HOUSING IN NEW HALOS

HOUSING IN NEW HALOS

A Hellenistic Town in Thessaly, Greece

Editors: H. Reinder Reinders & Wietske Prummel

with contributions by

Georgette M.E.C. van Boekel, Yvette Burnier, Margriet J. Haagsma,
Steven Hijmans, Colette Beestman-Kruyshaar, Bieneke Mulder,
Wietske Prummel, H. Reinder Reinders, Jaap Schelvis and Henk Woldring

A.A. BALKEMA PUBLISHERS / LISSE / ABINGDON / EXTON (PA) / TOKYO

Library of Congress Cataloging-in-Publication Data

Applied for

A previous publication 'New Halos, a Hellenistic Town in Thessalía, Greece' (Utrecht, 1988) is available from HES & DE GRAAF Publishers, Tuurdijk 19, 3997 MS 'tGoy-Houten, The Netherlands, fax (+31)(0)30-6011813, www.hesdegraaf.com

Cover design: <…>
Typesetting: Palm Produkties, Nieuwerkerk a/d IJssel, The Netherlands.
Printed in the Netherlands by Krips, Meppel.

ISBN 90 5809 540 1

CONTENTS

PREFACE

H. REINDER REINDERS

During the mapping of the surface remains of the lower town of New Halos in 1977 several well-preserved house foundations were discovered in a plot that was not in use as arable land. The relatively good preservation of these remains prompted negotiations with the owner, Christos Katrantzonis, an inhabitant of the village of Néos Plátanos, resulting in the purchase of a plot of four *strémmata* on behalf of the Greek state. Recently, more plots have been bought to ensure the preservation of all the excavated house remains.

The remains of six houses were excavated between 1978 and 1993 under the auspices of the Netherlands Institute at Athens. I am most grateful to the Greek Ministry of Culture for permission to work at Halos. My colleagues at the Vólos Museum played an important part in the project. In particular I would like to thank Roula Intzesiloglou, Anthi Efstathiou, Vasso Andrimi, Zoï Malakasioti, Litsa Skafida and Vasso Rondiri for their discussions, suggestions and assistance in the field, and also Nikos Manzoras, who died in 1996, for his support with the work in the depot of the Almirós Museum. I also would like to thank Susan Mellor who revised the manuscript, and Mette Bierma, Saskia Mulder, Hans Zwier and Jan Smit who prepared the manuscript and drawings for publication.

Until 1989 I was employed as a maritime archaeologist by the *Rijksdienst voor de IJsselmeerpolders* (IJssel-meerpolders Development Authority), which enthusiastically supported my work in Greece. The discussions I had with the planners and architects of Lelystad and Almere were most fruitful and my colleagues at the *Rijksdienst* provided a great number of worthwhile suggestions for my first book on New Halos. After 1989, I continued the investigation of the town in the context of a project launched by the Department of Archaeology of Groningen University and the Groningen Institute of Archaeology. This project was supported by the Netherlands Foundation of Scientific Research, which granted funds for the excavation of the houses in the lower town of New Halos. The Thessalika Erga Foundation and Reisburo Roever also supported the excavation with funds for the participation of the students.

Each campaign, a number of students and volunteers joined our team in Thessaly. Bakhuizen's pioneering work in Greece, on Gorítsa hill near Vólos, was continued in Aetolia by Bommeljé and Doorn, in Lávdha by Te Riele and Feije and at the archaeological site of Halos in the Almirós plain by our team; these excavations have offered generations of Dutch students a possibility of participating in fieldwork in Greece. In the years between the excavation campaigns at Halos much time was spent in the depot at Almirós, working on the abundant finds.

This preface ends with a survey of the campaigns at Halos with the names of the archaeologists, students and volunteers who participated each time. In particular I am indebted to Jan Jaap Hekman and Margriet Haagsma, who acted as field directors in 1987–1991 and 1993, respectively. The members of the excavation team stayed at Hotel Amalía in the small village of Amaliápolis until accommodation for the team became available in the same village in 1993.

The excavations of the house remain. The campaigns between 1978 and 1993:

1978 Initial investigation of the House of the Coroplast
Participants: Georgette van Boekel, Hildebrand de Boer, Niek Bosch, Cora Hoogcarspel, Ruud Olde Dubbelink, Dré van Marrewijk, Paulien de Roever, Vincent Rongen, Karel Vlierman, Joke Vlierman.

1979 Continuation of the excavation of the House of the Coroplast
Participants: Hans den Boer, Laura Bloemendaal, Jan Duim, Huib van der Hout, Petra Keppel, Ab Koelman, Bert Lanjouw, Paulien de Roever, Joke Vlierman, Karel Vlierman, Fons Vos.

1984 Excavation of the House of the Geometric *Krater*
Participants: Annie Bakker, Wim van Dijk, Jan Jaap Hekman, Kees Jan de Graaf, Huib van der Hout, Beatrice de Fraiture, Channah Nieuwenhuis, Monique Roscher, Ger Wieberdink.

1987 Excavation of the House of Agathon
Participants: Annie Bakker, Sandra Bakker, Yvette Burnier, Ido Dijkstra, Kees Filius, Henk Jan Hekman, Jan Jaap Hekman, Judith Hoogendijk, Huib van der Hout, Daan Lunsingh Scheurleer, Ange Nicklewicz, Barbara Stuart, Jaqueline Zoon.

1989 Excavation of the House of the Ptolemaic Coins
Participants: Annie Bakker, Corien Bakker, Gijsbert Boekschoten, Everhard Bulten, Yvette Burnier, Hella Grosfeld, Jan Jaap Hekman, Judith Hoogendijk, Hans Kok, Jona Lendering, Daan Lunsingh Scheurleer, Bieneke Mulder, René Oudhuis, Annechien Steendijk, Eric Wetzels.

1991 Excavation of the House of the *Amphorai*
Participants: Gerwin Abbingh, Annie Bakker, Yvette Burnier, Erie van Diemen, Ido Dijkstra, Kariene Erkelens, Maarten Grond, Margriet Haagsma, Joke van Haringen, Jan Jaap Hekman, Sarah Hines, Judith Hoogendijk, Steven Hijmans, Ruud Kok, René Oudhuis, Hilde Marie van Rozen, Sarah Tarlow, Eric Wetzels.

1993 Excavation of the House of the Snakes
Participants: Martine Bijlsma, Lizette ten Cate, Ido Dijkstra, Margriet Haagsma, Steven Hijmans, Michiel Huisman, Jacco Landskroon, Bieneke Mulder, Yvonne Oosthof, Wietske Prummel, A. Stamuli, Froukje Veenman, Michel John Versluys, Bethan Yates.

LIST OF CONTRIBUTORS

Georgette M.E.C. van Boekel, De Bréautélaan 9, 5263 GB Vught, the Netherlands.

Yvette Burnier, Van Houweningenstraat 3[1], 1052 TB Amsterdam, the Netherlands.

Margriet Haagsma, Department of History & Classics, University of Alberta, 2-1 Henry Marshall Tory Building, Edmonton AB T6G 2H4, Canada.

Steven Hijmans, Department of History & Classics, University of Alberta, 2-1 Henry Marshall Tory Building, Edmonton AB T6G 2H4, Canada.

Colette Kruyshaar, Bombardonstraat 91, 3822 CG Amersfoort, the Netherlands.

Bieneke Mulder, Prinseneiland 9–11, 1013 LL Amsterdam, the Netherlands.

Wietske Prummel, Groningen Institute of Archaeology, Poststraat 6, 9712 ER Groningen, the Netherlands (w.prummel@let.rug.nl).

H. Reinder Reinders, Groningen Institute of Archaeology, Poststraat 6, 9712 ER Groningen, the Netherlands (h.r.reinders@let.rug.nl).

Jaap Schelvis, Wirdumerweg 1, 9917 PA Wirdum, the Netherlands.

Henk Woldring, Groningen Institute of Archaeology, Poststraat 6, 9712 ER Groningen, the Netherlands.

INTRODUCTION

H. REINDER REINDERS

Archaeological knowledge of Greek towns in the Classical and Hellenistic world is restricted to a very small number of investigated towns. Studies focusing on Greek town planning and housing in general refer to these few examples. A number of these towns, like Priene and Olynthos, were excavated in the 19th and early 20th centuries. In recent years, however, a number of other towns have been excavated and plans have been presented which have greatly enhanced our knowledge of towns on the Greek mainland. Among these towns are Demetrias, Kassope, Gorítsa, Orraon and Halos. Nevertheless, the number of Classical and Hellenistic Greek towns of which the layout of the enceinte and the streets is known is very small. Even smaller is the number of towns in which public buildings and residential houses have been excavated and of which the results have been published.

The publication, in 1986, of *Haus und Stadt im Klassischen Griechenland* by Hoepfner and Schwandner aroused heated discussions on Greek town planning and housing. These authors re-evaluated the plans of cities like Milete, Rhodos, Peiraieus, Olynthos, Kassope, Priene, Abdera and Doura Europos and postulated the existence of *Typenhäuser* in the Classical Greek towns. Houses of the *pastas*, *prostas* and *Herdraum* types fill the housing blocks of their imaginative reconstruction drawings of Olynthos, Priene, Kassope and other towns. Hoepfner and Schwandner related the uniformity of the Classical towns to the ideology of *isonomia*, equality.

Hoepfner and Schwandner observed a trend towards the abandonment of the ideology of *isonomia* in Late Classical and Hellenistic times, in favour of the enlargement of houses with (parts of) neighbouring houses and their embellishment with mosaic floors and peristyles, indicating greater luxury. New Halos is an example of a small town founded in the late 4th or early 3rd century BC, a period in which Thessaly flourished and many new towns were founded and existing ones enlarged. Having been occupied for only a relatively short period of time, from 302 until 265 BC (Reinders, 1989), New Halos seemed to afford a good opportunity to find out to what extent, and how, views on town planning and housing changed after the Classical period. In most of the houses we observed only minor alterations of the original layout. The layout of the housing blocks seemed to have undergone very few changes, too.

Many of the Greek towns founded in the Classical and Hellenistic periods were reoccupied in later times. This resulted in an accumulation of occupation levels, making excavations in built-up areas very difficult. This was for example the case at the large Thessalian towns of Larisa and Pharsalos, where only small parts of the original layout have been uncovered. However, quite a few towns in Thessaly were not, or to a small extent only, built over in later times. Such towns, for example Pherai, offer better prospects for archaeological research. Unfortunately improved farming equipment and the use of pesticides and herbicides have in the past ten years caused serious damage to promising sites situated in agricultural areas. Forestation constitutes a similar threat to the cultural heritage of this region, for example on Gorítsa hill near Vólos.

The investigation of New Halos started in 1976 (fig. 0.1). The mapping of the surface remains of the upper and lower town in 1976 and 1977 resulted in the plan of a Hellenistic town with a rather rigid layout (fig. 0.2). In the twenty years that have passed since then, almost all the remains of the housing blocks, streets and houses of the lower town have been destroyed by ploughing. In 1977 many plots in the area of the lower town were in use as arable land. They lay fallow in the following two years. Since then, cotton, olives, lucerne and tomatoes have been grown in plots inside the town walls and the remains of housing blocks and houses have been removed as cultivation intensified.

To obtain a better understanding of the structure of the enceinte, the situation of the gates and the dimensions of the housing blocks and houses, we conducted several small-scale excavations. Between 1984 and 1993 we excavated several houses in the lower town. We initially intended to excavate houses of different housing blocks and in different quarters of the town. But when we started our excavations in 1984 we soon found that only a few houses were still undisturbed. Among them were four or five houses of housing block 6.4, situated in a plot which had not been cultivated but was in use as pastureland, so we focused our efforts on this block.

In the first publication on New Halos I discussed the results of the excavations carried out between 1976 and 1982 (Reinders, 1988). In those years only one house was excavated. For the sake of completeness the data relating to this house, the House of the Coroplast, have been included in the present volume. Between 1984 and 1993 we focused our attention on the housing remains and excavated the remains of five more houses. This volume contains the results of the excavations of the six houses investigated so far (fig. 0.3):

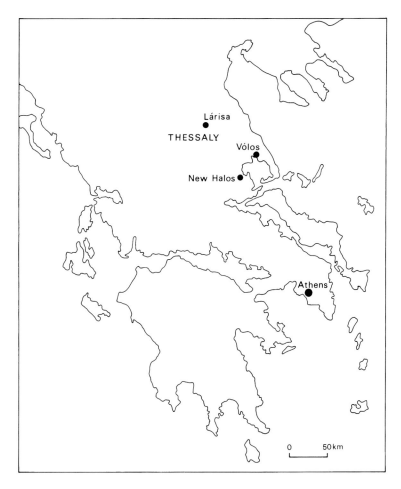

Fig. 0.1. Map of Greece, showing the position of the city of New Halos.

1. House of the Coroplast, excavated in 1978/79 (housing block 2.7, plot No. 11);
2. House of the Geometric *Krater*, excavated in 1984 (housing block 2.8, plot No. 17);
3. House of the Snakes, excavated in 1993 (housing block 6.4, plot No. 9);
4. House of the *Amphorai*, excavated in 1991 (housing block 6.4, plot No. 18);
5. House of the Ptolemaic Coins, excavated in 1989 (housing block 6.4, plot No. 21);
6. House of Agathon, excavated in 1987 (housing block 6.4, plot No. 22).
Unfortunately the House of the *Amphorai* proved to have been fairly badly disturbed: the house's outer walls were not intact. The greater part of the House of the Geometric *Krater* had also been destroyed, possibly during the construction of the old Almirós–Soúrpi road, which intersects the southwestern quarter of the house.

The six houses were excavated in seven campaigns. After each campaign we covered the entire plot with the excavated soil from which we removed as many large stones as possible. Our aim was to protect the structural remains from further damage. And fortunately our efforts were successful, for the remains of the isolated houses of the Coroplast and the Geometric *Krater* are still intact. This custom of covering up the plot each time did however involve the disadvantage that we were not able to determine structural relationships or check the measurements of previously excavated houses adjacent to those unearthed in later campaigns. Only in one campaign, in 1989, were the remains in several plots, notably those of the house of the Ptolemaic Coins and the house of Agathon, accessible throughout the excavation period.

The remains of the excavated houses are described and analysed in Chapter 2. All plot measurements quoted are approximate measurements. The same holds for the dimensions of the rooms, because the limestone blocks of the walls were not laid in an exactly straight line, which also made it difficult to determine the exact corners of the rooms. So everywhere where a room's measurements are specified, the reader should allow for a fault margin of 0.05 m. Time and

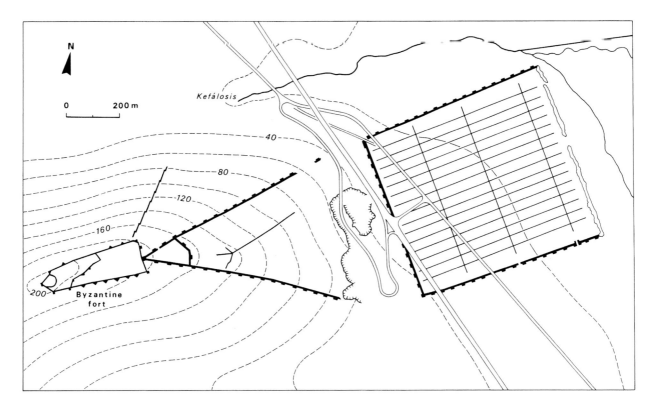

Fig. 0.2. Plan of the city of New Halos.

funds permitting, we intend to (re)excavate all the house remains to check our measurements and the structural relationships of the house foundations.

Leake (1835: pp. 336–337) described the lower town of New Halos as follows: 'the enclosed space, although thickly strewn with stones, the foundations of buildings and broken pottery, is sown with corn'. Apart from the foundations of the houses of New Halos there were no signs of building remains from later periods at the site, with the exception of a nineteenth-century watchtower. The excavation of the House of the Coroplast in 1978/79 led to the conclusion that the occupation remains comprised only one layer, which extended from only 5 cm to 40 cm beneath the surface. This single stratum was also observed during the excavations of other houses in 1984, 1987, 1989, 1991 and 1993.

During the excavation of the House of the Coroplast the lower parts of the house's walls were found to consist of two rows of blocks with a fill of small stones and pebbles. The way in which the walls of two adjacent houses were bonded together proved that the houses were integral parts of the housing blocks, and were not built on separate plots. No large blocks were observed inside the rooms, implying that the upper parts of the walls were made of mud bricks, their mud now forming part of the soil tilled by the local farmers. It is therefore more correct to speak of a stratum

consisting of four components rather than of a single occupation stratum. At the bottom of the stratum was the floor, consisting of compacted earth (1), on top of which many artefacts were found *in situ*, but crushed (2). In many rooms this layer of artefacts was covered with a layer of roof tiles (3) and, on top of that, a layer of dissolved mud bricks (4). This clear stratification was of course observed in only a few rooms. In addition, evidence of minor reconstruction work, artefacts from earlier periods and indications of the removal of artefacts were found.

The plan of the House of the Coroplast shows a large room with two smaller rooms on either side. This same layout was found in two more houses, excavated in 1987 and 1989; it was obviously the module of the city's house plans (Reinders, 1994: p. 219). The results of the excavation of the House of the Coroplast led to the hypothesis that New Halos was abandoned after an earthquake (Reinders, 1988). The discovery of coins struck during the reign of Ptolemaios II among the house remains suggests that this earthquake occurred around 265 BC.

In 1987, when we started excavating a series of houses in housing block 6.4, we set out to find answers to the following questions:
– how are the house plots arranged within the housing blocks;
– what are the dimensions of the house plots and what

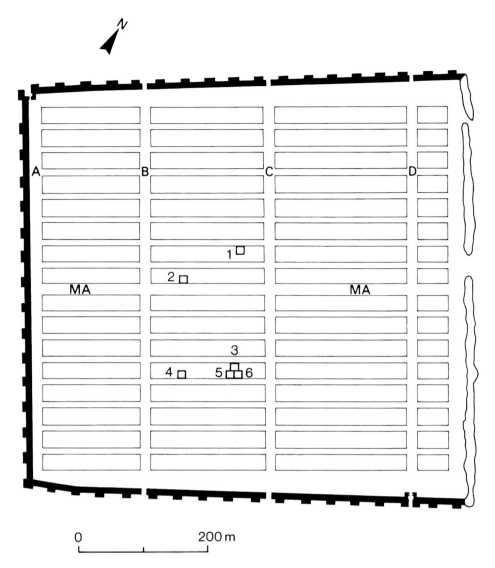

Fig. 0.3. Plan of the lower town of New Halos, showing the location of the six excavated houses (1-6), the Main Avenue (MA) and the main streets (A-D).

do their layouts look like;
- how many different types of houses can be distinguished;
- can the differences between individual houses be explained in terms of different functions;
- when and why was the city of New Halos abandoned;
- what, if any, evidence is there to suggest the city was destroyed by an earthquake;
- how do the layouts of the housing blocks and house plots of New Halos differ from those of cities in the Classical period?

We realize that it is not possible to draw decisive conclusions from evidence obtained for only six of the total of 1440 houses that once stood in the town of New Halos. However, since knowledge of Hellenistic

housing is almost non-existent, and only few house remains from this period have so far been excavated in towns on the Greek mainland, we find it worthwhile to present the data on these houses in a separate volume.

The present volume therefore contains the results of the excavations of house remains carried out between 1978 and 1993. It begins with an introduction discussing the archaeology of the Almirós plain and the coastal towns along the Pagasitikós and Maliakós gulfs, after which the layout of the city of New Halos, the various house plans and the distribution of artefacts are discussed. A long chapter is dedicated to the main categories of artefacts found in the six houses. Margriet Haagsma is going to use the results of these studies for a spatial analysis of the houses and a study of the households. She will discuss her results in her PhD thesis, which will be published separately. The chapter on the artefacts has been written by several archaeologists who participated in the excavation campaigns as students. Some of them have since then been granted funds enabling them to work on a PhD thesis while others have been fortunate enough to find jobs elsewhere. This created a few problems during the final phase of the completion of this volume.

A separate chapter focuses on environmental issues. The study of the animal remains prompted lengthy discussions about the nature of animal husbandry in New Halos and gave rise to the question whether a form of transhumance between the Almirós plain and the Óthris mountains may have been practised in Hellenistic times. The results of two botanical studies have therefore been incorporated in this volume. The investigation of the city of New Halos and its surrounding territory will continue in the years to come. Work has meanwhile started on the excavation of the Southeast Gate. The results of this and future excavations will be published in another volume, but some of the results so far obtained for the gate have been incorporated in this publication as the gate was presumably destroyed at the same time as the houses, and will consequently no longer have provided access to the city after that time. Evidence for the date of the destruction of the gate has been incorporated in the chapter on the abandonment of the city.

We have chosen to use the transliteration of ancient Greek names of people, places and artefacts for the spelling of Greek words in English, using Demetrios, Thebai and *krater* rather than the Latin or anglicised forms Demetrius, Thebes and crater. The same holds for modern Greek geographical names, which are distinguished from ancient ones by the use of an acute accent on the stressed syllable. Exceptions are Thessaly, Macedonia and Attica, used for the names of the districts in all periods. With respect to the transliteration of χ and φ, preference has been given to ch (f.i. Chlomón) instead of kh (Khlomón), and to f (f.i. Fársala) instead of Ph (Phársala), but to Pharsalos to indicate the ancient city. Exceptions are Athens and Corinth, used for the ancient as well as the modern cities.

REFERENCES

HOEPFNER, W. & E.-L. SCHWANDNER, 1986. *Haus und Stadt im klassischen Griechenland*. Deutscher Kunstverlag, München.

LEAKE, W.M., 1835. *Travels in northern Greece*. 4 vols. London.

REINDERS, H.R., 1988. New Halos, a Hellenistic Town in Thessalía, Greece. Hes, Utrecht.

REINDERS, R., 1994. Housing in Hellenistic Halos. In: R. Misdrahi-Kapon (ed.), *La Thessalie, Quinze années archéologiques, 1975–1990. Bilans et perspectives*, Volume B. Kapon, Athènes, pp. 217–220.

CHAPTER 1. NEW HALOS IN ACHAIA PHTHIOTIS

H. REINDER REINDERS

Classical Halos and Hellenistic New Halos were cities in Achaia Phthiotis. Like Perrhaibia and Magnesia, Achaia Phthiotis was a marginal district which was overshadowed by the powerful districts of Thessaly: Hestiaiotis, Thessaliotis, Pelasgiotis and Phthiotis. The marginal districts played no political role in the Thessalian League. In classical times two of the cities of Achaia struck coins: Melitaia and Halos (Reinders, 1988: pp. 236–37). This situation changed in Hellenistic times, when no fewer than five cities issued coins: New Halos, Phthiotic Thebai, Larisa Kremaste, Peuma and Ekkara. The cities of Achaia Phthiotis, and in particular the coastal towns Thebai, New Halos and Larisa Kremaste, obviously flourished in this period. The prosperity of Achaia Phthiotis was however not reflected in its political position.

1. ARCHAEOLOGICAL RESEARCH IN THE ALMIRÓS PLAIN

After the Treaty of London was signed on May 11th 1832, the border between the Greek kingdom and the Ottoman empire ran from the gulf of Árta in the west to the Pagasitikós gulf in the east. North of Lamía the border followed the crests of Mount Óthris and the course of the river Salamvrias down to the Soúrpi plain. The villages of Soúrpi and Amaliápolis were situated in Greece, while Almirós and the greater part of the Soúrpi plain were under Ottoman rule. A number of small Greek and Turkish forts lined the border on either side.

1.1. Previous research

After the liberation of Thessaly, in 1881, the inhabitants of Almirós showed a great interest in their past, which resulted in the founding of the Philarchaios Etaireia Almyrou 'Othrys' on April 28 1896. Three years later, on August 20th 1899, the Antiquarian Society published the first volume of its periodical; in 1911 the seventh and last volume was published. A local museum was built between 1910 and 1927 to house the archaeological finds recovered in excavations in the neighbourhood of Almirós (Doulyeri-Intzesiloglou, 1993). The archaeological interest of the citizens was actuated by Yannopoulos, who guided the members of the society to the archaeological sites in the plain. Many archaeologists, such as Tsoundas, Arvanitopoulos, Wace, Thompson and Vollgraff, visited the museum and the archaeological sites under his guidance. It is not surprising that Wace and Thompson (1912) and Vollgraff (1908) decided to carry out excavations in the Almirós plain.

Yannopoulos' knowledge of the archaeological sites was excellent. Most of the prehistoric sites known today were already known to him. They are mentioned by Wace and Thompson (1912: p. 10) in their list of prehistoric sites in northeastern Greece (fig. 1.1). "Round the Krokian plain are the following:

59*. The supposed site of Pyrasos at Néa Ankhíalos.
60*. Aidhiniótiki Maghúla, an hour north of Almiros.
61*. Zerélia, forty-five minutes southwest of Almiros.
62. Fifteen minutes north of Karatsadaghlí by a grove of oak trees.
63. Paleokhóri or Yiuzlár, half an hour west of the village of Daudzá.
64*. Maghúla Almiriótiki, fifteen minutes east of Almiros.
65. South of Almiros near the right bank of the Xeriás, and between the vineyards and the road to Karatsadaghli.
66. On the left bank of the Kholórevma, a few minutes downstream from the mill called Vaïtsi.
67*. The mound of Básh Mílos on the right bank of the Kholorevma just below the bridge on the Turkomusli road: cyst tombs have been found here.
68*. The acropolis of Phthiotic Thebes.
69*. The site identified as Phylake, just south of Kitík; by it is a good spring.
70*. The maghula of Surpi, half an hour south of the village by a mill: here Minyan ware is very common."

The asterisk "against the number implies that the mound is of the high type, or that at that site pottery of the Third (Chalcolithic) and Fourth (Bronze Age) Periods has been found". Yannopoulos published information on a great number of sites in the periodical edited by the 'Othrys' Antiquitarian Society in Almirós and other periodicals.

Among the visitors of the local museum at Almirós in the early 20th century was Stählin, who wrote an excellent book on the archaeological sites in Thessaly (Stählin, 1924). Stählin also mentioned the sites from Wace's catalogue, plus another site: *"Erwähnt sei auch das Kastro 3/4 Stunden nordöstlich von Turkomusli in Kaplage zwischen zwei zusammenmündenden Bächen. In den kyklopischen Mauern fand ich archaische und spätere Scherben, z.B. einen der attischen Vurwaart ähnlichen (Stück eines Tellers, schwarz auf rotbraun, eingeritzte Innen- und bemalte Umrisszeichnung)."*

Besides prehistoric sites, Stählin also mentioned sites from later periods. He presented plans of the two large towns in the Almirós plain, Phthiotic Thebai and New Halos (Stählin, 1924: figs 21 and 23). Unfortunately only the enceintes of the two cities are indicated in these plans, and not the grid of streets and housing blocks which must have been clearly visible at the surface in those years when farming methods were not yet as destructive as they were to become after the 1950s. Stählin (1924: fig. 22) located the site of Halos, the city destroyed by Philippos II in 346 BC, on the hill west of New Halos. A number of smaller fortified sites from Hellenistic times, remains of monasteries and other antiquities are also described in his work, which represents the first accurate survey of the archaeological remains in this plain.

For a long time Stählin's book on Thessaly was the most important reference work available to archaeologists and historians. Generally speaking, our knowledge of the occupation of the southern part of the plain remained largely unchanged from 1924, the year of the publication of his book, until 1976, when we started our investigation of New Halos. Throughout this long period this area was largely neglected and the activities of the 'Othrys' society declined.

Interest meanwhile shifted to the northern part of the plain, in particular to the village of Néa Anchíalos (fig. 1.2), where research efforts focused on the excavation of remains from the Byzantine period, when the city of Thebai flourished. Several excavation campaigns were carried out here after 1924, first by Sotiriou and later by Lazaridis. Five basilicas, dating from the 5th and 6th centuries AD, were excavated (Asimokopoulou-Atzaka, 1982). There was also some interest in the prehistory of this area; Theocharis excavated on the mound of Pyrasos. Less attention was paid to the city of Phthiotic Thebai west of Néa Anchíalos (fig. 1.2), which was inhabited from Classical to late Roman times. Arvanitopoulos excavated here in the early 20th century, but no further work was carried out until the 1990s, when the acropolis and the site of the theatre were excavated under V. Andrimi-Sismani.

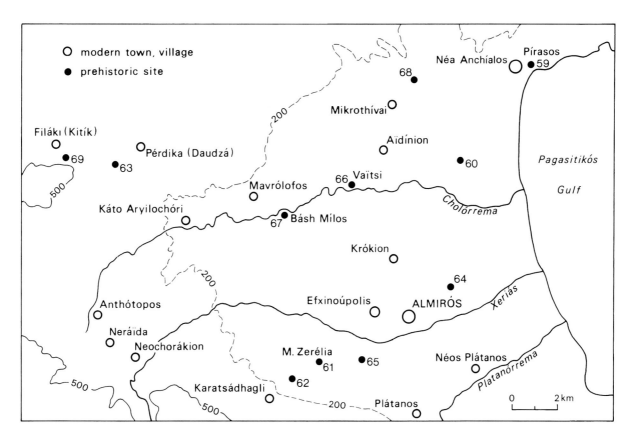

Fig. 1.1. Prehistoric sites on the Almirós plain, after Wace & Thompson (1912).

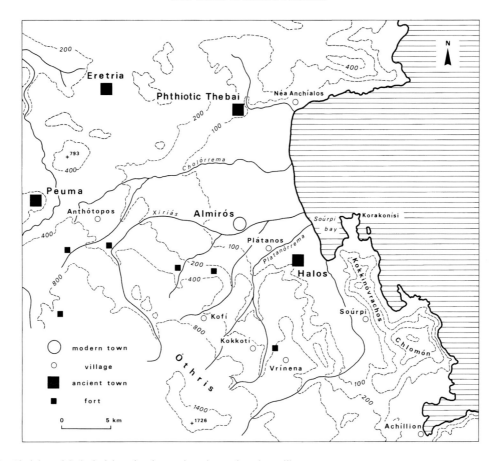

Fig. 1.2. The Almirós and Soúrpi plains, showing ancient sites and modern villages.

The investigation of the city of New Halos by a team of Dutch archaeologists started in 1976. From that year onwards interest in the archaeological sites in the southern part of the Almirós plain grew steadily. At first the investigation was restricted to the city of New Halos itself, but in the 1980s some Classical sites and Hellenistic forts within the city's former territory were investigated, too. A hypothesis concerning the location of the town of Classical Halos on *Magoúla* Platanió̈tiki was proposed (Reinders, 1988; 1993) and a number of forts were recorded (Reinders, 1988; Wieberdink, 1990).

1.2. Recent developments

Between 1982 and 1984 protest signs bearing slogans like *óchi anadasmós* (no reallotment) on bridges and buildings brought to my attention the Greek government's intention to reallot the Almirós–Soúrpi plains. Discussions with the geologist Van Andel and the archaeologists Jameson, Munn and Runnels of Stanford University resulted in a proposal to conduct a survey within the territory of the city of New Halos. This proposal was supported by Chourmouziadis, the

director of the Archaeological Museum at Vólos, and a permit was granted for the survey in 1984. Unfortunately it arrived too late for the Stanford archaeologists, who were unable to participate in the survey. Chourmouziadis meanwhile accepted a position as professor of prehistory at the Aristotelion University of Thessaloníki and plans for a survey were postponed until 1990.

The reallotment plans were carried out between 1985 and 1990. The topography of the area changed, as did the road network and the crops cultivated by the farmers. In the Voulokalíva plain, a short distance to the north of the city of New Halos, large areas of arable land had lain fallow for one or two years in the 19th and 20th centuries, during which time they were used as pastureland for the large flocks of transhumant pastoralists. After the reallotment, the cultivation of grapes, cotton, walnut and tomatoes intensified and started to constitute a threat to the Iron Age *tumuli*. Voulokalíva is the area where Wace and Thompson excavated an Iron Age *tumulus* in 1908. After 1985 a number of damaged *tumuli* were clearly recognisable by scattered boulders and cobbles, which do not occur in the plain's Pleistocene soil. The number of *tumuli* was found to have been many times larger than that indicated on a map drawn by Wace and Thompson (1911–12). In 1988 we appealed to the Greek government for a permit to conduct a survey in this area in order to determine the exact positions of these *tumuli* in the reallotted area.

In 1990 the Ministry of Culture granted us a permit for a survey within the territory of New Halos in cooperation with the 13th Ephorate of Prehistoric and Classical Antiquities at Vólos. The Ephorates of Byzantine Antiquities at Lárisa and Modern Monuments in Vólos also supported the survey, which was started in 1990 and continued in 1992, 1994 and 1996. For the preliminary results of this survey the reader is referred to the publications in Pharos, the periodical of the Netherlands Institute at Athens. The survey not only yielded answers to various questions concerning the history of the occupation of this area and a better understanding of the phenomenon of the large Hellenistic city of New Halos, but also led to the identification of many prehistoric sites, the most important of which now enjoy protection as ancient monuments.

The improved agricultural machinery used by present-day farmers and the construction of roads and factories constitute severe threats to this area's archaeological heritage. The construction of the national road in the 1970s led to the discovery of several Mycenaean tombs, only some of which could be excavated. Efforts aimed at preventing the construction of a canteen for an aluminium plant on top of *Magoúla* Plataniótiki were successful, but part of the site of Classical Halos was nevertheless fenced off (Efstathiou *et al.*, 1993), and large areas of the backswamp near the site were used as dumping grounds for the waste products of two large factories, which was of course disastrous for the environment. And last but not least, a conveyer belt was installed from a limestone quarry west of New Halos to a silo on the coast, passing by the southern wall of New Halos at a distance of 100 m. Apart from the deplorable impact on the skyline, large areas have become polluted with the fine dust particles released by the crushed limestone rock. Fortunately the land along the planned trajectory of the conveyer belt was surveyed before its construction and the remains of a Hellenistic cemetery from the 3rd century BC about 100 m south of the city's Southeast Gate were excavated, as well as a concentration of Byzantine amphoras on a beach ridge near the coast (Malakasiotis *et al.*, 1993).

Between 1970 and 1995 the Almirós–Soúrpi plain changed drastically – from a remote, neglected farming area into a busy part of Thessaly with a dense network of roads, industrial areas, oversized buildings and environmental pollution. In 1996 an archaeological survey was conducted in the area between Anthótopos and Plátanos (fig. 1.2). In the heat of June and July 1996, a team of Dutch students walked the stubble fields on the finger ridges between the villages of Neochorákion and Efxinoúpolis, where they located 50 archaeological sites. In 1997 a mining company presented plans to excavate this area of about 50 km². What will the future bring – protection or destruction without any chance of excavation?

Fortunately this same period saw a revival of interest in the past of Almirós and Achaia Phthiotis, and on August 31 and September 1 of 1991 a congress focusing on Almiriotic studies was organised (Kondonatsios, 1993), followed by a second congress in 1996. The 'Othrys' Antiquarian Society also received a fresh impetus and published the first volume of the second series of its periodical in 1997. The local museum, which had been damaged by an earthquake in 1980, was reopened in May 1980 by the Minister of Culture, E. Venizelos.

2. COASTAL TOWNS ALONG THE PAGASITIKÓS AND MALIAKÓS GULFS

2.1. Introduction

Until the 1970s little was known about the towns of Thessaly in Hellenistic times. Since then, however, a number of studies have been published which have greatly enhanced our understanding of town planning and settlement patterns in this period. In the 1970s Gorítsa hill (University of Utrecht) was surveyed, the site of Demetrias was surveyed and partly excavated (University of Heidelberg) and the first excavations were carried out at the site of the town of New Halos (University of Groningen). We now have accurate plans of three cities that were founded on the shore of the Pagasitikós gulf around 300 BC.

The investigation on Gorítsa hill was initiated by Bakhuizen in 1971. During three successive campaigns his team surveyed the surface remains visible on the hill. An enceinte was found to have enclosed an area of 32 ha, within which building remains were observed in an area of 17.5 ha. The site was occupied in the first half of the 3rd century BC. The absence of Megarian bowls indicates that it was abandoned before the middle of the century (Bakhuizen, 1992: p. 313). During the survey the vegetation was cleared and the remains were summarily cleaned. No excavations were carried out, so the date of the city's foundation and its period of occupation are not exactly known. The results of the survey of this small 4th-century city, however, yielded much information on Greek town planning.

The investigation of Demetrias by the University of Heidelberg also started with a survey of the visible remains. The greater part of this site was in use as arable land or contained orchards. Building remains were found in only a small part of the total area enclosed by the enceinte. Although only few house remains were identifiable as such, it was clear that an area of about 45 ha in the eastern part of the town had been occupied in Hellenistic times. In the western part of the town more or less the same amount of space was available on level ground as in the eastern part. Demetrias was founded around 290 BC by Demetrios Poliorketes and was occupied during Hellenistic and Roman times. The remains of some of the public buildings were excavated: an *anaktoron* and a citadel. Before the University of Heidelberg started its excavations Theocharis had started work on the excavation of the town's theatre. His efforts have been continued by the Archaeological Museum of Vólos in recent years. In addition, a Late Hellenistic house has been excavated in the eastern part of the city.

The strategic function of New Halos was evident from the very start (Reinders, 1988), not only from the town's strategic situation, at a narrow passage along the coastal route from northern Thessaly to central Greece, but also from its impressive 4.7-km-long enceinte, reinforced with at least 117 towers.

The investigations of these three cities focused on the cities' exact locations and their layouts, the public buildings and residential houses and the relationship between the cities and their territories. In the 3rd century BC Demetrias was the second capital of the Macedonian kingdom after Pella, and served as a naval harbour for the Antigonids (Marzolff, 1980; 1994). New Halos was in all probability also founded by Demetrios Poliorketes (Bakker & Reinders, 1996), in 302 BC, before he departed from Thessaly to Asia Minor to assist his father Antigonos in his struggle against Lysimachos and Seleukos. New Halos lay at a strategic point between a spur of Mount Óthris and a backswamp bordering the Pagasitikós gulf.

The city of New Halos may have been founded for military reasons. In our first publications on the investigation of New Halos we paid little attention to maritime aspects of the city. However, those aspects certainly deserve attention because maritime strategy was Demetrios' strong point. The excavation of the remains of six houses in New Halos yielded information on the period of occupation, but also evidence for overseas contacts.

In addition to agricultural implements, storage jars, amphoras, figurines and animal remains, some 150 coins were found at the site of New Halos, which showed that the city was abandoned around 265 BC (Reinders, Ch. 3: 5). The sources of these coins tell us with which cities New Halos was in contact. The coins point to contacts with neighbouring cities in Achaia Phthiotis, such as Peuma, Thebai, Ekkara and in particular Larisa Kremaste, but they also suggest contacts with cities on the island of Euboia, such as Histiaia and Chalkis, and with Lokris. As Furtwängler (1990) already demonstrated, most of New Halos' contacts were maritime, oriented towards the south; coins from other cities in Thessaly are rare (fig. 1.3). On the other hand, the Pagasitikós gulf was probably

used as fishing grounds, although only one fish bone was found in the excavated houses. Molluscs were gathered in Soúrpi Bay (Prummel, Ch. 4: 4).

Unlike New Halos, which struck its own coins, the nearby Hellenistic city of Demetrias (fig. 1.6) relied on the emissions of the city of Larisa in Pelasgiotis. The coins found during the excavations in Demetrias reflect contacts with the cities in the northern part of Thessaly and Macedonia (fig. 1.4). Virtually no coins from cities in Achaia Phthiotis or the island of Euboia were found. Furtwängler (1990: pp. 235–240) assumes that in the first quarter of the 3rd century BC Demetrias' contacts were restricted to Magnesia, the eastern part of the Pagasitikós gulf and, possibly, the city of Histiaia on the northern shore of the island of Euboia.

Trade and transportation are reflected in other artefacts, too. Stamps on amphora handles provide information on the importation of wine. The site of Demetrias yielded a large number of amphoras from Thasos and a relatively small number from Rhodos. Similar evidence from around 200 BC obtained in Athens reveals a preference for wine from Rhodos in that city; here, only a small number of amphoras from Thasos was found (Furtwängler, 1990). Unfortunately no amphora stamps were found at Halos. Although the exact nature of the contacts of New Halos is not clear at the moment, the coins suggest overseas contacts with the ports of Euboia and Lokris.

Does this mean that New Halos served as a port of call for part of Achaia Phthiotis? It was this question that led me to reconsider the situation of New Halos. In the following sections I will discuss the importance of the cities along the coast from a maritime point of view and try to find an answer to the question why there were so many ports in this area in Hellenistic times. I will first focus on the available written sources providing information on the coastline between Thermopylai and Demetrias, the sailing routes and the landmarks along the coast. Information on the shifting of the coastline in ancient times has been obtained in geological and palaeogeographical investigations. Maritime activity in this area continued until the Ottoman conquest of Thessaly in AD 1393.

2.2. Skylax, Strabon and Artemidoros

The oldest description of the coast between Thermopylai and Demetrias is a *períplous* (circumnavigation, a description of the coast with sailing directions) ascribed to Skylax of Karyanda, who lived around 500 BC (GGM 1 = Müller, 1965). The preserved version of this *períplous* (GGM 1, 15 ff.) is however a compilation of a later date, presumably written in the 4th century BC, possibly between 338 and 335 BC (Kretschmer, 1909: p. 153). Along the coast (GGM 1, 49–51) this *períplous* mentions, amongst other cit-

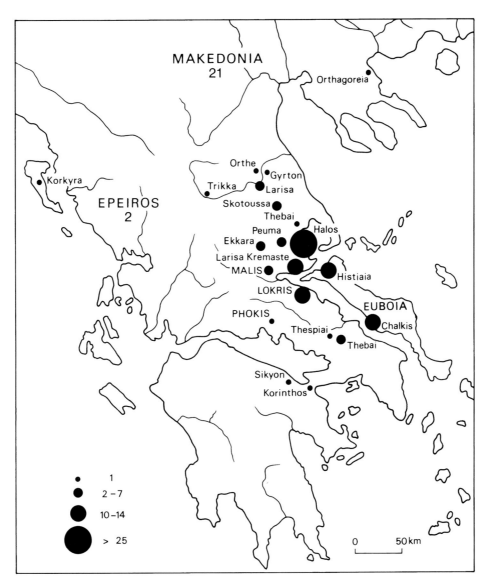

Fig. 1.3. Provenance of the coins found at the site of New Halos.

ies and features, Thermopylai and the river Spercheios and the cities of Lamia and Echinos in Malis; Echinos became part of Malis in 351 BC. Skylax continues with the towns of Achaia Phthiotis: Antron, Larisa, Melitaia, Demetrion, or Pyrasos, and Thebai (figs 1.6 and 1.10). The absence of Halos from this enumeration is not surprising, because the Classical city of Halos had been destroyed after a siege by a Macedonian army under Parmenion in 346 BC (Reinders, 1988; 1993). To the northeast of these towns, Amphanaion and Pagasai are mentioned along the shores of the Pagasitikós gulf; Amphanaion has not been located. In addition to these towns and rivers along the coast, Skylax mentions the distance of a voyage circumnavigating Attica, Boiotia and Lokris: the distance between Cape Sounion and the border of Malis near

Thermopylai (fig. 1.5) is said to be 1300 stadia. Skylax gives only a few names of Thessalian cities in the hinterland.

Another important source of information on the coast between Peiraieus and Thessalonike in Hellenistic times is Strabon's *Geography*. Strabon was born in Amasia in Asia Minor in 64 or 63 BC. He travelled to Rome many times and went on voyages to many other countries, but he never visited Thessaly. In his *Geography* he quotes passages from many authors, so his information comes from different sources and different periods.

One of the sources Strabon used for his description of Greece is Apollodoros' Commentary on the Homeric Catalogue of Ships. Strabon hence interwove the geography of his own time with that of Homeric

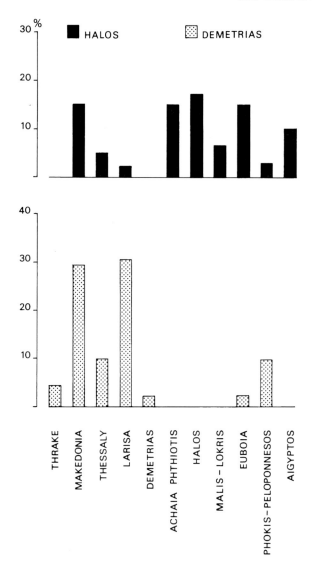

Fig. 1.4. Provenance of the coins found at New Halos and Demetrias (data relating to Demetrias after Furtwängler, 1990). Thessaly = cities of Thessaly, with the exception of Demetrias, Larisa in Pelasgiotis, Halos and the cities of Achaia Phthiotis.

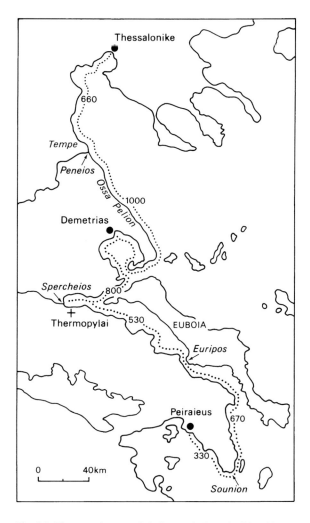

Fig. 1.5. The coast between Peiraieus and Thessaloniki, with distances indicated in *stadia*.

times. The information about Homeric times comes from the Catalogue of Ships in book B of the Iliad. This Ship Catalogue lists the towns that provided troops for the Troian War. It is generally accepted that the list to a certain extent refers to Late Bronze Age geography of shortly before 1200 BC. Some of the towns mentioned, such as Pylos, are known to have flourished until around that date. The site of Pylos has yielded no archaeological evidence for occupation after the Mycenaean period (Hope Simpson & Lazenby, 1970: p. 82). Some of the towns in certain regions, such as the Peloponnesos, can be associated with archaeological sites. Many of the towns mentioned in our area, however, cannot be identified on the basis of archaeological evidence.

The information that Achilles came from the area northeast of Lamia (fig. 1.6) has engendered, and indeed continues to engender, much speculation about his home town. A small number of 'Homeric' towns can be identified with more or less certainty. Excavations at present-day Iolkós, a short distance to the west of the city of Vólos (fig. 1.8), have yielded evidence for occupation in the Late Bronze Age, suggesting that the site may be identified as Iolkos (Hope Simpson & Lazenby, 1970: p. 136), although it has recently been proposed that the Mycenaean site near present-day Dhimíni may have been the Homeric site of Iolkos (Intzesiloglou, 1994). The southern part of the Almirós plain, the Krokion plain of Classical times, is known to have contained two large towns: Classical Halos and Hellenistic New Halos. Hope Simpson & Lazenby (1970: p. 126) placed Homeric Halos near the present-day city of Lamía, although another site that was also occupied in the Late Bronze

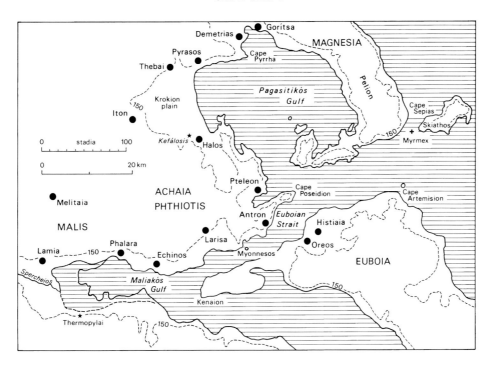

Fig. 1.6. Towns and landmarks along the coast between Thermopylai and Demetrias in Classical and Hellenistic times.

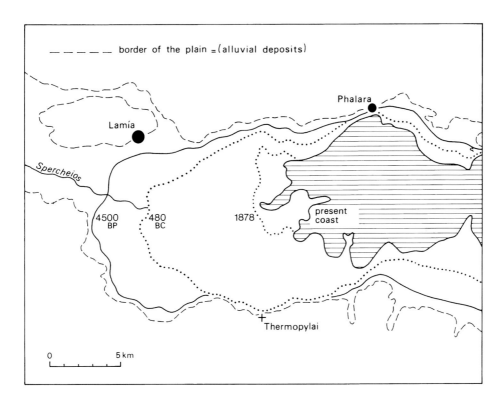

Fig. 1.7. Maliakós gulf. The shifting of the coastline near the outlet of the Spercheios (after Kraft *et al.*, 1987).

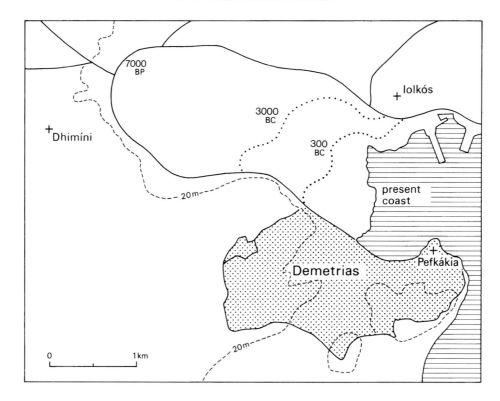

Fig. 1.8. Vólos Bay. The shifting of the coastline north of the Hellenistic city of Demetrias (after Zangger, 1991).

Age was found during an archaeological survey in Voulakalíva, a little to the north of Hellenistic Halos (Reinders, 1993). Smaller archaeological sites have been identified as Antron and Pteleon, but no indisputable evidence for Late Bronze Age settlements has yet been found.

If we omit the Homeric information Strabon borrowed from Apollodoros we are left with a less complicated account. The description of the geography of the coastal area of Thessaly is almost certainly based on Artemidoros. Around 100 BC Artemidoros of Ephesos wrote a work comprising eleven books. Strabon's description contains sections from Artemidoros' work or from similar *períplous*-like descriptions. The description of the coast from Thermopylai to Demetrias forms part of a longer account describing the route from Peiraieus to Thessalonike. Strabon, following Artemidoros (?), divided this route into six sections, each characterised by an important landmark, namely Cape Sounion, the strait of Euripos, the springs of Thermopylai, the city of Demetrias, the river Peneios and the city of Thessalonike (fig. 1.5). The length of each section is given in *stadia* (table 1.1). Strabon also mentions other landmarks that were important for sailors, such as the projecting headland of Kenaion, the small island of Myonnesos, a submarine reef, Cape Poseidion, the Krokion plain and Cape Pyrrha (fig. 1.6).

The springs of Thermopylai lie at the beginning of the section comprising the districts of Malis, Achaia Phthiotis and the western part of Magnesia. Various names of cities are given, with the distances between them expressed in *stadia*. The majority of these cities flourished in the third century BC. Along the Maliakós gulf, Strabon mentions the city of Lamia and its port Phalara. Next come the towns of Echinos, Larisa Kremaste, Antron and Pteleon, which formed part of Achaia Phthiotis. Along the western shore of the Pagasitikós gulf the cities of Halos and Thebai lay at either end of the Krokion plain; Demetrias was situated on the southern shore of Vólos Bay (fig. 1.6).

A few sites in the hinterland are mentioned, such as Homeric Iton and Phylake. It is doubtful whether the sites Iton and Phylake were still known in Hellenistic times. Distances in *stadia* from Iton and Phylake to Halos and Thebai are given, but no Hellenistic remains have been found at the implied sites. The information on the geography of the inland Thessalian plains is very poor in comparison with that on the coastal region. The geography of the coastal area was important for sailors. The town of Larisa Kremaste, *die schwebende genannt wegen ihrer vom Meere aus gesehen himmelhohen Lage* (Stählin, 1924: p. 182), was even named from a sailor's viewpoint. In his description of the section of the coast between Thermopylai and Demetrias, Strabon mentions Artemido-

Table 1.1. Distances along the coast between Peiraieus and Thessalonike in *stadia* (1 *stadium* is about 185 m, depending on the foot used).

From	To	Distance	Reference
Peiraieus	Sounion	330	Strabon 9.1.2
Sounion	Euripos	670	Strabon 9.2.8
Euripos	Peneios	2,350	Strabon 9.5.22
Euripos	Thermopylai	530	Strabon 9.4.17
Spercheios	Demetrias	800	Strabon 9.5.22
Demetrias	Peneios	1,000	Strabon 9.5.22
Peneios	Thessalonike	660	Strabon 8.8.5

ros (9.2.5). Assuming that Strabon based his description on Artemidoros, I have attempted to 'reconstruct' this part of Artemidoros' *períplous*.[1]

2.3. The shifting of the coastline

The greater part of the coast between Thermopylai and Demetrias is steep and rocky, here and there intersected by small bays offering sheltered anchorage. This part of the coast is covered in sea map 1556 of the British Admiralty, printed in 1890. From this map we may infer that the coastline has not changed that much since the 19th century. Palaeogeographical research has shown that further back in time, however, certain parts of the coast underwent dramatic changes.

The coastline between Thermopylai and Phalara has shifted eastwards under the influence of the river Sperchiós (figs 1.6 and 1.7). Strabon's description of this area, the distances between various landmarks and cities, the outlet of the Spercheios and the courses of the ancient rivers Dyras, Melas and Asopos have been closely studied (Radt, 1994: pp. 33–34). Strabon mentions that the city of Lamia lay 25 *stadia* from the sea, whereas the present shoreline lies more than 10 km from Lamia. Investigations in the area around the mouth of the Sperchiós (Kraft *et al.*, 1987) revealed that in 480 BC the shoreline lay close to Thermopylai. The site where in 480 BC a Greek army resisted a Persian army led by Xerxes nowadays lies buried beneath a 20-m-thick sediment. After 480 BC in particular the river transported enormous amounts of sediments into the Maliakós gulf. Deforestation of the mountains caused denudation of the soil and large amounts of sediments were washed downhill during heavy rainfall. This led to the formation of a wide mud plain, which also covered the original site of Thermopylai, situated on a narrow stretch of land between the mountains and the sea. Investigations by Kraft *et al.* (1987) have shown exactly how the coastline shifted between 2500 BC and AD 1972 and demonstrated the aforementioned rapid sedimentation between 480 BC and 1972.

The coastline of Vólos Bay, the northern part of the Pagasitikós gulf, has also shifted over the ages (fig. 1.8). In 3500 BC Vólos Bay extended up to the Late Neolithic site of Dhimíni. Between 3000 and 350 BC considerable sedimentation occurred here. The Late Bronze Age site of Iolkos and the city of Demetrias lay close to open water. According to Zangger (1991), the sedimentation in Vólos Bay was caused by deforestation, but this deforestation and the associated shifting of the coastline were on a much smaller scale than that at the mouth of the Sperchiós.

Smaller bays along the coast were closed off from the open sea by beach ridges, which created lagoons that later evolved into backswamps. This development has been studied in a backswamp along the shore of Soúrpi Bay, close to the area where the cities of Classical and New Halos lay (Reinders, 1988). The investigation showed that an inlet of Soúrpi Bay was gradually filled with marine sediments and that, around 3900 BC, a beach ridge was formed with an open lagoon behind it (Bottema, 1988; Van Straaten, 1988). It was on this beach ridge that the port of Classical Halos was situated (fig. 1.9). The lagoon gradually filled with sediments transported by the Amphrysos, the small river originating at Kefálosi spring. Nowadays a beach ridge constitutes an almost uninterrupted shoreline from the outlet of the Platanórrema to the outlet of the Salamvrias at the southern end of Soúrpi Bay (fig. 1.2).

It was not only perennial rivers like the Sperchiós that discharged sediments. In 1994 we observed that tremendous amounts of sediments are occasionally washed down from the mountains by rivers whose beds are dry in the summer. In October 1994, after a short period of heavy rainfall, the dry beds of the Xeriás and the Platanórrema in the Almirós plain (fig. 1.2) were filled with wild brown torrents. With a deafening noise, the 60-m-wide and 3–4-m-deep river Xeriás transported boulders, sand, pebbles, cobbles and the garbage of illegal dumps to the Pagasitikós gulf. In the upper part of the plain the river beds are quite wide, but along the rivers' lower courses they are narrow and shallow. Near the mouth of the Platanórrema large areas of arable land were flooded and covered with cobbles, pebbles and other sediments. The discharge of rainwater and sediments from the mountains led to sedimentation along the coast and to the creation of small underwater deltas beyond the mouths of the four rivers transecting the Almirós plain.

Geological and seismic research has shown that the Pagasitikós gulf is gradually silting up. In the central and eastern parts of the gulf, which are from 80 to over 100 m deep, there are Holocene marine sediments with thicknesses of up to 16 m (Mitropoulos & Michalidis, 1988). The greater part of the sea bed is covered with silt. Coarser sediments, sand, sandy silt and silty sand are to be found in the northern part of the gulf and at Vólos sill, the opening between the gulf and the open sea to the south between Tríkeri

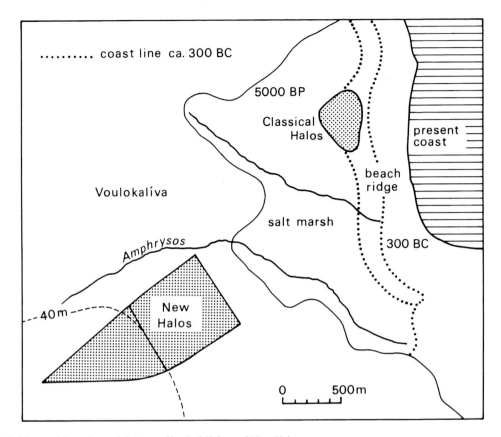

Fig. 1.9. Soúrpi bay and the salt marsh between Classical Halos and New Halos.

(fig. 1.6) and the mainland (Perissoratis *et al.*, 1988). The western part of the gulf, along the Almirós plain, is relatively shallow, with depths from 20 to 40 m. Coarser sediments, washed down from the mountains by the four rivers, are to be found along the west coast.

Generally speaking, the coastline has shifted only there where rivers like the Spercheios and Amphrysos empty into the Maliakós and Pagasitikós gulfs. Unfortunately, however, it is precisely in these areas that some important ports, such as Demetrias, Halos and Phalara, were situated in ancient times. The beach ridge on which Classical Halos was situated is now covered with 1–2 metres of sediment. Because of this, and the complicating factors of relative changes in sea level and relative land motion due to tectonic activity, we know fairly little about the ancient ports in this area.

2.4. Ports in Classical and Hellenistic times

Between Lamia and Demetrias there were many towns which, as we know from written sources and archaeological research carried out over the past 25 years, flourished in Hellenistic times in particular. The number of towns is indeed surprisingly high in compari-

son with what we know about other regions along the coast between Peiraieus and Thessalonike. Demetrias and New Halos were both founded in the early 3rd century BC (Marzolff, 1980; 1994; Reinders, 1988). Some of the other towns were founded at earlier dates, but were still occupied in Hellenistic times, in many cases with an enlarged enceinte. Only the towns of Antron and Pteleon are relatively unknown (fig. 1.6). The remains of the towns along the Maliakós gulf – Lamia, Phalara and Echinos – now lie buried beneath later buildings, but excavations have shown that these towns, too, flourished in Hellenistic times, and also in the Roman and Byzantine periods (Pantos, 1994).

Strabon describes the coast and the ports largely as they were in Late Hellenistic times, but he occasionally refers to Classical times, for example with respect to Pagasai, which information he may have borrowed from Apollodoros (figs 1.10 and 1.11). Some of the towns varied in importance in different periods. In Classical times Pagasai was the main port of Thessaly. In 377 BC Jason, the tyrant of Pherai, sent grain from Pagasai to the Boiotian city of Thebai (Xen. Hell. 5.4.56). It is also known from written sources that grain was transported from Thessaly to Kos in the 3rd century BC (Sherwin-White, 1978: p. 110).

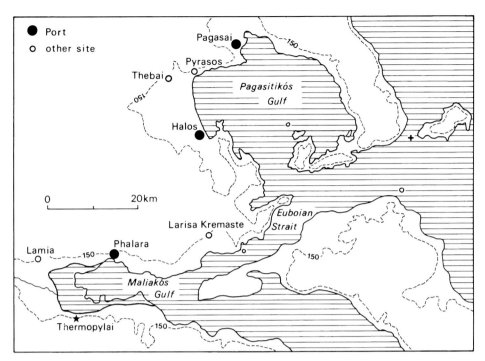

Fig. 1.10. Thessalian ports in Classical times.

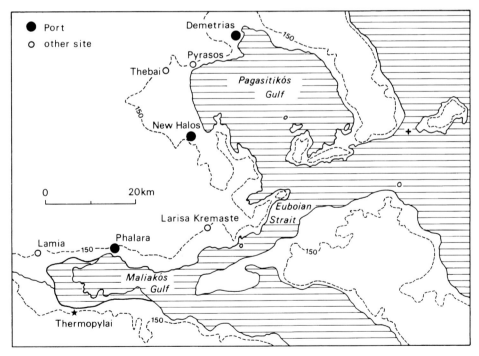

Fig. 1.11. Early Hellenistic ports.

Pagasai was in Classical times the port of the inland city of Pherai (Strabon 9.5.15). The town of Pagasai was probably situated on the Sorós, a conical rock, so a nearby stretch of beach will have been used to beach the vessels (Marzolff, 1994: p. 256). Intzesiloglou (1994: pp. 49–50), however, is of the opinion that the town on the Sorós was Amphanai, and that Pagasai was situated on the southern shore of Vólos Bay, near Pefkákia (fig. 1.8). The topography of the area around the gulf of Vólos is a much discussed topic. In Hellenistic times a newly founded city, Demetrias, was the naval station and residence of the Macedonian kings and was later to evolve into the chief port in this part of the Pagasitikós gulf. Pagasai became an insignificant town. The town of Demetrias must have had a good harbour near Pefkákia, where there is now a shipyard (Marzolff, 1980). Perhaps the southern shore of the small Vólos Bay was less silted up in Hellenistic times than it is today. Demetrias was undoubtedly an important port in Hellenistic times, although Strabon (9.5.15) mentions that the city's power had already begun to dwindle by his time.

The next town along the coast of the Almirós plain in Classical times was Pyrasos, the port of Phthiotic Thebai, situated near present-day Néa Anchíalos (fig. 1.10). Strabon (9.5.14) mentions the name of the port that succeeded Pyrasos in Late Hellenistic and Roman times: Demetrion or Demetreion. In the Byzantine period this town took over the position and the name of the inland city of Thebai and became the principal centre in this area (Asimakopoulou-Atzaka, 1982).

On the opposite side of the Almirós plain the important harbour of Classical Halos was favourably situated on the beach ridge south of the Platanórrema. Herodotos (7.173) reports that the Greek fleet disembarked an army of 10,000 men at Halos as Xerxes' army was approaching Tempe in 480 BC. In its sheltered position in Soúrpi Bay, the beach near Old Halos was evidently excellently suitable for beaching the vessels and disembarking such a large army. From Halos, the Greek army set off overland in the direction of Tempe, but returned to its ships when Xerxes avoided the route through the Tempe defile.

In 346 BC, an Athenian embassy en route to the Macedonian capital Pella disembarked at Halos while the city was besieged by a Macedonian army commanded by Parmenion (Demosthenes 19.163). The city managed to resist the siege of the Macedonian army thanks to the help of the Athenians, who were able to reach Halos overseas. The Athenians made peace, but excluded Halos from the peace treaty and left the Halians to their fate. The city was subsequently taken and its territory was given to the city of Pharsalos. The 11-ha-large *Magoúla* Plataniótiki has been identified as the site of Classical Halos (Reinders, 1993). At the end of the 4th century BC

New Halos was founded. Hellenistic Halos was situated at a strategic location, 2 km from the shore. We may assume that the beaches south of the Platanórrema were excellently suitable for beaching in Hellenistic times, too.

Harbours, or rather landing places, along the Maliakós gulf are also mentioned in written sources. In 302 BC Demetrios Poliorketes disembarked 32,000 men near Larisa Kremaste (Diodoros 20.110.3), which lay on a promontory, 2.5 km from the sea. The city of Echinos is not referred to as a port in written sources, but the town was in that respect ideally situated, at a distance of 1.5 km from a beach in a sheltered bay. Larisa Kremaste and Echinos both formed part of Achaia Phthiotis. At Echinos remains dating from the Classical period up to Roman times have been excavated (Fotini-Papakonstantinou, 1994: pp. 231–232). The next town along the coast of the Maliakós gulf was Phalara, the port of Lamia, situated in the district of Malis. At the beginning of the 20th century, Stählin was of the opinion that Phalara was situated near the mouth of the river Achelos. Although there is no sound evidence to support it, the present-day port of Stílis has been identified as the site of ancient Phalara, situated close to the sea (Pantos, 1994: p. 221). Phalara was destroyed by an earthquake in 476 BC, but excavations at Stílis have shown that the site was reoccupied from the second half of the 4th century BC onwards (figs 1.6 and 1.10).

Figures 1.10 and 1.11 show the main ports in Classical and Hellenistic times. Beaches near the aforementioned towns served as 'harbours', landing places or intermediate stations for voyages and troop movements; in those days there were presumably no harbours in the proper sense, with moles and quays. Ships could easily be hauled onto the beaches, or goods and people could be brought aboard and unloaded from vessels at an anchorage or roadstead with the help of smaller vessels, or simply by carrying the cargo through shallow water from the beach to a vessel and vice versa (Casson, 1971: fig. 191). Only occasionally does Strabon use the word 'port' for one of the towns mentioned along the coast from Thermopylai to Demetrias. The harbour of Demetrias he refers to as a *naustathmos*, which is translated as a 'naval base' in the Loeb translation. He uses different words for the harbours of Pagasai and Pyrasos, namely *epineion* and *eulimenos*, respectively, which are translated as 'seaport' and 'with a good harbour'. The words *naustathmos* and *epineion* are sometimes however translated as 'anchorage' or 'roadstead'. Considering the nature of the area, the latter translations are preferable in the case of sites like Demetrias, Pagasai and Halos.

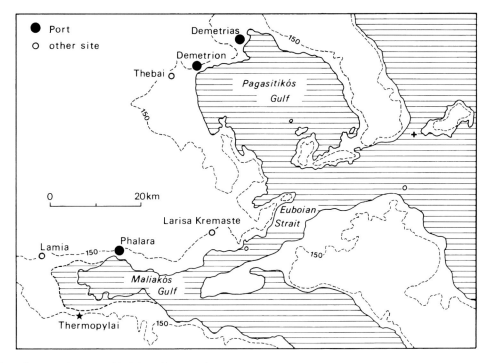

Fig. 1.12. Thessalian ports in Late Hellenistic and Roman times.

2.5. Maritime activity in the Roman and Byzantine periods

The coins recovered in the excavations of the remains of New Halos show that this Hellenistic city had overseas contacts. The city was abandoned in 265 BC, presumably after an earthquake (see sections 5.2 and 5.3). Phthiotic Thebai, in the northern part of the Krokion plain, then became the most influential city of this plain. Pyrasos, or Demetrion, the port of Phthiotic Thebai, is one of the three cities with harbours mentioned in an inscription of *c.* 151–150 BC. The towns of Thessaly undertook to deliver 430,000 *kophinoi* of grain to Rome: "That each of the cities arrange the transport of its allocated grain down to the harbour, whether that of the Demetreion or at Phalara or at Demetrias" (Garnsey *et al.*, 1984: p. 37). Stählin (1924: p. 173) assumed that the present-day harbour of Néa Anchíalos, in his days "*ein kleiner Weiher am Meere*", was a remnant of the old harbour of Demetreion. The position of this harbour and its long-distance trade are referred to in the inscriptions of many gravestones. The 2nd-century BC inscription mentioning the Thessalian grain transports shows that in Late Hellenistic times Demetrias, Demetrion and Phalara were the main ports of Thessaly (fig. 1.12).

The importance of Phthiotic Thebai gradually dwindled as the Byzantine city of Thebai rose to power. This city was situated at the site of the port of Demetrion, present-day Néa Anchíalos (figs 1.12 and 1.13). Remains of this city, including the ruins

of three basilicas, have been excavated at Néa Anchíalos. Near Aïdínion archaeologists found the remains of a small rural site with an olive press (Nikonanos, 1971: pp. 312–313). Byzantine Thebai flourished in the early Byzantine period, when it was the main centre in the Almirós plain. Surveys in the southern part of the Almirós plain led to the discovery of a large number of small rural sites dating from late Roman and Byzantine times. No other towns are mentioned in written sources and most of the archaeological remains from this period that were found during the surveys represent sites of only modest dimensions (Avramea, 1974; Malakasioti *et al.*, 1995).

Byzantine Thebai retained its powerful position until the end of the 7th century AD. Traces of a fire were found during excavations at Thebai. There is evidence for occupation in the 8th and 9th centuries, but by that time Thebai was no longer an important port (Koder & Hild, 1976: p. 271). Demetrias (Dimitriada), in the northern part of the Pagasitikós gulf, is mentioned as a port in written sources until the Ottoman conquest of Thessaly in 1393 (Koder & Hild, 1976: p. 271).

In mid-Byzantine times, before the Ottoman conquest, the city of Almiros (also called Almiro or Armiro) was an important port (fig. 1.14). In the 12th and 13th centuries Almiros and Saloniki were among the principal ports of the Byzantine empire. Communities of Jews and of people from Venice, Pisa and Genua lived in Almiros (Koder & Hild, 1976: pp. 170–171; Savvidis, 1993; Rizos, 1996). Within and

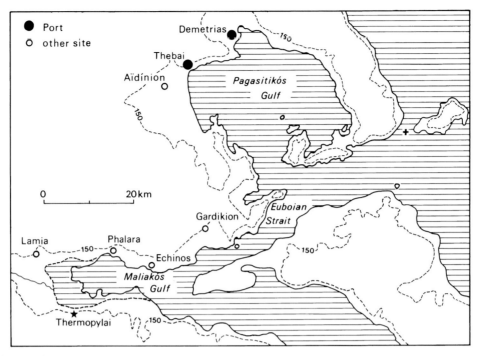

Fig. 1.13. Early Byzantine ports.

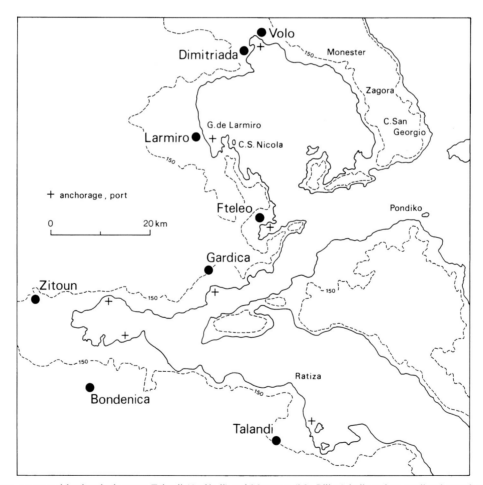

Fig. 1.14. Ports, towns and landmarks between Talandi (Atalándi) and Monester (Mt. Pílion) indicated on medieval portolan charts.

outside an enceinte close to the outlet of the river Xeriás, east of the present-day town of Almirós, glazed sherds dating from the 12th and 13th centuries have been found. In these centuries Almiros was the main port along the coast of the Pagasitikós gulf. Benjamin of Tudela, who visited Almiros, mentions that the town had some 400 Jewish inhabitants (Adler, 1907: p. 11). Archaeological surveys in the Almirós plain yielded evidence for the existence of a fort, monasteries and rural sites in this area in the same period. These sites indicate that medieval Almiros was not merely a commercial city, but presumably also had a hinterland in the Almirós plain (Dijkstra *et al.*, 1997).

Periploi were used in Byzantine and medieval times, too. Little is known about navigation in Byzantine times, but the *Stadiasmos i periplous tis Megalis Thalassis* provides specific data on the distances along the coasts of the eastern Mediterranean and also information on the nature of harbours, prevailing winds, the presence of anchorages and water, *etc.* (Delatte, 1947: p. XIX; Müller, 1965: p. CXXVI). This *periplous* is thought to be a Byzantine adaptation of a *stadiasmos* from the 3rd century AD. Unfortunately it contains no information on the coasts of the Greek mainland.

The medieval portolans, coastal descriptions and portolan charts were to a certain extent the successors of the *periploi* and *stadiasmoi*. The importance of the coastal stretch between Thermopylai and Demetrias is reflected in these portolans. After the fall of Constantinople in 1204, the territory of the Byzantine empire was divided amongst the crusaders. Although Thessaly was assigned to the Peregrini, the non-Venetian crusaders, the harbour of Fteleo remained under Venetian control (fig. 1.14). Several important ports and various other towns and landmarks are shown on portolan charts drawn in the 14th–16th centuries (Nordenskiöld, 1897). The contours of the coast

roughly indicate protruding capes and sinuosities; the coastline resembles a serrated holly leaf.

A number of towns between Talandi (Atalándi) and Monester (Mount Pílion) are indicated on almost all the maps from this period (fig. 1.14 and table 1.2). Many of the capes and ports can be identified without problems. Only the names near the river Sperchiós are not clear: placenames like Longiton, or Lo Giton, and Lambena, or Ladena, occur on some charts, sometimes in reversed order. The city of Lamia – also called Zeitoun or Zitouni in the Middle Ages – is not indicated under either of those names on the portolan charts. The name Zitouni was used to indicate the *Kastro* of Lamia: *Castrum de Situm super Ravenica*, which was rebuilt after 1204 (Fotini-Papakonstantinou, 1994b: p. 10). Ravenica was a medieval city that was probably situated south of Lamía (Koder & Hild, 1976: p. 251). According to Koder & Hild (1976: pp. 283–284), the names Gripton and Gitone were also used to indicate the city of Lamia between 1218 and 1259, when the city was again in Greek possession. The placenames Longito, Lomgito, lo Gittom, and also Lambena and Ladena occur on the Portolan charts in the neighbourhood of the outlet of the Sperchiós; they presumably refer to the city of Lamia (Kretschmer, 1909: p. 673).

An important stronghold in medieval times was the castle of Bondenitsa, present-day Mendenítsa, south of Thermopylai, defending a pass along the route from northern to central Greece. There was evidently also a port with the same name (Koder & Hild, 1976: p. 222) – *habemus portum et castrum Bondonicie* – which was important for the grain trade (Rizos, 1996). Among the other towns occurring on the portolan charts besides Almiros and Dimitriada are Gardikion (Gardikia hetera, Pelasyía) and Fteleo (Pteleó). Gardikion is said to have had a harbour in a written portolan. This harbour must have lain on the shore of Gardíkios Bay, because the city itself lay inland, at

Table 1.2. Place-names between Talandi and Monester on late medieval portolan charts.

	1318	1339	1375	1384	1384	1413	1467	XV	1500	1550	1552	1568
Talandi	+	+	+	+	+	+	+			-	+	+	+
Ratiza	+	+	+	+	+	+	-	+	-	-	-	-
Bondenica	+	+	+	+	+	+	-	+	+	+	-	-
Ladena, Lambena	-	+	+	-	-	+	+	-	-	+	+	+
Longito	-	-	-	+	+	-	+	+	+	+	+	+
Gardica	-	+	+	+	+	+	+	+	+	+	+	+
Feteleo	+	+	+	+	+	+	+	+	+	+	+	+
C. S. Nicolo	+	+	+	-	-	+	+	-	-	+	+	-
G. de Larmiro	+	-	-	+	-	-	+	+	+	+	-	-
Larmiro	+	+	+	+	+	+	-	+	+	+	+	+
Demitriada	+	+	+	+	-	+	+	+	+	+	+	+
Volo	-	-	-	-	-	-	-	+	+	-	+	+
C. S. Georgio	+	-	-	+	+	+	+	+	+	+	+	+
Monester	+	+	+	+	+	+	+	+	+	+	+	-

the site of ancient Larisa Kremaste (Koder & Hild, 1976: p. 161). Fteleo was an important Venetian stronghold until the fall of Negroponte in 1470, when the town was surrendered to the Ottomans and its inhabitants were deported to Constantinople (Koder & Hild, 1976: p. 241). Table 1.2 presents a survey of the towns and landmarks on the portolan charts.

Protruding capes give the coastlines on the portolan charts their peculiar shapes. Two capes were clearly important landmarks along the coast (fig. 1.14). Cape San Nicolo, or simply San Nicolo, is consistently mentioned after Fteleo, so it cannot be identified as the Cape Poseidion of ancient times or the Cape Stavrós of present-day maps. Along the coast north of Fteleo, Mitzéla Bay in the shelter of the island Áyos Nikólaos provided good anchorage. Byzantine remains have been found on this island (Iakovidis, pers. comm.), suggesting that it may well be San Nicolo of the portolan charts, as Kretschmer (1909: p. 637) already suggested.

Cape Sepias, on the southeast coast of Magnesia, was called Cape Ayos Yeoryos Zagora in medieval times, Zagora being the Slavic name for Mount Pílion. Almost all portolans indicate Monester, another medieval name for Mount Pílion. Mount Pílion was called *Kellia*, (monks') cells, in Greek, because of the great number of monasteries that were to be found on it (Koder & Hild, 1976: p. 186). A written portolan also mentions a Cape Monester situated along the east coast of Magnesia: "*Da cauo di moster a cauo di verliqui 35 miglia tra maestro e tramontana*" (Kretschmer, 1909: p. 323). If *cauo di verliqui* is Kávo Dhermatás (Cape Kíssavos), *cauo di moster* may be Cape Damoúchari. In addition to the capes and Mount Pílion, the outlet of the Sperchiós is indicated on many portolan charts.

The written Italian portolans prove that navigation was not restricted to the route between Euboia and the mainland, but also extended to the Aegean islands and routes further afield, if weather conditions permitted. The results of archaeological surveys in the Almirós plain suggest that most mid- and late Byzantine sites lie in the coastal zone (Dijkstra *et al.*, 1997), but little detailed information has so far been obtained in surveys covering the entire region. Many mid-Byzantine forts are moreover known along the east and west coasts of Magnesía, the east shore of Lake Kárla and to the west of New Halos. These strongholds and observation posts were intended to protect the Byzantine trade routes against piracy and incursions, rather like Frankish strongholds such as Bondenitsa and Platamona.

2.6. The Ottoman conquest of Thessaly

The Venetian documents of the 11th–13th centuries mention only Almiros as a port in Thessaly. Important Thessalian trade products were grain from the Thessalian plains and foothills, oil from the olive groves around the Pagasitikós gulf and wine from Fteleo (Rizos, 1996; Avramea, 1974: pp. 65–66). In the 13th century Venetian influence in Negroponte (Chalkis; figs 1.16 and 1.17) and Fteleo (fig. 1.14) increased and the latter took over Almiros' position as the most important port in the region. At the beginning of the 15th century the island of Euboia, also called Negroponte after the main city on this island, the Maliakós gulf and the gulf of Fteleo were still under Venetian control, whereas Thessaly and Macedonia had by this time already been incorporated in the Ottoman empire. After 1393 the town of Volos, Volo or Golos (fig. 1.14), which was first mentioned in 1333, took over Demetrias' position. A Turkish garrison was stationed at Golos, which was situated at the site of present-day Iolkós (Koder & Hild, 1976: pp. 165–166). During the Ottoman domination, from 1393 until 1881, it was the only important city and port in the area of the Pagasitikós gulf. In this period, too, grain was transported from Thessaly to Venice. The sultan's permission was needed for this transport and this led to grain smuggle (Rizos, 1996; Van Keulen, 1728). The portolan charts do not pay much attention to the political changes, although the Turkish port of Volos (Golos) is indicated on portolan charts from the end of the 15th century.

Besides portolan charts there are also written portolans. Delatte (1947) published a number of 16th-century Greek manuscripts containing coastal descriptions dating from the late Byzantine period, or perhaps from after the Ottoman conquest of Thessaly, but while the city of Negroponte and the island of Euboia were still under Venetian control. The choice of words reflects influences of both the Frankish and Venetian languages. In comparison with the *periploi* and the late-medieval portolan charts, these written portolans provide conspicuously little or no information on the ports along the coasts of the Maliakós and Pagasitikós gulfs; the ports of Almiros, Dimitriada and Fteleo along the Pagasitikós gulf are not mentioned at all. Instead, detailed information is given on the island of Euboia (Egripou) and the harbours of Negroponte (present-day Chalkís) and Karistos (Delatte, 1947: pp. 224–226). Unlike the portolan charts, the written portolans do mention the city of Lamia (Zeitoun or Zitoun, fig. 1.14). The distances between Lamia, Platamona and Saloniki are only roughly indicated, but much attention is paid to the passage between Skiáthos and the mainland (Delatte, 1947: pp. 296–297).

The lack of information on the Pagasitikós gulf is all the more surprising in view of the abundant, detailed information that is provided on the Aegean islands, for instance on the islands of Skiáthos, Skópelos and Alónnisos. This must mean that in late-Byzantine times, or perhaps in the period that Thessaly was under Ottoman rule while Negroponte (Euboia) was still under Venetian control, maritime

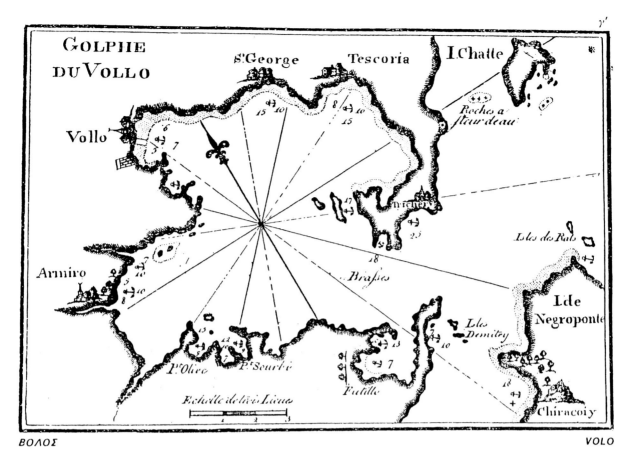

BOΛOΣ   VOLO

Fig. 1.15. Pagasitikós gulf with anchorages, shallows and landmarks such as trees, mosques, churches, a saltpan and a jetty; depths are indicated in fathoms (Roux & Allezard, 1800; reprint Maritime Museum of Greece, 1981).

activity shifted to the Aegean islands, which then started to play a prominent role in shipping. The written portolans mention the harbours and ports of every single island in the Aegean, even the tiniest. Under Ottoman rule the islands had a relatively independent status.

A corresponding amount of detail with respect to the islands and lack of information on the Maliakós and Pagasitikós gulfs is observable on the maps of 163 ports in the Mediterranean published by Roux and Allezard in 1800 (fig. 1.15). Among these maps there is however one of the Pagasitikós gulf, indicating depths in fathoms, shallows, and the rock between Skiáthos and Áyos Yeóryios on the mainland: *Roches a fleur deau*, which was called Myrmex in Classical times. Moreover, many anchorages in the gulf are indicated, for example near the Turkish towns of Armiro and Vollo and in Port Olive behind the island Áyos Nikólaos, and also various landmarks, such as the *Isles des Rats*, which in the late medieval portolans are indicated as Pondiko, present-day Pondikonísi. The map by Roux and Allezard is one of the two printed maps of the gulf of Vólos mentioned in Zacha-

rakis' catalogue (1992) – a very small number compared with the hundreds of maps available for the Aegean islands.

We also find evidence for maritime activity and an orientation towards the Aegean islands in the activities of the Greek communities on Mount Pílion (fig. 1.16). The occupants of the Greek villages on Mount Pílion engaged in shipping from the 16th century onwards; Zagoriana was the general term used for the vessels of these mountain villages (Makris, 1982: p. 182). One of these maritime villages without a harbour was Mitzéla on the east coast of Mount Pílion. When the Turks set fire to the village in 1828, the villagers fled to the Aegean islands; before that year many of them had fought in the struggle for independence as sailors or soldiers. Many families went to the nearby island of Skópelos; 70 families from Mitzéla are known to have been living on that island in 1829: a total of 295 persons, 147 male and 148 female. The heads of 37 families were sailors, others were soldiers or labourers; the heads of ten families were widows (Kalianos, 1984). In 1834 the Greek authorities granted the fugitives from Mitzéla permis-

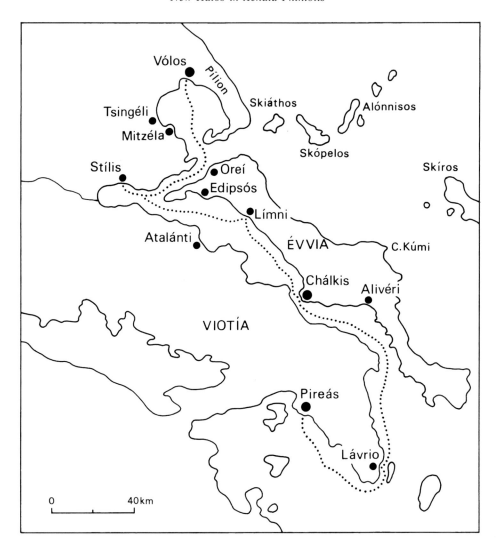

Fig. 1.16. Sea route and ports of call between Pireás and Vólos at the beginning of the 20th century (after Baedeker, 1910).

sion to found a new settlement, which they called Néa Mitzéla, also known as Amaliápolis, on the western shore of the Pagasitikós gulf (fig. 1.16).

During Ottoman rule, the Aegean islands were relatively independent. It was on the Greek islands that a Greek merchant navy developed (Sphyroeras *et al.*, 1985: pp. 9–21). Well-known Greek merchant vessels, such as *perama, trechandiri* and *bratsera*, can be traced back as far as the 17th century. These vessels originally weighed 3 tons. Later the maximum capacity ranged between 30 and 40 tons (Damianidis & Zivas, 1986), but larger vessels were also built. A network of maritime activity linked the individual islands, but the islanders also sailed to Smyrna and Constantinople and traded with European cities (Sphyroeras *et al.*, 1985). The Ottoman control of Thessaly and the Ottomans' dense network of land routes were not the only reasons why the sea route

between Euboia and the mainland fell into disuse; the islanders also had larger sea-going vessels, which were not restricted to sheltered passages and coastal navigation.

The importance of the Pagasitikós and Maliakós gulfs in maritime history is reflected in the number of ports that were to be found along their coasts through the ages. Phalara, Halos, Pyrasos, Pagasai and Demetrias in Classical and Hellenistic times; Phalara, Demetrion, Thebai and Demetrias in Roman and Byzantine times; Bondenitsa, Fteleo, Almiros and Dimitriada in the 12th–14th centuries AD. After the Ottoman conquest of Thessaly in 1393, all the ports along the Maliakós and Pagasitikós gulfs declined and lost their power to the Turkish port of Volos (Golos), the only port in Thessaly after the conquest of Negroponte and the deportation of the inhabitants of Fteleo to Constantinople in 1470.

2.7. Sea and land routes

Maritime sites along the Pagasitikós gulf played an important role in maritime history as ports of call for Thessaly, but what do we actually know about sea and land routes? Nowadays there are many land routes covered by buses and – less common – trains in Greece. Aeroplanes connect the capital Athens to remote parts of the country and a dense network of ferry boats links the major ports on the mainland with the islands in the Aegean and Ionian seas. Most of these connections are related to the tourist industry, but wooden merchant vessels still operate between the harbour of Vólos and the northern Sporádhes: Skiáthos, Skópelos and Alónnisos.

Until 1881, Thessaly and northern Greece formed part of the Ottoman Empire. The coastal section between Thermopylai and Demetrias had been split into two parts, divided between Greece and the Ottoman Empire. Between 1834 and 1881 the border ran from the gulf of Árta in the west to the Pagasitikós gulf in the east. Vólos and Almirós were in Turkish hands while the northern shore of the Maliakós gulf up to the peninsula of Amaliápolis formed part of the Greek kingdom. After the liberation of Thessaly, it was unusual to travel from Athens to Thessaly across land. From the schedules published by Baedeker (1910) we know that boats departed from Pireás to Vólos every week. The steamers of the early 20th century followed the route between the mainland and the island of Évvia, and halted at Lávrion, Alivéri, Chálkis, Límni, Atalánti, Edipsós, Stílis, Oreí and other places (table 1.3 and fig. 1.16). Baedeker (1910: pp. 213–217) describes this route as excursion No 16, *D'Athènes à Vólo par mer*, and provides detailed information on coasts, harbours and antiquities, continuing as it were in the tradition of the *periploi* of Classical and Hellenistic times.

Compared with travelling by train, a voyage by boat was relatively cheap and comfortable in those days. When the archaeologists Wace and Thompson excavated at *Magoúla* Zerélia in the Almirós plain they went by boat from Pireás to Vólos. "We left the

Piraeus by the steamer Ares at 8 P.M. on Wed. June 2nd 1908 and reached Volo on June 3rd at 2 PM. The voyage was without incident – the steward provided an excellent lunch, and Droop threw over board Ruskin's Art of Drawing" (Wace's diary). After they had visited the *ephoros* in the Museum of Vólos, they travelled overland to Almirós. Some steamers stopped at Néa Mitzéla and Tsingéli, on the shore of Soúrpi Bay, close to the site of Classical Halos (Baedeker, 1910: p. 214; Reinders, 1988: p. 161). From Pireás the steamer followed the route between Viotía (Boeotia) and Évvia (Euboia) through the narrow passage at Chálkis and from there to the Maliakós gulf and the strait of Tríkeri into the Pagasitikós gulf (fig. 1.16). The voyage from Pireás to Vólos took about 18 hours.

From the *periplous* of Skylax and Strabon's description of the coast between Peiraieus and Demetrias we may infer that ships in Classical and Hellenistic times followed the same sea route as those sailing in the early 20th century. This is confirmed by three other written sources. Herodotos (7.173) refers to the disembarkation of 10,000 men near Halos in 480 BC, while Demosthenes (19.163) mentions that in 346 BC the Athenian embassy sailed to Halos via the Euripos (fig. 1.5). In 302 BC Demetrios Poliorketes gathered his fleet and his troops at Chalkis, sailed along the coast and landed near Larisa Kremaste (Diodoros 20.110.2).

At first glance the sea route between Peiraieus and Demetrias offered sailors many landmarks for orientation and involved only few problems. The 19th-century sea maps of the British Admiralty show an abundance of protruding capes, towns and conspicuous trees for orientation. Near the shore were shallows and submarine rocks, but long stretches of the coast presented no problems, with only one exception. On their route to the north, the sailors had to negotiate an obstacle: the Euripos, a narrow channel between Chalkis and the mainland. In the Euripos a strong current changes direction at irregular intervals, a phenomenon already described by Strabon (9.2.8; quotations from Strabon's Geography have been taken from the edition of the Loeb Classical Library, translated by H.L. Jones): "Concerning the Euripus it is enough to say only thus much, that they are said to change seven times each day and night."

Under normal conditions the route from Chalkis northwards involved no problems, but the channel between Lokris, Malis and Euboia is bordered by two active tectonic faults. Earthquakes occur in this area. We know that earthquakes occurred here in Classical and Hellenistic times, too. In 426 BC the town of Phalara was destroyed by a *tsounámi*, a sea wave: "and as for Echinus and Phalara and Heracleia in Trachis, not only was a considerable portion of them thrown down, but the settlement of Phalara was overturned, ground and all." In Atalanta near Euboia "a

Table 1.3. Weekly Pireás–Vólos schedule of the John MacDowall & Barbour company: 'Tsón', Athens (Baedeker, 1910).

Day	Time	Harbour	Day	Time
Monday	19.00	Pireás	Friday	17.00
Monday	22.30	Lávrion	Friday	13.00
Tuesday	06.00	Chálkis	Friday	02.00
Tuesday	09.30	Límni	Thursday	22.30
Tuesday	12.00	Stílis	Thursday	19.00
Tuesday	20.00	Vólos	Thursday	12.00

Prices (excluding food; first and second class): Pireás-Lávrion, 6 dr., 4 dr.; Pireás-Chálkis, 8 dr., 5 dr; Pireás-Vólos, 15 dr., 10 dr.

trireme was lifted out of the docks and cast over the wall." This information Strabon (1.3.20) derived from Demetrios of Kallatis, who wrote a treatise on earthquakes.

These circumstances were of course exceptional. In 480 BC the Greek army sailed without problems to the Pagasitikós gulf, from where it travelled overland to its destination. The army had reason to travel partly overland instead of going all the way to Tempe by ship: the sea route was safe for only part of the year. "The whole voyage along the coast of Pelion is rough, a distance of about eighty *stadia*; and that along the coast of Ossa is equally long and rough" (Strabon 9.5.22). With eastern winds, the route along the east coast of the Pelion and Ossa was particularly dangerous. The voyage along the east coast of Pelion and Ossa was however considerably longer than Strabon suggested. In 480 BC Xerxes' Persian fleet "was lying in wait (on a small beach between Kasthaniá and Cape Sepias) when, a violent east wind bursting forth, some of the ships were immediately driven high and dry on the beach and broken to pieces on the spot" (Strabon 9.5.22). Another obstacle on the route to the north along the east coast of Magnesia is a rock between the mainland and the island of Skiathos (fig. 1.6). The Persians left a stone column on the rock, called Myrmex, as a sign for the other vessels (Herodotos 7.183).

On account of storms and poor visibility in the winter, the sailing season in Classical and Hellenistic times was generally restricted to the summer; the season lasted from 10 March until 10 November at best (Casson, 1971: p. 270; Vegetius, *mil.* 4.39). The sailors had to rely on landmarks and especially with poor visibility the route between the island of Euboia and the mainland was too dangerous. In wintertime the seas were 'closed' (Vegetius, *mil.* 4.39). Even nowadays sailing along the east coasts of Évvia and Pílion in wintertime is not without danger. In 1996, on the 29th of December, the *Dístos*, carrying a cargo of cement, was hit by two large waves on the broadside and capsized off Kími on the east coast of Évvia (fig. 1.16).

In early and mid-Byzantine times the route to Thessaly between Évvia and the mainland was still used for the transportation of Thessalian products to Venice, Pisa, Genua and other maritime cities (Koder & Hild, 1976: pp. 101–102). In the 12th century a city like Almiros also had trade connections with Saloniki and Constantinople. The written portolans, although dating from a later period, show that the route to Saloniki ran from Zeitoun to Gardika, Ayos Yeoryos Zagora, along the east coast of Magnesia, via Seta (Cape Pourion), Verliqui and Platamonas to Thessaloniki (Delatte, 1947: p. 226). Koder and Hild believe that there were also connections between the ports along the Pagasitikós gulf and the northern Sporádhes, as suggested by the wreck of a Byzantine vessel that

was found near the island of Alónnisos in 1970. The cargo of this vessel, consisting of over 1500 plates and amphoras, yielded a mid-12th-century date for the wreck (Asimakopoulou-Atzaka, 1982: pp. 168–169; Kritzas, 1971). The glazed plates were beautifully decorated with birds, mammals and geometrical and floral patterns. The same kind of pottery was found during surveys in the Almirós plain (Dijkstra *et al.*, 1997). According to Throckmorton (1971), the vessel "resembles a modern caique rather than a tenon fastened Roman ship."

Strabon's Geography, the *periploi* and the portolans give us a sailor's view of the coastal area between Thermopylai and Demetrias. From the 4th century BC until AD 1393 Thessaly, especially the districts of Malis and Achaia Phthiotis, had a maritime orientation; Thessaly was reached overseas from the south via a great number of ports of call along the coasts of the Maliakós and Pagasitikós gulfs. After the Ottoman conquest the old ports disappeared; their role was taken over by a dense network of land routes (Rizos, 1996).

The network of land roads, cobbled roads, or *kalderimis*, and bridges covered the whole of northern Greece and the Balkans. European travellers in the 19th century made use of these land roads and mentioned long caravans along their routes. At regular intervals were *khans* with overnight facilities for human and animal travellers. Small sites called *to chani* are still to be found here and there today, like the *khan* at the Foúrka pass north of Lamía and that at the pass near Brálos. There was also a *khan* just west of Kefalosi spring in the Almirós plain (fig. 1.6), as we know from Wace's diary, but it lay in ruins by the time Wace and Thompson carried out their excavations in the area east of Plátanos.

In Turkish times goods were transported by oxdrawn carts, by pack animals and also by camels, to be traded at annual fairs (Ritzos, 1996). These annual fairs more or less took over the function of the ports of call in previous periods. Although no documentation is currently available on land roads, we may assume that before the Ottoman conquest there were land roads between the various districts of Thessaly and a nearby harbour, or between an inland town and its harbour, as in the case of, for instance, Pherai–Pagasai, Thebai–Demetrion and Lamia–Phalara. These land roads will however not have constituted a dense network. The Ottomans provided the infrastructure for this network and Christian communities enhanced it by building bridges (Rizos, 1996). The network was not restricted to Thessaly, but connected Thessaly to central Greece, Macedonia, Constantinople, Anatolia and central Europe.

We know almost nothing about the role of the harbour of Volos and the sea route between Euboia and the mainland during Ottoman rule. Two Turkish portolan charts in the Khalili collection (Soucek,

1996) show accurate, detailed geographical representations of the coast of Volos (Koloz) and the island of Euboia (Igriboz). Both charts, which are in the tradition of Piri Reis, indicate that the gulf of Volos and the passage along Negroponte were important for the Ottomans, at least for the Ottoman navy, for which Piri Reis' 'Book of the sea' was intended. A portolan chart of the Aegean by Mehmed Reis, dated 1590, shows the same attention to detail for the whole Aegean (Biadene, 1990: pp. 94–95). The dates of the first two portolan charts are not known with any certainty, but the Euboia chart closely resembles the chart of Euboia from the *isolario* of Bartolomeo dalli Sonetti (1485) and the maps of Benedetto Bordone (1528); the Turkish portolan charts are obviously based on the Italian printed atlases of the Aegean islands (Sphyroeras *et al.*, 1985: p. 28).

West European sailors never used the old sea route to Thessaly through the Euripos. Trade in the Mediterranean began to flourish after 1590. A passage in Van Keulen's *Zeefakkel* shows that it was simply impossible for large vessels to pass the Euripos; it was only just wide enough for a galley without oars: "*maer van 't Kasteel tot de stad is een windbrug om op te halen als 'er kleyne scheepen en Galeyen door halen, want t'is er niet wy'er, als dat er een galey zonder riemen deur kan*" (Van Keulen, 1728). West European sailors occasionally called at the port of Volos. Van Keulen describes the sea route from Makronisos, the island east of Cape Sounion, via the islands of Schiro (Skíros), Schopelo (Skópelos) and Sciatta (Skiáthos) to the entrance of the Pagasitikós gulf (table 1.4 and fig. 1.17). He describes the anchorages in the gulf of Vólos, among which were anchorages near the village of Seigne (Fteleo) in Pteleós Bay, in a bay near *Moordenaars Eyland* and between the island of Tríkeri and the mainland (fig. 1.15). Although Van Keulen's map is not accurate, *Moordenaars Eyland* is without doubt Áyos Nikólaos Island, which is indicated on the portolan charts as Cape San Nicolo (figs 1.14 and 1.17).

What had West European sailors to look out for in this part of the Aegean? Van Keulen provides one

warning: in describing an anchorage off the west coast of Skópelos, in Panórmos Bay, he mentions that although this anchorage is good, he nevertheless advises sailors to anchor behind Dhásia Islands, two small islands north of Panórmos Bay. He explains that it is difficult to leave Panórmos Bay with western winds and warns sailors against the risk of the sudden appearance of Turkish galleys in search of vessels loading grain without permission: "*om niet beset te worden van de Turcksche Galeyen die de scheepen hier kooren ladende somtijds komen betrappen.*" From this we may infer that the Turkish tsiflíkia in the Almirós plain produced a surplus of grain and that it was worthwhile to attempt to export the grain without obtaining the necessary Turkish permit.

2.8. Conclusion

The section of the coast between Thermopylai and Demetrias formed part of the sea route from Peiraieus to Thessalonike. In Classical, Hellenistic and Byzantine times there were many ports along the shores of the Maliakós and Pagasitikós gulfs. Written sources inform us about the sea route from Peiraieus via the Euripos channel to these ports, which served as ports of call for the large Thessalian plains. They were also important as intermediate stations for voyages and troop movements to Tempe and Macedonia, because the sea route from Demetrias to Thessalonike was dangerous. The coast between Cape Sepias and the river Peneios was rough; only few beaches provided shelter in the event of gales.

During a long period, spanning seventeen centuries, the predominance of maritime activity shifted from one port to another, but the ports of call along the Maliakós and Pagasitikós gulfs retained their important positions until Thessaly, Lokris and Euboia came under Ottoman control in the 15th century (fig. 1.18). With the exception of Golos (Vólos), almost all the ports lost their former importance. After 1393, when Thessaly came under Ottoman rule, a dense network of land roads connected Thessaly to central

Table 1.4. Sea route from Macronisi (Makrónisos) to Negroponte (Chalkís) and Volo (Vólos), according to Van Keulen (1728).

From	To	Course	Distance
Macronisi (Makrónisos)	Asturi Isl. (Petalí Isl.)	northeast	4–5 miles
Asturi Isl.	Suhi de Basa (Stíra, Kavalianí)	northwest	3 miles
C. Negro (C. Alivéri)	Volie	northwest	3–4 miles
Volie	Strait S. Marco (Euripos)	west	1 mile
Strait S. Marco	Negroponte (Chalkís)	-	-
C. Martelo (C. Mandíli)	Schiro (Skíros)	north	11 miles
P. S. Georgio de Schira	Schoppelo (Skópelos)	northwest	6 miles
Schoppelo (north point)	Sciatta (Skiáthos)	west	2 miles
Sciatta (southcoast)	Gulf of Volos	west-southwest	3–4 miles

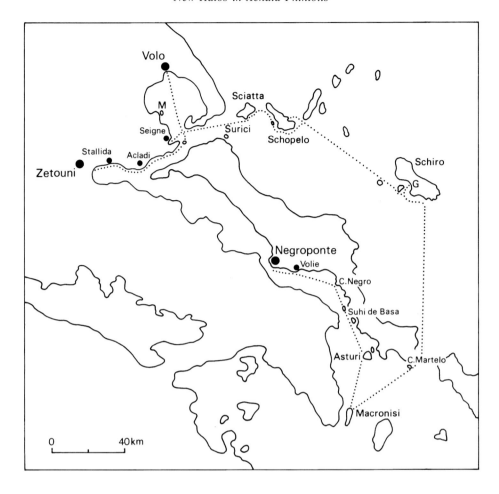

Fig. 1.17. Sea routes from Macronisi (Makrónisos) to Negroponte (Chalkis) and Volo (Vólos) after Van Keulen (1728); M = Moordenaars Eyland (Cape San Nicolo).

Greece and, via northern Greece, to the Balkans and Constantinople.

In the previous section the long-term development, extending beyond the Classical–Hellenistic period, of the coastal area of Magnesia and Achaia Phthiotis has been discussed in order to determine the positions of the coastal towns, and in particular those of Classical and Hellenistic New Halos. Classical Halos was situated close to the sea and was mentioned as a port in classical times. No written evidence is available for New Halos. Despite its strategic position at a short distance from the sea, we may assume that New Halos was also a port which had contacts in particular with the towns on Euboia and probably served as a port of call for Achaia Phthiotis. At the beginning of the 3rd century BC at least, Magnesia and Achaia Phthiotis had four important coastal towns; it is questionable whether we should see Achaia Phthiotis as a marginal district of Thessaly around this time.

2.9. Note

1. Strabon's description of the coast between Thermopylai and Demetrias (quotations have been taken from the edition of the Loeb Classical Library, translated by H.L. Jones):

Thermopylae, then, is separated from Cenaeum by a strait seventy stadia wide; but, to one sailing along the coast beyond Pylae, it is about ten stadia from the Spercheius; and thence to Phalara twenty stadia; and above Phalara, fifty stadia from the sea, is situated the city of the Lamians ... (Strabon 9.5.13). The Spercheius is about thirty stadia from Lamia, which is situated above a certain plain that extends down to the Maliac Gulf (Strabon 9.5.9).

... and then next, after sailing a hundred stadia [from Phalara] along the coast, one comes to Echinus, which is situated above the sea; and in the interior from the next stretch of coast, twenty stadia distant from it, is Larisa Cremastê (it is also called Larisa Pelasgia) (Strabon 9.5.13).

Then one comes to Myonnesus, a small island; and then to Antron ... Near Antron, in the Euboean strait, is a submarine reef called "Ass of Antron" (Strabon 9.5.14).

... and then one comes to Pteleum and Halus (Strabon 9.5.14).

And Artemidorus places Halus on the seaboard as situated outside the Maliac Gulf, indeed, but as belonging to Phthiotis; for proceeding thence in the direction of the Peneius, he places Pteleum after Antron, and then Halus at a distance of one hundred and ten stadia from Pteleum (Strabon 9.5.8).

The Phthiotic Halus is situated below the end of Othrys, a mountain situated to the north of Phthiotis, bordering on Mount Typhrestus and the country of the Dolopians and extending from there to the region of the Maliac Gulf. ... Halus is called both Phthiotic and Achaean Halus, and it borders on the country of the Malians, as do also the spurs of Othrys Mountain .... Halus (either feminine or masculine, for the name is used in both genders) is about sixty stadia distant from Itonus... It is situated above the Crocian Plain; and the Amphrysus River flows close to its walls (Strabon 9.5.8).

... and then to the temple of Demeter; and to Pyrasus, which has been rased to the ground.... Pyrasus was a city with a good harbour; at a distance of two stadia it had a sacred precinct and a holy temple, and was twenty stadia distant from Thebes (Strabon 9.5.14).

Below the Crocian plain lies Phthiotic Thebes (Strabon 9.5.8). Thebes is situated above Pyrasus, but the Crocian Plain is situated in the interior back of Thebes near the end of Othrys; and it is through this plain that the Amphrysus flows.... Now Phylacê is near Phthiotic Thebes ... it is about one hundred stadia distant from Thebes, and it is midway between Pharsalus and the Phthiotae (Strabon 9.5.8).

... and then to Cape Pyrrha, and to two isles near it, one of which is called Pyrrha an the other Deucalion. And it is somewhere here that Phthiotis ends (Strabon 9.5.14).

The seaport of Pherae is Pagasae, which is ninety stadia distant from Pherae and twenty from Iolcus.... Now Pherae is at the end of the Pelasgian Plains on the side towards Magnesia; and these plains extend as far as Pelion, one hundred and sixty stadia (Strabon 9.5.15).

Demetrias, which is on the sea between Nelia and Pagasae, was founded by Demetrios Poliorcetes, who named it after himself .... Furthermore, for a long time this was both a naval station and a royal residence for the kings of the Macedonians (Strabon 9.5.15). Iolcus is situated above the sea seven stadia from Demetrias.

3. THE INVESTIGATION OF NEW HALOS

3.1. Introduction

The investigation of New Halos, a city in Achaia Phthiotis, started in 1976. It was, however, not the first investigation carried out by Dutch archaeologists in Thessaly. In 1906, Vollgraff (1908), who was then a professor at the University of Groningen, had conducted a number of small-scale excavations in this region. He dug a number of trenches at, successively, *Magoúla* Zerélia, a hill near the village of Karatsádhagli and at *Magoúla* Plataniótiki. His main aim was to locate the sites of the temples of Athena Itona and Zeus Laphystios. In all three cases he gave up his attempts after a few days only. In later years, after he had been appointed professor of archaeology at the University of Utrecht, Vollgraff (1908) conducted successful excavations at Argos.

In the early seventies a group of Dutch archaeologists started to survey the area of Gorítsa hill, east of Vólos, under the supervision of Bakhuizen (Bakhuizen *et al.*, 1973). The investigation, conducted in four campaigns, of the remains of the enceinte, quarries and house walls resulted in the plan of a late-4th-century town of moderate dimensions with a regular grid of perpendicularly intersecting roads and housing blocks of various lengths (Boersma, 1983). No further excavations were carried out to study the period of occupation or details of the houses. Nevertheless, the plan of Gorítsa, which was obtained with moderate means in a short period of time, is of great interest to any student of the history of Greek town planning (Bakhuizen, 1992).

Initially, the aim of the fieldwork at Halos was the same as that of Bakhuizen's survey of Gorítsa. The archaeological remains visible at the surface were recorded in 1976 and 1977 (Reinders, 1988). Later the emphasis shifted to smallscale excavations in order to obtain a better understanding of the plan of the town and investigate the layout of the houses and the period of habitation. In 1990 an archaeological survey of the territory of Hellenistic Halos was started. The results obtained in the excavations carried out at the site of New Halos between 1976 and 1982 were published in 1988. The reader is referred to the first publication for detailed information (Reinders, 1988). In this section a general survey of the results will be given to provide a context for the results of the excavation of six houses presented in this volume.

3.2. The town's enceinte

The town of New Halos was situated in the southern part of the Almirós plain on a narrow stretch of flat land between Mount Óthris and a salt marsh along the Pagasitikós gulf (Reinders, 1988). Halos was in all probability founded in 302 BC, by Demetrios Poliorketes. The fieldwork carried out in 1976 concentrated on mapping the remains of the enceinte, the buildings and houses that were visible at the surface in the plain and on a spur of Mount Óthris.

This survey of the visible remains revealed that the site of New Halos comprised a lower part in the plain and an upper part on the hill west of the lower town (fig. 1.18). The lower town comprised a built-up area of about 40 hectares enclosed by an almost square enceinte measuring 700×700 m. This enceinte had a total length of 2.8 km and was reinforced with 68 towers. Two large gates and a number of posterns provided access to the town. The city's main gates were of the courtyard type. The Northwest Gate, situated in the northwest corner of the town, could be reached from the Almirós plain via a small corridor between the flanks of a hill and the Amphrysos, the small river running along the north wall of New Halos. The Southeast Gate opened onto the Soúrpi plain.

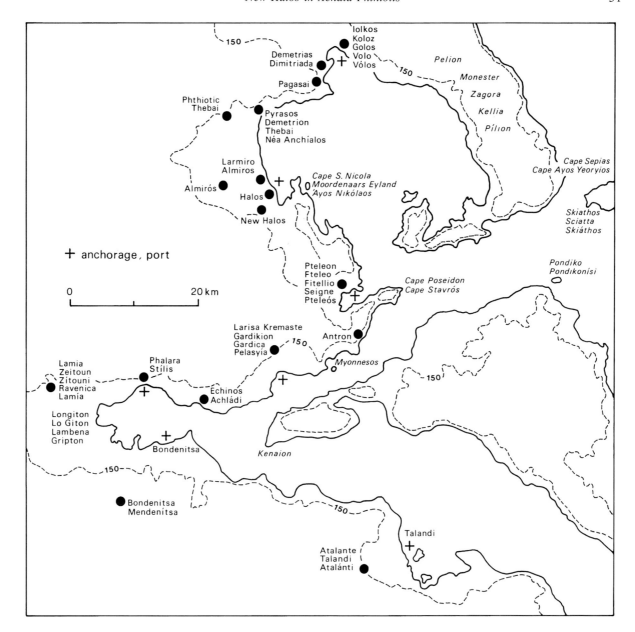

Fig. 1.18. Names of towns, islands and landmarks which are mentioned in Chapter 1.

Two walls, each with a length of over 1 km, extended uphill to a small acropolis at an altitude of about 200 m. These walls were reinforced with at least 50 towers. The small acropolis at the apex was accessible from the north side via a small gate, a simple opening in the line of the wall, besides a tower. Presumably a battery was situated at the apex. If so, its remains now lie buried beneath those of a tower which formed part of a fort that was built to the west of the acropolis of New Halos in the 12th century.

In many places two to five courses of the limestone blocks of the enceinte surrounding the lower town have survived. In other places the wall has been completely destroyed. The entire eastern wall, for example, was torn down in 1837. Some of the limestone blocks of the walls disappeared in lime kilns. In 1988, in spite of the absence of the upper courses, I concluded that the walls and towers had been built entirely from limestone blocks. However, since then Greek archaeologists of the Vólos Museum have cleaned part of the southern wall and found it to be reasonably intact and have a fairly level top. Not a single fallen block was found among the debris on either side of the wall, nor were any coping stones found at the foot of the wall. This, and our discovery of mudbricks among the debris of the Southeast Gate,

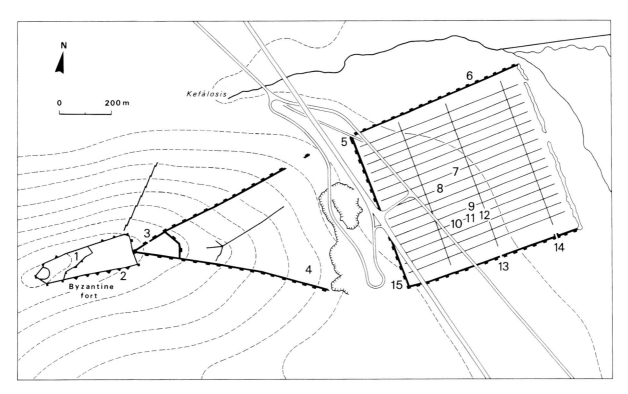

Fig. 1.19. Plan of the city of New Halos showing the sites which were investigated by the University of Groningen. 1. Building in the Byzantine fort; 2. Tower of the fort; 3. Acropolis Gate; 4. Sepulchral building; 5. Northwest Gate; 6. Part of the enceinte with three towers; 7–12. Houses in the lower town; 13. Small gateway; 14. Southeast Gate (excavation in co-operation with the 13th Ephorate of Antiquities at Vólos); 15. Western wall of the lower town and the connection between the southern wall of the upper town and the southeast corner of the lower town, excavated by archaeologists of the 13th Ephorate of Antiquities at Vólos.

makes it more likely that the upper parts of the walls were built from mudbricks. Marzolff (pers. comm.) also postulated a mudbrick upper structure for the enceinte of New Halos.

Between 1976 and 1982 we focused on the gates as part of the fortification system. We investigated four gates by measuring the remains visible at the surface after we had removed the overlying debris and bushes (fig. 1.19). In addition, we dug trial trenches to obtain information on the plan and the structure of the gates. By 1982 we had gathered detailed information on the plan and structure of the Acropolis Gate (No. 3), the Northwest Gate (No. 5), the Southeast Gate (No. 14) and a small gate in the southern wall (No. 13). The gates, which lay in line with the city's main streets, formed integral parts of the rigid fortification system of New Halos.

The results of the preliminary investigation of the gates were published in 1988. At first we paid little attention to the observation that the passages of the minor gate in the southern wall and the Southeast Gate had been blocked off with *poros* blocks. The investigation of the Southeast Gate was resumed in 1995 (Reinders *et al.*, 1996). By the end of the excavation

campaigns of 1995, 1997 and 1999 it was clear that the entrance to the city via the Southeast Gate had been completely blocked off by transverse walls and that people had lived around, in and on top of the former gate complex during the reign of Antigonos II Gonatas (276–239 BC). Other remains of buildings were observed when the Vólos Museum cleaned the southern wall in 1997; people had evidently lived in buildings constructed against the southern wall after the gates had been destroyed.

A project of large-scale excavations was started in 1998 along the Athens–Thessaloníki national road which is to be widened in the near future. A long stretch of the impressive enceinte of New Halos was excavated by the 13th Ephorate of Prehistoric and Classical Antiquities in Vólos. In some places five courses of the limestone blocks of the western city wall had survived. Evidence obtained in this excavation confirmed that there had indeed been a square tower at the southwestern corner of the enceinte, as previously assumed. One of the main problems concerning the plan of the enceinte of New Halos was whether or not the two walls enclosing the upper town had been connected to the square enceinte of the lower town. In 1998 indisputable evidence of a connection was found just north of the square tower at the south-west corner.

3.3. Remains of housing blocks and houses in the lower town

The housing blocks were separated from one another by fourteen streets running from east to west and three streets running from north to south. The housing blocks of New Halos were oblong and varied in length. They were 100 ancient feet wide and 250, 600 or 700 feet long: a striking difference with respect to the housing blocks at Priene, which measured 160× 120 feet, and those at Olynthos, which measured 300×120 feet (Reinders, 1988: p. 112). After the survey we tried to reconstruct the arrangement of the houses of New Halos within the housing blocks but our efforts were not successful. We assumed that the houses were all of the same size, just like those of Olynthos and Priene. However, we were unable to fit round numbers of identically sized houses in the housing blocks. We therefore concluded that houses of 15×15 m must have stood at either end of the blocks and that houses of smaller sizes were built in between.

The first house within the enceinte of New Halos was excavated in 1978 and 1979 (fig. 1.19, No. 7). Only one stratum containing artefacts and roof tiles was observed, beneath a thin layer of topsoil. Other houses were excavated in 1984, 1987, 1989, 1991 and 1993 (fig. 1.19, Nos 8–12). During the period spanned by those years we observed a dramatic change in land use in the area we were investigating (fig. 1.20). In the late 1970s the majority of the plots within the enceinte had lain fallow and had been used as pastureland. A row of plots inside the eastern wall of the enceinte had been used for cultivating cotton and lucerne. These plots were irrigated with water from a ditch which already existed in Stählin's time. No remains of Hellenistic houses were found in these plots; all foundation stones had already been removed by the farmers.

In the early 1980s this situation changed. Plots of pastureland and arable land were bought or rented by farmers who started to grow olives or cotton, lucerne and tomatoes. The remains of the foundations of the houses and housing blocks were removed with heavy machinery and the land was irrigated with water from the powerful Kefálosis spring. The scattered blocks of the destroyed stretches of wall were removed by the farmers, but not the blocks of the intact houses and intact stretches of wall. We recorded the remaining foundations in 1976 and 1977. By that time, the land on either side of the house foundations was already under cultivation, so the archaeological context had been disturbed and the artefacts destroyed. Foundations that were found to lie along the edge of a plot of arable land had in several places been used to mark

the plot's boundaries. As a consequence of this, we were able to measure only relatively few intact house plans in 1976. Six houses were assumed to be intact, with well-preserved outer walls, but when we excavated these houses, between 1978 and 1993, we found that two of the houses and the four outside walls of another house had been partly disturbed. Some of the rooms of the latter house were moreover found to lie in plots that had already been taken into cultivation. This house was later unfortunately destroyed by the farmers.

Between 1987 and 1989 we bought a plot of land from Christos Katrantzonis, an inhabitant of the village of Néos Plátanos. This plot contained the remains of six houses, four of which have since then been excavated. The remaining two will be excavated in the near future. Unfortunately the area within the city walls has been divided into a large number of small plots – 171 in total. Determining the exact limits of the plots and tracing the names and addresses of their 150 owners proved very difficult. Until the 1980s many of the owners were inhabitants of the small village of Néos Plátanos who were willing to come along to the excavation to discuss matters of boundaries and ownership, but in later years plots were sold or rented to more enterprising farmers or cooperatives of farmers from nearby Almirós and other villages, which made matters more difficult for us.

Additional problems were the discovery that one house lay on two different plots of land and along the border of a third plot, and that a single house of 15×15 m lay at the centre of a large plot measuring 17 *strémmata* from which all other archaeological remains had already been removed. The farmers argued that the plot of land would become useless if they were to sell an area of 1 *stremma* (=1000 m^2) containing the house remains, whereas I of course argued that the remains would be destroyed if we would not be able to buy the 1 *stremma*. Unfortunately, my efforts to protect the remains only accelerated the removal of the limestone foundation blocks of the houses in question. It was hard to convince the local population of the need to protect the archaeological heritage. Only pastoralists showed some understanding, because if the land were to remain uncultivated, that would be in their interest, too, as they would then be able to use it as pastureland.

During the second campaign of the excavation of the Southeast Gate in 1997 we took measures to protect the remaining house foundations. On March 3 and 4, 1997, during a visit to the site of New Halos, we saw serious attempts were being made to remove all archaeological remains from the site. On March 6 we therefore advised a representative of the Vólos Museum to buy several scattered plots of land containing house remains and to fence off the remains already excavated. The Greek Ministry of Culture, which had already granted us permission to conduct excavations

Fig. 1.20. Remains of the city of New Halos at the transition of the Almirós to the Soúrpi plain, seen from the northwest.

in 1997, approved these measures and in the autumn of 1997 the most important house remains were fenced off. So now at least the remains of seven houses will be preserved for later generations – a deplorably small number in comparison with the original number of 1440 houses that once stood in New Halos, but together with the plan obtained in the 1976–77 survey this is not a poor result, considering how little is known about the lower towns of cities like Pelinna, Tithorea, Dhrimaia and Alea, to mention only a few towns with similar layouts (Reinders, 1988).

The threat of intensifying use of the arable land in the lower town area was the reason why we focused our efforts on excavating the remains of the houses. Unfortunately, the land containing the scanty remains

of what may have been an *agora* had already been disturbed when we investigated this area in the late 1970s. Rectangular carved blocks, a large threshold and a few column drums were, and still are, observable along the borders of an olive yard. There may have been public buildings along the city's main avenue, but we have no evidence to prove it.

3.4. The upper town

Between 1984 and 1993 we paid no attention to the public buildings whose remains we had found near the 80 m contour line in the upper town. We had excavated one of these buildings in 1981 and 1982 (fig. 1.19, No. 4). This building, which comprised four rooms, the two rear rooms containing what we assumed to be a double grave and votive gifts, we had interpreted as a sepulchral building, although the occurrence of graves inside the enceinte and the absence of human skeletal remains were surprising. An alternative interpretation was suggested in a lecture at Rethymnon University: the female figurines and the small *hydriskes* found among the remains could indicate that the building was a Demeter sanctuary. What we had originally interpreted as a 'double grave' covered with stones may have been a symbolic entrance to the underworld or a sacrificial table.

3.5. An early Hellenistic city

The results of the excavation of the House of the Coroplast in 1978 and 1979 led to the hypothesis that the city of New Halos was founded around 302 BC and was abandoned around 265 BC. The evidence obtained in the excavation of the other houses seems to confirm these dates. Additional information on the time at which the city was abandoned was obtained in the excavation of the Southeast Gate (Reinders *et al.*, 1996). The evidence concerned will be presented in Chapter 5.

Evidence obtained in the excavation of a cemetery 100 m south of the Southeast Gate (Malakasiotis *et al.*, 1993: pp. 93–103) shed some doubt on our hypothesis that the entire city was abandoned following some catastrophe around 265 BC. This cemetery contained only 66 graves. Although one of the graves contained an excellently preserved coin of the city of New Halos, placed in the mouth of a dead man, the majority of the artefacts dated from the second half of the 3rd century BC. Perhaps these were the graves of the people who had lived along the southern wall of the city after the gates and the rest of the enceinte had been destroyed. It will be interesting to compare the artefacts found near the gate with those recovered from the cemetery after they have been studied and published. So far, no cemetery representing the period 302–265 BC has been discovered.

In my first book on New Halos (Reinders, 1988) I also discussed general aspects of Greek town planning. The main characteristics of a fourth-century-BC Greek town, as clearly illustrated by the plan of the city of Priene, are:
– an enceinte following a natural line of defence;
– no relation between the enceinte and the grid layout of the housing blocks;
– housing blocks of identical shapes and dimensions, constituting the module of the layout;
– public buildings incorporated in the grid layout.
These characteristics can be observed in a number of cities of the Classical period, such as Olynthos, but it should be added that only few cities from this period have so far been thoroughly investigated.

The survey of the town area and the small-scale excavations conducted between 1976 and 1982 re-sulted in a plan of New Halos. Although large parts of the city have been destroyed, we were able to reconstruct the layout of the city's street plan and housing blocks. We then observed several differences in layout between the plan of New Halos and those of Priene and Olynthos:
– the enceinte of New Halos does not follow a natural line of defence;
– at New Halos the enceinte and the housing blocks were laid out relative to one another;
– the housing blocks of New Halos were not all of the same shape and size and did not constitute the module of the layout;
– public buildings were incorporated in the grid layout to a certain extent only.
One of the main objectives in writing the present publication was to determine to what extent the houses of New Halos differed from or resembled the houses in other cities of the Classical period.

4. REFERENCES

ADLER, M.N., 1907. *The itinerary of Benjamin of Tudela*. London.

ASIMAKOPOULOU-ATZAKA, P., 1982. Palaiochristianikí kai Bizandiní Magnisía. In: G. Chourmouziadis (ed.), *Magnisía, to chronikó énos politismoú*, pp. 105–175.

AVRAMEA, A.P., 1974. *I Bizandiní Thessalia mechri tou 1204*. Athina.

BAKHUIZEN, S.C. *et al.* 1972. Goritsa, a new survey. *ADelt* 27: Chronika, pp. 347–374.

BAKHUIZEN, S.C. (ed.) 1992. *A Greek city of the fourth century B.C.* Bretschneider, Roma.

BAKKER, A. & R. REINDERS, 1996. Tetrobolen en tetradrachmen. Een muntvondst in Nieuw Halos. *Paleo-aktueel* 7, pp. 66–69.

BIADENE, S. (ed.), 1990. *Carte da navigar. Portolani e carte del Museo Correr 1318–1732*. Marsilio Editori, Venezia.

BOERSMA, J.S., 1983. Goritsa: the residential district. *BABesch* 58, pp. 61–82.

BOTTEMA, S., 1988. A reconstruction of the Halos environment on the basis of palynological information. In: H.R. Reinders, *New Halos, a Hellenistic town in Thessalía, Greece*. Hes, Utrecht, pp. 216–226.

CASSON, L., 1971. *Ships and seamanship in the Ancient World*. Princeton University Press, Princeton & New Jersey.

DAMIANIDIS, K. & A. ZIVAS, 1986. *To trechandiri stin Elliniki navpiyiki techni*. Eommex, Athína.

DELATTE, A., 1947. *Les Portulans Grecs* (= Bibliothèque de la Faculté de Philosophie et Lettres de l'Université de Liège, Fascicule 107). Liège.

DOULYERI-INTZESILOGLOU, A., 1993. To archaioloyiko Mousio tou Almirou. In: V. Kondonatsios (ed.), *Achaiophthiotika* 1. Ores, Almiros, pp. 17–47.

DIJKSTRA, Y., H.R. REINDERS, V. RONDIRI & Z. MALAKASIOTIS, 1997. Van Duivelsberg tot Rode Rots. *Paleo-aktueel* 8, pp. 89–92.

EFSTATHIOU, A., Z. MALAKASIOTI & R. REINDERS, 1990. Halos archaeological field survey project, Preliminary report of the 1990 campaign. *Newsletter of the Netherlands Institute at Athens* 3, pp. 31–45.

FOTINI-PAPAKONSTANTINOU, M., 1994a. To notio kai to ditiko tmíma tis Achaías Phthiotídas apo tous klasikous mechri tous romaikous chronous. In R. Misdrahi-Kapon (ed.), *La Thessalie, Quinze années de recherches archéologiques, 1975–1990, Bilans et perspectives* 2. Athènes, pp. 229–238.

FOTINI-PAPAKONSTANTINOU, M., 1994b. *The Kastro of Lamia.* Archaeological Receipts Fund, Athens.

FURTWÄNGLER, A.E., 1990. *Demetrias, eine Makedonische Gründung im Netz Hellenistischer Handels und Geldpolitik.* Habilitationsschrift, Saarbrücken.

HOPE SIMPSON, R. & J.F. LAZENBY, 1970. *The catalogue of the ships in Homer's Iliad.* Clarendon Press, Oxford.

INTZESILOGLOU, B.F., 1994. Istoriki topografia tis periochis tou Kolpou tou Volou. In: R. Misdrahi-Kapon (ed.), *La Thessalie, Quinze années de recherches archéologiques, 1975–1990, Bilans et perspectives* 2. Athènes, pp. 31–56.

KEULEN, J.G. VAN, 1728. *De nieuwe groote ligtende Zeefakkel,* deel 3. Amsterdam.

KODER, J. & F. HILD, 1976. *Hellas und Thessalia* (Österreichische Akademie der Wissenschaften, Philosophisch-Historische Klasse Denkschriften, Band 125). Wien.

KALLIANOS, K., 1984. O katalogos ton Thessalomagniton paroikon tis Skopelou sta 1829. *Thessaliko Imeroloyio* 7, pp. 31–38.

KONDONATSIOS, V. (ed.) 1993. *Achaiophthiotika, praktika tou A' sinedriou Almiriotikon spoudon.* Ores, Almiros.

KRAFT, J.C., G. RAPP, G.J. SZEMLER, C. TZIAVOS & E.W. KASE, 1987. The Pass at Thermopylae, Greece. *Journal of Field Archaeology* 14, pp. 181–198.

KRETSCHMER, K., 1909. *Die italienischen Portolane des Mittelalters. Ein Beitrag zur Geschichte der Kartographie und Nautik* (= Veröffentlichungen des Instituts für Meereskunde und des Geographischen Instituts an der Universität Berlin, Heft 13). Mittler und Sohn, Berlin.

KRITZAS, Ch., 1971. To vizandinon navayion Pelagonnisou-Alonnisou. *AAA* 4.2, pp. 6–182.

MAKRIS, K.A., 1982. Metavizandini kai Neoteri Magnisia. In: *Magnisia, to Chroniko enos Politismou.* Athina.

MALAKASIOTIS, Z., V. RONDIRI & E. PAPPI, 1993. Prosfates anaskafikes erevnes stin archaia Alo: mia proti parousiasi ton evrimaton. In: V. Kondonatsios (ed.): *Achaiophthiotika* 1. Ores, Almiros, pp. 93–101.

MALAKASIOTIS, Z. V. RONDIRI & R. REINDERS, 1995. Groninger bijdrage aan Griekse monumentenzorg. *Paleo-aktueel* 6, pp. 71–74.

MARZOLFF, P., 1980. *Demetrias und seine Halbinsel* (= Demetrias III). Bonn.

MARZOLFF, P., 1994. Antike Städtebau und Architektur in Thessalien. In: R. Misdrahi-Kapon (ed.), *La Thessalie, Quinze années de recherches archéologiques, 1975–1990, Bilans et perspectives* 2. Athènes, pp. 255–276.

MITROPOULOS, D. & S. MICHALIDIS, 1988. Seismic stratigraphy and structure of Pagasitikos and Maliakos Gulf and the surrounding areas, Aegean Sea, Greece. *Rapp. Comm. int. Mer Médit.* 31.2.

MÜLLER, K., 1965. *Geographi Graeci Minores* (Volume 1; reprint of the edition of 1855). Georg Olms, Hildesheim.

NIKONANOS, N., 1971. Aïdínion. *Arch. Deltion (Chronika),* pp. 312-313.

NORDENSKIÖLD, A.E., 1897. *Periplus, an essay on the early history of charts and sailing directions.* Stockholm.

PANTOS, P., 1994. La vallée du Spercheios – Lamia exceptée –

aux époques Hellénistique et Romaine. Quinze années de recherches, 1975–1990. In: R. Misdrahi-Kapon (ed.), *La Thessalie, Quinze années de recherches archéologiques, 1975–1990, Bilans et perspectives* 2. Athènes, pp. 221–228.

PERISSORATIS, C., P. ZACHARAKI & A. ANDRINOPOULOS, 1988. Texture and composition of the bottom sediments of Pagasitikos Gulf and Trikeri Strait, Thessaly (Greece). *Rapp. Comm. int. Mer Médit.* 31.2.

RADT, S.L., 1994. Aus der Arbeit an der neuen Strabonausgabe. *Pharos* 2, pp. 31–35.

REINDERS, H.R., 1988. *New Halos, a Hellenistic town in Thessalía, Greece.* Hes, Utrecht.

REINDERS, H.R., 1993. I topothesia tis Alou. In: V. Kondonatsios (ed.), *Achaiophthiotika* 1. Ores, Almiros, pp. 49–59.

REINDERS, R., Y. DIJKSTRA, V. RONDIRI, S.J. TUINSTRA & Z. MALAKASIOTIS, 1996. The Southeast Gate of New Halos. *Pharos* 4, pp. 121–138.

REINDERS, R., 1998. Earthquakes in the Almirós Plain and the abandonment of New Halos. In: E. Olshausen & H. Sonnabend (Hrsg.), *Stuttgarter Kolloquium zur historischen Geographie des Altertums 6, 1996 'Naturkatastrophen in der Antiken Welt'.* Stuttgart, pp. 198–210.

RIZOS, A., 1996. *Wirtschaft, Siedlung und Gesellschaft in Thessalie im Übergang von der Byzantinisch–Frankischen zur Osmanischen Epoche.* Hakkert, Amsterdam.

SAVVIDIS, G.K., 1993. I emporiki akmi tou Almirou kata to 12o aiona m. Chr. In: V. Kondonatsios (ed.), *Achaiophthiotika* 1. Ores, Almiros, pp. 203–211.

SHERWIN-WHITE, S.M., 1978. *Ancient Cos. An historical study from the Dorian settlement to the Imperial period* (= Hypomnemata, Untersuchungen zur Antike und zu ihrem Nachleben, Heft 51). Vandenhoeck & Ruprecht, Göttingen.

SOUCEK, S., 1996. *Piri Reis & Turkish mapmaking after Columbus. The Khalili Portolan Atlas.* Nour Foundation, Azimuth Editions, Oxford University Press, London.

SPHYROURAS, V., A. AVRAMEA & S. ASDRAHAS, 1985. *Maps and mapmakers of the Aegean.* Olkos, Athens.

STÄHLIN, F., 1924. *Das Hellenische Thessalien.* Engelhorn, Stuttgart.

STRAATEN, L.M.J.U. VAN, 1988. Mollusc shell assemblages in core samples from ancient Halos. In: H.R. Reinders, *New Halos, a Hellenistic town in Thessalía, Greece.* Hes, Utrecht, pp. 227–235.

THROCKMORTON, P., 1971. Exploration of a Byzantine wreck at Pelagos Island near Alonnessos. *AAA* 4.2, pp. 183–185.

VOLLGRAFF, W., 1908. Notes on the topography of Phthiotis. *BSA* 16, p. 224.

WACE, A.J.B. & M.S. THOMPSON, 1911–1912. Excavations at Halos. *BSA* 18, pp. 1-29.

WACE, A.J.B. & M.S. THOMPSON, 1912. *Prehistoric Thessaly.* University Press, Cambridge.

WIEBERDINK, G., 1990. A Hellenistic fortification system in the Othrys Mountains (Achaia Phthiotis). *Newsletter Netherlands Institute at Athens* 3, pp. 47–76.

ZACHARAKIS, Ch. G., 1992. *A catalogue of printed maps of Greece 1477–1800.* Samourkas foundation, Athens.

ZANGGER, E., 1991. Prehistorical coastal environments in Greece: The vanished landscapes of Dimini Bay and Lake Lerna. *Journal of Field Archaeology* 18, pp. 1–15.

CHAPTER 2. THE HOUSES OF NEW HALOS

MARGRIET J. HAAGSMA

1. INTRODUCTION

Six houses were excavated within the enceinte of the lower town of New Halos in campaigns in the years 1978, 1980, 1984, 1987, 1989, 1991 and 1993. The excavation of each of the houses lasted one campaign, except that of the House of the Coroplast, which was excavated in two campaigns (1978 and 1979). Most of the excavation campaigns lasted five to eight weeks. The size of the excavation teams varied from ten to twenty members. The teams consisted of archaeologists and students.[1]

In this chapter I will give a general description of the architecture and principles of construction of the houses in Halos, followed by a detailed description of each excavated house. In the final sections I will discuss the houses of Halos in the broader context of early Hellenistic Greek housing in general, and house types and the role of *Typenhäuser* (Hoepfner & Schwandner, 1994)[2] in particular.

The following houses – each named after a characteristic artefact or artefacts found during the excavations – will be discussed *in extenso*:

1. House of the Coroplast (housing block 2.7, plot 11; 13.75×15 m). This house was excavated in 1978 and 1979; the results of the excavations were published by Reinders in 1988. Since then, studies of the artefacts and architecture have provided new insight into the history of the occupation of this house and the activities that took place in it. This new information will be included in the discussion of this house below;
2. House of the Geometric *Krater* (housing block 2.8, plot 17; 12.70×8.80 m). The foundation stones of the southwestern part of this house were found to have disappeared, obviously because the Almirós-Soúrpi road once ran across this part of the house. The preserved northeastern part was excavated in 1984;
3. House of the Snakes (housing block 6.4, plot 9; 15.10×15.45 m). This house was selected for excavation, in 1993, because until then mainly houses in the southern parts of housing blocks had been studied; Considering the house's position, we expected that its plan would differ from the plans of the House of Agathon and the House of the Ptolemaic Coins bordering this house on its southern side;
4. House of the *Amphorai* (housing block 6.4, plot 18; 15.50×14.30 m). This house was excavated in 1991 in order to gain insight into the plan of a house

measuring 15×15 m.[3] Unfortunately we found the house remains had been severely disturbed;
5. House of the Ptolemaic Coins (housing block 6.4, plot 21; 12.70×14.80 m), excavated in 1989. This house bordered the House of Agathon (fig. 2.1), but was less well preserved as the southeastern part had been partially disturbed;
6. House of Agathon (housing block 6.4, plot 22; 12.50×15.20 m), excavated in 1987.

At times I will also refer to one other house, House A, in housing block 6.2 (fig. 2.1). The foundations of this house were cleaned in 1978 in order to gain insight into the house's structure and plan. The house was however not excavated, and virtually no artefacts were collected during the fieldwork. As the results of the investigation of this house have been published elsewhere (Reinders, 1988), this house has been left out of the description of the individual houses (section 2.3).

1.1. Excavation methods and find processing

The students and the members of the excavation teams shared responsibility for every aspect of the fieldwork. Specific supervisory tasks were divided among the senior members of the team. For didactic purposes, the students were usually involved in the digging, recording, measuring, drawing and processing of the finds and were obliged to keep daily records of their results.

The actual excavation work was performed with the help of pickaxes, mattocks and trowels. Layers with thicknesses of about 0.05–0.10 m were removed until the virgin soil was reached. After each layer had been removed, the exposed surface of each unit was drawn. Units usually coincided with the rooms of the individual houses. Finds were recorded per room and per layer. The precise positions of all diagnostic finds – pottery, metal objects, organic matter and coins – were recorded and were indicated in the drawing of each unit.

The sherds recovered in the first three excavation campaigns (the House of the Coroplast, the House of the Geometric *Krater* and the House of Agathon) were partially sorted in the field. The roof tiles were discarded but most of the other pottery was kept, as were all other finds (organic remains, loomweights, metal objects). All of the ceramic objects encountered in the last three campaigns (House of the Ptolemaic Coins, House of the *Amphorai*, House of the Snakes) were

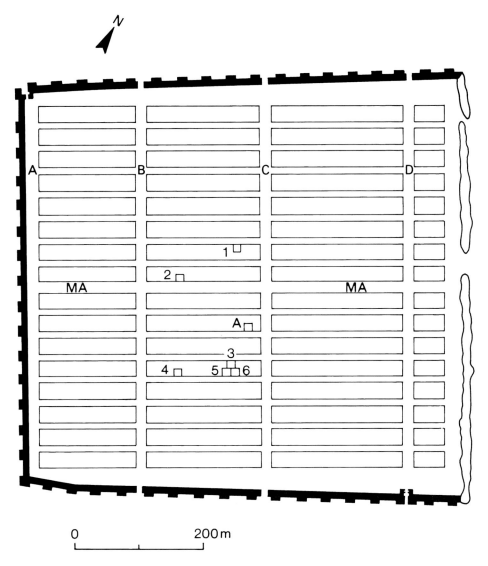

Fig. 2.1. Lower town of Hellenistic Halos with housing blocks and the location of the excavated houses.

recovered and processed. All finds were recorded per layer on forms, on which the type of ware, number and weight were specified, after which they were taken to the *apothiki* in Almirós to be washed. Finds of particular significance were taken to the Vólos museum for consolidation and restoration. In separate study campaigns the sherds were sorted into different categories on the basis of their shape and the type of ware, and all finds were individually described. The diagnostic sherds were drawn by members of the team and – where possible – restored either by us or by the Vólos museum. A significant proportion of the metal finds were also drawn, and the most important metal objects were restored by the Vólos museum. For descriptions and the results of the analyses of the artefacts and faunal remains the reader is referred to Chapters 3 and 4 of this volume. In 1998 a selection

of the finds was included among the archaeological remains displayed at the newly restored museum of Almirós.

This detailed registration and analysis of the finds of New Halos was deemed worthwhile in view of the nature of the deposits and the siteformation processes. As a result of the sudden, catastrophic destruction of New Halos and its subsequent abandonment without any rebuilding attempts, a significant percentage of the household inventory – essentially objects which were broken, of little intrinsic value or beyond recovery – was left *in situ*. The straightforward stratigraphy associated with the houses moreover implies that the site was not occupied for very long. The numismatic evidence yielded a very accurate date for the destruction (*cf.* Chapter 5). These find conditions definitely called for a detailed system of registration and description.

Fig. 2.2. Shattered ceramics: House of Agathon.

1.2. Post-depositional processes and stratigraphy

The occupation layer containing the remains of the Hellenistic town of New Halos was found to be extremely shallow (fig. 2.2). Until recently, many foundations of the houses that once stood in the area enclosed by the enceinte could be identified in the field without much difficulty. In the cropping system employed by the local farmers, the plots were left fallow for one or two years after a crop of wheat had been grown in them. In those years the land was used for pasturing flocks of sheep and goats. These activities caused virtually no damage to the stratigraphy. Thirty years ago, however, intensive agriculture was introduced in the Almirós area and this had major consequences for the preservation of the site. The change in land use led to widespread damage not only to the archaeological stratigraphy, but also to the architectural remains. The owners of the plots of land within the enceinte of Halos used heavy machinery to remove the foundation stones of most of the houses and to plough the stony, but fertile soil. At present (summer of 1999), about 98% of the land within the city walls is in use for the cultivation of olives, wheat, cotton or tomatoes. In recent years, most of the house remains have consequently been destroyed, except those in the plots of land that were expropriated for this research.

During our excavations we reached the virgin soil at an average depth of 0.40 m. The stratigraphy of each individual house or area within a house will be discussed in detail below. Generally speaking, the following layers were encountered:

0–0.15 m: topsoil; disturbed; mixed with worn sherds and other artefacts dating from the Hellenistic period and recent times;

0.15–0.35 m: layer containing destroyed remains; sherds (not worn) and other artefacts and structural remains dating from the Hellenistic period

0.35–0.40 m: floor level(s); soil mixed with a few small, worn sherds dating from the early Hellenistic period and a few artefacts dating from the Geometric period;

0.40– ...: virgin soil, consisting of Pleistocene hardpan. The foundation stones had been dug into the virgin soil.

2. PRINCIPLES OF THE CONSTRUCTION OF THE HOUSES

In this section I will discuss the basic principles of construction used for the houses at New Halos. These principles have been deduced primarily from the archaeological record, with supplementary information derived both from analogous sites and from the study of subrecent houses constructed along similar lines.

2.1. Foundations and walls

The foundations of both the internal and the external walls of the houses at New Halos consisted of large, roughly hewn blocks of limestone, usually laid out in foundation trenches dug into the Pleistocene soil. Most of the blocks were quarried on the hill of the upper town (Reinders, 1988: p. 60). Although the outer faces of some of the blocks of the foundations of external walls show signs of having been worked

Fig. 2.3. House of Agathon, Halos: two rows of foundation stones. The lower row consists of flattened limestone blocks.

and smoothed to some extent, most blocks were un-worked. Conglomerate stone was used sparsely. *Poros* was used for some elements, primarily thresholds (see below).

The majority of the foundations of the external walls consist of two rows of blocks with an overall width of 0.45–0.50 m. Internal walls were somewhat less carefully constructed, the majority consisting of two rows of blocks and some of a single row of larger blocks. The width of these foundations is 0.30–0.35 m. The gaps between the elongated foundation stones and between the foundations and the courses of the wall proper were filled with soil mixed with small pieces of limestone.

In most cases only one course of foundation blocks had survived. Some foundation blocks were found to rest on flat limestone slabs. The upper walls were most likely built from perishable materials such as mud-brick, wood and/or wattle and daub. No remains of these walls were found in the excavated houses, so we have no direct evidence as to which of these materi-als was used at New Halos. The gaps between the foundation stones were wide enough to have contained the vertical timber beams that would have been re-quired for wattle-and-daub walls. Indirect evidence and evidence obtained at similar sites however sug-gests that the walls were of mudbrick. Foundations such as those found at New Halos were quite suitable for supporting mudbrick walls, and in recent excava-tions of the remains of a small building adjacent to the Southeast Gate of New Halos the outlines of burnt intact mudbricks were observed (Dijkstra *et al.*, 1996: p. 63). Evidence obtained elsewhere in Greece sug-gests that mudbrick was used more commonly than wattle and daub.[4]

The foundations of the front and rear walls of houses at New Halos have a few characteristics in common. The stones of the external rows are gener-ally broader and longer than those of the internal rows. The outer parts of a building's external walls had to bear more pressure from the upper part of the wall and the roof than the inner parts (fig. 2.4). The row of larger stones thus prevented the risk of the external wall collapsing outwards. This implies that the rear walls of the south-facing houses bordered an open area. This we were able to check in one case only: the rear walls of the House of the *Amphorai* and the House of Agathon indeed bordered the courtyard of the House of the Snakes.

There where only one course of foundations had survived, the average height of the stones was 0.30 m. Only in the House of the Coroplast and parts of the rear wall of the House of Agathon had two courses of the foundations been preserved. As the foundation stones had been dug into the hardpan, the height of the preserved foundations measured from floor level was 0.15 m on average. In my opinion this cannot have been their original height. Observations made during a study of present-day mudbrick housing in the Almi-rós area (see Haagsma, in prep. a) show that the height of most foundations nowadays is between 0.40 and 1.00 m, measured from floor level. Such a height is necessary to prevent the risk of water seeping up or spattering the lowest row of mudbricks, which would affect the stability of the upper wall. That this risk was considered at New Halos, too, is suggested by the two courses of foundation stones preserved in the House of the Coroplast and the House of Agathon, but also by various stones found during our excavations which are assumed to derive from collapsed upper courses

Fig. 2.4. House of the Coroplast. Foundation stones of the external wall. Outer stones are larger.

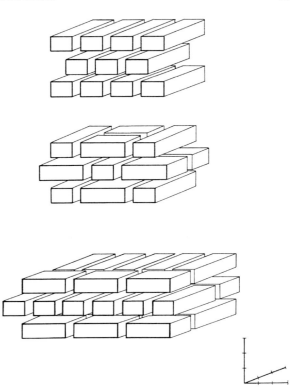

Fig. 2.5. Various techniques of laying mudbrick.

of foundation stones (*e.g.* House of the Snakes, room 2). In some houses (notably the House of the *Amphorai*) in which parts of collapsed upper walls were discovered, these parts were moreover found to comprise the fill of part of the wall foundations. I therefore believe that the foundations of the houses at Halos originally comprised at least one course more than is preserved in the archaeological record, and that their height above floor level was at least 0.40 m.

Mudbricks, which were still being made in the Almirós area in recent times, usually measure 0.12× 0.15×0.30 m (Skafida, 1994; Haagsma, in prep. a). These measurements coincide with those known from the Southeast Gate of New Halos. The bricks known from Olynthos were a little flatter, longer and wider, measuring 0.08–0.10×0.17–0.19×0.39–0.49 m (Robinson & Graham, 1946: p. 185). We may assume that the method used to produce mudbricks in antiquity was similar to that employed in recent times. The soil used for the mudbricks was usually obtained from an area very close to the site where the new building was to be erected. It was mixed with organic temper such as

hay to improve its mouldability and was then placed in rectangular *kaloupia,* wooden moulds. When the soil had dried sufficiently, the mudbricks were removed from the moulds and were left to dry in the sun. Mortar was made from purified soil, free of stones and grit.[5]

As already mentioned above, the widths of the wall foundations in New Halos varied. Several brick-laying techniques are consistent with the varying widths. These techniques, which are illustrated in figure 2.5, all resulted in stable, flexible upper structures. Most of the techniques have been inferred from the above-mentioned subrecent houses in Almirós and its surrounding area and from other archaeological evidence (Skafida, 1994; Haagsma, in prep. a). From ethno-archaeological research we know that rooftile fragments are sometimes inserted between mudbricks to strengthen the walls (fig. 2.6a). Tile fragments are sometimes found on top of foundation stones, too (fig. 2.6b). Whether this same custom was practised at New Halos we do not know, as we assume that the upper course of foundation stones has not survived. Large quantities of tile fragments were found in the excavations, but it is impossible to say with certainty whether any of those fragments derive from walls rather than from roofs.

A technique which was frequently used to increase the flexibility of walls in areas susceptible to earth-

Fig. 2.6. Mudbrick wall in subrecent housing in Almirós. a. Sherds within the mudbricks; b. Tile-fragments separate the foundation stones and the bricks; c. Straw used as temper in the mudbricks.

quakes in subrecent times involved incorporating crossties at regular distances in the walls (fig. 2.7), held in place by iron nails. This technique was used in all subrecent houses in Almirós and surrounding areas, whether they were built from mudbrick or consisted partially or entirely of stone (Haagsma, in prep. a). It is likely that a similar technique was also used in the houses of New Halos. Although we found no evidence for the use of crossties in the walls, we do have some evidence in the form of postholes for vertical beams, which may have formed part of a framework ensuring the required strength and flexibility, for instance in the southwestern foundation wall of room 3 of the House of Agathon.

The finished walls were presumably plastered on both sides. Plaster prevented the risk of insects and other animals housing in the bricks and mortar. At New Halos, the most probable form of plaster will have consisted of a mixture of powdered limestone and water. Such plaster is still used in Greece and other countries today (De Baïraicli Levy, 1981). In antiquity, it could easily have been obtained from the upper town of New Halos. However, no traces of any plaster were found in the houses of Halos, so we cannot entirely rule out the possibility that the houses were left unplastered.[6] It is unlikely that they were plastered with the kind of preconcrete plaster found in the houses of Olynthos, Kassope and elsewhere, because traces of such plaster would almost definitely have been preserved in the archaeological record.

2.2. Floors

As already mentioned above, the soil in the area of Hellenistic Halos is very hard and dry. The shallow depth of the occupation layer and the great uniformity in the colour and texture of the soil made it difficult for us to distinguish archaeological *strata*. Features were difficult to recognise even in the exposed horizontal surfaces. In most of the rooms we excavated we left baulks at the centre of the room in the hope that they would help us analyse the stratigraphy. In most rooms and houses we were able to distinguish only one floor level in the sections. In room 2 of the House of the Ptolemaic Coins we encountered two floor levels. Here a small part of the floor was found to consist of a layer of larger pebbles, beneath which was the usual floor of trodden earth encountered elsewhere.

Most of the floors consisted of a very thin layer

Fig. 2.7. Wall of house in Almirós, partially constructed in stone, partially in mudbrick. Note the horizontal beams protecting the wall from collapsing during an earthquake.

(0.005 m) of trodden earth containing the odd sherd, animal bone and mollusc shell overlying the Pleistocene hardpan. The hardpan, in fact, constituted a perfect floor foundation. In some houses and rooms, however, a thin layer of limestone grit had been incorporated in the trodden earth to consolidate the floor. In some cases earth had also been used to level the floor. This was most clearly observable in the House of the Ptolemaic Coins, whose room 5, especially the southwestern part, contained a layer of earth with a thickness of 0.05 m that was virtually devoid of finds and had evidently been deposited to level the floor in this part of the room.

Floor layers found at other archaeological sites were less thin. The floors of House A at Eleutherna (Kalpaksis *et al.*, 1994: p. 55), which also consisted of trodden earth, were 0.03–0.20 m thick. The latter thickness was measured in areas where the irregular floor had been levelled. But in other areas the floor layer was at least 0.03 m thick. At Olynthos, the majority of the floor layers were 0.05 m thick.

I observed how earthen floors are treated in present-day Almirós (Haagsma, in prep. a). Every year, shortly before Easter, the inhabitants of the mudbrick houses cover their earthen floors with a fine red slip of clay obtained from a riverbed near the town. This *kókkino chóma* is purified with water before being spread over the floors and then left to dry. I found no evidence of a similar custom in the houses of New Halos, but it should be borne in mind that such a layer of purified clay is usually only a few millimetres thick,

and by the time a new layer is applied the previous layer will have been worn away completely.

2.3. Doorways

Doorways in external walls at New Halos were all of the same type. Doorways that provided access to a courtyard were designed as a so-called *prothyron*, a roofed entrance. This *prothyron* consisted of a limestone threshold with two pivots and usually a double door which opened inwards. The door was protected by a small tiled roof (fig. 2.8). Only one doorway with a threshold was found in an external wall in the excavations, in the House of the Coroplast. In most of the other houses evidence of a *prothyron* was found in the form of a concentration of roof tiles near a doorway in an external wall and the foundations of the doorjambs of such a structure. The majority of these doorways were between 1.20 and 1.60 m wide. An exception is the doorway in an external wall of the House of the Snakes, which was much narrower and provided access to the roofed area of the house. The smaller width of this doorway indicates that it contained a single door. The doorway in an external wall of the House of the *Amphorai* also seems to have been too narrow to have contained a double door, but this house most probably did have a *prothyron*.

It is generally assumed that carts could enter through the *prothyron* to be parked in the courtyard, but the doorways in the external walls of the houses at New Halos – with the probable exception of that of

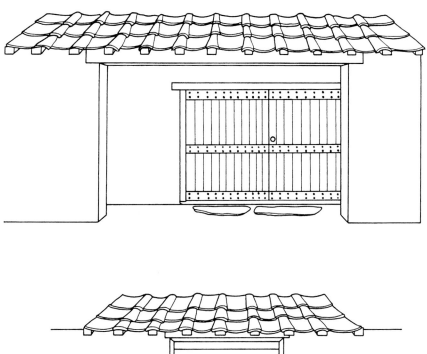

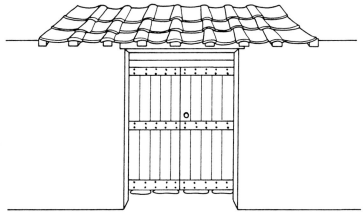

Fig. 2.8. *Prothyra* in Halos. a. House of the Coroplast; b. House of Agathon.

the House of the Coroplast – seem to have been too narrow to have allowed carts to enter.

There is a lot of information on what outside doors from the Classical and Hellenistic periods looked like. The archaeological record, representations of doors in Attic vase painting (fig. 2.9) and marble doors preserved in Macedonian tombs (fig. 2.10) show us doors made of vertically arranged wooden planks held together by cross-latches and large clenched nails which were sometimes decorated with bronze bosses or had large, boss-like iron heads (fig. 3.38, M155). At New Halos there is evidence from the House of the Coroplast and the House of the Snakes that outside doors were of this design. In both houses concentrations of large numbers of clenched nails with large heads were found near the doorways in the external walls. In the case of the House of the Snakes the nails were even

found in association with a lock, a key and a bolt.

Doorways in internal walls were of a simpler design. The doors were mostly single, made of wood and had an average width of 1.0 m. Some incorporated clenched nails and turned in a pivot hole hacked into a stone threshold.[7] A few thresholds have survived. They were about 0.15 m high, measured from the original floor level, and were made from various materials. Some, such as those in the House of the Snakes, were made from *poros* stone, while those in the House of the Coroplast were of limestone. One of the latter thresholds had supported a double door (between rooms 4 and 5). Some of the doorways in the House of Agathon and the House of the Ptolemaic Coins contained a neat row of limestone pebbles. Whether these pebbles served as a foundation for a large threshold block is not clear. There is evidence that internal

Fig. 2.9. Door of a house as depicted on an Attic *Pyxis*.

doors could also be locked: in the House of Agathon and the House of the *Amphorai* keys and keyholes were found in the vicinity of internal doorways.

2.4. Windows

No direct evidence for the existence of windows was found during the excavation.[8] Many iron fragments – including nails – came to light which may have been parts of the mountings of windows or shutters, but they may also derive from ceilings, furniture or other wooden items. As the houses excavated at New Halos so far shared most of their external walls, they can have had windows only in walls bordering the courtyards. For reasons of privacy, windows in Greek houses rarely opened directly onto the street. This means that direct daylight will have been admitted to, on average, only three rooms in a house. Apparently two or more rooms had no daylight whatsoever. No evidence of so-called *opaia* (Robinson & Graham, 1946; Svoronos-Hadjimichalis, 1956), openings for letting daylight enter a house and smoke escape, was found.

2.5. Ceilings

It is not clear whether the houses at New Halos had ceilings. Usually single-floor houses from Classical and Hellenistic times are reconstructed without ceilings, with the wooden structure of the roof exposed (*e.g.* Hoepfner & Schwandner, 1994: pp. 107–108). There are, however, no reasons why (single-floor) Classical and Hellenistic houses should not have had ceilings. From ethnoarchaeological evidence (Haagsma, in prep. a) we know that ceilings are common in subrecent single-floor mudbrick houses. They insulate the house, which is necessary in the harsh winters as well as in the hot summers in Greece. Ceilings moreover prevent draughts and dust from entering the house.

Ceilings of traditional mudbrick and stone houses in Almirós and surrounding areas were usually constructed by arranging reeds longitudinally across the rooms in the house. The reeds were held in place by iron mountings attached to the horizontal beams of the timber framework of the roof. They were covered with purified earth or mortar and then plastered. Whether the same principle was applied in the houses at New

Fig. 2.10. The marble leaves of the Heuzey Tomb, Vergyína, Late 4th century BC, Paris, Musée du Louvre.

Halos is far from certain. A small number of fragments of iron mountings have been found, but it is not clear from what objects they derive. Reeds may moreover very well have been held in place by means which have left no trace in the archaeological record, such as lashing, pegging, *etc.*

Ceilings may also have been created by placing wooden boards longitudinally on top of the horizontal beams of the timber framework, but such a design would have been more costly.

2.6. Roofs

In view of the large numbers of roof tiles found at New Halos, the houses most likely had gently sloping roofs consisting of a timber framework covered with large roof tiles. The principles used to construct such timber roof frameworks in Classical and Hellenistic buildings are not well attested in the archaeological record.[9] The framework may well have resembled that used in subrecent houses in Almirós and surrounding areas (Haagsma, in prep. a) and also in traditional housing in a wider Thessalian context (Skafida, 1994). These frameworks consisted of a horizontal beam with a vertical beam attached to it in the middle. Two smaller beams were added to form a triangle, supported by two small crossbars. Several of such triangles were arranged on top of the walls at

intervals of on average 1 m. It is not clear whether the roofs at New Halos sloped on all four sides.

The roofs were covered with large, slightly concave rectangular roof tiles whose upturned edges were covered with elongated convex tiles. The concave tiles of this 'monk and nun' roofing system measured 0.40×0.80 m; the convex tiles measured 0.20×0.80 m. Although very large numbers of rooftile fragments were found at New Halos, only one reasonably complete tile of each type had been preserved in an archaeological context. No other types of roof tiles, such as antefixes, spouts or tiles with gutters, were found.

2.7. Upper floors

Examination of ethnoarchaeological evidence (Skafida, 1994; Haagsma, in prep. a) shows that mudbrick houses are sturdy enough to support an upper floor. Ancient sources frequently refer to upper floors in Classical and Hellenistic houses as the *gynaikonitis*, the woman's quarter. Upper floors are however difficult to infer from archaeological remains. Usually, the only indications are bases of wooden stairs such as those found at Olynthos, or parts of stairs made of stone like those found at Thorikos (Robinson & Graham,1938; 1946), but in neither of these cases is it clear whether the stairs led to a flat roof or roof terrace or to an upper floor. Other indications of upper floors are holes for beams supporting the first floor like those found in rear walls of houses in Priene (Hoepfner & Schwandner, 1994: p. 218). It is however difficult to prove the former existence of upper floors on the basis of stratigraphic evidence alone, except in cases involving certain architectural features such as mosaics.[10]

When a house with an upper floor collapses, household items from the upper and the ground floor usually become mixed on the ground floor along with architectural debris from the upper structure. In ancient times, such architectural debris will usually have consisted of organic, and hence perishable, material such as wood and reed, which leave no traces in the archaeological record.

The stratigraphic evidence obtained at New Halos suggests that the houses had no upper floors. No indisputable evidence of collapsed upper floors was found. Certain stones found in some of the houses, such as the House of Agathon, room 3, may have been bases of wooden stairways, but this is far from certain. No other finds or evidence suggest the existence of an upper floor in any of the houses we excavated.

2.8. Water supply, storage and drainage

During our excavations we found no remains of devices or structures for supplying or draining water. This is very unusual for a town from the Hellenistic period. A section dug across the street between housing blocks 6.4 and 6.5 in 1993 revealed no traces of gutters, drainage pipes or other drainage systems, nor did we find any remains of watersupply systems. This lack of evidence may in part be due to the limited scope of our excavations. It could well be that all traces of waterworks like public cisterns and drainage systems such as gutters along the main streets of Halos were destroyed in ploughing.

Most houses were similarly devoid of drainage pipes, wells and the like. In 1978, when we cleaned part of a foundation wall of a house in the lower town of Halos, we found some evidence of a drainage system in the form of a roof tile inserted between two large foundation stones – probably the remains of a gutter for draining excess water from the house to the street (fig. 2.11). In the courtyard of the House of the Snakes we found a feature which proved to represent a well that was secondarily used as a rubbish pit (Haagsma, 1998).

Drinking water was probably drawn from public cisterns, which were presumably widely distributed across the town, and stored in *pithoi* inside the houses. Some houses may have had wells. The most likely sources of water were the Kefálosis spring and rainwater.

3. DESCRIPTION OF THE HOUSES

In this section I shall discuss the evidence obtained in the excavation of each house separately. For each house I shall provide general information on the excavation, descriptions of the foundations, individual rooms, areas and units, the finds recovered in them and the associated find conditions, traces of rebuilding/renovation, *etc.* and general conclusions concerning the nature of the house, its ground plan, any changes it underwent, *etc.* A 'room' will here be understood to be an area enclosed by four walls. An 'area' is part of a house that was enclosed by fewer than four walls, *i.e.* usually a courtyard or a corridor. For practical reasons I have subdivided some areas into 'units'.

3.1. The House of the Coroplast (Housing block 2.7, plot 11)

The House of the Coroplast was excavated in 1978 and 1979. Although the results of these excavations have already been published (Reinders 1988: pp. 117–134), we have decided to include this house in the present publication because closer examination of its foundations has yielded new information on the house's building phases.

The dimensions of the House of the Coroplast were 13.75 by approx. 15.00 m. The house lay in the northern part of housing block 2.7, along street No. 6. It was entered through a *prothyron* with a double door with a width of 2.25 m. This door provided access to the courtyard which comprised two separate areas: areas 1 and 2. To the left of the entrance was a small room (No. 9) measuring 2.70×4.40 m, which was not connected to any of the other rooms (figs 2.12 & 2.13).

From the courtyard, the covered part of the house could be entered via what we have called room 3. This

Fig. 2.11. Gutter for excess water.

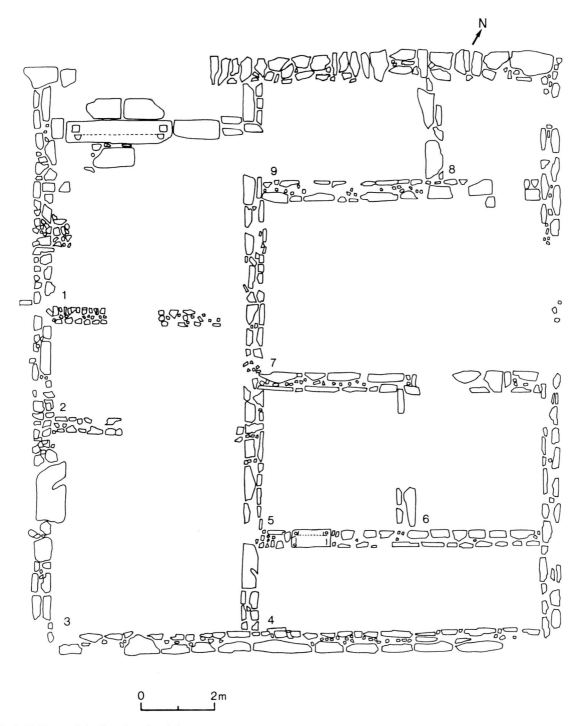

Fig. 2.12. House of the Coroplast: foundations.

room, which measured 5.20×5.20 m, has been identified as the workshop of a coroplast on the basis of the relatively large number of moulds and figurines that were found in it. This room provided access to room 4, measuring 2.15×7.65 m, which was probably used for weaving and/or storage. A double door in room 4 provided access to room 5 (3.60×3.60 m), which is thought to have been the cooking area. Room 5 provided access to room 6 (3.60×3.60 m), which was devoid of finds. From room 6 one entered the largest room of the house, room 7 (4.55×7.60 m), which has been identified as the main living room. Room 8, a small room (2.65×2.75 m) which yielded very few finds, could only be entered via this main living area.

Fig. 2.13. House of the Coroplast, view from the north.

The most remarkable feature of the House of the Coroplast is the single route running through the house. There seems to have been no circular route. Neither did any one of the rooms provide access to several other rooms as in the House of Agathon or the House of the Ptolemaic Coins. To get to, say, room 8, one had to walk through almost all the rooms in the house. Whether this feature formed part of the original design of the house or perhaps evolved as a result of the use of the house we do not know.[11]

Closer examination of the foundations of the west wall of room 7 revealed a stretch with a width of 1.20 m consisting of smaller stones breaking the pattern of the rest of the wall foundations. The other wall foundations comprise two rows, the outer (western) row consisting of large limestone blocks and the inner (eastern) one of significantly smaller stones. The different pattern of the aforementioned stretch suggests that there was a doorway here originally, which was later walled up. Photographs of this stretch support the assumption that it was built at a later stage. If this is correct, this access from the courtyard would transform room 7 into the central room of the house, via which various other rooms could be entered, making it comparable with all the other main living areas encountered in the houses we excavated: room 3 in the House of Agathon, room 3 in the House of the Ptolemaic

Coins, room 3 in the House of the Geometric *Krater* and room 8/11 in the House of the Snakes.

In this context it is worth noting that the foundations of the two east-west walls in the area at the western end of the house identified as the courtyard also differ significantly in character from the other wall foundations in that they consist of significantly smaller stones. The stones of these walls were moreover laid against the north-south walls of the courtyard, instead of being incorporated in them. This, and the walled-up doorway of room 7, suggests that the house was redesigned at least once.

We may conclude that in phase 1 (fig. 2.14) the house had an open courtyard at its western end and that the roofed part of the house could be entered through rooms 7 and 4. In phase 2 (fig. 2.15) the courtyard comprised three areas, one of which, the coroplast's workshop, was roofed (room 3). Walling up the doorway between the courtyard and room 7 will have ensured the privacy of the living quarters, closing them off from the more public area of the workshop. The function of the wall between areas 1 and 2 remains unclear as there is no evidence of either of these areas having been roofed.

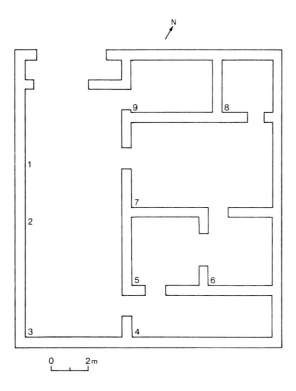

Fig. 2.14. House of the Coroplast: phase 1.

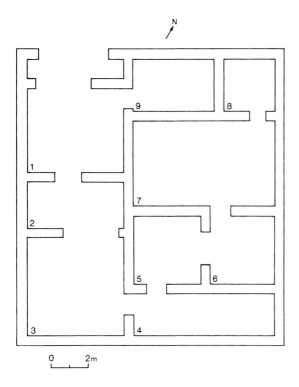

Fig. 2.15. House of the Coroplast: phase 2.

3.2. The House of the Geometric *Krater* (Housing block 2.8, plot 17)

Alongside the old road running from Almirós to Soúr-pi clear remains of wall foundations were visible *in situ* at the surface. Although this area had been used for arable farming, its situation near the elevated road had evidently saved it from extensive ploughing. We decided to excavate this undisturbed area in order to obtain insight into the structure and layout of the architectural remains. The excavation was carried out in the autumn of 1984 by a team of eleven people who worked for four weeks using pickaxes, mattocks and trowels. They dug right down to the virgin soil. Diagnostic finds were plotted and kept for further investigation. The majority of the undiagnostic sherds were described and then discarded.

In the course of the excavation the observed concentration of wall foundations proved to be part of a house (figs 2.16 & 2.17). This house, which we called the House of the Geometric *Krater*, lay in the southern part of housing block 2.8, bordering the Main Avenue. The wall foundations are the same as most of the others at Halos, consisting of double rows of differently sized limestone blocks. Stretches of 12.70 m and 8.80 m of the northeastern and northwestern wall, respectively, had survived. The average width of the northeastern wall was 0.50 m, that of the northwestern wall 0.55 m. The inner walls were built from smaller stones with widths of approximately 0.45 m. The widths of the northeastern and rear walls and the absence of any doorways in them suggest that these walls separated the House of the Geometric *Krater* from the adjacent buildings. As the remains were uncovered it was found that the northeastern wall had suffered the consequences of recent agricultural activities. The stones at the southern end had been disturbed and parts of the wall were missing. Inspection of the assumed courses of the outer walls of the southern part of the house showed that the foundation stones in this part had disappeared in ploughing. This means we were unable to determine the house's plan with 100% certainty (but see below).

Three main rooms were distinguished in the excavated area. Gaps with widths of between approximately 1.00 and 1.15 m in the foundations indicate doorways. No threshold stones were found. A double row of small limestone blocks found just above floor level in the southern entrance to room 3 represents a threshold or the foundations of a threshold stone.

Inside the rooms were dissolved mudbrick, small stones, roof-tile fragments, pottery and other finds. The stratigraphy and dates of the finds are the same as described in section 2.1, with the exception of a few finds encountered at a greater depth which date from the Geometric period. Differences in the structure of the walls however suggest that the house underwent renovation work at least once. The southwestern walls

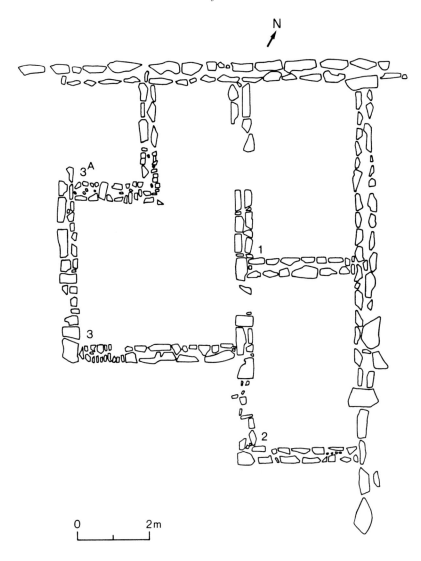

Fig. 2.16. House of the Geometric *Krater*: foundations.

of rooms 2 and 3 both contain a doorway which was evidently walled up at a later stage. The presence of the internal wall in room 3, creating what we have called room 5, the level at which it was founded and the way it was linked to the other walls also indicate renovation activities. Starting in the northeast, I shall now describe the main features of the various rooms and the finds discovered in them.

Room 1. This room was situated in the northeastern part of the house. Its inside measurements were 4.65×2.85 m. The room was bordered on two sides by the house's external walls. In the northwestern wall was a 1.00-m-wide doorway. The room's floor consisted of trodden earth. It was covered with a thin layer of grit whose density increased towards the entrance. This layer may have served to consolidate the floor. The many small roof-tile fragments indicate that the room was roofed, and also suggest that it was not

immediately buried beneath debris after the final earthquake. There is a strong possibility that looting and scavenging took place during and after the site's abandonment.

Some body fragments of a *pithos*, a *chytra* and a *lekane* were found, but the majority of the sherds found in Room 1 derive from vessels used for consumption: a plate, four *kantharoi* and three bowls, all of black gloss ware (see section 3.1). In addition, parts of three lamps were found, all near the entrance to the room. The bulk of the pottery was found in the southeastern corner of the room. This location and the distribution of the sherds suggest that the pottery fell from a higher level, probably a shelf on the wall (*cf.* Svoronos-Hadjimichalis, 1956).

No fewer than 22 coins were found in this room. They all date from the 4th and early 3rd centuries BC, the latest being a coin of Antigonos Gonatas dating

Fig. 2.17. Excavation of the House of the Geometric *Krater*, view from the north.

from the beginning of his rule, *i.e. c.* 270–265 BC (see section 3.5). The level at which this coin was found suggests that it dates from the time when the house was in use. Interestingly, the majority of the coins found in close association with one another derive from the same sources. Coins from Larisa were found mostly in the northeastern corner of the room and coins from Chalkis in the southwestern part of the room.

Few metal finds and animal remains were recovered here. Most of the artefacts, especially pottery and coins, were found in the southern part of the room; the central part of the room was devoid of finds.

Room 2. Situated directly south of room 1, room 2 had more or less the same dimensions: 4.60×2.85 m, measured inside the walls. The room's northeastern wall was found to have been severely damaged by recent ploughing. The other walls were reasonably intact. The shapes and arrangement of the foundation stones suggest that there were originally two doorways in the southwestern wall: a northern one, with a width of 0.85 m, providing access to room 3 and a southern one, with a width of 0.90 m, which probably provided access to the courtyard. The stones at both ends of the middle wall stretch were clearly worked. They both have straight corners. Small limestone blocks found in the southern doorway suggest that it was walled up during later rebuilding work.

A thin layer of grit, probably deposited to consolidate the floor, was found at the same depth as in room 1 (-0.12 m at the centre of the room measured from surface level). The large number of roof-tile fragments supports the assumption that room 2 was roofed.

The number of finds recorded in this room is considerably smaller than that in room 1 despite the fact that a layer of large roof-tile fragments was found to cover most of the finds, suggesting an undisturbed context. The small number of pottery sherds is attributable to the fact that most of the body sherds found in this room were discarded during the excavation, resulting in an incomplete picture. Fragments of *lekanides*, cooking pots and *lopades* were found here, together with fragments of bowls, cups and a plate plus part of an *amphora*. The only coin found in this room came to light just beneath the surface in the northern doorway in the foundation wall. It came from Histiaia. The only animal remains found in this room were shells.

Rooms 3 and 3A. Room 3 was the largest of the three excavated rooms. Its dimensions were 5.85×4.35 m (including room 3A). Its situation at the centre of the living quarters and its shape and size bring to mind the *oikoi*, the main living rooms, identified in houses at Olynthos (Robinson & Graham, 1938; 1946), Kassope (Hoepfner & Schwandner, 1994) and elsewhere.

Fig. 2.18. House of the Geometric *Krater*: threshold room 3 to south room.

Two partition walls in the northwestern corner of the room designate room 3A as a separate area within room 3. Its contents will be described separately.

Evidence of rebuilding activities was observed in the structure of the wall foundations. The southwestern foundation wall consisted of the usual double row of limestone blocks, except at its southern end. As in room 2, two stones, with neatly carved corners, marked a doorway which was walled up at a later stage (fig. 2.19). From its northern end onwards the wall was disturbed, but the remains suggest that it originally extended to the rear wall.

The walls partitioning the area designated as room 3a did not form part of the original layout of the house. The way in which they were connected to the northwestern and southwestern walls and the level at which they were founded show that they were added at a later stage. The northwestern foundation wall of room 3A originally contained a doorway with a width of approx. 0.85 m, evident from the arrangement of the stones near its connection to the southwestern wall (fig. 2.16). This doorway provided access to a room, now lost, similar to room 2. The foundations of the southeastern wall of room 3 were completely undisturbed. A fairly wide doorway with a width of 1.17 m provided access to room 3, with small limestone cobbles serving as a threshold or its foundation (fig. 2.18).

The layer containing the remains dating from the time of the destruction of the Hellenistic town was in room 3 covered with large fragments of roof tiles, indicating that this part of the house was roofed. The soil was mixed with small bits of charcoal. The floor of this room consisted of trodden earth mixed with large cobbles. Finds consisted mainly of domestic pottery like fragments of cooking pots, an *amphora* and fragments of bowls. Fragments of shells and bones were found, too. Some bones represented astragali, used in games or in divination.

Room 3A was defined by a wall with a length of 2.50 m extending from the northwestern external wall and a wall with a length of 1.80 m extending from the southwestern wall of room 3. As already mentioned above, these walls must have been built in a later occupation phase. Room 3A was most probably entered through a room lying to the west of it, as no doorway providing access to room 3 was observed. The southwestern wall of room 3A had not survived.

In view of the situation and size of this room and the evidence for food-processing activities found in room 3, room 3A has been interpreted as a kitchen flue of the kind also encountered in the houses at Olynthos (Robinson & Graham, 1938; 1946), in spite of the absence of evidence such as ashes or an abundance of charcoal.

Finds recovered from room 3A consist mostly of storage pottery. In addition, part of a *pythos* came to light and a medium-sized storage jar was found *in situ*. A remarkable find was a bronze bracelet in the form of a snake.

Reconstruction. Although only part of the House of the Geometric *Krater* has been preserved, the four excavated rooms provide enough information for us to attempt to reconstruct the plan of the house. We know that the lengths of the houses at New Halos varied only little. The average width of a housing block was 31.5 m, so the average length of the houses was 15.75 m. About 13.00 m of the length of the House of the Geometric *Krater* had survived, so it must have

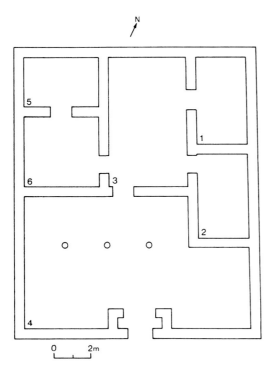

Fig. 2.19. House of the Geometric *Krater*: phase 1.

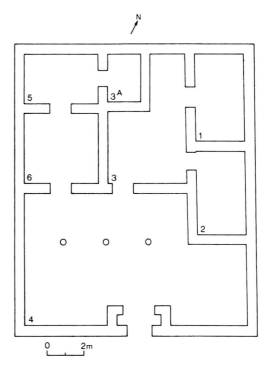

Fig. 2.20. House of the Geometric *Krater*: phase 2.

been about 2.75 m longer originally. The houses at New Halos had three standard widths: 12.50, 13.75 and 15.00 m. As will be discussed at the end of this chapter, the basic living quarters in the houses of New Halos consisted of a large living room bordered by two smaller rooms on either side. When we apply this principle to the House of the Geometric *Krater* we arrive at a width of approx. 12.50 m.

The reconstruction illustrated in figures 2.19 & 2.20 shows the various spatial components of the house. The house compares well with two other south-facing houses with widths of 12.50 m excavated at New Halos: the House of Agathon and the House of the Ptolemaic Coins (see below). Rooms 5 and 6 may have been comparable with those in the latter two houses, and what we have reconstructed as area 4 can be interpreted as the courtyard, with a *prostas* or a *pastas*. These components and their consequences for the house typology of New Halos and for house typology in early Hellenistic town-planning in general will be discussed at the end of this chapter.

Geometric Finds. The stratigraphy of the House of the Geometric *Krater* was more complex than that of the other houses. In room 3, a layer beneath the Hellenistic floor level yielded some finds dating from the Geometric period. These finds consisted of three coarse *pithos* bases, one fragment of an unknown terracotta object, a two-handled cup and a decorated *krater* (Dyer & Haagsma, 1993). The *pithos* bases, the terracotta fragment and the *krater* were found in three pits with depths of approx. 0.25 m, measured from the

Hellenistic floor level, which had been dug into the virgin soil. The two-handled cup was found at the Hellenistic floor level. It is interesting to compare these finds with the results of the excavations of the Hellenistic cemetery of New Halos conducted by the 13th Ephoria of Prehistoric and Classical Antiquities in the area southeast of the town, outside the city walls. These excavations yielded remains of a reused Iron Age *tholos* and some *pithos* burials, which probably date from the Iron Age (Malakasiotis *et al.*, 1993).

The finds discovered in room 3, the *pithos* burials and the *tholos* most likely represent the remains of one or more Geometric cemeteries situated south of the river Amphrysos, which were partially destroyed during the construction of the Hellenistic town. Of a somewhat similar date but different in nature is the well-known Iron Age cemetery north of the Amphrysos (Wace & Thompson, 1911–1912; Reinders *et al.*, 1992), which consisted of burial mounds containing both inhumations and cremation remains.

3.3. The House of the Snakes (Housing block 6.4, plot 9)

The House of the Snakes formed part of the same housing block as the House of the *Amphorai*, the House of Agathon and the House of the Ptolemaic Coins. It lay to the northwest of the latter two houses and was hence situated at the northwestern end of the housing block. The house was excavated by a team of fifteen excavators in eight weeks in the summer of

1993. The excavation of a well in the house's court-yard was not completed that year, but was finalised in the summer of 1997.

Parts of the house's wall foundations were visible at the surface. The foundations of this house were also damaged in some places, especially near the external northwestern wall, where all the foundation stones of a stretch of approx. 0.7 m of the internal walls were found to have been removed (figs 2.21, 2.22 & 2.25). The foundation stones consisted of large, irregular blocks of limestone, usually arranged in two rows with smaller stones used to fill the gap between them. One course of the foundations (with a height of approx. 0.45 m) had survived. As elsewhere, the external walls were a little wider (0.5 m) than the internal walls (0.4 m). The house measured 15.10×15.45 m.

As usual, the topsoil (0.08–0.1 m) had been ploughed and contained worn pottery and some present-day artefacts. The layer beneath (approx. 0.25 m thick) was more compact and seemed to have been disturbed in a few places only. The virgin, Pleistocene soil, which contained many cobbles and pebbles, con-stituted the floor level in both the roofed and the open parts of the house.

Room 1. Room 1 lay in the northeastern part of the house and bordered room 2, area 4 and the external northeastern wall. The room was square and measured 2.80×2.80 m. It could be entered from area 7 through a doorway whose exact width could no longer be de-termined as the wall foundations had been disturbed at this point. No threshold stone was found. A few large limestone blocks were found in the destruction layer, especially in the northern area of the room. Some may have come from a gap that was found in the wall between rooms 1 and 2 (which was not a doorway but the result of disturbance), while others may derive from a second course of foundation stones.

A number of stones were found arranged in a rec-tangle next to the northeastern external wall. They rested on the floor and had surrounded a pit measur-ing 0.7×0.6 m with a depth of 0.07 m. The pit's fill consisted of dark soil mixed with a few small sherds and organic matter. No ashes were found and the finds were not burnt. The pit may have been an old cook-ing pit or hearth, but was not in use as such at the time when the house was destroyed.

Diagnostic finds recovered from this room are frag-ments of a *lagynos*, two *lopades*, two *chytrai*, an am-phora, a *pithos*, a storage vessel, a plate and a lid. A few iron artefacts like nails and a ferrule came to light as well as many shells and bone fragments.

Room 2. This room lay immediately northeast of room 1 and was of the same size and shape. It was likewise entered from area 7. The doorway to room 2 was found to be intact and had a width of 0.8 m. It had a threshold made of volcanic stone which meas-ured 0.75×0.45 m. The stone was worn at the centre due to use. No pivot holes were observed.

A trial pit was dug in this room in order to exam-ine the relationship between the floor level and the foundation trenches of the northeastern external wall. This pit showed that the room contained only one, thin floor layer, consisting of compacted soil mixed with fine grit, at the same level as the top of the founda-tion trench. The trench was filled with somewhat softer, mostly virgin soil mixed with cobbles but no finds. The absence of a later floor layer on top of the foundation trench confirmed our assumption that most of the houses at Halos were occupied in a single con-tinuous phase.

No specific features were found in this room. The destruction layer contained the usual finds mixed with roof-tile fragments. The following diagnostic finds were recovered in room 2: an *amphora*, a *lopas*, a kind of jug, a *kantharos*, a lid, a bowl, a *lekane*, a storage jar and animal remains.

Room 3. Room 3 lay northeast of room 2 and ap-pears to have been rectangular. The surviving part of the room measures 4.70×2.80 m. The foundations of the northeastern wall and those of the entire southwest-ern wall had been destroyed by ploughing in recent times; all the foundation stones in these parts had dis-appeared. The trench that had contained the founda-tion stones of the southwestern wall was however still visible over what we took to be its entire length.[12] We found a few other physical remains of this wall in the form of a foundation stone in the northwestern corner of room 2 and two stones lying approximately 2 m further northwest in the presumed course of the wall. It is not clear whether room 3 was subdivided or not.

Significant differences are observable in the nature and distribution of the finds recovered in the northeast-ern and southwestern parts of the room. The south-western part contained a clear destruction layer with many roof-tile fragments, pottery and other finds, whereas only few finds, mixed with many small, worn roof-tile fragments came to light in the northeastern part. As a similar difference was observed in areas 7 and 8, we assume that it is attributable to recent agri-cultural activities affecting part of the northwestern area of the house (fig. 2.21) rather than to different uses of the areas concerned or to some sort of inter-nal subdivision in the room.

The southwestern part of the room contained a compact layer of pottery fragments mixed with roof-tile fragments. Characteristic finds are five *chytrai*, two *lekanides*, a *lopas*, a plate, some *kraters*, jugs and lamps. Among the finds were also the remains of a door, consisting of a large number of clenched nails similar to those found in the House of the Coroplast. A bolt, lock and key which most probably belonged to the same door were found opposite room 3, near the doorway of room 8.

Area 4. This area lay southwest of room 1. It was open on the courtyard side of the house (area 9) and

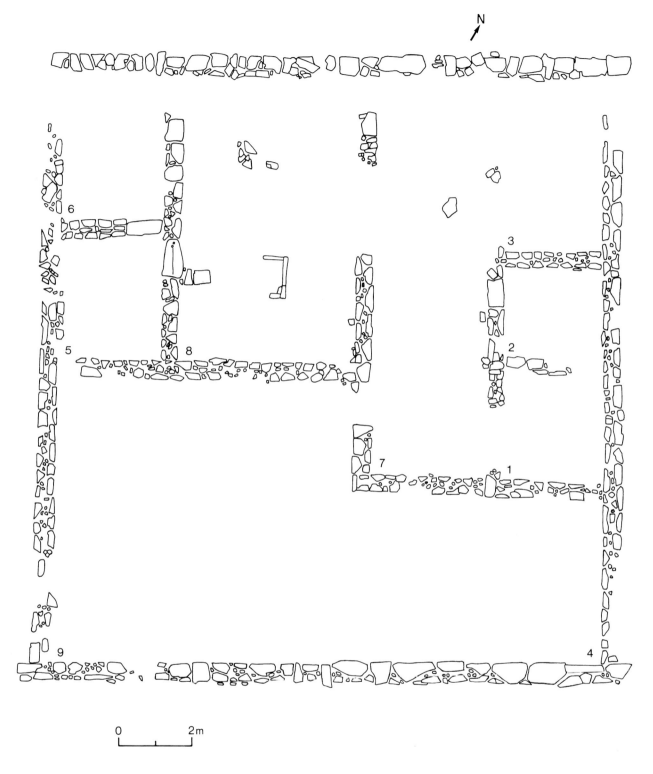

Fig. 2.21. House of the Snakes: foundations.

Fig. 2.22. Excavation of the House of the Snakes, view from the west.

could probably be reached from the street through the corridor (area 7) and the courtyard. The area was rectangular and measured 4.35×6.90 m. The area's situation and the fact that it yielded only small, worn roof-tile fragments indeed suggest that it formed part of the courtyard. So this part of the courtyard was not covered, or at least not with roof tiles. It may have been covered with some organic material like reed, but there is no archaeological evidence to support this.

Interestingly, the floor layer contained cobbles of a larger size than those found in the covered parts of the house. Also, the floor level was at least 0.1–0.2 m higher in area 4 and the rest of the courtyard than in

the other parts of the house. The floor sloped gently upwards towards the west. The foundation trenches of the northeastern and southwestern external walls were less clearly visible in this area, probably as a result of disturbance by the roots of a tree that had grown at the southeastern corner of the house.

As already mentioned above, pottery fragments and other finds were found scattered all over this area. Only a few diagnostic objects were found: an *amphora*, a *lekane*, four coins, some nails and a keyhole.

Room 5. This room lay in the western part of the house and was accessible from room 8. The room, which provided access to room 6, was rectangular and measured 3.30×2.70 m. Its wall foundations were reasonably intact, although parts of the southeastern and southwestern walls had been disturbed. The doorways in the northeastern and northwestern walls were 1.0 m and 1.1 m wide, respectively, and each had a threshold made of limestone. The threshold in the doorway to room 8 contained two small pivot holes and was neatly finished. The surface of the western part of the threshold was lower than that of the eastern part, indicating that the door swung open towards the west.

The topsoil of room 5 contained only a few roof-tile fragments and no sherds of other pottery whatsoever. The destruction layer however yielded an abundance of pottery mixed with large roof-tile fragments. The pottery consisted mainly of storage ware like *amphorai* and *pithoi*. In addition, the features of two *pithos* holes were found along the northwestern wall. They were approximately 0.4 m deep (measured from the original floor level) and had a diameter of 0.8 m. Both were filled with soil mixed with pottery from the destruction level. The floor consisted of a compact layer of small pebbles mixed with a few small, worn sherds and many metal objects. Traces of a walking route were observed by examining the colour of the soil of the floor: near the two doorways the soil was darker than elsewhere in the room.

The diagnostic pottery found in this room consists of five *amphorai*, two lids, a *chytra*, a jug, a storage pot, a bowl and fragments of a *pithos*. A relatively large number of nails were found, plus what may have been the bolt of a door. Room 5 also yielded a large number of molars and bone fragments, most of which were found close together in the northeastern part of the room.

Room 6. Room 6 could be entered from room 5. The doorway between the two rooms was 1.10 m wide and contained a threshold made of limestone without any traces of pivot holes. The room was slightly larger than room 5, measuring 2.50×3.95 m. The wall foundations were partially disturbed, especially in the northern part of the room, where foundation stones appeared to have been removed in ploughing across the entire length of the house. The rest of the room was however reasonably intact.

Room 6 contained three features interpreted as *pithos* holes along its southwestern wall. They lay at equal distances from one another and were 0.7–0.8 m wide and about 0.3 m deep. The holes were filled with loose yellowish earth mixed with pebbles and a few finds. Large *pithos* remains were found scattered across the room's floor. The room also yielded fragments of other vessels, such as a *pyxis*, storage jars, lids, a bowl and a plate. Metal finds included several iron implements and organic finds consisted of shells. In addition, a few pieces of a bathtub were found scattered across the floor.

Area 7. This area was a corridor running from the entrance of the house to the courtyard, which provided access to rooms 1, 2, 3 and 8 and to the courtyard itself (area 9). The remains of the original entrance of the house consisted of a 1.55-m-wide doorway in the external wall with a threshold made of a large block of limestone measuring 1.30×0.45 m, which contained no pivot holes. The northern part of area 7 lies in the aforementioned zone affected by ploughing. Consequently, only one layer, containing mostly small, worn roof-tile fragments, was observable in the disturbed stratigraphy. The stratigraphy of the southern part of this area was intact. Large roof-tile fragments indicated that the area was roofed. Other finds include a figurine, *hydriai*, *lekanides*, *chytrai* and a storage jar.

Room 8. Room 8 was situated at the centre of the house, between rooms 5 and 6 and area 7. It was the largest room of the house, measuring 7.50×4.70 m. The room could be entered from area 7 and provided access to room 5. No clear indications of any internal subdivisions were observed in the room, apart from two foundation stones south of the entrance to room 5 and some foundation stones at the centre of the northern half of the room. The latter may perhaps be interpreted as the remains of a base for a wooden roof support; the former may have formed part of a short stretch of a wall shielding the southern corner of room 8.

Except in the northeastern part of the room (fig. 2.21; see above), the stratigraphy was reasonably intact. A layer of roof-tile fragments mixed with pottery was found in most parts of the room. A few parts of the room had however been disturbed in recent ploughing. This was most clearly observable near the hearth that was found at the centre of the southern half of the room. A few of the blocks of volcanic stone of this hearth, which had been dug into the floor, had been removed in the ploughing; the damaged remaining blocks were found in the vicinity of the hearth. The furrow made by the plough could still be followed across the room.

The hearth was originally rectangular and measured 1.75×0.4 m. Its blocks protruded slightly (about 0.07 m) above the level of the floor. The hearth contained ashes and burnt pottery fragments. Many remains of cooking vessels were found in the vicinity of the hearth, as well as remains of an iron chain and hook, which undoubtedly formed part of an implement for hanging a kettle and/or food above the fire.

An exceptional find came to light next to the hearth, near its eastern corner (figs 2.23 & 2.24). It consisted of a vessel with a lid, both made of local volcanic stone, which had been dug into the virgin soil. The hole into which the vessel had been placed was dug some time after the blocks for the hearth had been laid. The vessel contained reddish soil, a bone fragment, a shell, a sherd, and two small serpent-shaped pieces of metal, one of iron and one of silver. This assemblage, which is to be interpreted in the context of a house cult, *i.e.* a sacrifice to Zeus Ktesios, will

Fig. 2.23. House of the Snakes: hearth and stone vessel in room 8.

Fig. 2.24. House of the Snakes: stone vessel in room 8.

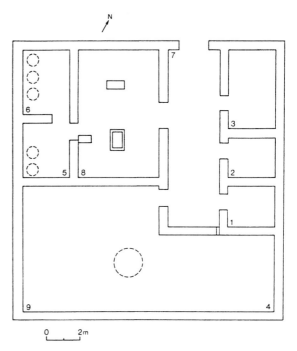

Fig. 2.25. House of the Snakes: reconstruction.

be discussed in greater detail in a separate publication (Haagsma, in prep.).

Immediately to the south of the doorway to room 5 was a rectangular limestone block inscribed with the text ΕΠΙΔΟΣΙΣ. This inscription will also be discussed in a separate publication. Its presence in a household context is very odd and we therefore assume that the block in question derives from some other source and was secondarily used as building material in this house.

Room 8 yielded a wide variety of finds: cooking pots such as *chytrai*, *lopes* and other pottery like plates, *hydriai*, bowls, juglets, lids, storage jars, *kraters*, an *unguentarium*, a lamp, loomweights and figurines, metal objects such as nails and parts of furniture (mountings) and stone objects such as a rubber and a grinding stone. Few organic remains were found here, the most noteworthy being a large cask shell *(Tonna galea)* that came to light near the hearth.

Area 9. Area 9 can be interpreted as the house's courtyard. It could be reached from the entrance of the house through area 7. The courtyard consisted of a large rectangular area, measuring 7.50×8.10 m, which was open, and a narrower area, measuring 4.35×6.90 m in the northeast, which was most likely covered, perhaps not with roof tiles, but with reed or some other perishable material, thus forming a kind of *apothike*. The courtyard contained two features. The first, feature a, came to light a short distance to the southeast of the entrance to the courtyard and consisted of a heap of large cobbles mixed with fragments of pottery, roof tiles, bone and charcoal plus some 40 oyster shells. This feature, which was 0.8 m wide, was found at the courtyard's floor level. It was interpreted as a rubbish heap. Feature b came to light a short distance to the southeast of feature a. It consisted of a deep pit which became slightly narrower towards the bottom. The pit was lined with roof-tile fragments and filled with soil devoid of pottery fragments.

Five layers were distinguished in the pit's fill. Layer 1 (0.15 m thick) represents the destruction layer, containing roof-tile fragments and pottery. Layer 2 (0.15–0.27 m) consisted of brown crumbly soil, containing a few limestone pebbles. Layer 3 (0.27–0.78 m) consisted of a greenish black clay-like soil mixed with many small limestone fragments. Some larger pieces of limestone seemed to mark the boundary between layers 4 and 5 (0.78–1.04 m), which consisted of a greyish loose soil with a distinctive smell. At a depth of 0.89 m from the surface a horizontal layer of roof-tile fragments appeared (0.15 m), overlying a soft brownish soil (layer 5). The pit was obviously full by the time when the houses of New Halos were abandoned. In my opinion it was therefore not a cistern or a well, as was first thought, but a *kopron* (Haagsma, 1998).

The rest of the courtyard floor was covered with scattered small fragments of pottery and roof tiles, evenly distributed across the area. The floor consisted of a compact layer of large cobbles.

History of occupation and reconstruction. The remains of the House of the Snakes seem to represent a single building phase, as only one floor level was found and there was no evidence for any changes in the house's layout during its occupation. The roofed part of the house bordered the street directly, in contrast to the roofed parts of the 'south-facing houses', which were entered through an open area, and to the roofed part of the only other 'north-facing' houses: the House of the Coroplast, which was entered through a *prothyron* and an elongated courtyard. The House of the Snakes lacked a *prothyron*, having instead a large door in its facade, consisting of wooden beams held together by cross-beams and clenched nails resembling the nails found in the House of the Coroplast.

As with most of the other houses, one room (room 1) protruded into the courtyard, thus providing a space for a *pastas* over the remaining width of the house (fig. 2.25). But, as with the other houses, no indisputable evidence of any *pastas* was found, such as column bases or a distribution of large fragments of roof tiles distinctly differing from the distribution pattern in the rest of the courtyard.

No remains of staircases indicating an upper floor were found. The number and weight of the roof-tile fragments (see the next chapter) moreover indicate that the entire roof was tiled and that no parts of the building had a flat roof.

Although this is a 'north-facing house', the same canon of a large room flanked by four side rooms encountered in all the other houses excavated so far (see below) was observed here, too, be it in a somewhat different configuration. The corridor (area 7) leading from the entrance of the house to the courtyard separated the living quarters into two parts, one consisting of two or three rooms (rooms 1 and 2 and perhaps 3), and the other of the main living room flanked by two smaller rooms (rooms 5, 6 and 8). There was a hearth in the main living room (room 8) and probably also in room 1 and in the courtyard. Holes for large storage jars such as *pithoi* were found in rooms 5 and 6. A well that was secondarily used as a rubbish pit and a *kopron* was found in the courtyard.

3.4. The House of the *Amphorai* (Housing block 6.4, plot 8)

The House of the *Amphorai*, which formed part of the southern row of houses in housing block 6.4 (figs 2.26 & 2.27), was excavated in its entirety between May and July of 1991. A team of seventeen people used pickaxes, mattocks and trowels to clear away the soil and debris to an average depth of 0.40 m, at which the virgin soil was reached.

Fig. 2.26. House of the *Amphorai*: foundations.

We decided to excavate this house because its foundations seemed to be in a reasonable condition at first sight. A second argument was the house's width of 15.00 m. By this time, houses with all three of the widths encountered at New Halos (12.50, 13.75 and 15.00 m) had been excavated or studied (House of the Coroplast – 13.75, House A – 15.00 (Reinders, 1988: p. 117) and two houses with widths of 12.50 m during the 1987 and 1989 campaigns). We were interested in finding out whether the differences in width correlated with differences in plans. If this were the case, the next question would be to what extent the differences in house plan reflected differences in the use of space.

After the topsoil had been removed the house's external wall foundations became more clearly visible, but it also became clear that the remains had been disturbed, and that large parts of the foundations were

Fig. 2.27. House of the *Amphorai*, view from the north.

missing. The layer of debris that had to be removed turned out to be only 0.20–0.25 m deep on average. Below this debris no clear destruction layer was observable. The northwestern part of the house in particular had been badly damaged by agricultural activities. Large parts of the northwestern and southwestern walls were missing as well as parts of the internal walls. The plan of the house could therefore be reconstructed in part only. Nevertheless, it was clear that the plan as a whole differed in several important respects from the plans of the other houses hitherto excavated.

The structure of the intact stretches of wall foundations did not differ from that of the foundations found elsewhere at New Halos. But differences were observed between the eastern and western parts of the house, reflecting different building stages (see below). The house's dimensions were 15.50×14.30 m.

Room 1. This room was the most easily recognisable when the excavation was started. As the surface of the land on which housing block 6.4 was built slopes downwards slightly to the northeast, the surface was highest at the northwestern wall. The same holds for the floor level, which decreased approximately 0.04 m from the northwest to the southeast.

For practical reasons, the room was divided into two units, with a baulk left at the centre of the room to enable us to study the stratification. The room's soil fill was found to vary. Part of the fill, especially that

in the northwest corner of the room, consisted of clayey soil, for which there is no clear explanation, whereas other parts consisted of soil of a yellowish sandy texture. The latter most probably represents the remains of the house's original mudbrick walls. The greater part of the fill however consisted of dark red *rendzina* mixed with small particles of charcoal and tiny bits of limestone.

The stratification observable in the baulk left at the centre of the room yielded no new information; it comprised the usual 0.10-m-thick top layer of worn remains followed by the destruction layer and a layer containing diagnostic finds. The floor consisted of the usual large cobbles of the Pleistocene hardpan intermingled with finds, some deriving from the destruction layer, others, such as worn, small sherds, embedded in the original floor. A few present-day objects were encountered among the finds of the destruction layer, such as a button and a bullet cartridge. These finds indicate that the greater part of the area of this house was severely disturbed. The lower layers of room 1 however seemed to be more or less intact.

The finds discovered in room 1 are a large rim of a *pithos*, an *amphora*, two or three *chytrai*, part of a juglet, a strainer, a *pyxis*, some nails, six coins and some shells.

Room 2. This room was situated northwest of room 1 and measured 4.50×2.90 m. It could be entered from area 8. Its southeastern wall probably contained an entrance to room 1. The doorways could not be located due to severe disturbance. As in room 1, a number of subrecent artefacts were found in the upper layers. The lower layer of room 2 again seemed to be reasonably intact. And as in room 1, the destruction layer and the original floor came to light at a deeper level at the wall foundations as the surface of the land sloped slightly downwards (approx. 0.03 m) towards the centre of the room. Fewer finds were recovered at the centre of the room than at the sides. The majority of the finds came to light in the southern part of the room. Many lumps of burnt soil were found all over the room, especially in the western corner, together with small pieces of charcoal, indicating that a small fire burned here after the site was abandoned. Roof-tile fragments indicate that the room was originally roofed.

The main finds recovered from this room are *amphorai*, a *kantharos*, a juglet, two miniature cups, a *lagynos*, a plate, a *hydria*, a lower and upper grinding stone, several nails, clamps and coins, plus organic remains like bone fragments and shells.

Area 3, units A, B and C. Area 3 was situated northwest of room 2. To the north and west it was bordered by the house's external walls. To the east, a wall comprising two courses of foundation stones extending westwards over a length of 6 m constituted the third boundary. Further to the east, the land had been severely disturbed and no traces of any wall foundations had survived.

For practical reasons we divided the area spanning the entire width of the house into three equal units, which we called A, B and C, from the west to the east. The western part of the southern boundary of area 3 is consequently artificial. Each of the units thus distinguished measures 5.00×4.22 m. Due to the disturbance it is not certain how area 3 was originally entered, but the area was probably open to the courtyard.

Area 3 had clearly been disturbed. Many worn sherds and small stones were encountered in the top layer of 0.2 m and in some places it was even difficult to identify the original floor level. In units B and C, which seemed to be the most severely disturbed, the original floor levels were totally absent. Instead of an even surface of small pebbles and cobbles, we found large boulders mixed with worn pieces of pottery and roof tiles. During the excavation of area 3 we found a second sign of disturbance in unit A in the form of a pit with a depth of approx. 0.4 m containing large pieces of charcoal near the southern wall. This pit had once held a medium-sized bush or a small tree. The finds recovered from this area consist of many small, worn sherds of both household vessels and roof tiles, some iron fragments and objects, a coin and some organic remains. Only one more or less complete piece of pottery was found: a *kantharos*. The roof-tile fragments were all rather small and worn. Due

to the disturbance it is difficult to say whether or not this area was originally roofed.

Room 4. Room 4 was situated in the southern corner of the house and was therefore bordered by the south-western and southeastern external wall foundations. An internal wall running in a northwesterly direction from the southeastern wall formed the third boundary. To the east, room 4 was bordered by another internal wall. The room was entered from area 6 in the northeastern corner. During the excavations the remains of a previously unobserved internal wall were discovered, running parallel to and between the southwestern external wall and the northeastern internal one.

Room 4 consequently originally comprised two parts, which we called units 4 and 4A (fig. 2.26). These units do not exactly coincide with the original division of the room into two different parts due to the late discovery of the internal wall, part of which was concealed from sight under the baulk. This wall was found to contain a doorway allowing passage from one part of the room to the other. The foundation stones of the internal walls differ remarkably in size and regularity from those of the other walls of this house and those of the other houses at Halos studied so far. These units may have been built at a later stage (see below).

A remarkable feature was found in unit 4 in the form of a 0.40-m-wide strip of soil alongside the southwestern wall whose surface appeared to lie about 0.15 m higher than that of the floor (fig. 2.28), suggesting that it was a small platform or bench. It did not seem to have a support in the form of a row of stones separating it from the floor beneath (see room 4 of the House of the Ptolemaic Coins). The fill did not contain any diagnostic finds. We do not know for certain what purpose this platform served, but it may well have been used to support storage pots.

Many traces of burning were observed in both units in the form of small and large charcoal particles and seven pieces of lead which seemed to have melted on the spot. However, neither the pottery nor the other finds recovered from the room were severely burnt. Among the finds in question are fragments of several *pithoi*, two *pithos* lids, one bearing an inscription, a *kantharos*, a *chytra,* a *krater*, a juglet, some lead used for repairing pots and some organic remains. The many tile fragments found indicate that this area was roofed. One tile fragment bore part of an inscription: *ΚΑΛΛΙ...*

Room 5. The northwestern part of the house is poorly defined due to recent agricultural activities. Room 5 is defined by the house's external southwestern wall foundations, which had partially collapsed inwards, onto the destruction layer, the northwestern wall foundations of room 4 and the remnants of the foundations of an internal wall in the southeast. Together, these walls formed an almost square room. The southeastern wall foundations had been recently dis-

Fig. 2.28. House of the *Amphorai*: platform in room 4, unit 4; see fig. 2.26.

turbed. It is likely that room 5 was entered through area 8. There may have been a doorway in the eastern corner of the room, but, if so, the width of the gap observed in the foundations probably exceeds the doorway's original width. Despite these disturbances the stratigraphy in this room was straightforward: the first 0.10 m of soil contained mostly worn sherds, the majority of which were roof-tile fragments. The lower layer contained a large number of artefacts that were fragmented but reasonably complete and included a consistent layer of roof-tile fragments indicating that the room was roofed. The bulk of the finds was buried under the rubble of the mudbricks from the northwestern external wall. The floor level was easily distinguishable and did not seem to have been disturbed. The floor itself consisted of a layer of very small pieces of limestone measuring about 0.05 m directly on top of the virgin soil, and contained only few finds, mostly metal objects and small worn sherds.

No features were observed in this room, nor were there any signs of the burning evident in the adjacent room 4. The finds that came to light just above floor level consisted of fragments of *amphorai*, a *lagynos*, lamps, *chytrai*, a cooking pot, a beaker, a juglet, *lopades*, nails and other metal fragments, a coin, a large number of bone fragments and several loomweights.

Room 6. This room was situated between room 4 and area 7 in the southeastern part of the house and was defined by the house's external southeastern wall foundations, the northwestern foundations of room 4 and the foundations of the partially destroyed northeastern wall of room 6, separating it from area 7. No remains of a northwestern wall separating room 6 from

area 8 were found. We cannot be certain whether there ever was such a wall or whether room 6 was completely open to area 8, as this part of the room was severely disturbed.[13] The northwestern boundary of room 6, postulated as an extension to the extant – albeit damaged – foundations of the wall separating rooms 4 and 5, is therefore artificial. It is most likely that room 6 could be entered from area 8, either directly, with the latter serving as an extension to the room, or through a doorway in a wall now lost, defining it as a separate room.

Despite the widespread disturbance in the wall foundations, the southeastern and southwestern parts of the room did have a fairly clear, undisturbed stratigraphy. After the first 0.1 m of soil had been removed a layer of pebbles came to light. They probably derived from an upper course of the wall foundations. The next layer consisted primarily of roof-tile fragments. Between and beneath the roof tiles were pottery and other finds. In the northwestern part of the room the entire deposit was mixed with large pebbles, pieces of charcoal and large quantities of bone fragments and sherds which were all misfirings, recalling the so-called blister ware from Corinth (Edwards, 1975).

Room 6 yielded a very large quantity of sherds. The diagnostic pottery consisted of two *lekanai*, a bowl, two plates, a *kantharos*, a lid, a *hydria* and a jug. The aforementioned misfirings comprised five or six juglets, none of which were complete. Metal finds consisted of a bronze pin (for clothing) and nail fragments. As already mentioned above, a large number of bone fragments was also found in this area, a fairly large proportion of which was burnt. Other finds consisted of a few loomweights and a fragment of a grinding stone.

Area 7. This area was situated immediately behind the entrance to the house. The entrance itself was not designed as a *prothyron*, as in the case of the House of Agathon, but consisted merely of a 1.1-m-wide doorway in the wall. No threshold had survived. The northwestern part of the area seemed to have been severely disturbed. Owing to the absence of wall foundations or a foundation trench, the boundary between areas 7 and 8 could not be exactly located. Roof-tile fragments and other pottery sherds were found only in the upper layer and were all badly worn. The layer of soil overlying the floor yielded only few finds and very few roof-tile fragments, suggesting that the area was open. The floor itself consisted of medium-sized cobbles and pebbles. One feature was found in the southwestern corner of this area, next to the wall foundations separating area 7 from room 6. It seemed to be a shallow, oval, north–south oriented pit. It was 1.45 m long, 0.80 m wide and 0.10 m deep. Its fill consisted of pottery and roof-tile fragments. The feature is too shallow to represent a rubbish pit. It could represent the trench of a wall whose stones have disappeared, though the proximity of the wall of room 6 makes this fairly unlikely. Finds consisted of a *kantharos*, a bowl and a fairly large number of bone fragments.

Area 8. This large rectangular area lay at the centre of the House of the *Amphorai*. It was bordered by rooms/areas 2, 3, 4, 6 and 7. No meaningful measurements can be given due to the fact that some of the boundaries of this area are artificial, notably those in the west. Neither is it certain whether the area had boundaries in the form of walls separating it from areas 6 and 7. We have therefore defined area 8 as a separate unit for practical purposes only. The discovery of subrecent artefacts in the upper layer and part of layer 2 suggests that this area was disturbed. The layer just above the original floor level was intact, but yielded few finds, mostly small roof-tile fragments and worn sherds. The floor itself consisted of a compact layer of large cobbles and pebbles and did not resemble the floors of for example rooms 5, 1 and 2, but seemed similar to those in the courtyards of the House of the Ptolemaic Coins and the House of Agathon (area 6 in both houses). No specific features were found in area 8. The absence of large, unworn roof-tile fragments indicates that the area was open.

The area yielded only few finds, among which were surprisingly no diagnostic ceramic items: some nails, two coins, some bone fragments and a few loomweights.

History of occupation and reconstruction. From the descriptions of the different parts of the house it will be clear that the remains of the House of the *Amphorai* were too badly damaged to allow a reliable

reconstruction. It is even difficult to say whether the house was occupied in a single phase or perhaps several phases. The foundation stones of the southwestern part of the house (rooms 4 and 5) seem to have been arranged less regularly than those of the southeastern part (rooms 1 and 2) and room 6/7 and were bonded to the external wall foundations, like the foundations of the walls of room 1 and area 7. The sections created during the excavation however revealed no evidence to suggest that this may be due to different phases of occupation or rebuilding activities.

We originally interpreted this house as a building with a central courtyard surrounded by rooms (Haagsma & Reinders, 1991). This interpretation was however not satisfactory as it was based entirely on negative evidence and was formed at a time when no comparable houses had yet been studied. Two years later, the excavation of the House of the Snakes, which also measured approx. 15×15 m, showed that the House of the *Amphorai* could also be reconstructed with a courtyard at the back, away from the street. This would mean that the covered part of the house bordered the street directly, which would explain why there was no *prothyron* at the entrance to the house. To conclude, the scanty evidence available for the House of the *Amphorai* does not permit a reliable reconstruction.

3.5. The House of the Ptolemaic Coins (Housing block 6.4, plot 21)

The House of the Ptolemaic Coins formed part of the southern row of houses of housing block 6.4 (figs 2.29 & 2.30). It was excavated in 1989 by a team of eighteen people. The method of excavation was the same as in the other campaigns: small teams excavated the individual rooms of the house with pickaxes, mattocks and trowels down to the virgin soil.

It was clear from the start that the house's external dimensions were the same as those of the adjacent House of Agathon, the greater part of which had been excavated two years earlier. The house measured 12.70×14.80 m on the outside.

The wall foundations of the House of the Ptolemaic Coins did not differ from those found elsewhere in the town. They consisted of blocks of local limestone and conglomerate which were usually arranged in two rows. The gap between these rows was filled with small cobbles and pebbles. The outer faces of some of the foundation stones had been worked and smoothed. In most cases only one course of the wall foundations had survived. The top of this course lay approx. 0.35 m on average from the Hellenistic floor surface. The stones had been laid in a shallow (approx. 0.05 m) foundation trench.

In the course of the excavation it became clear that large parts of the western and southern walls had been destroyed by deep-ploughing. Stretches of the inter-

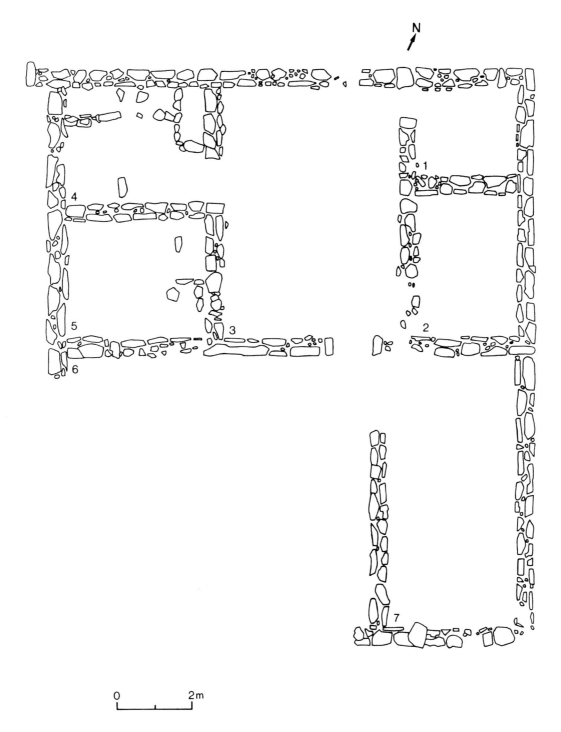

Fig. 2.29. House of the Ptolemaic Coins: foundations.

Fig. 2.30. Excavation of the House of the Ptolemaic Coins and the House of Agathon, view from the northwest.

nal wall foundations were also damaged, as was a large part of the courtyard (area 6). It was therefore decided to excavate only part of the courtyard.

Area 6. Despite the disturbance, the dimensions of this area could be measured: 8.15×7.15 m, inside the wall foundations. Area 6 was indisputably the court-yard of the house. This was also evident from the lower percentage of tile fragments found here. On account of the damage, the entrance from the street could only be inferred by comparing the layout of the house with that of the House of Agathon. The court-yard must have been of the same shape as that of the adjacent House of Agathon. The northern part of the courtyard may have been roofed, thus constituting a *pastas*, but no column bases were found to prove this. Finds consisted mainly of worn sherds and roof-tile fragments. Among the few diagnostic finds were a *chytra* and a *kantharos*. No specific architectural re-mains or features were found in area 6.

Room 7. This room was situated in the southeast-ern part of the house. The large roof-tile fragments found here indicate that this area was roofed. The room measured 3.55×7.15 m. It could be entered from the courtyard through a wide doorway with a width of 2 m. No specific architectural remains or features were found in room 7. Noteworthy finds were two *lekanai* and a *hydria*. One *lekane* was found on top of the *hydria*. The *hydria* was more or less complete and was

found upside down in a pit against the room's north wall. A few small undiagnostic iron objects were found in this room, too.

Room 3. During the excavation room 3 was divided into four equal units enabling one person to work in each. This room was the central, largest part of the roofed area of the house. Its measurements were 6.60× 4.75 m. The room's northern wall was partly damaged, as was the southeastern corner. The room could be entered from the courtyard via a doorway with a width of 1.07 m. The room could also be entered from the four side rooms. The doorway to room 5 had a thresh-old; the same probably holds for that to room 2. The soil was loose and mixed with sherds and tiny bits of limestone. Along the southern wall of room 3 parts of the soil however seemed to be more compact and of a brighter colour. These parts could represent the re-mains of partially dissolved mudbricks.

Roof tiles were found scattered across the floor, covering pots and other finds. Some of the roof tiles were more or less complete (fig. 2.31). Finds consisted of a wide range of domestic pottery and other items. Several *amphorai* were found scattered across the floor, together with *lopades, chytrai*, jugs, bowls, *kan-tharoi* and parts of a brazier. In addition, a small bronze *pyxis* came to light, together with a conspicu-ous number of metal finds and organic remains, plus two grinding stones and one rubbing stone.

Fig. 2.31. House of the Ptolemaic coins: room 3 finds (almost complete roof tile).

Room 4. This room was situated in the northwestern part of the house and could be entered from room 3 through a doorway with a width of 1.10 m. The room's measurements were 3.00×3.80 m. The remains of some remarkable architectural structures were found in room 4. These structures originally comprised three compartments formed by courses of pebbles along the northwestern external wall, which were filled with earth, thus forming some kind of platforms. The soil of which these platforms consisted had a soft sandy texture, especially lower down, indicating that it represented a fill. The platform was a little (0.1 m) higher than the level of the floor at the centre of the room. The partition separating compartments 1 and 2 seemed to have been disturbed. There were some indications that the platform extended along the southwestern and southeastern walls of the room, too (fig. 2.32). The compartments were on average 0.75 m wide and 1.60 m long.

These compartments can be interpreted as the remains of platforms for either *klinai* or storage vessels. Exploring the first interpretation, if we assume that the *klinai* were 1.60 m long, exactly five *klinai* would have fitted on the platforms (fig. 2.33).[14] Evidence of platforms with indications of *klinai* along the walls of a room in a domestic context usually implies that the room was used as the men's dining room, the *andron*. However, *andrones* are usually square and have an entrance offcentre due to the positioning of the *klinai*, and this is not the case in room 4. The artefacts found in this room do moreover not support an interpretation as an *andron*, as they comprise no remains of vessels intended for consumption, but mainly sherds of storage vessels, notably the remains of at least three

pithoi and one *pithos* lid. In addition, a concentration of eleven loomweights came to light here, plus a relatively large number of coins, three of which were made of silver. These finds in association with the platforms suggest that part of the room, at least, was used for storage and make it fairly unlikely that the room was an *andron*.

Room 5. This room was situated directly south of room 4 and measured 3.15×3.75 m. It could be entered from room 3 through a doorway with a width of 1.05 m which had a threshold consisting of loosely arranged pebbles with no signs of any pivot holes. A few more stones, seemingly arranged in a row, were found in front of the doorway. They may have formed part of platforms like those in room 4. The floor of room 5 lay at a slightly lower level than that in room 3 and the soil was of a sandy texture. The small number of finds and (platform) stones found here however constitute insufficient evidence to conclude that the room was an *andron*. The stones moreover came to light at a comparatively high level, overlying other finds, which makes it unlikely that they actually formed part of any structure. Their presence here is probably attributable to recent agricultural activities. The finds recovered from room 5 consisted of many worn fragments of roof tiles, small iron artefacts, an *amphora* and some shells and bones.

Room 1. Room 1 could be entered from room 3 via a narrow doorway with a width of only 0.55 m. The room could also be entered from room 2 originally, as indicated by a gap in the foundations of the southwestern wall of room 1. This gap was originally 0.50 m wide and was later walled up with blocks of a greyish limestone differing from the white limestone and

Fig. 2.32. House of the Ptolemaic Coins: room 4. Platforms for possible *klinai* or storage jars.

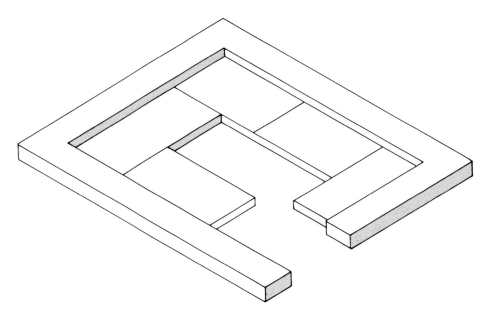

Fig. 2.33. House of the Ptolemaic Coins: room 4. Reconstruction position platforms for possible *klinai*.

conglomerate used for the foundations of the south-western wall of room 1. The room was small, measuring only 2.80×2.30 m. Apart from an abundance of roof tiles, few finds were recovered from this room. A small *lagynos* was found, an arrowhead and six loomweights. The upper layer of soil in this room was somewhat disturbed and contained some present-day artefacts.

Room 2. This room was situated southeast of room 1 and was slightly larger than the latter, measuring:

2.75×3.80 m. The doorway in the southwest was severely disturbed. A few threshold stones were however still visible and part of the rubble of the core of the foundations of the original southwestern wall was still partly *in situ*, indicating that the doorway was originally 1.05 m wide. In an earlier stage of occupation room 2 could also be entered from room 1 (see above). An abundance of roof tiles was found in this room, all fragmented but the fragments were quite large.

Fig. 2.34. House of the Ptolemaic Coins: room 2 with remains of pebble floor.

The room's floor lay immediately beneath these remains. The northeastern part of the room was paved with fairly large white limestone pebbles (fig. 2.34). The fact that the roof tiles were found all over the room, both on top of this pebble floor and in the part of the room which was not paved, implies that the pebbles were deposited to consolidate part of the floor and are not to be seen as the remains of an originally completely paved floor which was damaged some time after the town was abandoned.

Both the colour and texture of the soil and the discovery of pottery beneath the pebble floor indicated the presence of an earlier floor at a lower level, 0.05–0.10 m beneath the pebble floor. This lower floor yielded no specific finds, only very small fragments of pottery. The soil of the destruction layer contained six coins, a *kantharos* and other pottery fragments plus a stone shaped like a dice.

History of occupation and reconstruction. The House of the Ptolemaic Coins yielded indisputable evidence for at least two building phases in the form of the doorway between rooms 1 and 2 which was later walled up and the second, partially paved floor in room 2 (figs 2.35 & 2.36).

The internal division of the House of the Ptolemaic Coins is largely the same as that of the House of Agathon and also closely resembles that of the House of the Geometric *Krater.* The house comprised an – originally roofed – northern part and a partially open area in the south (fig. 2.35). The northern part consisted of five rooms arranged in the usual layout observed in the houses at New Halos, with one large room (room 3) at the centre flanked by two smaller rooms on either side (rooms 1 and 2; 4 and 5). A door-

way in the southern wall of room 3 provided access to this part of the house from the courtyard. The southern part of the house was apparently partially roofed. Room 7 resembles room 7 of the House of Agathon in both its position and its shape. This spatial arrangement suggests that part of the courtyard was covered – as in the House of Agathon – forming a *pastas*. An entrance in the southern wall originally provided access to the house from the street, but no traces of this entrance (probably a *prothyron)* were found due to the disturbance in this area.

3.6. The House of Agathon (Housing block 6.4, plot 22)

In 1987, after inspecting the surface remains of the lower city as a whole, we decided to concentrate our excavation efforts on the central part of housing block 6.4, which appeared to have suffered less from the ploughing than other areas inside the city walls (see Chapter 1). Some of the foundation stones in this area were visible at the surface and they seemed to be undisturbed. The House of Agathon, which formed part of housing block 6.4, was excavated down to the level of the Pleistocene soil in the summers of 1987 and 1989 by teams of thirteen and eighteen people using mattocks, pickaxes and trowels.

The foundation stones of the house's external walls were clearly visible before the excavation was started, indicating that the house had had a width of approx. 12.50 m. A house of this type had not yet been excavated at New Halos. The house's outside dimensions were 13.50×15.20 m.

The House of Agathon belonged to the southern

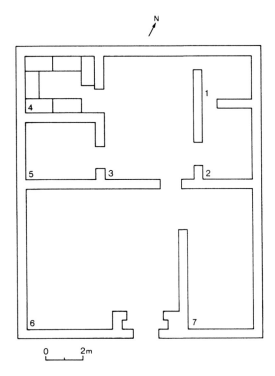

Fig. 2.35. House of the Ptolemaic Coins: phase 1.

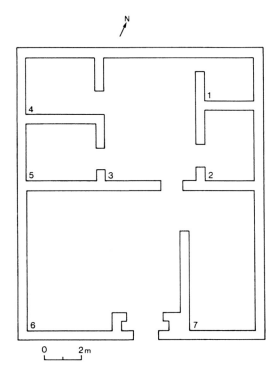

Fig. 2.36. House of the Ptolemaic Coins: phase 2.

row of houses in housing block 6.4 (figs 2.37 & 2.38). It lay adjacent to the House of the Ptolemaic Coins and south of the House of the Snakes. The wall foundations were the same way as those of the other houses excavated at New Halos, consisting of two rows of irregular blocks with the space between the two filled with cobbles and pebbles. One course (with a height of 0.30–0.40 m) of the foundation stones had survived, but it is assumed that the foundations were originally higher. Only the wide foundation wall separating the northern and southern rows of houses and the foundations of the housing block's external walls consisted of a single row of large limestone blocks. These walls were wider than the internal walls, measuring approximately 0.60 m.

The excavation revealed that the building had been only slightly disturbed. Some of the foundation stones of the northeastern part of the southeastern wall had disappeared in ploughing and the remains of the *prothyron* in the middle of the southeastern wall had been disturbed.

Area 6. This large area was excavated in 1989. For analytical purposes and the sake of efficiency it was decided to divide this area into six equally sized units. When the excavation was started, the internal foundation walls defining the area that was later designated room 7 were not yet visible.

Area 6 was entered from the street through a *prothyron*, indicated by architectural remains and an abundance of large roof-tile fragments. Its internal width

– insofar it could be measured – was 1.60 m. No threshold stone was found. It is not sure whether there ever was one or whether perhaps the threshold consisted of several flat slabs of limestone. The soil in the area of the entrance was soft and humid, indicating that it had been disturbed by agricultural activities. The pieces of limestone that were still *in situ* showed no signs of any pivot or bolt cavities. They may therefore have been the foundations of a large threshold block which was removed after the site was abandoned.

Area 6, which extended over the entire width of the house, sloped slightly downwards towards the southeast. More earth consequently had to be removed in the southeast than in the southwest. This suggests that the latter part was affected by post-depositional processes like ploughing to a greater extent than the former. Very few artefacts were found in the western half of area 6. A large bronze Geometric *fibula* was found here among the cobbles of the Pleistocene subsoil. This find, together with the Geometric finds recovered from the House of the Geometric *Krater* (Dyer & Haagsma, 1993) and the Geometric finds found in the Hellenistic cemetery south of the city (Malakasiotis *et al.*, 1993), support the hypothesis that there was a cemetery, or cemeteries, in this area in the Geometric period, before the city was built.

Northeast of the house's entrance a concentration of several loosely arranged cobbles and pebbles was found. It is not clear what it represents. Remains of a

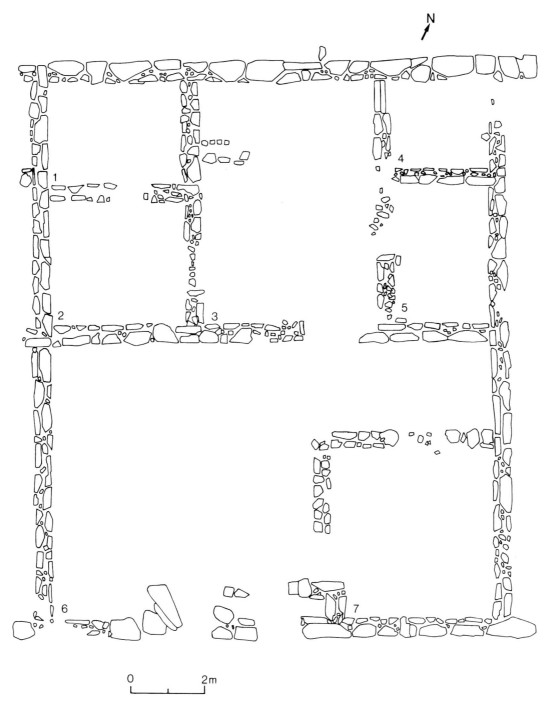

Fig. 2.37. House of Agathon: foundations.

Fig. 2.38. Excavation of the House of Agathon: view from the southeast.

pithos were found outside the western corner of room 7. Its position possibly indicates that it served to collect rainwater, running from either the roof of the *pastas* or the roof of room 7 itself. A layer of ashes came to light 1.70 m west of this *pithos*, representing the remains of a hearth or perhaps a place of sacrifice.[15] A knife was found in the vicinity. The part between room 5 and room 7, which probably represents a *pastas,* yielded more finds. Several pots were found among the pebbles of the floor, such as a *kantharos* and a bowl, plus roof-tile fragments, indicating that this part of the area was roofed.

The area's position and size and the small number of roof-tile fragments found in most parts indicate that area 6 constituted the house's courtyard; it probably contained a roofed *pastas.* The courtyard provided access to rooms 7 and 3.

Room 7. This room was situated in the eastern part of the house. It was a square room, measuring 4.45× 4.40 m. Part of the northeastern wall of the room was damaged, a stretch of about 1.30 m of the middle part of the wall having disappeared. That this gap did not represent a former doorway was evident from the softness of the soil in the area where the foundation stones originally lay. The room could be entered from the courtyard (area 6) through a doorway in the southwestern wall represented by an 0.8-m-wide gap. Room 7 contained two interesting features. One was discov-

ered in the southwestern corner of the room and consisted of a square patch of burnt soil mixed with ashes measuring 1.0×0.9 m. The feature was on one side bordered by the external south-eastern wall of the house and on another by one of the walls of the *prothyron.* A few roof-tile fragments had been placed vertically against the foundation stones of these walls, probably in order to protect them from the heat. In spite of the absence of any observable boundaries in the form of for example stones along the northeastern and northwestern sides, this feature may be interpreted as a fireplace.

Another feature came to light a little to the west of the centre of the room. It consisted of two rows of pebbles which probably originally formed a rectangle of about 0.6×0.8 m. Its position and shape suggest that this, too, is the feature of a hearth, but no ashes or burnt material were found here. Instead, the fill of the feature consisted of earth with a high concentration of organic matter, evident from its colour and smell. A large *pithos* rim was found at the centre of the feature. The feature is difficult to interpret as it was severely disturbed. It may represent a hearth, but, if so, it was certainly not in use as such at the time when New Halos was abandoned. Perhaps there was originally a hearth here and the remains of the hearth were later used to support a *pithos* containing *e.g.* olive oil or other foodstuffs. The absence of further *pithos* fragments (removed in ploughing?) however makes this interpretation conjectural.

Other finds recovered from room 7 comprise a wide range of domestic earthenware vessels, the majority of which are associated with the processing and consumption of food, such as cooking pots, *lekanides*, a strainer and bowls, but also an *amphora*. Some metal objects were found, too, including a key, a piece of lead and a few coins. The abundance of roof tiles found in room 7 indicates that the room was roofed.

Room 3. Room 3 lay at the centre of the roofed part of the house. It was the largest of all the rooms, with dimensions of 4.80×6.30 m. Parts of the foundations of the room's walls were disturbed. Some of the foundation stones of the eastern end of the southeastern wall were missing, as were some of those of parts of the northeastern wall. Most of the foundations consisted of the usual two rows of limestone blocks. The greater part of the northwestern wall, however, which was also the boundary between the southern and northern houses of the housing block, consisted of a single row of large limestone blocks with smooth outer faces, the gaps between the blocks having been filled with smaller stones. Room 3 could be entered from the courtyard via a doorway whose northeastern side was disturbed. Its width could therefore not be measured. The room provided access to three side rooms. An 0.80-m-wide threshold was found *in situ* in the doorway providing access to room 4. The width of the doorway providing access to room 5 could not be determined due to the severe disturbance of the wall foundations at this point. The southwestern corner of room 1 was also disturbed. A 1.25-m-wide doorway in the southwestern wall of room 3 provided access to room 2. In the doorway was a neat row of small cobbles whose top sides were worn. They may have served as the threshold themselves or they may have supported a threshold stone. A number of clenched nails with large heads indicated that there was once a wooden door in this doorway.

Room 3 contained some remarkable structural remains: an almost semi-oval arrangement of foundation stones extending from the northern part of the northwestern wall, a short distance to the east of room 1, to the centre of the room. It was 1.35 m long and 0.60 m wide. The stones had been deposited on top of the Pleistocene soil instead of embedded in it like the foundations of the house itself. The function of this arrangement of stones must be inferred from the finds discovered in its vicinity: two large grinding stones, both of volcanic rock (fig. 2.41). The upper stone had a concave top side and a slot in its bottom side, the lower stone was flat. Both had incised grooves. They represent two millstones of the hopper rubber type (Moritz, 1958; White, 1963; Amouretti, 1986), which was in common use in the Classical and Hellenistic periods. The structural remains are therefore most probably the foundations of a platform on

which the grinding equipment rested (Robinson & Graham, 1938: pp. 326–336). Another possibility is that the feature represents the foundation of a staircase (Robinson & Graham, 1938; Hoepfner & Schwandner, 1994: p. 102), but no convincing parallels are known to support the latter interpretation. All known foundations of staircases consist of a single large elongated flat slab, not of several small cobbles.

At the centre of room 3, 3.30 m from the southwestern wall and 2.50 m from the northwestern wall, was a small, more or less round elevation built from pebbles, with a diameter of 0.75 and a height of about 0.10 m, measured from the Hellenistic floor level. The absence of traces of a fire makes it unlikely that this feature represents the remains of a hearth. Its function remains unknown.

The number of finds recovered from room 3 is rather small considering the room's size. The finds are a lid and a large number of pyramidal loomweights. The latter were found in the southeastern part of the room. A few coins and some organic remains were found here too.

Room 1. Room 1 lay in the eastern corner of the house. It could only be entered through room 2, to the south of this room. The room's dimensions were 3.60×2.75 m. The southeastern side of the 0.90-m-wide gap representing the door was slightly disturbed. The wall foundations defining this room consisted of the usual double row of stones, partly embedded in the Pleistocene soil. The majority were of the local limestone, but some of the stones of the northeastern wall consisted of conglomerate. During the excavation, remains of mudbrick were found in the form of irregularly fired amorphous lumps of terracotta. No other indications of fire were found in the room.

This room contained no specific architectural remains or features. An important observation, however, is the fact that the pebbles marking the floor level were not encountered in all parts of the room. A small circular area in the southeastern corner with a diameter of about 0.60 m was clear of pebbles. Considering the character of the other finds recovered from this room (see below), among which are many fragments of a *pithos*, we now believe that the absence of pebbles in this particular area may well imply that a hole was here dug into the Pleistocene soil for a large container such as a *pithos*. The feature was however not identified as such a hole during the excavation.

This rather small room yielded a large amount of pottery and other finds: a large part of a *pithos* with a lid bearing the inscription ΑΓΑΘΟΝΟΣ (hence the house's name), four *krateres*, a cooking pot, an *amphora*, three *chytrai* and some tableware like a *kantharos*, a plate and a bowl. In addition, quite a few loomweights and several metal objects such as a knife and some coins were found.

Room 2. Room 2 lay southwest of room 1. It had two doorways: one with a width of 0.90 m in the northwestern wall providing access to room 1, and one with a width of 0.80 m in the northeastern wall providing access to room 3. The latter doorway had a threshold that has already been described under room 3 above. The room's dimensions were 3.65×3.35 m. Its walls were founded on a double row of limestone blocks. A number of blocks found in the southern corner of the room probably derive from a second course of wall foundations.[16]

Beneath the layer of roof tiles, which were particularly abundant in this room, the following finds came to light: an *amphora*, a *hydria*, a jug, a *chytra*, a miniature cooking pot and a small jug, plus a large number of loomweights and some iron objects, including a sickle and a few coins.

Room 4. This room lay in the northern corner of the house and was accessible from room 3 through a 0.90-m-wide doorway. In the doorway was a threshold consisting of two rectangular stones set at different heights longitudinally in the gap in the wall. The stones were of a grey, harder kind of limestone than that quarried in the direct vicinity of the town. The higher stone contained a pivot hole at its southern end.

The room measured 2.75×2.40 m. As already mentioned above, the foundations of the northeastern wall and the southern corner of room 4 were disturbed. The other wall foundations were of the usual kind. The foundations were all of the white local limestone. The floor of this room lay at a slightly higher level than the floors of the neighbouring rooms: 0.1 m higher than the floor of room 3 and 0.2 m higher than that of room 5.

No architectural remains or features and remarkably few artefacts were found in this room. A semicircular patch of ash with a width of 0.50 m and a maximum depth of 0.01 m came to light a short distance to the southeast of the entrance, against the southwestern wall. There was no evidence to suggest that the ash derived from a formal fireplace. Finds from this room consisted mainly of roof tiles and some nails.

Room 5. Room 5 lay to the east of room 3 and southeast of room 1. Its dimensions were 2.60×3.70 m. The room was entered from room 3. The foundation walls in the vicinity of the gap representing the doorway were severely damaged, so no reliable width can be given. The blocks of the foundations of room 5 were all of local limestone and were laid in the usual double row with the gap between filled with smaller stones and pebbles.

Room 5 contained no architectural remains or features. In some parts of the room the soil was darker in colour and harder, indicating that a fire had burned here. The occurrence of fire in this part of the house during or after the abandonment of the town was further confirmed by the remains of an iron spade comprising large chunks of charcoal – what was left of the wooden handle. Pieces of melted lead objects were moreover found here, too. Most of these pieces were beyond recognition, but two of the lead artefacts had indisputably melted on the spot. One is a roughly oval plaque which still bears the impression of woven reed – probably from a basket or a mat – and the other is a partially melted lead weight.

Room 5 contained the largest number of artefacts per room and the largest density of artefacts per m² encountered at New Halos to date. The finds came to light along the walls of the room, where they had originally stood on the floor or a shelf or hung from the wall. Beneath a layer of roof tiles were the following artefacts (see the next chapter for details): a *krater*, parts of a *pithos*, *amphorai*, *chytrai*, bowls, juglets, a *lagynos*, a miniature bowl, a miniature cooking pot, lids, plates, *lekanides*, an *unguentarium*, two sets of burnt *astragalia*, two burnt roe deer antlers, over a hundred loomweights, figurines, agricultural equipment such as a *dikella*, a hoe, a spade, sickles, other tools and implements such as a bronze *strigil*, a *stylus*, a bronze spoon, a knife, parts of a pair of scissors, a meat hook, a keyhole, pot repairs, a lead weight, a lead lid, molten pieces of lead, nails, coins and many undiagnostic artefacts, plus organic remains such as shells and bones.

History of occupation and reconstruction. Housing block 6.4 slopes downwards in an easterly direction. Nevertheless, the floors in the House of Agathon, and also those of the House of the Ptolemaic Coins, were only slightly levelled. Only one floor level was observed in all the house's rooms. No evidence for renovation activities, such as the walling up of former doorways or the addition of new foundations, was found either. We may therefore assume that the remains of the House of Agathon represent a single building phase and that the house's plan at the time of its destruction and abandonment was the same as that at the time of its construction, about 40 years earlier (fig. 2.39).

The house's dimensions and plan are very similar to those of the House of the Ptolemaic Coins and the House of the Geometric *Krater*. The plan included a courtyard in the southeast, with an entrance from the street through a *prothyron*. Only part of this area yielded large quantities of roof-tile fragments. Hardly any roof tiles were found in the other units, where an elongated *pastas* could be postulated. It is therefore likely that only between rooms 7 and 5 the part had a tiled roof, representing a structure that could be described as a short version of a *pastas*. The area may have been roofed across its entire length with some perishable material such as reed, but there is no proof for this in the form of the bases of columns that may have supported such a roof. The courtyard probably contained a hearth or an altar.

The rest of the house was roofed with tiles as can

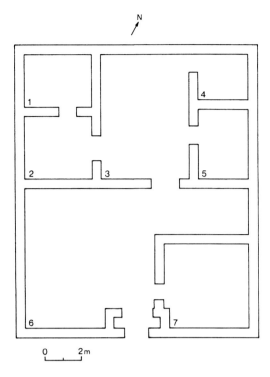

Fig. 2.39. House of Agathon: reconstruction.

be inferred from the abundance of tile fragments that were found scattered across all the rooms. There was a fireplace in room 7 and a hole in the floor of room 1 which may have held a *pithos*.

4. CONCLUDING REMARKS

The houses of New Halos provide important new information on the planning and nature of houses in newly planned cities in the early Hellenistic period. In comparable cities of the Classical period the houses, as observed by Hoepfner & Schwandner (1994), were completely integrated in the plan of the city or district as a whole. The measurements of the houses, their positions within the housing blocks and all architectural elements were part and parcel of the overall plan of the city. Examples of such planned cities are Peiraieus, Rhodos, Olynthos and Kassope. As a consequence of this rigorous planning, resulting in the same spatial elements in all of the houses of these cities, the houses can be divided into several 'types', whose origins lie in housing traditions of earlier (Archaic) times. Adhering to uniform house types in planning a city implies obvious utilitarian advantages.

However, Hoepfner & Schwandner (1994) argue that the uniformity of the Classical houses in the aforementioned cities was so all-embracing that it

would be incorrect to speak simply of 'house types', i.e. houses whose layouts were based on a common set of measurements and a general set of principles. They therefore coined the term *Typenhäuser* to denote houses that formed an integral part of a planned city, both in size and in their layout. Each newly planned city had its own *Typenhäuser*. The houses in question were strikingly similar, and Hoepfner and Schwandner are of the opinion that this is not attributable to utilitarian considerations alone; they associate the emergence of *Typenhäuser* with the popular political ideology of *isonomia*, equality, for the citizens.

More difficult to explain, according to Hoepfner and Schwandner, is the existence of *Typenhäuser* in the Hellenistic period. Referring to the layout and planning of the city Doura Europos, whose construction was commissioned by Seleukos in 289 BC, they state:

Wir sind nicht sicher daß noch im Hellenismus den Bewohnern neugegruendeter Städte Typenhäuser empfohlen wurden. Der praktische Nutzen ist jedenfalls so groß, daß man mit einem vom Architekten vorgeschlagenen Typus rechnen kann. Es kennte allerdings auch nur ein Schema einer bestimmten Raumfolge und damit mehr ein Haustypus als ein Typenhaus in dieser spaeteren Zeit die Regel gewesen sein (Hoepfner & Schwandner, 1994: p. 277).

The authors claim that if it is indeed true that *Typenhäuser* continued to be built in Hellenistic times, this was done merely out of a practical need for such houses and for reasons of planning, and no longer out of any desire to implement the (Classical) ideology of *isonomia*.

A major problem for Hoepfner and Schwandner was the lack of adequate information on early Hellenistic house plans. In cities like Doura Europos a long history of occupation has almost entirely obscured the original layout of the earliest houses. One of the greatest merits of the New Halos project is hence the insight it is providing into houses and planning principles in the early stage of the crucial transition from Classical to Hellenistic city planning.

An analysis of the architectural and social aspects of the houses of New Halos in the wider context of Classical and Hellenistic housing in general is currently in preparation. A brief discussion of the implications of the architecture of the houses at New Halos, specifically in relation to the evolution of the *Typenhäuser* postulated by Hoepfner and Schwandner, will follow below.

4.1. Spatial elements in houses at Halos

The sizes and layouts of the excavated houses at Halos differ to varying extents. To summarise, the dimensions of the houses are as follows:

House A: approx. 15.00×15.00 m (measured from the centres of the foundation walls)

House of the Coroplast: approx. 13.75×15.00 m (measured from the centres of the foundation walls)

House of the Geometric *Krater*: 12.70×15.00 m (reconstructed)

House of the *Amphorai*: 15.50×14.30 m

House of Agathon: 13.50×15.20 m

House of the Ptolemaic Coins: 12.70×14.80 m

House of the Snakes: 15.10×15.45 m

So the lengths of the houses range from 14.30 to 15.45 m and their front widths from 12.70 to 15.50 m. The town and the houses consequently show a systematic, but irregular layout. The differences in the front widths of the houses are not attributable to mere coincidence. These differences are mainly responsible for the differences in the individual houses' surface areas. There is a difference of 45 m^2 between the smallest excavated house, which measures 188 m^2, and the largest house, measuring 233 m^2. The front width of the houses was consequently deliberately varied, to create houses of different sizes for inhabitants with different means and/or needs. The houses do not differ in size according to some regular system applied to the individual housing blocks as at for instance Olynthos and Peiraieus. Neither were the housing blocks laid out according to a regular system of the kind used for former newly planned cities whose houses all had comparable layouts, formed integral parts of housing blocks and sometimes had dimensions relating to the city's overall plan. These differences imply a clear break in planning traditions between the Classical and Hellenistic periods.

The layouts of the houses themselves do not show any clear regularity either, like those of the houses of the newly planned cities of the Classical period. The arrangement of the rooms within the houses differs considerably from one house to another, although certain similarities are observable. The spatial arrangements of the rooms of the House of the Geometric *Krater*, the House of the Ptolemaic Coins and the House of Agathon are very similar. These are all south-facing houses which could be entered directly from the street. The entrances provided access to a courtyard which covered a large part of the house's total area. This courtyard in turn provided access to the living quarters, which consisted of a rectangular main living room with two side rooms on either of its long sides.

The part of the courtyard close to the entrance to the living quarters may have been covered, in which case it constituted a *pastas*. No sound archaeological evidence was found to indicate that such a *pastas* extended across the entire width of any of the houses. Part of the courtyards of the House of the Ptolemaic Coins and the House of Agathon was definitely covered: the part between the main living quarters and another room which could be entered from the courtyard, to the right of the entrance (rooms 6 and 7, respectively). No evidence for such a covered area was found in the House of the Geometric *Krater,* but here room 2 protruded into the courtyard, thus creating extra space for the main living quarters.

The two excavated north-facing houses, the House of the Coroplast and the House of the Snakes, both have a different layout. These houses could likewise be entered directly from the street. The entrance to the House of the Coroplast, however, provided access to a smallish courtyard, whereas the covered living quarters of the House of the Snakes could be entered directly from the street. In the latter house a corridor led to several rooms and areas inside the house: the main living room, two or three smaller rooms and the courtyard, which lay in the southern part of the house. The courtyard of the House of the Coroplast provided access to a workshop, but had formerly also provided access to the main living area. The House of the *Amphorai* was unfortunately too damaged to provide reliable information on its layout.

The houses' covered living quarters (that of the House of the Geometric *Krater* as reconstructed) show close similarities in shape and in access between the individual rooms and areas. The large living room usually provided access to four side rooms, resulting in a layout comprising five rooms. In the case of the House of the Snakes, the corridor leading to the courtyard separated the main living room from the two eastern side rooms. This arrangement of five rooms can be seen as the 'canon' of the houses of New Halos, which is observable not only in the layouts of the houses, but also in the domestic activities carried out in them as reflected by artefact assemblages.[17]

4.2. Discussion

The 'canon' of five rooms was not universally applied as in the arrangement of the living quarters in cities such as Olynthos and Priene. At New Halos, the orientations and dimensions of the individual houses and the ways in which the various rooms provided access to one another varied considerably (table 2.1).

The close similarities observable in the layouts of the houses of newly planned (parts of) cities of the Classical period led scholars to associate specific 'house types' with specific sites, such as the *pastas* house with Olynthos (Robinson & Graham, 1938; Drerup, 1967; Krause, 1977, Cordsen 1995), the *prostas* house with Priene and Piraeus (Drerup, 1967; Hoepfner & Schwandner, 1994), the peristyle house with Eretria and the *herdraum* house with Kassope (Hoepfner & Schwandner, 1994). The significance of the definition of house types of the Hellenistic period

Table 2.1. Dimensions of the houses of New Halos (in metres).

House of the Coroplast	
Large living area	7.60×4.55
Room 8	2.55×2.75
Room 9	2.70×4.40
Room 6	3.60×3.60
Room 5	3.60×3.60
House of the Geometric *Krater*	
Large living area	5.85×4.35
Room 2	2.85×4.60
Room 1	2.85×4.55
House of the Ptolemaic Coins	
Large living area	6.60×4.75
Room 4	3.00×3.80
Room 5	3.15×3.75
Room 1	2.80×2.30
Room 2	2.75×3.80
House of Agathon	
Large living area	6.30×4.80
Room 1	3.60×2.75
Room 2	3.65×3.35
Room 4	2.75×2.40
Room 5	2.60×3.70
House of the Snakes	
Large living area	7.50×4.70
Room 6	3.95×2.50
Room 5	3.30×2.70

is still a matter of discussion; I am of the opinion that there is little sense in attempting to include the New Halos houses in a typology.

The *Typenhäuser* defined by Hoepfner & Schwandner (1994) are a different matter. *Typenhäuser* are houses that are found in planned cities, forming integral parts of the city's overall layout in terms of size and internal arrangement. Each newly planned city had its own *Typenhäuser*. These houses showed a striking similarity, which Hoepfner and Schwandner associate with the popular political notion of *isonomia*, equality, for the citizens.

The houses of New Halos are not *Typenhäuser*. This is evident from the differences in the sizes of the houses, but also from the variations observable in their layouts. The latter variations represent another break with the house and city planning traditions of the Classical period. It is difficult to explain the wider significance of this break. Few other houses of early Hellenistic cities have so far been excavated. We know the general layouts of a few other late Classical-early Hellenistic cities in the neighbourhood of New Halos, such as Gorítsa (Bakhuizen, 1992), Demetrias (Marzolff, 1978) and Dion (Stephanidou Tiveriou, 1998). The first city had an orthogonal layout, comprising housing blocks of different sizes. These differences in

size are thought to reflect adaptations to the local terrain. Little is known about the individual houses of this city. Demetrias seems to have had a regular layout, with regular housing blocks, but again little is known about individual houses, and the same holds for Dion. In their discussion of Doura Europos (see the above quotation), Hoepfner and Schwandner argue that in the Hellenistic period cities were no longer planned according to the former ideological principle of *isonomia*, and that the former system based on *Typenhäuser* had by that time become a practical means for planning a city's occupation area. We now know that this was not the case at New Halos at least, but we have too little information to claim a widespread change in planning practices.

To summarise, at New Halos both the ideological and the utilitarian principles underlying the use of *Typenhäuser* in newly planned cities were evidently abandoned in favour of different considerations. Closer analysis of these considerations must await the forthcoming study of the houses of Halos in the wider context of Greek and Hellenistic housing in general.

5. NOTES

1. For the composition of the teams, see the preface to this volume. Two workmen were hired in the 1989 and 1991 campaigns.
2. I will discuss the social and socioeconomic aspects of the houses of Halos more extensively in a monograph I am preparing on the houses of Classical and Hellenistic Greece (Haagsma, in prep.b).
3. The houses in Halos had three basic sizes, with a width of either 12.50, 13.75 or 15.00 m and a uniform length of 15.00-15.75 m. All quoted dimensions were measured inside the walls. *Cf.* this volume, Chapter 1.
4. For Classical and Hellenistic houses with mudbrick walls on stone foundations, *cf.* Olynthos (Robinson & Graham, 1938; 1946), Kassope (Hoepfner & Schwandner, 1994) and many other sites.
5. For mudbricks and their use *cf.* Skafida (1994).
6. *Cf.* Hoepfner & Schwandner (1994: p. 155) on the fact that *Kalkpfutz* plastering leaves no traces in the archaeological record.
7. It is interesting to note that the majority of the internal doors at Olynthos were double doors, even those with widths of only 1.00 m or less (Hoepfner & Schwandner, 1994: p. 106)
8. Evidence for windows in houses has been obtained at for example Olynthos in the form of the capitals of pillars which were so small that they can only have been part of windows (Hoepfner & Schwandner, 1994: p. 106).
9. In the Villa of the Bronzes at Olynthos the remains of a wooden beam (1×0.26×0.10 m) were found. They are thought to derive from a roof beam (Robinson & Graham, 1946; Hoepfner & Schwandner 1994: p. 107).
10. Other houses which probably had an upper floor are House MC at Olynthos, whose upper floor was in all probability paved with mosaics (Robinson & Graham, 1946: p. 204), and a Classical House at Kallipolis (Themelis, 1979: p. 245 ff.).
11. I would like to thank Todd Whitelaw and Lisa Nevett, with whom I first discussed this anomaly, for their helpful comments

and suggestions, which led me to re-examine the architecture of this house.

12. This is remarkable because usually the foundations of the internal walls of houses at New Halos were laid directly on top of the Pleistocene cobble soil.

13. The deposit was also significantly shallower (0. 25 m) than elsewhere.

14. Measurements of *klinai* found at other sites range from 1.60 m (Kassope) to 2.00 m (Priene).

15. Altars dedicated to Zeus Herkeios, the protector of the house, have regularly been found in courtyards, at least at Olynthos (Robinson & Graham, 1938: pp. 159, 321 ff.; Hoepfner & Schwandner, 1994: p. 97).

16. The blocks in question seemed to roughly describe a quarter of a circle, extending from the southeastern to the southwestern wall. Although not all of the stones adjoined one another, it was first thought that they represented the remains of a structure with sides of approx. 0.85 m – possibly an aborted wall or a well – but no evidence was found to support either of these interpretations. Besides a single sherd, some charcoal and a loom weight, no specific finds were discovered in association with the stones. The colour and texture of the soil surrounding (and beneath) the blocks were moreover much the same as elsewhere. Furthermore, the Hellenistic floor lay at the same level here as in other parts of the house, implying that the stones lay about 0.05 m above the Hellenistic floor level at the time of their discovery. The most likely conclusion, therefore, is that they derive from a collapsed upper course of wall foundations.

17. The functions of the various rooms will be discussed in a separate study (Haagsma, in prep. b).

6. REFERENCES

AMOURETTI, M.-C., 1988. La viticulture antique: contraintes et choix techniques. *Revue des études anciennes: paraissant tous les trois mois* (1988) (= Communications présentées à la réunion de la SOPHEAU, Bordeaux 1986), p. 5.

BAÏRACLI-LEVY, J. de, 1909. *De Horizon en daar Voorbij: Reizen en Trekken in de Landen rondom de Middellandse Zee*. De Driehoek, Amsterdam [Dutch translation: 1981].

BAKHUIZEN, S.C. (ed.), 1992. *A City of the Fourth Century B.C. by the Gorítsa Team* (= Bibliotheca archaeologica 10). "L'Erma" di Bretschneider, Roma.

CORDSEN, A., 1995. The Pastas House in Archaic Greek Sicily. *Acta Hyperborea* 6, pp. 104–121.

DIJKSTRA Y., et al., 1996. De zuidoostpoort van hellenistisch Halos. *Paleo-aktueel* 7, pp. 62–65.

DRERUP, H., 1963. Zum geometrischen Haus. *Marburger Winckelmann-Programm* 1963, pp. 1–9.

DRERUP, H., 1967. Prostashaus und Pastashaus. Zur Typologie des griechischen Hauses. *Marburger Winckelmann Programm* 1967, pp. 6–17.

DYER, C. & M.J. HAAGSMA, 1993. A Geometric Krater from New Halos. *Pharos* 1, pp. 165–174.

EDWARDS, G.R., 1975. *Corinthian Hellenistic Pottery* (= Corinth 7, pt. 3). American School of Classical Studies at Athens, Princeton, N.J.

FUSARO, D., 1982. Note di architettura domestica greca nel periodo tardogeometrico e acaico. *Dialoghi di Archeologia* 4, pp. 5–30.

GRAHAM, J.W., 1966. Origins and interrelations between the Greek house and the Roman house. *Phoenix* 3, pp. 3–31.

HAAGSMA, M. & H. REINDERS, 1991. Verwoesting door een aardbeving. *Tijdschrift voor Mediterrane Archeologie* 8, pp. 16–25.

HAAGSMA, M.J. (in prep. a). House construction and abandonment in present-day Greece: a comparative study.

HAAGSMA, M.J. (in prep. b). *Creation and Use of Domestic Space: an Archaeological Study of Domestic Economy and Social Organisation on the Greek Mainland in the Classical and Hellenistic Periods*. PhD. thesis, University of Groningen.

HOEPFNER, W. & E.-L. SCHWANDNER, 1994. *Haus und Stadt im Klassischen Griechenland* (= Wohnen in der klassischen Polis 1), 2nd ed. München.

KALPAKSIS, T., A. FURTWÄNGLER & A. SCHNAPP (eds), 1994. *Eleutherna, tomeas II, 2. Ena ellinistiko Spiti ("Spiti A") sti thesi Nisi* (contributions in Greek, German and French), Ekdoseis Panepistimiou Kritis, Rethymno.

KRAUSE, C., 1977. Grundformen des Griechischen Pastashauses. *Archäologische Anzeiger*, pp. 164–179.

MALAKASIOTIS, Z., V. RONDIRI & E. PAPPI, 1993. Prosfates anaskafikes erevnes stin archaia Alo: mia proti parousiasi ton evrimaton. In. V. Kontonatsios (ed.), *Achaiophthiotika* 1. Ores, Almiros, pp. 93–101.

MAKRIS, K.I., 1980. *H seismikotita tis periochis tis Thessalikis Larisas*. Congress paper, International Congress of Thessalian Studies.

MARZOLFF, P., 1978. Bürgerliches und herrscherliches Wohnen im Hellenistischen Demetrias. In: W. Hoepfner (ed.), *Wohnungsbau im Altertum* (= Diskussionen zur Archäologischen Bauforschung 3). Deutsches archäologisches Institut, Berlin, pp. 129–144.

MORITZ, L.A.., 1958 *Grainmills and Flour in Classical Antiquity*. Clarendon Press, Oxford.

ROBINSON, D.M. & J.W. GRAHAM 1938. *The Hellenic House: a Study of the Houses Found at Olynthus, with a Detailed Account of Those Excavated in 1931 and 1934* (= Excavations at Olynthos 8, The Johns Hopkins University studies in archaeology 25). John Hopkins Press, Baltimore.

ROBINSON, D.M. & J.W. GRAHAM, 1946. *Domestic and Public Architecture* (= Excavations at Olynthos 12, The Johns Hopkins University studies in archaeology 36). John Hopkins Press, Baltimore.

SKAFIDA, E., 1994. Kataskevastika ilika, techniki kai technologyia ton plinthinon spition sti Neolithiki Thessalia: mia ethno-archaioloyiki prosengisi. In: R. Misdrahi-Kapon (ed.), *La Thessalie: quinze années de recherches archéologiques, 1975-1990: bilans et perspectives: actes du colloque international, Lyon, 17-22 Avril 1990, Volume A*. Ministère Grec de la Culture, Athens, pp. 177–188.

STÄHLIN, F., 1924. *Das Hellenische Thessalien*. Stuttgart.

STEFANIDOU-TIVERIOU, Th. 1998. *Anaskafi Dio*. Tomos 1, I Ochirosi. Aristotelou Panepistimiou Thessalonikis, Thessaloniki.

SVORONOS-HADJIMICHALIS, 1956. L'Évacuation de la fumée dans les maisons grecques des Ve et IVe siècles. *Bulletin de correspondance hellénique* 80, pp. 483–506.

THEMELIS, P., 1979. Ausgrabungen in Kallipolis (Ost-Aetolian), 1977–1978. *Athens Annals of Archaeology* 12, pp. 245–279.

THEOCHARIS, D.R., 1973. *Neolithic Greece*. National Bank of Greece, Athens.

WACE, A.J.B. & M.S. THOMPSON, 1911–12. Excavations at Halos. *Annals of the British School at Athens* 18, pp. 1–29.

WHITE, K.D., 1963. Wheatfarming in Roman times. *Antiquity* 37, p. 207.

Many artefacts were found in each of the six houses that were excavated in the city of New Halos. In this chapter the main categories of artefacts – pottery, figurines, loomweights, metal objects and coins – are presented and discussed. In a separate publication attention will be paid to a small number of artefacts that have not been included in this volume, for instance a stone vessel with an iron and silver snake that was found in the House of the Snakes, millstones and other stone objects. A selection of the artefacts from the six houses is on permanent exhibition in the Archaeological Museum of Almirós.

1. POTTERY

COLETTE BEESTMAN-KRUYSHAAR

1.1. Introduction

The history of Hellenistic pottery, and especially the pottery of the early Hellenistic period (late 4th, first half of the 3rd century BC), is still much discussed among scholars of Greek pottery. One of the major problems in this field of study is the lack of accurately dated pottery assemblages. Only two fixed reference points have, after a long discussion, now been accepted for this period by most scholars: 348 BC, the year when Olynthos, a city in Chalkidike, was destroyed by Philippos II, and 280–261 BC, when there is known to have been a Ptolemaic military camp at Koroni on a promontory along the east coast of Attica.[1] Besides the lack of accurately dated assemblages, there are also other aspects causing problems in dating early Hellenistic pottery, such as a lack of reference material, changes in political and economic life in the Hellenistic period and changes in production methods.

Hellenistic domestic pottery, and domestic pottery in general, has hitherto attracted very little interest. Only few detailed studies of Hellenistic domestic pottery have so far been published. The publications covering the pottery of well-known and/or large sites like Olynthos (Robinson, 1933; 1950), Athens (Thompson, 1934; Rotroff, 1997), Corinth (Edwards, 1975), Koróni (Vanderpool et al., 1962; 1964), Pergamon (Schäfer, 1968) and Eretria (Metzger, 1969) are all of relatively early dates. In a recent study, Susan Rotroff (1997: p. 4) was forced to restrict herself to the fine wares from among the large quantity of pottery found on the Athenian Agora, although she had initially intended to give a full account of the Agora's Hellenistic pottery. Her study is nevertheless an important reference work for the fine ware found at New Halos.

Another dating problem concerns the political, social, economic and artistic changes that occurred after 350/40 BC, when the Classical period in Greece came to an end. Though these changes are to varying extents related to the growing power of the kings of Macedonia after Alexander the Great's conquests of Greece and large parts of the eastern mediterranean world, the exact influence of these events on the production and use of pottery cannot always be estimated. Around this time the former supremacy of Attica as a pottery-production centre decreased, without this centre however altogether losing its role in pottery production, as is demonstrated by the wide distribution of West Slope ware and 'Megarian' bowls. Meanwhile, new pottery production centres emerged in Greece and the eastern mediterranean world, for example in Alexandria in Egypt (Adriani, 1936; 1940; 1952), and in Pergamon in Asia Minor. And outside these major centres more pottery began to be produced locally, for instance on Thasos (Blondé, 1985) and at various local centres on the Peloponnesos like Elis and Argos (Bruneau, 1970).

Changes took place in the production of the pottery itself, too. The lack of painted figurative decoration for instance makes it difficult to date pottery in isolation, and implies a strong dependence on closed pottery assemblages. A complicating factor is that similar pottery types and wares were used in different parts of the mediterranean area, including the western parts. In some cases the quality of the pottery varies considerably, but on the whole the vessels are all of the same types. Until more information is obtained on distribution patterns, technology and artistic values, it is difficult to estimate the time it will have taken for a finished vessel or a pottery type to 'move' from one place to another.

In the past few decades Hellenistic pottery has begun to attract more attention, and many studies covering typological, chronological and even technological aspects of the pottery have been published. But in spite of this increased interest there is still no illustrated survey of dated pottery types, systematically described and arranged for use in the field. As the reasons for this are closely related to the characteristics of the material, the employed production methods and, as mentioned above, a lack of accurately dated pottery assemblages, excavators often have to make do with surveys covering small specific regions only.

The present survey of the pottery of Hellenistic New Halos may help solve some of the aforementioned problems to some extent. Firstly, and most importantly, the pottery assemblage presented in this chapter can be soundly dated to the early Hellenistic period, from about 302 to 265 BC, on the basis of associated coins. Secondly, the survey presents a collection of pottery from a region not well known elsewhere in Greece in the Hellenistic period: Thessaly. Unlike New Halos, several sites in Thessaly have not yet been precisely dated, such as Demetrias, Gorítsa and Phthiotic Thebai. And, finally, thanks to the fact that New Halos was suddenly and abruptly destroyed, we are able to form a fairly good impression of conditions in the town just before the catastrophe. We know that the site was occupied for only a short period of time prior to its destruction and that no new buildings were built here in later times. The site has consequently survived practically unchanged since that event (Dijkstra et al., 1996; Reinders & Kloosterman, 1998). These conditions granted us an opportunity to study a more or less complete household inventory and to reconstruct at least some aspects of daily life in Hellenistic Halos, even though only six randomly selected houses out of an estimated total of 1440 have been excavated (Reinders, 1988: pp. 110–113).

The pottery considered below was recovered from among the remains of six houses in the lower town of New Halos:

1. House of the Coroplast (housing block 2.7, plot 11);
2. House of the Geometric *Krater* (housing block 2.8, plot 17);
3. House of the Snakes (housing block 6.4, plot 9);
4. House of the *Amphorai* (housing block 6.4, plot 18);
5. House of the Ptolemaic Coins (housing block 6.4, plot 21);
6. House of Agathon (housing block 6.4, plot 22).

All the pottery encountered during the excavations was collected as we assumed we would be able to reconstruct some of the vessels. In some cases, sherds from the same piece of pottery were found at different levels due to disturbance caused by ploughing. All the pottery from houses 1, 4, 5 and 6 was analysed and sorted. Part of House 2 was found to have been disturbed, so the pottery from this house does not provide a complete survey of the inventory. Of the pottery recovered from house 3 I have chosen to present only complete vessels representing types not encountered in the other houses owing to lack of time to describe the pottery in its entirety. Part of the pottery found in house 1 has already been discussed elsewhere (Reinders, 1988: pp. 110–113), but it will be included below as further research has enabled us to improve the descriptions and some of the drawings.

Generally speaking, the pottery is in very poor condition. Much of the fine and medium-coarse wares is powdery. Some of the medium-coarse ware, especially the locally produced New Halos ware, readily crumbles. This is probably attributable to small fires that burned in the town during or after its destruction and to the effects of water and soil over the past two millennia. For example, in some cases a single sherd of an *amphora* or other large vessel was found to be blackened, whereas the rest of the vessel was clean, and some glazed sherds show bands of well-preserved black glaze where other sherds had apparently lain on top of them since the time of the town's destruction. Various other processes may have affected the condition of the pottery, too, such as the preparation of the clay and firing techniques, but, if so, we do not yet know how. Almost all the pottery is covered with a lime scale, which can only be removed with great difficulty or not at all. This sometimes made it very difficult to describe the ware and draw the types. The deep-ploughing of parts of the site in the past thirty years has also caused a lot of damage, in some cases even resulting in the destruction of the sherds in parts of the houses or their dislocation across distances of several metres.

During the excavations, a layer of at most 30–40 cm, measured from ground level to the cobbles and loam of the virgin soil, was removed. Because of the lack of a well-defined stratigraphy, we are unable to discuss and illustrate variations in shapes and types in any chronological order in this chapter. Only there where a small proportion of the pottery was found to be more or less intact were we able to draw some conclusions concerning the typology of the pottery at or shortly before the time of the town's destruction. The reconstruction of some pottery types, especially jugs and other large containers, proved very time-consuming and difficult, so not all types have been reconstructed. A large part of the pottery was very fragmentary, representing household ware that was in use for an indefinite period of time within a span of about 40 years. Some profiles are based on only a very small part of the original vessel.

First a survey of the wares will be given. The temper and colour were determined with the naked eye, the colour by using Munsell's Soil Color Chart. It should be noted that various people have worked on the descriptions of the pottery.[2] After that, the pottery will be discussed according to class and function. The following classes have distinguished: cooking pots (for preparing food), closed vessels (for storing and serving liquids), open vessels (for grinding, mixing liquids and preparing solid food), storage vessels (for storing liquids and solid food), fine glazed tableware (for eating and drinking) and miscellaneous vessels (not in any way associated with food). The individual pottery types and their variations in shape and ware will be discussed within this general classification.

1.2. Wares

The excavated pottery represents a wide range of wares. The various wares have been defined on the basis of temper and colour as perceived by the eye, though it should be borne in mind that the colour of the pottery may have changed as a result of use and post-depositional processes. The following main wares have been distinguished:

- medium-coarse orange and red wares (local);
- New Halos ware (local);
- medium-coarse light ware (imported);
- glazed orange ware (local and/or imported);
- fine glazed ware (local and/or imported);
- fine unglazed ware (imported);
- coarse ware (local and imported).

Medium-coarse orange and red wares: The red ware has been classified as 10 R and 2.5 YR, the orange ware as 5 YR and 7.5 YR. Due to – most probably – irregularities in the firing process, the colours of individual sherds or vessels of these wares occasionally vary between 2.5 YR and 5 or 7.5 YR, or the colours of the in- and outside are sometimes dark red to black as a result of burning in an oven or over a cooking fire. This could imply that these wares should in fact be seen not as two separate wares, but as one and the same, represented by a wide range of shades. The fabric is powdery, containing a medium-coarse temper of mostly quartz and lime. The thickness of the sherds ranges from 2 to 5 mm in the case of small vessels, and from 5 to 12 mm in the case of larger vessels. The greater part of the pottery (an estimated 60 to 70%) is of medium-coarse orange ware. A rare feature of the medium-coarse red and orange pottery is an incised decoration consisting of a horizontal line with thin lines on either side creating a floral pattern. This form of decoration was observed on four of the red-ware vessels (P521–524) and two of the orange-ware vessels (P302, P525) and also on some of the New Halos-ware pottery (P336–337, P340) (table 3.1).

The vessels of medium-coarse orange and red ware were most probably produced locally. They comprise many different types of domestic vessels like *chytrai* (cooking pots), jugs, *hydriai* (water jugs), *krateres* (mixing bowls), jars, *lekanides* (wide bowls), *lopades* (casseroles), lids and stewpots.

New Halos ware: The New Halos ware is dark red to reddish brown, ranging from 10 R to 2.5 YR. New Halos ware is characterised by a medium-coarse temper, including some lime and a little mica or glimmer. The fabric is hard and cracky. Most of the sherds have slightly smoothed in- and outsides. The sherds are of more or less the same thicknesses as those of the medium-coarse red and orange wares. In some cases this ware is hard to distinguish from the medium-coarse red ware. New Halos ware is macroscopically identical to the wares of some of the terracottas and moulds found in the House of the Coroplast, suggesting that this pottery was made locally. Three vessels of this ware (P336–337, P340) show an incised floral decoration of the same kind as observed on the vessels of medium-coarse red and orange ware.

Like the medium-coarse orange and red wares, Halos ware is represented mostly by domestic types such as *chytrai*, jugs, *krateres*, jars, *lekanides*, *lopades* and lids and occasionally by types like bowls and *amphoriskoi* (miniature *amphorai*, e.g. P517).

Medium-coarse light ware: This ware ranges from the lighter shades of 2.5 YR to 10 YR, but 5 YR 6/6 or 7/6 reddish yellow is also represented. Some *amphorai* made of this ware were cracked and scaled, probably as a result of secondary burning. The thickness of the sherds ranges from 4 to 16 mm. The temper is medium coarse and occasionally contains mica. Besides *amphorai* (large wine-storage containers), a Chian *lagynos* (wine container; P186) and a *lekane* (P234) are made of this ware. An analysis of the petrochemical characteristics of this ware (Jager & Van der Meulen,1989) revealed a wide variety of minerals, including biotite, leucite, sandine or nephaline and aenigmatite. This medium-coarse light ware was probably imported from Chios and/or other wine-production centres.

Glazed orange ware: The fabric is reddish yellow to bright orange and the temper is medium to medium coarse. The thickness of the sherds ranges from 2 to 5 mm. The fabric is basically the same as that of the medium-coarse orange ware, except for the glaze, which is of very poor quality and mostly worn, surviving in a few traces only. Only a few vessels are made of this ware. They all had a specific function: two *squat lekythoi* (oil containers; P173 and P174), one oil lamp (P475) and one *lagynos* (P187).

Fine glazed ware: The fabric is mainly light pink and yellow (between 5 YR and 7.5 YR), varying to

Table 3.1. Local pottery with a decoration of floral pattern.

	Ware	House 1	House 2	House 3	House 4	House 5	House 6
Lid	NHW	-	-	P337	P341	-	P336
?	MCR	-	-	-	-	P522–523	P521
Jar	MCO	-	-	P302	-	-	-
Jar?	MCR	-	-	P524	-	-	-
Jug?	MCO	P525	-	-	-	-	-

light brown. The temper ranges from very fine to medium-coarse. The fine-medium wares occasionally show white lime particles. Both thick (4–6 mm) and egg-shell thin (2–3 mm) sherds were found, but the thicker sherds are the most common. The glazed pottery from New Halos is completely glazed (with only one exception, bowl P358, which was dipped) but the quality of the glaze is mostly very poor, with dark grey being the predominant colour.[3] The glaze of many vessels had disappeared almost completely, probably due to soil conditions. Some pots were irregularly glazed or fired. The glaze of these pots is of a more reddish hue. Two bowls (P398, P405) show stacking marks, which are particularly common in Attic pottery of the late fourth and early third century (Rotroff, 1997: p. 11). The greater part of the fine-tempered Attic pottery found at Hellenistic Halos has a glaze of good quality.

Only a few bowls have stamps on the inside of the base, are decorated with rouletting, or show a combination of the two (table 3.2). The stamps are mostly palmettes, but they also include horse-shoe (P360) and tree-like (P399) motifs. On complete bases the stamps are arranged in groups of four or occasionally five around the centre. One early specimen (P397), which must have had about ten palmette stamps linked by interlocking arcs, possibly dates from before c. 300 BC (Rotroff, 1997: p. 37). Almost all the stamps are fairly large and were carelessly impressed. Fine rouletting, common in Attic bowls until c. 275 BC, and coarser rouletting consisting of thicker, more widely spaced lines, common in Attic bowls after c. 275 BC (Rotroff, 1997: p. 37), are both represented.

The four vessels decorated with the West Slope technique show incised curls containing traces of worn paint (P511), a wavy line (P486) and simple ivy and grape garlands painted in a brown slip, now almost faded (P371, P435). Similar designs are also known from Athenian fine-ware vessels dating from

about 300–275 BC, on which they were however usually combined with other designs (Rotroff, 1997: p. 3). Two very fine vessels, a cup-*kantharos* (P435) and a *pyxis* (P491), are decorated with limy red paint or *miltos* in grooves in the base and body; in the case of P435 this was combined with West Slope decoration.

The pottery also includes one piece of Attic red-figure ware, a *pyxis* lid (P498) decorated with two facing heads. At Athens, red-figure ware was also encountered in Hellenistic assemblages, suggesting that this kind of pottery was still being made in Hellenistic times[4], or had been in use for a long time. One fragment of a *hydria* or *pelike* (P182) shows white painted ovules on top of the rim.

The greater part of the tableware, including the various drinking and serving bowls and wine cups such as *kantharoi*, luxury items like *pyxides* (lidded boxes), and oil lamps, is made of fine glazed ware.

Fine unglazed ware: The first of the two most important types of fine unglazed pottery found at New Halos is Blister ware, represented by a small *aryballos* (P512) imported from Corinth. This particular vessel is wheel-made and has an egg-shell-thin (1–3 mm) wall of a hard, almost vitrified, fabric typical of Hellenistic Blister ware. The temper consists partly of large quartz particles. The colour of the fabric, bright orange with a grey shine caused by induction, is however more reminiscent of pottery from the Classical period (Edwards, 1975: pp. 7 and 10). Having characteristics of both periods, this vessel could consequently date from around the transition from the Classical to the Hellenistic style of pottery production. The *aryballos* is decorated with incised ivy leaves and a band of incised crescent moons ('lunate design') around the shoulder – a decorative pattern that can be dated to the last quarter of the 4th century.

The grey ware of *unguentaria*, perfume jars, has a reddish orange core and a thin grey internal and external surface, and contains a fine to medium-fine

Table 3.2. Decoration on black glazed pottery.

	House 1	House 4	House 6	House 5	House 5/6
Bowl, base fragments					
stamped	-	-	P359–361	-	-
rouletting	-	P362–363	P364	P365	P366
stamped + rouletting	-	P367–369	-	P370	-
Bowl, incurved rim					
stamped + rouletting	P397	-	-	-	-
rouletting	-	-	P351	-	-
Bolsal					
rouletting	-	P375+380	P377+387	-	-
stamped + rouletting	-	-	P376	-	P374
painted (West Slope)	-	-	P371	-	-
Bowl, everted rim					
stamped + rouletting	-	-	P405	-	-

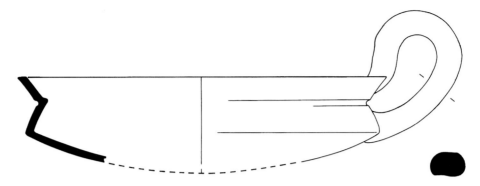

Fig. 3.1. *Lopas* P84 (scale 1:2).

temper consisting of lime. The fabric is hard and the thickness of the sherds varies from 2 to 4 mm. The body, shoulder and neck of some of the vessels are decorated with bands painted in a dark grey or dark red slip. It is not certain where this ware was produced, but it is likely that it came from Attica (Hellström, 1971: pp. 25–26; Rotroff, 1997: p. 177).

Coarse ware: The temper of the coarse ware consists mainly of grog and small quartz particles. The fabric is cracky and full of small holes, which were probably formed when organic inclusions burned during the firing. The thickness of the sherds ranges from 15 to 40 mm. Most of this pottery has a light reddish yellow colour, varying between Munsell 10 R and 10 YR. The most common types of vessels made of this ware are *pithoi* (large storage jars) and tubs. They were probably produced locally. One mortar (P271) and one *louterion* (P272) made of a light pinkish yellow coarse ware came from Corinth.

1.3. Types

Next, the types of vessels will be described (see also Appendix 1) and a survey of their distribution in the six houses will be presented in Appendix 2. The order in which the types will be described in each section is based on the numbers of vessels of the individual types found in the various houses, so the most common types will be presented first.

Cooking vessels (P1-147)

Chytrai (cooking pots, P1–73; fig. 6.1): A fairly large number of complete profiles of this common type of cooking pot could be reconstructed. Most *chytrai* have a convex base, a globular body, a wide concave neck, an everted rim and a convex or bevelled lip. Some variation is observable in the occasionally angular transition from the body to the neck (P2; P3 and P60). The base is moreover sometimes a little flattened (P2). The complete *chytrai* all have a single small vertical strap handle attached to, but never arching above the

lip. The ware of most of the vessels is medium-coarse, orange and red; some vessels are of New Halos ware. The common type of medium-sized *chytra*, with a height of about 10 to 15 cm and a body diameter of 9 to 23 cm, was found in all six of the houses (P1–66). In addition, a few smaller *chytrai*, with heights of 6 to 9 cm, were found in five of the houses (P67–72). One specimen, P72 from house 5, is quite a bit larger than the others, with a height of about 29 cm, but it is nevertheless of the same shape. Their frequencies suggest that *chytrai* were more common cooking pots than *lopades* and stewpots.

Lopades (P74–98; figs 3.1 and 6.1): *Lopades* are wide, squat cooking pots or casseroles. The heights of the vessels found at New Halos vary from 4.4 to 5.8 cm. The rim is about 14 to 24 cm wide and has a flange for a lid. All *lopades* have a convex base, but four main types can be distinguished on the basis of the shapes of the body, rim and handle. The two most common types (P74–78), which were encountered in four of the houses, have a straight or slightly curved profile, a flaring rim and two horizontal ring handles just below the rim. A variation on these types (P83–84), found in houses 3 and 4, has a vertical strap handle arching above the rim (fig. 3.1). Another distinctive type (P79–82) has an angular profile, a small upright rim with a convex or bevelled lip, and horizontal ring handles. In house 2 a variation on the latter type was found (P85), with a splayed wall and a vertical strap handle attached to the rim, but not arching above it. Most rim fragments (P86–98) do not have handles, so it is impossible to assign them to a specific type. The walls of most of these fragments are straight or slightly curved. Most of the *lopades* are of medium-coarse orange ware, some are of medium-coarse red ware or New Halos ware. *Lopades* were found evenly distributed in all six houses, but in smaller numbers than the *chytrai*.

Stewpots (P99–108; fig. 6.1): A small number of rim and body fragments were identified as sherds of stewpots with a convex base, a wide globular body, occasionally with an angular profile formed by a

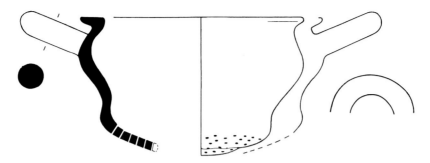

Fig. 3.2. Strainer P109 (scale 1:2).

sharply inturning shoulder, and a flaring rim with a flange for a lid. Some rim and wall fragments show the attachments of horizontal ring handles (P102, P103, P106). Fairly small rim fragments, with diameters of about 10 to 15 cm, are most common; they were found in houses 1, 2 and 5 (P99–104). Other rim fragments derive from large stewpots with rim diameters of 20 to 24 cm, which were found in houses 1, 3 and 5 (P105–108). The stewpots are of medium-coarse red and orange ware and New Halos ware. Cooking pots of this type were less common than *lopades* and *chytrai*.

Strainers (P109–114; figs 3.2 and 6.1): Strainers may have been used for purifying oil or wine, lined with a piece of cloth, or for straining dairy products. Two more or less complete strainers are of two different types. P110, made of medium-coarse red ware, has a wide convex base with 3–4-mm-wide perforations, a broad tapering body, an everted rim and an overhanging, convex lip. No handles have survived. The strainer is about 12 cm high. The other strainer (P109) is about 7 cm high and is made of medium-coarse orange ware (fig. 3.2). It has a convex base with 2-mm-wide perforations, a splayed wall, an angular transition to the shoulder and an everted rim with a flattened top. This strainer has (at least) one horizontal ring handle attached to the shoulder and arching a short distance above the rim. Base fragments of strainers with several perforations were found in houses 5 and 6. The perforations are either large (5–6 mm) and set at short distances from one another (P111) or small (3–4 mm) and spaced far apart (P112–114).

Lids (P115–146; fig. 6.1): Small cooking pot lids have simple, slightly domed profiles and small knobs of three different shapes. The most common is the small conical knob. Other shapes are a little wider, and have a concave or flattened top. The majority are of medium-coarse red and orange ware; a few are of New Halos ware. One lid and a small knob are made of a very attractive fine pink ware (P124, P125). The diameters of these lids vary from 12 to 24 cm. The lids may have covered the cooking vessels discussed

above or other vessels such as jars, jugs, *hydriai* or *krateres*.

Brazier (P147; fig. 6.1): A large rim fragment of a deep brazier or stove which was refitted from sherds found in rooms 3 and 5 of house 5, has a rim with a flattened top with a diameter of about 22 cm, a shoulder with an angular profile and an attachment of a horizontal strap handle. The brazier is made of medium-coarse red ware.

Closed vessels (P148–195)

Closed vessels were containers for liquids: wine, water, milk or oil. Many may have been used for either wine, water or milk:

Jugs (P148–164; figs 3.3 and 6.2): Dozens of base, rim and handle fragments of various jug-like vessels were found at New Halos. It is impossible to say to exactly which type of vessel they once belonged, as only a few vessels could be completely reconstructed. The rim fragments with handles derive from either jugs, whose handles were attached to the lip, or the slightly larger *hydriai*, whose handles were attached below the lip. The jugs, which were intended for holding and pouring all kinds of liquids (except oil), are made of medium-coarse red and orange ware and New Halos ware.

Two similar jugs were found in houses 1 (P148) and 4 (P149; fig. 3.3). They are of medium size, about 18 cm high, and have a raised flat base, a wide globular body with a slight angle at the level of the widest diameter, a profiled lip and a vertical strap handle arching above the rim.

Several jugs with differently shaped bodies which could be completely reconstructed were found in room 5 of house 6. They are all about 30 cm high. Jug P150 has a slightly convex base, a globular body and a narrow neck with a wide everted rim; P151 has an oval body and a wide neck; P152 has a raised flat base, a globular body and a slightly offset wide neck. A collection of the best-preserved rim fragments shows the variation in the diameters and shapes of the lips (P153-164).

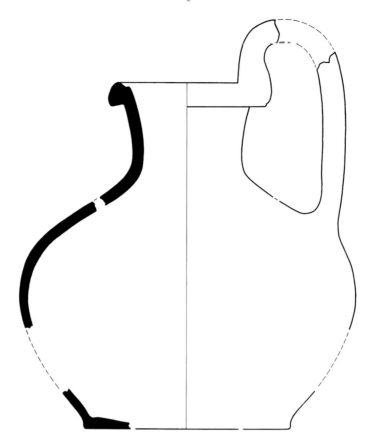

Fig. 3.3. Jug P148 (scale 1:2).

The large vessels with heights of about 15 to 25 cm presented below contained oil used for preparing food or filling oil lamps. Several small jars of fine ware contained perfumed oil for cosmetic purposes. They will be discussed later (see under *Miscellaneous*):

Lekythoi (P165–174; fig. 6.2): One *lekythos* (P171) with a height of about 18 cm is made of fine black-glazed ware. It has a splayed base ring, a wide ovoid body, a narrow neck with a vertical double strap handle, an everted, carinated rim and a thickened lip with a flattened top. A small group of *lekythoi* from houses 4 and 5 (P165–168) of slightly larger dimensions are made of medium-coarse orange and red ware and New Halos ware. Insofar as can be ascertained they have a raised flat base and a vertical ring handle. As far as their shape is concerned, these locally made vessels are very similar to the aforementioned black-glazed specimen, if somewhat simpler. House 6 yielded a distinctive type of *lekythos* without a carinated rim (P169 and P170). The shape of the vessel's body and neck is however similar to that of the other *lekythoi*.

A fragment comprising the base, shoulder and rim of a large *lekythos* (P172) with an estimated original height of about 30 cm was found in house 4. The vessel had a raised flat base, a convex shoulder, a narrow neck, a wide everted carinated rim with a thick bevelled lip and a vertical double strap handle. The lip was decorated with a brown slip on the in- and outside.

In addition, two different types of squat *lekythoi*, probably of local origin, were found in houses 1 and 5. P173 is about 9 cm high and has a convex base, a broad tapering body, a sloping shoulder, a narrow neck and an everted rim. The strap handle is attached to the rim. The bright reddish yellow medium-coarse ware tempered with lime was coated with a dark grey glaze, the greater part of which has now disappeared. Squat *lekythos* P174, which has a preserved height of 6 cm, is made of the same ware. The base of this vessel is also convex, but the body is much squatter than that of P173, and it had a vertical ring handle on the shoulder.

Some vessels were used as containers for water:
Hydriai (P175–182; fig. 6.2): Five *hydriai*, large vessels for storing, carrying and pouring water, with an estimated total height of about 40 to 46 cm, could

be partly reconstructed from parts of the base, body and rim. The best-preserved New Halos *hydriai* have a raised flat base (P181) or a ring base (P175), an ovoid, almost globular body and three handles. One vertical strap handle is attached to the shoulder and the neck, below the rim, and two horizontal ring handles are attached to the shoulder. The shoulder is convex and the neck is straight or slightly splayed. The majority are of a particularly bright shade of the medium-coarse orange ware, but some are of medium-coarse red ware or New Halos ware.

A small rim fragment (P182) deriving from a late Classical Attic *pelike* or *hydria* with a wide overhanging rim was found in house 1. The rim has a diameter of 17.5 cm and was painted with ovules in white slip. As no complementary fragments were found, this vessel was probably imported, used and discarded in the early days of the town's occupation.

Water pitchers (P183–185): Only a few rim fragments of this type of jug were found; they have diameters of 12 to 13 cm. The vessels were intended for serving water (or other liquids) and have a wide straight neck and a broad everted rim with a broad groove on top. The most distinctive fragment (P183) was found in house 3. The other rim fragments (P184 and P185) are from houses 4 and 5. The ware is fine and unglazed, of a reddish yellow colour.

A few specific wine containers are represented among the pottery:

Lagynoi (P186–187; fig. 6.2): Two different types of these wine containers were found. One large vessel (P186), with a height of about 27 cm, has a ring base, a wide angular body and a long narrow neck. The vessel's size and shape and the reddish yellow, medium-coarse ware suggest it came from Chios. Parallels of this type of *lagynos* that have been dated to the 3rd century BC were found at Koróni (Edwards, 1963: No. 50) and Athens (Grace, 1961: No. 50). A fragmentary, smaller specimen (P187) from house 1 has a flat base and a wide angular body with a preserved height of about 12 cm. The neck and rim broke off at some stage. The vessel is of the local medium-coarse orange ware tempered with lime and was coated with a black glaze, the greater part of which has disappeared, like the glaze on the squat *lekythoi* P173 and P174 mentioned above.

Olpai (P188–190; fig. 6.2): Three slender jugs may have been used as measuring cups for wine, or *olpai* (Rotroff, 1997: p. 128). Two flat base fragments with a diameter of 6 cm from room 5 of house 4 are glazed on the inside, but bare on the outside. They are of an asymmetrical shape and show broad ridges on the inside, suggesting that the jugs were thrown on a slow wheel. The fabric of *olpe* P188 is soft and powdery with a medium-coarse temper and a reddish yellow colour. That of P189 is a little finer. A fragmentary shallow jug or *olpe* found in house 3 (P190)

has a low ring base and a long, irregularly shaped, slender body. The shoulder, rim and handle are missing. The vessel is of exceptionally soft, gritty, medium-coarse ware of a light reddish brown colour.

Pelikai (P191–194; fig. 6.2): An almost complete *pelike*, a wine jug with a raised flat base, a globular body, a wide neck, two vertical strap handles and a carinated rim, was found in room 5 of house 6. It is about 18 cm high. Three other rim fragments of this type of vessel were found in other rooms in this house. The ware has a medium fine temper and a pale brown colour.

Amphora (P195; fig. 6.2): One very small *amphora* with a height of about 10 cm which was found in house 1 was used for serving wine instead of storing liquids (storage *amphorai* will be discussed below): it has a broad raised flat base, an almost globular body with a slightly angular shoulder, a wide neck with an everted rim and an overhanging lip. Two vertical strap handles are attached to the shoulder and below the lip. The baggy shape is similar to that of the Attic West Slope *amphorai* found on the Athenian Agora and elsewhere – a popular shape that was introduced in Athens around the beginning of the second quarter of the 3rd century BC (Rotroff, 1997: p. 121). The latter vessels are however about twice as high and have twisted handles and a moulded ring base. The New Halos vessel is made of New Halos ware, and may therefore be a local version.

Open shapes, for serving and processing food (P196–272)

Lekanides (P196–236; fig. 6.3): *Lekanides* are common household basins that were used for general, and occasionally heavy-duty purposes. Complete specimens are between 7 and 21 cm high, their rim diameters ranging from about 16 to 60 cm. Most common is a rim diameter of between 27 and 36 cm. The vessels have a raised flat base (only P221 from house 5 has a ring base) and a broad, slightly splayed body. Two horizontal ring handles are attached just below the rim, touching the lip for greater strength. The handle of one rim fragment (P233) from house 4 was attached to the lip with a broad strip of clay to give it even greater strength. The rims of three of the smaller vessels (P205–206 and P213) contain small perforations, drilled before firing, from which the bowls could be suspended from the wall. The vessels show a lot of variation in the shape of the rim, but overhanging rims with a flattened top are the most common (type 1). The majority of the vessels are of medium-coarse orange and red ware, a few are of New Halos ware. One vessel (P234), from house 1, is made of medium-coarse light ware.

Krateres (P237–252; figs 3.4 and 6.3): *Krateres* (kraters) are large bowls that were used for mixing wine and water. One more or less complete profile

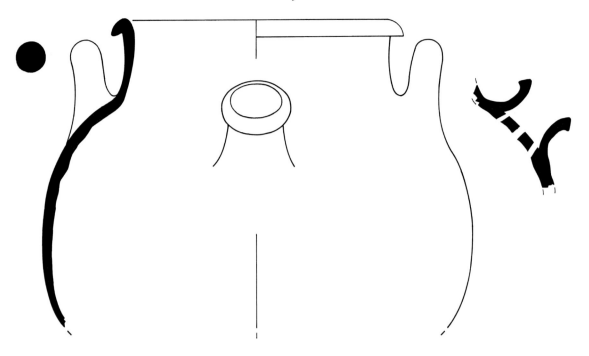

Fig. 3.4. *Krater* P252 (scale 2:5).

(P237) has a height of about 27 cm. The New Halos *krater* has a raised flat base, an ovoid body and a wide, straight or slightly flaring neck with an overhanging rim. The shape of the overhanging rim shows minor variations. The ring handles on the shoulder are set either vertically (P237, P252), or inclining slightly inwards (P251) or outwards (P238). Two fragmentary *krateres* (P251 and P252; fig. 3.4) have a spout on the shoulder; one is open, the other contains a strainer in the wall of the pot. The majority of the vessels are made of medium-coarse orange and red ware; only a few are of New Halos ware.

Tubs (P253–267; fig. 6.3): Large vessels intended for heavy-duty domestic and industrial purposes such as pressing grapes or olives. At New Halos (fragments of) tubs were found in houses 1, 4, 5 and 6. One tub (P253) could be completely reconstructed. The fragments are mainly rims and a few bases. The complete tub, from house 6, has a 46-cm-high basket-like shape with a raised flat base, a splayed profile, an everted rim with a flattened top and a diameter of about 60 cm and lugs beneath the rim. Rim fragments of presumably basket-like shapes have diameters of between 40 and 76 cm (P254, P256, P257, P259-260). Rim fragment P257 shows several small stamped ovals with a curly sign in relief on the rim and wall and (illegible) incisions on the lip, applied before firing, probably a name or sign indicating the potter. Large raised flat base fragments probably deriving from tubs have diameters of between 19 and 20 cm (P263–265). A smaller ring base (P262), from either a tub or a large *lekane*, has a diameter of about 12 cm.

Parts of two distinctive tubs were found in house 5: a rim fragment of a rectangular vessel (P258) and a rim fragment with a spout (P266) deriving from a tub that was also rectangular, but most probably had a rounded base. The two fragments are of medium-coarse to coarse red and reddish yellow ware.

A few thick wall fragments (P267; with thicknesses of 13 to 19 mm) of a vessel of coarse red ware that was at some stage repaired with strips of lead attached to the wall with pins inserted through drilled holes may also derive from a large tub. The fact that it was repaired implies that this vessel was used for a long time and that such large vessels were precious objects.

Lebetes (P268–270; fig. 6.3): Fragments of two almost complete *lebetes* or *dinoi* and one rim fragment (P269) were found in houses 3 and 4. The large vessels had a globular body with a round (P268) or slightly pointed (P270) base, a wide rim and two large horizontal ring handles attached to the shoulder and arching above the rim. Both vessels had an estimated height of 32 cm and a rim diameter of about 22 cm. They were used for mixing wine with water, like *krateres*, or as wash bowls. The vessels are of medium-coarse red ware and New Halos ware.

Mortar (P271; fig. 6.3): A small rim fragment of a mortar with an overhanging rim and a bolster hand grip was found in house 6. The rim, with a diameter of 44 cm, is decorated with a strip of clay with regular round incisions. The ware of this fragment, which is pale yellow and contains medium-coarse temper,

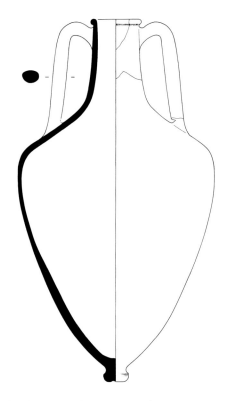

Fig. 3.5. Storage amphora P276 (scale 1:8).

suggests a Corinthian provenance. Numerous parallels are indeed known from Corinth, (especially) the Rachi settlement, and elsewhere.[5] On account of the similar shape of stone mortars it is often assumed that earthenware mortars were used for grinding roots or herbs, but besides that also the use as milk basins or even cheese moulds is suggested (Edwards, 1975: p. 110). The rough inner surface may have been intended to accelerate the coagulation of milk (cf. section 4.4), after which the whey would have been poured out through the long narrow spout. The fact that the coarse granular inside surface of many of the vessels of this kind found at Corinth was not worn, as would be expected if the vessels had been used for grinding for long periods of time, indeed suggests that these 'mortars' were used for other purposes.

Louterion (P272; fig. 6.3): One rim fragment of a shallow type of supported wash basin or bread board (Sparkes & Talcott, 1970: p. 218) came to light at New Halos. Some wall fragments may derive from the same vessel. The overhanging rim had a diameter of 56 cm. The rim fragment was broken at the point where the lip, which was made separately, was attached to the edge of the basin. The shapes of the rim and the lip, which is ribbed on the outside, are characteristic of Corinthian types, whose ribs were usually painted brown and red. Some Corinthian vessels, which were found in the West Cemetery of Isthmia near Corinth, have been dated to the 5th and 4th cen-

turies BC (Iozzo, 1987: pp. 356–357). Rim fragments of this type and date are also known from the Athenian Agora (Sparkes & Talcott, 1970: p. 219). The louterion fragment found at New Halos may have been part of a 5th- or 4th-century vessel which had been kept for its practical or decorative value.

Storage vessels (P273–347)

Amphorai (P273–301; figs 3.5 and 6.4): Large wine storage amphorai came to light in all of the houses. Some could be completely reconstructed. Four of these large, almost complete vessels were found together in room 5 of house 4, which was consequently called the House of the Amphorai. None of the complete amphorai or the surviving handles is stamped.[6] The shapes of the bases and bodies of the amphorai vary, as do their colours and sizes, suggesting that the vessels were imported from several wine-production centres. It is however very difficult to say where the vessels originated as no satisfying identically shaped parallels are known. All quoted capacities were calculated up to the rim.[7]

The largest amphorai, with heights of about 74 to 80 cm (type 1, P273–283), have a wide ovoid body with a slight angle at the transition to the shoulder, a slightly concave shoulder, a long cylindrical neck and a rolled rim. Other typical features of these vessels are the slightly diverging handles and the short button knob on the base, with a rounded edge and a small nipple on the underside. The ware of these amphorai is macroscopically similar to that of the Chian lagynos P186, but no good parallels of this type are known. The capacities of the best-preserved vessels, P275, P276 (fig. 3.5) and P273, are about 29.1 l plus an estimated 1.5 l for the largely missing neck, about 37.8 l and about 59.4 l, respectively. Such a range of sizes and capacities may be quite remarkable for amphorai from one and the same production centre.

P284 is a pointed amphora of much smaller dimensions, with a height of about 47 cm, a long, slightly splayed body with an angular transition to the shoulder, a slightly concave shoulder, a cylindrical neck and a rolled rim. The handles incline slightly inwards. The capacity of this amphora is about 8 l. The base knob has a broad cylindrical band at the edge and a concave underside with a small nipple. Base fragments P285 and P286 are of the same shape.

The high lip of P287 and P288 is a distinctive feature. The amphorai are probably of Rhodian origin. The complete amphora P287 has an ovoid body, a short cylindrical neck and small vertical ring handles. The base knob has a concave underside and a straight edge. The capacity of this amphora is about 10 l, a little more than that of P284, which is however of a comparable height.

The size and most aspects of the shape of P290, a fragmentary amphora with a height of about 80 cm,

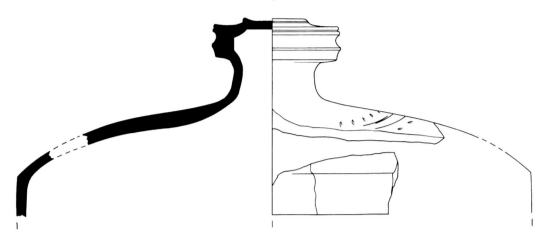

Fig. 3.6. Large lid P336 (scale 1:2).

are reminiscent of those of the *amphorai* of type 1, for instance the angle at the transition from the body to the shoulder, the slightly concave shoulder, the (probably) slightly diverging strap handles and the shape of the rim. The lower part of the body is however more pointed, so the capacity of this vessel, without the neck, is about 29 l. Another 1.5 l may be added for the volume of the neck, which is largely missing.

Three fragmentary *amphorai* (P291, P292 and P293) with ovoid bodies and slightly convex shoulders have an estimated height of about 74 cm. The capacity of P293 is about 20 l. The two other *amphorai* have slightly wider bodies, but they are too fragmentary to allow us to estimate their volumes.

The following are mainly neck, rim and handle fragments of large *amphorai*. Rim fragment P289, with a diameter of about 13 cm and part of a distinctive Koan double barred handle, suggests that wine from Kos was imported (Grace, 1961: Nos. 58 and 59). For some rim fragments no parallels can be quoted. Rim fragment P296 has a very long cylindrical neck and a rolled rim. The shoulder is slightly concave and the handles incline inwards. P297 is a vessel represented by a neck and rim fragment with a height of 21 cm and a base knob. The vessel cannot be reconstructed. The neck and rolled rim have a diameter of about 11 cm. The base knob has a flat underside and a rounded edge. Fragmentary amphora P298, represented by rim and neck fragments, probably broke some time before the town's destruction, because the fractures appear worn. The rim is of a different, everted type with a broad, bevelled lip. The handle shows slightly raised vertical ribs.

Various types of solid base knobs were found (P294–301). Two base fragments of pointed *amphorai* with similarly shaped knobs with an angular edge (P294 and P295) are similar to the knob of P292, but the lower parts of the vessels' bodies seem to have been more pointed.

As far as the distribution of the *amphorai* in the houses of New Halos is concerned, fragments of type 1 were found in all six houses, so this was the most common type at New Halos. An *amphora* of type 3 was found only in house 3, and house 1 is the only house where a Koan *amphora* was found. Three types were represented in house 4: type 1, one *amphora* of type 6 and one fragment of an *amphora* of type 2. Besides three *amphorai* of type 1, house 2 yielded *amphorai* of two other types and house 5 contained only one other type besides type 1. House 6 yielded only one fragmentary vessel of type 1, which was so common in the other houses, but also fragments of another eight types, so this house contained the most varied range of *amphorai*.

Jars (P302–326; fig. 6.4): Rim fragments of jars were found in all the houses. None of the storage jars could be completely reconstructed, but two fragmentary vessels (P309, P311) have a ring base, a large globular body, a wide opening with a narrow everted rim and horizontal ring handles attached to the shoulder. The jars are of different sizes, those of medium size, with rim diameters of 14 to 26 cm, being the most common (P302–308, P318–319). A few rim fragments have larger diameters, of about 30 to 37 cm (P309–317), or very small diameters, of about 6 to 8.5 cm (P320–321). One fragment, P302, made of medium-coarse orange ware, is decorated with an incised floral pattern on the shoulder. Rim fragments of three other types of jars were found. Fragments of the first type, P322–323, came to light in house 3. They derive from large jars with rim diameters of about 20 and 34 cm and an angular shoulder. The second type, which was found in house 1 (P325), is a jar with a large globular body with a diameter of 50 cm, a thick wall (about 1.6 cm), and a relatively narrow opening with a rim diameter of 23 cm. And the third type, represented by rim fragment P326 from house 4, is of a similar shape, but the rim itself is thicker, about 1.6 cm. The majority of the jars are made of medium-coarse orange and red ware.

Fig. 3.7. Pithos lid P342 (scale 1:4).

Pithoi (P327–335): Relatively small rim and wall fragments of these large, coarse storage vessels were found in all the houses except house 1. The heavy rims are everted with a flattened top and bevelled lip. The *pithos* is the largest type of storage jar, with rim diameters ranging from 42 to 77 cm. It was used for storing large quantities of food like olives or grain. The *pithos* had a round base and was partly dug into the floor for better access, to save storage space and to keep the contents cool. In rooms 5 and 6 of house 3 and room 1 of house 6 several round pits which once contained such vessels were found in a line along a wall. These pits contained nothing besides a little rubble, suggesting that the vessels were removed shortly before or after the house was destroyed. This implies that they were of high (practical) value.

Large lids (P336–341; figs 3.6 and 6.4): Remains of four large lids, intended to cover jars, *lebetes* or other large vessels, were found in houses 3, 4 and 6. P336 is almost complete; it is domed and has a straight rim with a diameter of 28 cm (fig. 3.6). The dome is decorated with an incised floral pattern around the knob. The knob is elaborately decorated with several ridges created on the wheel. House 3 also yielded a rim fragment of this type (P337) of a slightly smaller size. The knob is missing. These lids are all made of New Halos ware.

In house 6 another large rim fragment of a domed lid was found (P338). The rim has a diameter of 29 cm and is slightly flared. This lid shows no decoration and is made of medium-coarse red ware. The knob is missing, but a large knob also made of medium-coarse red ware (P339) which was found in this same house may well have belonged to this lid. The knob is less elaborately decorated than that of P336; it has a concave top with a nipple at the centre and a simple, slightly rounded edge. A smaller version of this type of lid, with a width of about 18 cm, is represented by P340 of New Halos ware. This lid was found in house 3.

Some very large wall fragments (P341), probably deriving from a large, very slightly domed lid with a diameter of at least 40 cm, were found in house 4. This lid is made of New Halos ware and is decorated with bands of an incised floral pattern, but with additional finger-drawn wavy lines between the bands.

Pithos or tub lids (P342–347; figs 3.7 and 6.4): Two types of large *pithos* or tub covers were found.

The first, which was represented by vessels with diameters of about 28 to 46 cm in houses 4 and 6 (P342–345), is shaped like a flat disk with a hole at the centre surrounded by a flange. This device probably made it possible to take small amounts of food (or oil or wine) from the tub without exposing too much of the rest of the contents. Two of these lids, P342 and P343, show the inscription of a name on the top side, around the aperture. The complete inscription of P343, which was found in house 6 and gave this house its name, reads 'αγαθονος' anticlockwise around the aperture (fig. 6.4). The incomplete inscription on lid P342, which was found in house 4, reads 'αγ[...]' (perhaps also from 'αγαθονος'?) clockwise around the aperture. As the inscriptions were drawn in the wet clay by finger before the vessel was fired, they probably represent the potter's name.

The two large lids of the second type consisting of a simple flat disk have diameters of 53 and 57 cm (P346 and P347; fig. 6.4). They were both found in house 5.

Fine glazed tableware (P348–462)

Different types of bowls were found in the various houses: *echinos* bowls, small bowls, large bowls with everted rims, *bolsals* and one hemispherical bowl.

Echinos bowls (P348–369; fig. 6.5): Plain *echinos* bowls (bowls shaped like the *echinos* of a Doric capital), which were used for serving food, constitute the largest part of the bowl-shaped tableware found at Hellenistic Halos. Numerous rim and base fragments were found, but only complete *echinos* bowls or profiles and decorated base fragments will be presented below. The bowls are all completely glazed, except for P355, which was dipped. The resting surfaces of the bases are not grooved. The complete bowls are not stamped, but one is decorated with rouletting (P351). Some base fragments show rouletting and/or palmette stamps (for a discussion of the decoration of the bowls, see Ch. 3: 1.2).

The development of the shapes of bowls seems to reflect changing fashions in tableware. These changes may serve as useful references for dating (Rotroff, 1997: pp. 161-162; Edwards, 1975: pp. 30-31). As the complete *echinos* bowls of New Halos were probably in use around 260 BC, they should have certain features in common, assuming they were all made at the same production centre. The bowls are 4.4 to 7.6 cm high and have curved (P351, P354, P358) or almost straight (P348–P350, P352–353, P356–357) walls and rim diameters of 10 to 12.5 cm. P349 is the only bowl with a slightly angular profile. The widths of the bodies of many of the bowls are a little over twice to almost 2.8 times the vessels' heights (table 3.3). P358 has the highest profile: its width is 1.8 its height. At the other end of the spectrum is bowl P396, whose width is 3 times its height (see *Large echinos bowls*). In the 3rd century the ratio of Athenian *echinos* bowls was 1:3.1, in the second half and up to the end of the 2nd century it was 1:2.6 to 1:2.4 (Rotroff, 1997: p. 162). As the many *echinos* bowls found at New Halos cannot be a century ahead of their time, they are most probably not Attic.

Table 3.4 shows the ratio of the diameters of the bases and bodies of the complete *echinos* bowls. The bodies of most of the bowls are between 2 and 2.5 times wider than the bases. P351 has a very wide base and a body that is almost 1.7 times wider. P357 represents a type with a very small base, just over one third of the body. It moreover has a very elongated profile.

Most common are bowls with thick walls (4–6 mm). Two *echinos* bowls with almost eggshell-thin walls (3 mm) were found in house 3 (P356 and P357).

The shapes of the bowls do not seem to reflect a single stage in the development of this type. As the bowls were most probably all in use at the same time, it is likely that they derive from different (local?) sources, as is indeed also suggested by the similar proportions of an *echinos* bowl of local ware (P403; see also under *Large unglazed bowls*).

Bolsals (P371–387; figs 3.8 and 6.5): Three Attic *bolsals*, drinking bowls with two horizontal strap handles, could be completely reconstructed, although large parts of the handles are missing. Other finds of this

type of bowl are ring base fragments with diameters of 6.5 to 11.5 cm. A common feature is a wide base ring with a rounded, bevelled or slightly upturned edge and a groove in the resting surface. The transition from the base to the body is concave. The lower part of the body is slightly convex, the upper part and rim are straight. P371 is a rather squat, wide *bolsal* with a height of 6.6 cm and a rim diameter of 16 cm. The transition from the base to the body is marked by a groove. The upper part of the body is decorated on the outside with ivy leaves painted in a buff slip, in the West Slope style while the inside of the base is decorated with rouletting. A parallel from the Athenian Agora dates from the second half of the 4th century BC (Rotroff, 1997: No. 169 (P 443)). With its height of 7.8 cm, P372 (fig. 3.8) is a little taller than P371, but they both have the same rim diameter. The transition from the lower to the upper part of the body is angular. This *bolsal* is not decorated. P374 consists of non-fitting rim and base fragments. The *bolsal* from which these fragments derive was smaller than the other two, with a rim diameter of about 11–12 cm and a base diameter of 8.7 cm. The base ring has a bevelled edge and the transition from the base to the body is marked by a groove. The handles are upturned, arching above the rim. The vessel has four nine-petalled palmette stamps and rouletting on the inside of the base.

One bolsal base fragment (P376) shows four carefully positioned impressed thirteen-petalled palmette stamps and rouletting. A parallel of the New Halos bolsals P381 and P385 dated to the last quarter of the 4th century was found on the Athenian Agora (Rotroff, 1997: No. 168 (P 29119)). In Athens this type continued to be made until the end of the 4th century BC (Sparkes & Talcott, 1970: p. 108; Rotroff, 1997: p. 97).

Table 3.3. *Echinos* bowl, ratio between the diameter of the body and the height.

Cat. No.	Height	D. body	D. body/height
P358	7.6	13.6	1.79
P357	5.5	10.8	1.96
P401	7.3	15.6	2.14
P351	5.1	11.3	2.22
P400	7.5	17.4	2.32
P353	5.1	12.0	2.35
P349	4.5	11.0	2.44
P354	4.9	12.1	2.47
P350	4.8	12.0	2.50
P355	4.7	12.0	2.55
P356	4.9	13.0	2.65
P352	4.4	12.0	2.73
P348	4.6	12.6	2.74
P403	5.0	13.8	2.76
P396	5.3	16.2	3.06

Table 3.4. *Echinos* bowl, ratio between diameters of base and body.

Cat. No.	D. body	D. base	D. base/d. body
P351	11.3	6.8	1.66
P352	12.0	6.1	1.97
P403	13.8	6.6	2.09
P353	12.0	5.7	2.11
P396	16.2	7.5	2.16
P350	12.0	5.4	2.22
P355	12.0	5.4	2.22
P354	12.1	5.3	2.28
P349	11.0	4.7	2.34
P401	15.6	6.6	2.36
P358	13.6	5.6	2.43
P400	17.4	7.0	2.49
P356	13.0	4.9	2.65
P357	10.8	3.8	2.84

Table 3.5. Small bowl, ratio between height and diameter of the body.

Cat. No.	Height	D. body	Height/d. body
P390	3.6	6.8	1.89
P391	3.5	7.0	2.00
P393	3.5	7.2	2.06
P388	3.7	8.2	2.22
P389	3.1	7.0	2.26
P392	3.4	8.8	2.59
P394	2.8	9.2	3.29

Table 3.6. Small bowl, ratio between diameter of body and base.

Cat. No.	D. body	D. base	D. base/d. body
P389	7.0	4.0	1.75
P388	8.2	4.5	1.82
P392	8.8	4.7	1.87
P391	7.0	3.5	2.00
P394	9.2	4.5	2.04
P390	6.8	3.2	2.13
P393	7.2	3.3	2.18

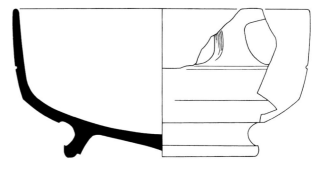

Fig. 3.8. *Bolsal* P372 (scale 1:2).

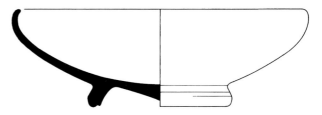

Fig. 3.9. Large *echinos* bowl P396 (scale 1:2).

Small bowls or '*saltcellars*' (P388–395; fig. 6.5): This type of bowl is generally referred to as a 'saltcellar' on account of its small size. It was probably indeed used for serving salt or other condiments. Eight more or less complete specimens were found at New Halos. Seven are shaped like small *echinos* bowls, with body heights of 3.1 to 3.7 cm and rim diameters of about 6–6.6 cm (P392 is slightly larger, with a diameter of 8 cm). They have relatively wider bases than full-size *echinos* bowls (table 3.5). P394 has a squatter, slightly angular profile, the width of its body being more than three times wider its height (table 3.6). Probably this bowl is from Boiotia (Vatin *et al.* 1976: No. 115*bis*4).

Three of the eight more or less complete small bowls were found in house 6, two in house 4 and one each in houses 2, 3 and 5. Rim fragments were found in houses 4, 5 and 6.

Large echinos bowls (P396-399; figs 3.9 and 6.5): The finds of this type of bowl are mostly fragmentary. The size of several ring bases and rim fragments of *echinos* bowls suggest that they had diameters of 20 to 25 cm. Two bowls could be partly reconstructed. P397 is a base fragment of an Attic bowl found in room 3 of house 1. The ring base broke off at some stage, after which the base was repaired with strips of lead, attached through drilled holes with the aid of a lead nail. In spite of the fact that they cannot be fitted to the base fragment, parts of a turned-in rim that were found in room 1 most probably belonged to this bowl, as they are of the same ware and one of the fragments contains a small hole pierced through it for repair purposes. The base is decorated with stamps

and rouletting. It must have had about ten palmette stamps linked by interlocking arcs originally, considering the positions of the two stamps and arcs on the preserved base fragment. The decoration, dated before *c.* 300 BC (see Ch. 3: 1.2 *Wares*), and the repair work suggest that this bowl was used and re-used for some time.

Another large bowl (P396) with a turned-in rim and a ring base with a grooved resting surface was found in house 3 (fig. 3.9). This fine-quality Attic bowl is not decorated. Its proportions are similar to those of 3rd-century Attic bowls (see table 3.3). Two large Attic bowls are represented by stamped ring base fragments. P398 is a simple ring base with a straight edge and a flat resting surface which is decorated with four seven-petalled palmettes with interlocking arcs. The bottom side of the bowl also shows a stacking mark. P399 has a slightly rounded edge and a groove in the resting surface and is decorated with a large stamp resembling a fir tree and rouletting.

Large unglazed bowls (P400–403; figs 3.10 and 6.5): Three bowls of local origin were found in room 5 of house 6. They are unglazed and have a concave raised base with a rounded edge, a splayed body and a turned-in rim. P400 (fig. 3.10) and P401 are large vessels with heights of about 7.5 cm and an almost hemispherical profile. They are made of New Halos ware. A base fragment of this type of bowl (P402) was found in house 5.

P403 is squatter than the other two, resembling an average *echinos* bowl in shape and size. The bowl is about 5 cm high and is made of medium-coarse orange ware. The height and size of the base relative to

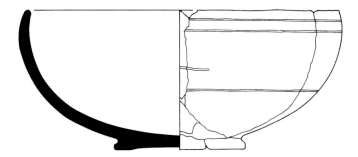

Fig. 3.10. Bowl P400 (scale 1:2).

the diameter of the body are also reminiscent of the average *echinos* bowl (table 3.4).

Bowls with everted rims (P404–405; fig. 6.5): Two complete specimens were found of this type of bowl, which was not suitable for drinking from, but was used for serving dry foods like nuts or olives or cooked food. The vessels differ in size and quality. P404 is a small bowl with a rim diameter of 9.5 cm. The base ring has a rounded edge. The lower and upper parts of the body are straight and are separated by a distinctive angle. The bowl has an everted rim. The base is rather narrow, with a diameter of about 4.2 cm. P405 is a large, fine-quality Attic bowl with a rim diameter of 18.6 cm. The base ring is almost straight and has a groove in the resting surface. The lower part of the body is straight, the upper part is slightly concave. The inside surface is decorated with four palmette stamps set within a carefully made band of rouletting defined by a line drawn with the aid of a ruler. Another bowl was apparently stacked on top of this bowl while it was fired, as the centre of its base is red (evidence of a reducing atmosphere) on the inside. The shapes of the base ring and the body, the grooved resting surface and the decorative pattern are typical of Attic bowls from around 300 BC or a little later (Rotroff, 1997: p. 157).

Hemispherical bowl (P406; fig. 6.5): One Attic hemispherical drinking bowl was found in room 3 of house 1. The bowl is made of very fine glazed ware and has a thin wall and three feet shaped like *Cerastoderma glaucum* valves. Beneath the rim are two grooves. Large parts of the bodies of Attic hemispherical bowls are usually covered with painted West Slope decorations. The specimen from New Halos shows no traces of painting, possibly because the exterior is badly worn. The glaze on the wall of part of the bowl which was covered by a sherd of another vessel had survived intact, but no traces of painting are observable on this part either. So if the vessel ever had any painted decoration it must have been restricted to the upper part of the body. This type was first produced in Athens shortly before about 275 BC (Rotroff, 1997: p. 108). The specimen found at New Halos, presumably in use at about 265 BC, could therefore be a rather contemporary vessel.

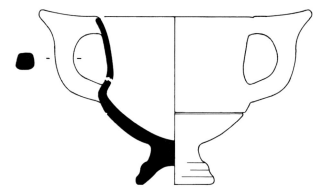

Fig. 3.11. *Kantharos* P415 (scale 1:2).

Kantharoi (P407–434; figs 3.11 and 6.5): The *kantharoi*, wine-drinking cups, which were found in all of the houses, constitute a varied group in terms of wares and design, but they all seem to have had more or less the same, common classical Attic shape, with a plain body, a slightly everted plain rim, a convex lip and moulded spur handles. Only a few ribbed wall fragments were found (P432–434). The body diameters of these small cups range from 7.2 to 8.5 cm. The diameter:height ratios of most of the complete cups range from 0.83 to 0.93 (table 3.7), which, if they were made in Athens, would date them in the late 4th century BC (Rotroff, 1997: p. 84). The *kantharoi* of types 1 and 3, and possibly also those of type 2, are made of the same ware as other Attic vessels,[8] and were therefore probably produced in Attica. Attic cup P424 has a diameter:height ratio of 0.76, dating it to the first quarter of the 3rd century BC. One cup of a distinctive fine orange ware of unknown provenance (P423) is quite a bit squatter and has a diameter:height ratio of exactly 1. Although some, at least, of the cups seem to be of Attic ware, it is difficult to assign them to a specific Attic workshop on the basis of their shape as their shapes vary considerably, especially the shapes of the bases; the majority of the *kantharoi* (fragments) have high ring bases (types 2, 3, 4, 5, 7, 9, 10 and 13), a few have high stemmed bases (types 1 and 11). Two *kantharoi* have

Table 3.7. *Kantharos*, ratio between height and widest diameter of the body.

Cat. No.	Height	D. body	H/d	D/h
Attic				
P424	9.6	7.3	1.32	0.76
P421	9.6	8.2	1.11	0.90
P415	8.8	8.5	1.07	0.93
Other				
P416	10.2	8.5	1.20	0.83
P423	7.8	7.8	1.00	1.00

Table 3.8. Fish plate and saucer, ratio between diameters of rim and base.

Cat. No.	D. rim	D. base	Ratio
Attic			
P443	17	8.4	2.02
P442	19	10.3	1.84
P444	25	14	1.79
Other			
P452	16	7.4	2.16
P447	18	5.2	3.46
P449	19.2	6.7	2.87
P448	21	6.5	3.23
P451	22	6	3.67
Saucer			
P461	17	5	3.4

low stemmed bases (P415 and P424) and P426 has a low ring base. The ring bases and most of the stemmed bases are not, or virtually not, profiled, except stemmed bases P412 and P429, and P430, which has a profiled low ring base. The bases are also narrower than those of the contemporary *kantharoi* that were found on the Athenian Agora.

The bases of two *kantharoi* dated to 290–275 BC that were found on the Athenian Agora (Rotroff, 1997: Nos. 109 (P 7765) and No. 110 (P 6948)) are exactly the same as those of P429 and cup-*kantharoi* P435 and P436, while a parallel of base fragment P412 has been dated to 300–280 BC.[9] A parallel of base fragment P429, dated to 285–275 BC, is also known (Rotroff, 1997: No. 31, P 6311). These close parallels indicate that some, at least, of the *kantharoi* of New Halos were contemporary vessels.

Cup-kantharoi (P435–437; figs 3.12 and 6.5): The cup-*kantharoi* are on the whole a little larger than the *kantharoi* proper. The shape of one Attic cup-*kantharos* (P435; fig. 3.12) with a rim diameter of 11.3 cm which originally had high-swung handles matches that of a group of painted cup-*kantharoi* from the Athenian Agora dated between 300 and 260.[10] As already mentioned above, the base can be dated to about 290–275 BC, like that of P436. The profile could not be reconstructed in its entirety, making it difficult to give the precise shape, but it is likely that the cup-*kantharos* is a contemporary vessel. The rim was probably originally decorated with a simple painted garland of ivy leaves in the West Slope style; only one leaf was still visible, the rest of the decoration having faded. Grooves in the resting surface and the lower part of the body bear traces of a limy red *miltos*. P437 may also have been part of an Attic cup-*kantharos*, con-

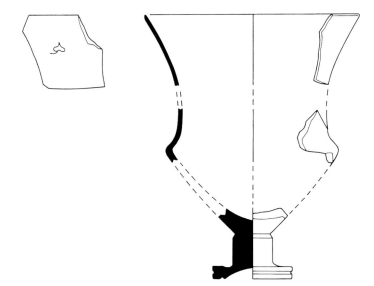

Fig. 3.12. Cup-*Kantharos* P435 (scale 1:2).

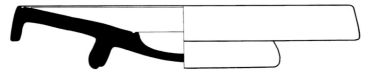

Fig. 3.13. Fish plate P442 (scale 1:2).

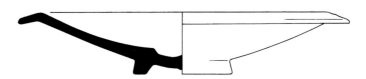

Fig. 3.14. Fish plate P447 (scale 1:2).

sidering its rim diameter of 10.7 cm and the sharply offset upper body.

Bowl-kantharoi (P438–439): Only two handle fragments of this type of *kantharos*, shaped like a wide, almost hemispherical bowl on a high stem, were found. They both came to light in room 6 of house 5 and probably belonged to the same vessel. The thumb rests shaped like an ivy leaf are attached quite close to the lip; they are characteristic of the Hellenistic type of bowl-*kantharoi*. Comparable vessels found on the Athenian Agora were presumably made for a short period of time, from about 290 to 275 BC (Rotroff, 1997: p. 93). This timespan seems a little too narrow, as one comparable, complete specimen of such a vessel found in the Sepulchral Building in the upper town of New Halos was most probably in use from around 306 until 297 BC (Reinders, 1988: p. 146, No. 54.14). Another possibility is that the dates of the Sepulchral Building should be revised.

Skyphoi (P440–441): Very few remains of *skyphoi*, or wine-drinking cups, were found at New Halos. They all derive from vessels of the Attic type. The fragments in question are a rather wide, presumably 4th-century[11] base fragment (P440, house 6, room 3) with a straight edge and a diameter of 8 cm and some rim fragments, the best-preserved of which is P441. The small number of – small – fragments suggests that this type of drinking cup was not very popular at Halos. This is in accordance with the small numbers of *skyphoi* found in Menon's Cistern and at Koróni, suggesting that the type had gone out of general use by the 260s (Rotroff, 1997: p. 94).

Fish plates (P442–460; figs 3.13, 3.14 and 6.5): Various types of plates with a circular central depression for sauces or collecting juices were found at New Halos. A parallel of P452, which has a rim diameter of 16 cm and a down-turned rim, is known from Labraunda. The latter vessel was dated to the second half of the 4th century BC.[12] The base is narrow and the depression fairly deep (table 3.9). The badly worn, fragmentary plate was found in the floor of room 1

of house 3. This findspot also suggests that the plate is old and was probably used by the town's earliest inhabitants. The three complete Attic fish plates (P442–444; fig. 3.13) are fairly flat and have deeply grooved resting surfaces. They have rim diameters of about 17 to 25 cm and broad bases (rim:base ratios: 2.02, 1.84 and 1.79; see table 3.8). These specimens were found in houses 1, 4 and 6. The other houses yielded only small fragments of fish plates, except house 5, where this type was absent. Comparable plates found on the Athenian Agora have been dated to about 310–290 BC (Rotroff, 1997: No. 713 (P 23538) and No. 715 (P 20318)).

The other fish plates (P447–460; fig. 6.5) are made of different wares. They have base rings with splayed or straight edges and the shape of the central depression varies from very shallow to about 0.5 cm deep (table 3.9), and from rounded to angular in profile. Common features of the more or less complete plates are the thin spreading walls and the slightly offset rims. One exception is P453 which may be from Boiotia (Heimberg, 1982: No. 336). Though very fragmentary it shows an angular profile. The rim diameters vary from 14 to 22 cm. The bases are quite

Table 3.9. Fish plate, ratio between diameters of floor and depression.

Cat. No.	D. floor	Depression	Ratio
Attic			
P443	16	4.4	3.64
P442	18	4.8	3.75
P445	20	5.2	3.85
Other			
P452	16	5.2	3.08
P447	15.6	3.0	5.2
P449	16.8	2.4	7.0
P448	18	2.4	7.5
P451	18.8	2.6	7.23

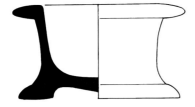

Fig. 3.15. Spool saltcellar P462 (scale 1:2).   Fig. 3.16. Oil lamp P469 (scale 1:2).

narrow, too, in comparison with the rim diameters (table 3.8). A parallel of P449, but with a slightly smaller rim diameter, was found in the Sepulchral Building (Reinders, 1988: No. 52.04). A few of the fish plates are indisputably Attic, but we have not yet been able to identify the production centres of the other specimens.

Saucer: P461 (fig. 6.5) is the only complete saucer found at New Halos. It has a base ring with a straight edge, a splayed wall and a slightly offset rim. The rim has a diameter of 17 cm, the relatively narrow base ring is 5 cm wide (table 3.8). The saucer is made of the same ware as fish plates P447 and P448.

Spool saltcellar (P462; figs 3.15 and 6.5): This Attic vessel was probably intended for serving salt. It has a wide base with a rounded edge, a cylindrical body and a wide everted rim with a convex top. Vessels of this type found on the Athenian Agora have been dated to *c.* 325–295 BC (Rotroff, 1997: No.1070 (P 20908)).

Miscellaneous ceramics (P463–520)

a) *Oil lamps* (P463–479; figs 3.16 and 6.6): The oil lamps found at New Halos are all wheel-thrown and totally glazed. Three related types can be distinguished. The first, represented by P463, P464, P467 and P468, match Howland type 30B, which was made in Athens from about 325 to 275 BC (Rotroff, 1997: p. 500). It has a slightly concave base and a convex body with a narrow turned-in rim and a simple convex lip. The lamps have wide apertures and the short spouts extend smoothly from the body. Most of the other lamps seem to be of a type representing a variation on type Howland 30B with a broader, more overhanging rim (P471, P473 and P474), occasionally with a flattened top (P469, fig. 3.16; P470). One rim fragment of an Attic lamp (P476) shows two incised parallel lines running around the lower part of the body. This lamp, of Howland type 30A, was dated to 410–370 BC[13] by Howland and Scheibler.

P475 is a lamp of the same type as the Attic lamps, but made of medium-coarse orange ware. It may have been made locally, in imitation of the Attic lamps. A few small rim fragments show different profiles of

unknown types (P477–479), that may be of a fairly early date; probably they have been incorporated in mudbricks.

Various types of lidded boxes were used for storing cosmetic powder or small precious objects like ornaments. Five such types were represented at Hellenistic Halos:

Lekanides (P480–490; fig. 6.6): A *lekanis* is a bowl-like *pyxis* with a ring base, a splayed wall, a slightly incurved rim and a flange for a lid on the outside. One complete specimen (P488) from house 3 has a rather squat profile, with a height of 4.7 cm and a rim diameter of 12 cm. Rim fragments of this type with diameters ranging from 13 to 20 cm were found in houses 1, 4, 5 and 6 (P480–486). One of these fragments (P486) was decorated with a wavy line painted in slip, of which only a shadow now remains. One fragment (P489) has a fairly straight upper wall. *Lekanides* with straight upper walls found on the Athenian Agora (Rotroff, 1997: Nos. 1254–1255) have been dated to the first half of the 3rd century BC. A contemporary parallel of Boiotian origin is known from a tomb in Medeon is Phokis (Vatin *et al.*, 1976: No. 63.7). Two small rim fragments, P483 and P487, have small lugs or attachments of lugs or handles just beneath the flange.

Pyxides (P491–494; figs 3.17, 3.18 and 6.6): An unusual, large, baggy *pyxis*, or perhaps a *lekanis* (P493), with a height of 9 cm and a rim diameter of 8.2 cm, has a globular body with a slightly incurved rim and two horizontal ring handles recurved to the body (fig. 3.17). This vessel, which was found in the same room in house 3 as *lekanis* P488, is decorated with grooves filled with a red limy paint or *miltos* on the widest part of the body and at the transition from the ring base to the body. No parallels of this type are known, but the ware is very fine and hard like that of hemispherical bowl P406 and cup-*kantharos* P435, and the decoration is also similar to that on the latter vessel, suggesting that this *pyxis* came from Attica, although no parallels of this type are known from Athens.

A globular *stamnoid pyxis* (P493) of a type with two horizontal ring handles arching above the lip and a high ring base (missing in the case of this vessel)

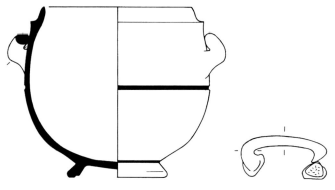

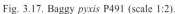

Fig. 3.17. Baggy *pyxis* P491 (scale 1:2).

Fig. 3.18. *Pyxis* P492 (scale 1:2).

was found in room 5 of house 5. It is made of medium-coarse red ware. A *pyxis* of the same type, also made of medium-coarse red ware, was found in a grave in the nearby Hellenistic cemetery. This vessel is now in the Almirós Archaeological Museum (BE 9244).

P492 is a badly worn, 5-cm-high Attic *pyxis* with a rim diameter of 4.5 cm. The transition from the base to the body is marked by a wide moulded ridge (fig. 3.18). The wall is upright and the rim has a flange for a lid. A parallel found on the Athenian Agora has been dated to the 4th century BC (Sparkes & Talcott, 1970: No. 1312 (P 7396)). P494, another small, fragmentary *pyxis*, has a flat base and an upright wall. Its surviving height is 4 cm. The shape is similar to that of an Attic Type D *pyxis* from the Athenian Agora, which has hesitantly been dated to the mid-3rd century BC.[14] Rotroff (1997: p. 191) suggests that *pyxides* of this size were too small for storing ornaments and were probably used for ointments or cosmetics.

Glazed pyxis lids (P495–502; fig. 6.6): Fragments of glazed lids may have originally formed part of the lids of *pyxides*. Four different types of lids can be distinguished on the basis of differences in the rims. Simple domed lids with straight edges and rim diameters of 8.5 to 23 cm were found in houses 1, 3 and 4 (P495–498). One of these lids (P498) shows a decoration of dark painted ovules at its edge, two facing heads on the dome and a floral pattern in between in red-figure. A parallel from Thasos has been dated to about 350 BC (Blondé, 1985: p. 283, fig. 4, No. 34), but recently obtained evidence suggests that red-figure vessels continued to be produced in Athens until the early Hellenistic age (Rotroff, 1997: p. 3) Other fragments of lids of this type show no traces of decoration.

A lid with a wide ridge at the edge of the rim was found in house 1 (P502). A rim fragment of this type was also found in the Sepulchral Building (Reinders, 1988: No. 54.20). House 5 contained two different types of *pyxis* lids: a domed lid with a turned-in rim with a groove at the edge and a diameter of 22 cm

(P499, P500) and a small domed lid with a thickened lip and a rim diameter of about 8 cm (P501).

All of the excavated houses contained one or more small jars which originally contained perfumed oil for personal hygiene or medicine. They are of various types, namely *unguentaria, aryballoi, guttoi, amphoriskoi* and squat *lekythoi.*

Unguentaria (P503–510; figs 3.19 and 6.6): Small jars, mostly of grey ware, were used for storing perfumed oil, but also *silphium* (resin), honey, vinegar and *garum* (fish sauce) and possibly also medicine (Hellström, 1971: p. 24). P503, a complete *unguentarium* of grey ware, has an almost bulbous body and a hollow foot. Base fragment P505 derives from a slightly larger jar of the same bulbous type. Base fragment P504 (and possibly also wall fragment P506) seems to derive from a somewhat slenderer type.

Part of a large grey *unguentarium* or *amphoriskos* (P507; fig. 3.19) with a hollow foot and two vertical ring handles on the shoulder was found low down in a pit in the courtyard of house 3 (Haagsma, 1997). A parallel of this large handled *unguentarium* is known from a cemetery in nearby Phthiotic Thebai. This vessel, dated to the early 3rd century BC, is now in the Archaeological Museum of Almirós (K 2951 VII). Another parallel, dated shortly after 250 BC, is known from Olympia.[15]

The bulbous type can be dated to the late 4th century BC and is thought to represent the earliest stage in the development of the *unguentarium*, which was later to become shallower and more elongated (Hellström, 1971: pp. 23–27). Rotroff (1997: p. 178) observed that the lower part of the body became concave after 275 BC. The first signs of this development are observable in the grey *unguentaria* found at New Halos, but not yet in the only black-glazed specimen P508. The latter more or less complete *unguentarium* (only the rim is missing), has a small base, an ovoid body and a solid foot.

Aryballoi (P511–513; fig. 6.6): Three small luxurious perfumed oil jars, all with missing necks and rims, were found at Hellenistic Halos. The preserved

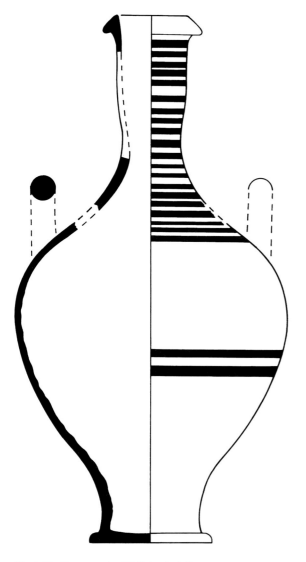

Fig. 3.19. *Unguentarium* P507 (scale 1:2).

far as both the shape and the decoration are concerned, close parallels dated to the last quarter of the 4th century are known from Corinth.[15] The peculiarity of the ware of this vessel has been discussed in Ch. 3: 1.2.

The third *aryballos* (P513) is fairly small, with a preserved body height of about 3 cm and a diameter of 5 cm. It has a raised flat base, a wide, slightly squat globular body and a narrow neck. The temper is very fine and it is coated with a black glaze.

Guttoi (P514–516; fig. 6.6): *Guttoi* were also containers for fine perfumed oil. Three fragmentary vessels of this type were found at New Halos. One (P514), from house 3, is probably Attic, and has a splayed base ring and a groove in the resting surface, an attachment of a round vertical strap handle, a narrow neck and a wide overhanging rim. A reference from the Athenian Agora (Rotroff, 1997: No. 1142) was dated *c.* 285 BC, which gives a contemporary date. The ware of a rim fragment of a similar type (P515) found in house 2 is a little finer.

Base and body fragment P516, probably from a *guttus* or an *askos*, has a splayed base ring edge and a squat rounded body with a slight angle at the level of the widest diameter. The vessel is made of soft, powdery, medium-fine orange ware with a lime temper and was probably not glazed. Traces of a reddish slip were visible on the outside of the base.

Amphoriskos (P517): A pointed base fragment of an *amphoriskos*, another perfume jar, of New Halos ware was made from the same clay as mould F9 found in house 1 (see section Ch. 3: 2 and Appendix 3). This vessel was therefore possibly made locally.

Squat lekythos (P518; fig. 6.6): A fragmentary black-glazed squat *lekythos* with an almost flat base with a profiled edge and a globular body with a diameter of 4.4 cm was found in house 6. It was probably made in Athens and can be dated to the 4th century BC.[17]

The last type of miscellaneous pottery is probably associated with children, which, as far as the pottery is concerned, are almost invisible:

Miniature bowls (P519–520; fig. 6.6): Two very small bowl-like vessels of fine black-glazed ware were found in two separate rooms in house 4. The domestic context suggests that they were toys.

1.4. Discussion

The large number of different pottery types and their distribution among the six houses indicate that cooking, storage, serving, eating, drinking and personal care took place in all of the houses. The most common shapes are large wine-storage *amphorai*, *chytrai*, *echinos* bowls, *kantharoi*, *lekanides*, jugs, *lopades*, fish plates and small domed lids, which were found in large quantities in all of the houses. Slightly less common, represented in five houses, are shapes like small bowls or saltcellars, large *echinos* bowls, *kra-*

height of the bodies of the two larger *aryballoi* is about 8 cm. They are both decorated on the shoulder. An Athenian black-glazed *aryballos* (P511) shows incised curls on the shoulder – remnants of a West Slope decorative motif, probably originally comprising ivy leaves or grapes. The vessel has a base ring with a rounded edge and an almost globular body with a slight angle at the transition to the shoulder. A double strap handle is attached to the shoulder and the neck.

House 6 yielded another *aryballos* (P512) of Corinthian Blister ware, so far the only object of this ware to have been found at New Halos. The base is flat, the body is slightly splayed with a slight angle at the transition to the shoulder. A strap handle is attached to the shoulder. The slightly convex shoulder is decorated with two bands of crescent moons or 'lunate design', with an ivy garland in between. As

teres, jars, oil lamps, *pithoi* and *lekanides*. *Bolsals*, *hydriai*, large lids, large *lekythoi*, stewpots, strainers, tubs, *unguentaria* and glazed *pyxis* or *lekanis* lids were found in four of the houses. Rarer are *guttoi*, *lebetes*, lids of *pithoi* or tubs and water pitchers, which were found in small numbers in three of the houses. Even rarer, represented in only two of the houses, are *aryballoi*, bowls with everted rims, cup *kantharoi*, *lagynoi*, *olpai*, squat *lekythoi* and unglazed bowls. A few miniature bowls and *skyphoi* were found, but all in the same house. Unique vessels are a small serving *amphora*, an *amphoriskos*, a brazier, a hemispherical bowl, a painted *hydria*, a glazed squat *lekythos*, a *louterion*, a miniature jug, a mortar, a *pelike*, four different types of *pyxides* and a spool saltcellar.

The cooking vessels are all simply shaped, showing few variations. Most are of medium size (up to 20 cm high or wide), some are small (about 10 cm high); large cooking vessels are rare. The almost complete absence of braziers or stoves may imply that the food was cooked over open fires. Another possibility is that cooking stands were removed from the houses before or after the town's destruction. The pouring vessels show a wide range of shapes and functions. Most of the vessels are about 20 cm high, the *hydriai* being about twice as high as the other vessels. The smallest jug-like vessel is a baggy *amphora*. The open shapes also show a wide variety and are somewhat larger than the jugs and cooking vessels. The storage *amphorai* all have a height of either about 80 cm or about 40 cm, but their volumes vary considerably, even those of vessels of the same type. The jars show about the same range of sizes as the *chytrai* and stewpots. The tableware includes a high proportion of Attic shapes, besides some types from a local, as yet unidentified source (Thessaly, Boiotia or Euboia?). Some local versions of Attic shapes are recognisable. The majority of the bowls and fish plates are non-Attic. Some bowls that were definitely produced locally are of an almost hemispherical shape. The majority of the cups and large glazed bowls are Attic. So are most of the miscellaneous vessels, but one *aryballos* is of Corinthian origin. The stamnoid *pyxis* and the *amphoriskos* are local vessels.

The pottery seems to represent varied early Hellenistic household inventories. The number of (more or less) complete vessels of each type, supposedly representing the inventories at the time of the town's destruction, ranges from one to at most five. The range of house 2 is definitely distorted as a result of disturbances, but the types which were found in only small numbers in this house were also represented in smaller proportions in the other houses. It is possible that refugees removed vessels from the houses just before or after the disaster, as has been ascertained in the case of metal objects (Ch. 3: 4). Large *pithoi* were for example probably removed after house 3 had collapsed. Other large vessels may very well have survived the disaster and likewise have been removed afterwards. This could account for the almost complete absence of vessels like braziers or cooking stands.

As far as the distribution of the imported fine wares is concerned, house 5 yielded the most varied range, in terms of both numbers and types, followed by houses 3 and 6. Houses 4 and 5 contained considerably larger numbers of stamped or rouletted bowls and fragments of such vessels than the other houses; houses 2 and 3 yielded no stamped bowls whatsoever. A few rare, worn vessels with West Slope decoration were found in houses 1 and 5. The latter house also contained the widest range of *amphora* shapes. This difference could reflect a difference in wealth, but it is more likely that it is attributable to the different state of preservation of the house remains.

Three dates of the pottery of New Halos are relevant within the scope of this study: the date of manufacture, the date when the vessel was purchased by the customer and the date when it was discarded or deposited. The last date of the complete vessels is about 265 BC. Part of the pottery, especially the fine ware, was manufactured in Athens and Corinth and traded to, or brought by the inhabitants of New Halos. These vessels are important references as the types in question have been fairly accurately dated in both absolute and relative terms on the basis of evidence obtained in Athens and Corinth and surrounding areas. Their dates provide information on the trade in pottery to Hellenistic Halos and on the timespan at which the pottery was in use.

Dated parallels of no more than 34 of the 520 chosen vessels of New Halos are known from Athens (Agora), Corinth and other sites (table 3.10). The early types included among the pottery, dated to the second half of the 4th century BC or even earlier, such as oil lamps, *louterions*, *lekythoi*, *skyphoi*, *pyxides* and *pyxis* lids, probably represent highly durable vessels. Vessels of similarly early dates were also encountered among the pottery found at Koróni (Vanderpool *et al.*, 1962: p. 59). Most of the datable pottery is however from the last quarter of the 4th century BC. Only a few types can be dated to the first quarter of the 3rd century BC, such as Attic fish plates, *kantharoi* and hemispherical bowls, which date from no later than 275 BC (table 3.11). Vessels of new well-dated types introduced in Athens around 260 BC such as the large baggy Hellenistic *kantharos* and vessels of well-dated later stages in the development of particular types such as deeper fish plates and fusiform *unguentaria* are not represented in the pottery of New Halos.

The varying provenances, among which are Athens and Corinth, indicate that no single pottery centre ruled the market in Hellenistic Halos. The determination of the provenance of the wares, especially the medium-coarse light ware (*amphora* ware) and the fine glazed wares, will require closer study.

Table 3.10. Dated parallels to pottery from New Halos.

Type	Cat. No.	Publication	Site	Date (BC)
Provenance: Athens				
1. oil lamp	P476	Scheibler, 1976 (type 30A)	Agora	410–370 (HT-date)
		Howland, 1958: No. 418 (L 1873) (type 30A)		last quarter 5th cent.
2. squat *lekythos*	P518	Sparkes & Talcott, 1970: No. 1137 (P 876); 1138 (P 765); 1139 (P 7606);	Agora	410–300
		1140 (P 14811) and 1141 (P10436)		325–300
3. *pyxis*	P492	Sparkes & Talcott, 1970: No. 1312 (P7369) (type D)	Agora	4th cent.
		Jones *et al.*, 1962: No. 37	Dema House, Attica	last quarter 5th cent.
4. *pyxis* lid	P498	Knigge, 1966: No. 4 125 (hS127)	Eridanos-*Nekropole, Grab*	
		Blondé, 1985: No. 34	Thasos	*c.* 350
5. *bolsal*	P371	Rotroff, 1997: No. 169 (P 443)	Agora	350–300
6. *unguentarium*	P508	Edwards, 1975: No. 584 (C-48-119)	Corinth	*c.* 325
7. *lekanis*	P489 (P 25688)	Sparkes & Talcott, 1970: No. 1275	Agora	325–310
		Vatin *et al.*, 1976: No. 63.7	Medeon Phocidis	300–250
8. *bolsal*	P380, P382	Rotroff, 1997: No. 168 (P 29119)	Agora	325–300
9. *unguentarium*	P503	Vatin *et al.*, 1976: No. 75.9-12; 115.4	Medeon Phocidis	*c.* 300
		Schlörb-Vierneisel, 1966: No. 166.1; 170.2; 174.1	Eridanos-*Nekropole*	325–300
10. spool saltcellar	P462	Rotroff, 1997: No. 1070 (P 20908)	Agora	325–295?
11. oil lamp	P463	Rotroff, 1997: (type 30B)	Agora	325–275
	P464	Howland, 1958: No. 419 (L 2890)		350–275
	P467	Howland, 1958: No. 420 (L 3530)		310–250
	P468	Howland, 1958: No. 421 (L 1899)		375–300
		Broneer, 1977: No. 58 (IP 716) (type IVB)	Isthmia	2nd half 4th–1st half 3rd cent.
		Henninger, 1976: No. 184 (DP 71-162, 37)	Demetrias	before end 3rd cent.
		Scheibler, 1976: (type 30B)		360–270 (HT-date)
		Howland, 1958: Nos. 419–422 (type 30B)	Agora	mid 4th–first quarter 3rd cent.
12. oil lamp	P469–471	Henninger, 1976: No. 186 (DP 71-199, 56); No. 187 (DP 71-202, 1)	Demetrias	before end 3rd cent.
13. fish plate	P450	Rotroff, 1997: No. 713 (P 23538)	Agora	310–290
		Sparkes & Talcott, 1970: No. 1073 (P 23538)		*c.* 325
14. fish plate	P451	Rotroff, 1997: No. 715 (P 20318)	Agora	310–290
15. bowl *kantharos*	P438–439	Reinders, 1988: No. 54.14	Halos, Sepulchral Building	*c.* 300
		Rotroff, 1997: No. 141 (P 27970)	Agora	290–275
16. cup *kantharos*	P435–436 (base)	Rotroff, 1997: No. 109 (P 7765)	Agora	290–275
		Reinders, 1988: No. 54.14	Halos, Sepulchral Building	*c.* 300
		Rotroff, 1997:	*Abschnitt* I	300–280
		Braun, 1970: No. 10 (KER 9722)	Kerameikos, Spring of Dypilon	*c.* 300
		Rotroff, 1997:	Menon's Cistern	280–275
		Miller, 1974: No. 15 (P 27970)	Agora, Menon's Cistern	
17. *unguentarium*	P507	Arch. Museum Almiros K 2951 VII	Phthiotic Thebai, cemetery	beginning 3rd cent.
		Hausmann, 1996: No. 58 (K 1110)	Olympia, *Brunnen N. von Bau C*	shortly after *c.* 250
18. *guttus*	P514 (base)	Rotroff, 1997: No. 1142 (P 28028)	Agora	*c.* 285
19. hemisph. bowl	P406	Rotroff, 1997: No. 311 (P 27986)	Agora	285–275
		Smetana-Scherrer, 1982: No. 523	Aegina, city	first half 3rd cent.
20. *pyxis*	P494	Rotroff, 1997: No. 1249 (P 6282)	Agora	250–240?
Provenance: Korinthos				
1. *louterion*	P272	Sparkes & Talcott, 1970: No. 1861 (P 24252)	Agora	5th–4th cent.
		Iozzo, 1987: No. 126 (IPG 1970-69)	Corinth	mid 5th cent.
		Iozzo, 1987: No. 131 (IPG 1970-74a,b)	Corinth	5th or 4th cent.
		Blondé, 1989: No. 252	Thasos, *puits public* (TAQ 330)	2nd–3rd quarter 4th cent.
2. *aryballos*	P512	Pemberton, 1989: No. 388 (C-61-154)	Corinth, sanct. Demeter and Kore	4th quarter 4th cent.
		Pemberton, 1989: No. 387 (C-69-186)		late 3rd–4th quarter
		Edwards, 1975: No. 777	Corinth	last quarter 4th cent.

Table 3.11. Chronology of a selection of the ceramics.

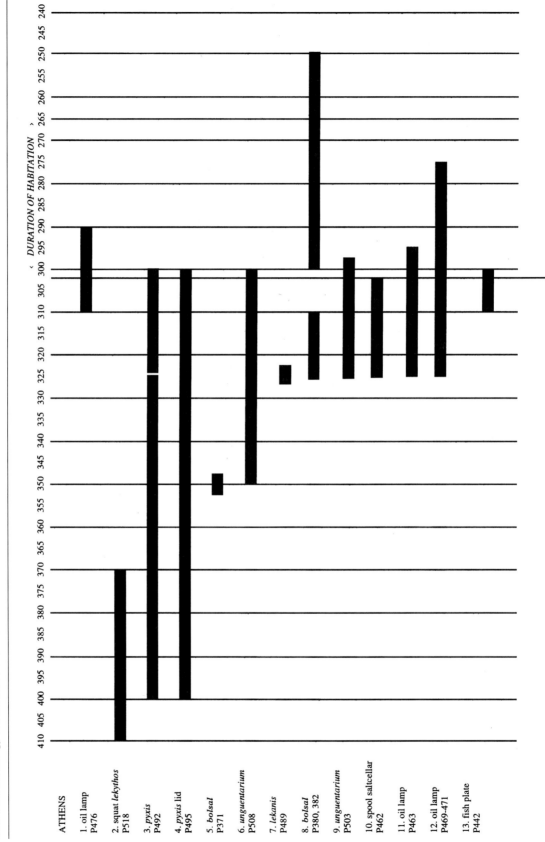

DURATION OF HABITATION

ATHENS

1. oil lamp
P476

2. squat *lekythos*
P518

3. *pyxis*
P492

4. *pyxis* lid
P495

5. *bolsal*
P371

6. *unguentarium*
P508

7. *lekanis*
P489

8. *bolsal*
P380, 382

9. *unguentarium*
P503

10. spool saltcellar
P462

11. oil lamp
P463

12. oil lamp
P469-471

13. fish plate
P442

Table 3.11 (continued).

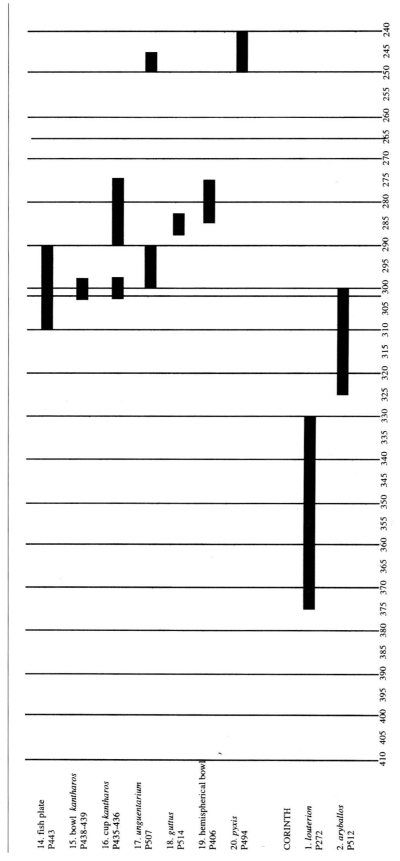

The range of vessels made of the medium-coarse orange and red wares and the New Halos ware suggests that these wares were locally produced. Mineralogical and petrographical analysis of terracottas, moulds and pottery may confirm this. Further investigation of the orange, red and New Halos wares may show whether the composition of the clays of these wares differs or whether the pottery was treated differently, for instance during the firing. Important evidence is provided by four pots showing the same incised floral decoration as observed on vessels of the medium-coarse red and orange wares. Unfortunately, no kilns or clay pits, which would certainly help in identifying local wares, have so far been found.

1.5. Notes

1. For a recent discussion of these reference points, which in the case of Olynthos should probably be changed to 316 BC, see Rotroff, 1997: pp. 18–23 and 31–32.

2. I would like to thank the following people for reconstructing the pottery and preparing the descriptions and drawings of most of the material: G.A. Abbingh, E. Akkerman, A.J. Bakker, G. Boekschoten, M. Bijlsma, Y. Burnier, H. Grosfeld, B.J. Haagsma, M. Haagsma, J.J. Hekman, M. Huisman, S. Hijmans, W. Jansen, J. Lendering, D. van der Meulen, B. Mulder, R. Oudhuis, E. Seiverling, F. Veenman, E. Wetzels, G. Wilmink, Chr. Williamson and B. Yates. Special thanks are due to G.A. Abbingh and M. ten Haaf for designing a database program enabling us to effectively describe this large amount of pottery.

3. The glaze of the pottery picked up during a topographical survey at Gorítsa, near Vólos in Thessaly, was also found to be of poor quality (Vermeulen Windsant, 1973: p. 371). This similarity could imply that the pottery of these two sites comes from the same source, but the poor quality cannot be classified as a typical provincial feature, as pottery of inferior quality was produced elsewhere, too, for instance in Athens (see Rotroff, 1997: p. 11).

4. Rotroff, 1997: p. 3.

5. Istmia: Anderson-Stojanović, 1993: p. 276 No. 21 (IP 883), dated before 275 BC.

6. Two handle fragments found in 1977 during a survey of the city area actually do show stamps. One represents an amphora (not Corinthian), the other is a round stamp depicting a monogram with a combination of the letters A or Λ, and T. Cf. Coja, 1986:, p. 446, No. 184 (Inv. V20869, in a square stamp). The provenance of this particular stamp is not clear, but most of the stamps from this site are from Thasos.

7. The capacities were calculated by dividing the amphora body in half and then into sections with almost straight outside walls. The volumes of these sections, which are in fact halves of cylinders with one diagonal side, were calculated using the following formula:
$V = \pi h (r^2 + r.dr + \frac{1}{3}.dr^2)$. The volumes of the sections were then added up to obtain the capacity of the entire vessel.

8. Hemispherical bowl P406 and cup-*kantharos* P435.

9. Rotroff, 1997: No. 166 (P 7760), actually a base of a bolster cup.

10. Rotroff, 1997: No. 72, (P 7358), dated 300–275 BC; No. 73 (P 28049), dated *c.* 275 BC and No. 77 (P 5811), dated 275–260 BC.

11. Changes in the shape of this type reflect a development towards an increasingly angular profile and a smaller foot. The bases of the specimen from the Athenian Agora and related vessels dated between *c.* 325 and 260 BC are all about 4 to 4.5 cm wide. See Rotroff, 1997: p. 94, Nos. 151–154.

12. Hellström, 1971: No. 56. According to Hellström the poor quality of the ware indicates that this vessel was not produced in Attica.

13. Howland, 1958: p. 97; Scheibler, 1976: p. 191, adhering to the date suggested by Howland in 1958. This type of lamp is not mentioned in Rotroff, 1997.

14. Rotroff, 1997: p. 191, No. 1249. This specimen may also date from the 4th century BC.

15. Hausmann, 1996: pp. 28, No. 58, 32–34; though the neck is somewhat longer. Rotroff, 1984: p. 258.

16. See Pemberton, 1989: No. 388, and Edwards, 1975: No. 777, for decoration; see Pemberton, 1989: No. 387, for the shape of the base and the body, dated to the late third to fourth quarter of the 4th century BC.

17. Sparkes & Talcott, 1970: Nos. 1137–1141, dated as a group to 410–300 BC.

1.6. REFERENCES

ADRIANI, A., 1936. La nécropole de Moustafa Pacha. *Annuaire du Musée Gréco-Romain (Alexandrie)* 1933/34–1934/35.

ADRIANI, A., 1940. Fouilles et découvertes. *Annuaire du Musée Gréco-Romain (Alexandrie)* 1935–1939 (1940), pp. 15–163.

ADRIANI, A., 1952. Nouvelles découvertes dans la nécropole de Hadra. *Annuaire du Musée Gréco-Romain (Alexandrie)* 1940-1950 (1952), pp. 1–27.

ANDERSON, J.K., 1958. Excavations on the Kofinà Ridge, Chios. *Annual of the British School at Athens* 49, pp. 138–181.

ANDERSON-STOJANOVIĆ, V.R., 1987. The chronology and function of ceramic Unguentaria. *American Journal of Archaeology* 91, pp. 105–122.

BLONDÉ, F., 1985. Un remblai Thasien du IVe siècle avant notre ère. A. La céramique. *Bulletin de Correspondance Hellénique* 109, pp. 281–344.

BLONDÉ, F., 1989. Le comblement d'un puit public à Thasos. *Bulletin de Correspondance Hellénique* 113, pp. 467–545.

BRAUN, K., 1970. Der Dipylon-Brunnen B1: Die Funde. *Mitteilungen des Deutschen Archäologischen Instituts, Athenische Abteilung* 85, pp. 129–269.

BRONEER, O. 1977. *Terracotta Lamps* (= Isthmia, Vol. III). American School of Classical Studies at Athens, Princeton, New Jersey.

BRUNEAU, P., 1970. Tombes d'Argos. *Bulletin de Correspondance Hellénique* 94, pp. 437–531.

COJA, M., 1986. Les centres de production amphorique identifiés à Istros Pontique. *Bulletin de Correspondance Hellénique* Suppl. 13.

DOULGÉRI-INTZESILOGLOU, A. & Y. GARLAN, 1990. Vin et Amphores de Péparétos et d'Ikos. *Bulletin de Correspondance Hellénique* 114, pp. 361–389.

EDWARDS, G.R., 1963. Koroni: The Hellenistic pottery. *Hesperia* 32, pp. 109–111.

EDWARDS, G.R., 1975. *Corinthian Hellenistic Pottery* (= Corinth

7: III). The American School of Classical Studies, Princeton, New Jersey.

GRACE, V.R., 1974. Revisions in Early Hellenistic chronology. *Mitteilungen des Deutschen Archäologischen Instituts, Athenische Abteilung* 89, pp. 193–203.

HAAGSMA, M.A., 1997. Baubo in de put. Resultaten van de opgravingen in huis 9 in Halos, 1997. *Paleo-aktueel* 9, pp. 36–39.

HAUSMANN, U. 1996. *Hellenistische Keramik, eine Brunnenfüllung Nördlich von Bau C und Reliefkeramik Verschiedener Fundplätze in Olympia* (= Olympische Forschungen, Band 27). Walter de Gruyter, Berlin/New York.

HEIMBERG, U., 1982. *Die Keramik des Kabirions* (= Das Kabirenheiligtum bei Theben, Band 3). Walter de Gruyter & Co, Berlin.

HELLSTRÖM, P., 1971. *Pottery of Classical and Later Date. Terracotta Lamps and Glass. Labraunda, Swedish Excavations and Researches* (= Skrifter utgivna av svenska institutet i Athen 4, V, II:1). Stockholm.

HENNINGER, F., 1976. Lampen. In: V. Milojčić und D. Theocharis (Hrsg.), *Demetrias I, Die Deutschen Archäologischen Forschungen in Thessalien* (= Beiträge zur Ur- und Frühgeschichtlichen Archäologie des Mittelmeer-Kulturraumes 12). Rudolf Habelt Verlag GmbH, Bonn, pp. 133–136.

HOWLAND, R.H., 1958. *Greek Lamps and their Survivals* (= The Athenian Agora 9). The American School of Classical Studies at Athens, Princeton, New Jersey.

IOZZO, M., 1987. Corinthian Basins on High Stands. *Hesperia* 56, pp. 355–416.

JAGER, J. & D. VAN DER MEULEN, 1989. *Halos-Thessalië-Griekenland.* Unpublished paper, State University of Groningen (Netherlands).

JONES, J.E., L.H. SACKETT & A.J. GRAHAM, 1962. The Dema house in Attica. *Annual of the British School at Athens* 57, pp. 75–114.

KNIGGE, U., 1966. Eridanos Necropole: II. Gräber hS 205–230: 3. Die hellenistischen Gräber. *Mitteilungen des Deutschen Archäologischen Instituts, Athenische Abteilung* 81, pp. 122–134.

METZGER, I.R., 1969. *Die Hellenistische Keramik in Eretria* (= Eretria, Ausgrabungen und Forschungen 20). Francke Verlag, Bern.

MILLER, S.G., 1974. Menon's Cistern. *Hesperia* 43, pp. 194–239.

PEMBERTON, E.G., 1989. *The Sanctuary of Demeter and Kore; the Greek Pottery* (= Corinth 17: 1). American School of Classical Studies at Athens, Princeton, New Jersey.

REINDERS, H.R., 1988. *New Halos, a Hellenistic Town in Thessalía, Greece.* Hes Publishers, Utrecht.

REINDERS, R. & A. KLOOSTERMAN, 1998. Hellenistische muntvondsten uit Nieuw Halos. *Paleo-aktueel* 9, pp. 40–45.

ROBINSON, D.M., 1933. *Mosaics, Vases and Lamps of Olynthus Found in 1928 and 1931* (= Olynthus 5). Baltimore.

ROBINSON, D.M., 1950. *Vases Found in 1934 and 1938* (= Olynthus 13). Baltimore.

ROTROFF, S.I., 1997. *Hellenistic Pottery. Athenian and Imported Wheelmade Tableware and Related Material* (= The Athenian Agora 29). The American School of Classical Studies at Athens, Princeton, New Jersey.

SCHÄFER, J., 1968. *Hellenistische Keramik aus Pergamon* (= Pergamenische Forschungen, Band 2). Walther de Gruyter & Co., Berlin.

SCHEIBLER, I., 1976. *Griechische Lampen* (= Kerameikos, Ergebnisse der Ausgrabungen, Band 11). Walter de Gruyter & Co, Berlin.

SCHLÖRB-VIERNEISEL, B., 1966. Eridanos Necropole: I. Gräber und Opferstellen hS 1-204: 6. Spätklassisch–frühhellenistische Gräber nach der Schotteraufschüttung ab 338 v. Chr. *Mitteilungen des Deutschen Archäologischen Instituts, Athenische Abteilung* 81, pp. 75–107.

SMETANA-SCHERRER, R., 1982. Spätklassische und hellenistische Keramik. In: H. Walter (Hrsg.), Alt-Ägina II, 1. Verlag Philipp von Zabern, Mainz am Rhein, pp. 56–91.

SPARKES, B.A. & L. TALCOTT, 1970. *Black and Plain Pottery of the 6th, 5th and 4th Centuries B.C.* (= The Athenian Agora 12). The American School of Classical Studies at Athens, Princeton, New Jersey.

THOMPSON, H.A., 1934. Two centuries of Hellenistic pottery. *Hesperia* 3, pp. 311–480.

VANDERPOOL, E., J.R. MCCREDIE & A. STEINBERG, 1962. Koroni: a Ptolemaic camp on the east coast of Attica. *Hesperia* 31, pp. 28–61.

VANDERPOOL, E., J.R. MCCREDIE & A. STEINBERG, 1964. Koroni: The date of the camp and the pottery. *Hesperia* 33, pp. 69–75.

VATIN, C., P. BRUNEAU, C. ROLLEY & T. HACKENS, 1976. *Tombes Hellénistiques – Objets de Métal –Monnaies* (= Médéon de Phocide, 5). Diffusion de Bocard, Paris.

VERMEULEN WINDSANT, C.TH.F., 1973. Goritsa, single finds. *Αρχαιολογικόν Δελτιον* 28, p. 371.

2. TERRACOTTA FIGURINES

GEORGETTE M.E.C. VAN BOEKEL AND BIENEKE MULDER

2.1. Introduction

During the excavations at New Halos, 46 intact and fragmentary terracottas were discovered in the upper and lower towns.[1] The five statuettes and four moulds that were found in the lower town will be considered here. Three of the figurines and all four moulds came to light in the House of the Coroplast in housing block 2.7 (fig. 3.20). The House of Agathon in housing block 6.4 yielded two statuettes. The terracottas represent four standing female figures, Helle riding a ram, a banqueter, a horse and a relief showing Aphrodite riding a billy-goat (see also Appendix 3).

2.2. Date and distribution of the terracottas

The terracottas are of early Hellenistic date, based on the period of Halos' existence. The town was probably founded in 302 BC, after the destruction of classical Halos in 346 BC. The new town was destroyed by an earthquake, most probably in 265 BC. Artefacts discovered in the houses that have so far been excavated have therefore been generally dated to the period from *c.* 300 to 265 BC. The terracottas from the House of the Coroplast and the House of Agathon must hence also date from this period, their latest possible date being 265 BC. The three female figu-

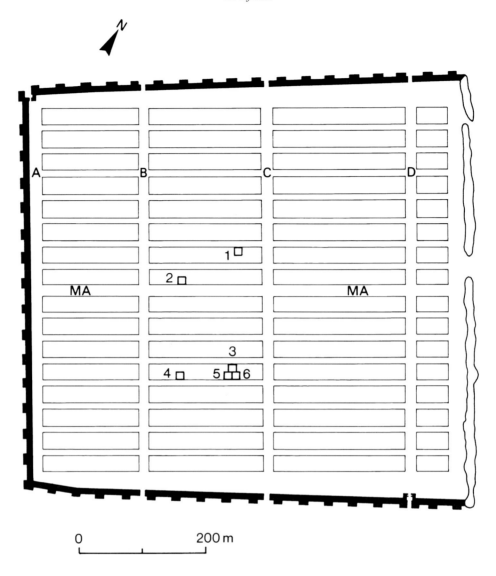

Fig. 3.20. Plan of the lower town of New Halos, showing the situation of the House of the Coroplast (1) and the House of Agathon (6).

rines (F1–3) can be dated to the first quarter of the 3rd century on stylistic grounds, too. The mould of a female figure (F4) has a 4th-century type of plinth, making this the oldest representative of this group. The representation of a banqueter (F6) similarly shows older, traditional features but its use in the early Hellenistic period is proved by the fact that the statuette was found with its mould.

The terracottas from the House of the Coroplast were recovered from rooms 1, 3 and 5 (fig. 3.21). One figurine of a standing woman (F3) was found near the post of the door between rooms 1 and 2, another came to light in room 5 (F2). Two partial moulds for a horse (F8), a disc-shaped relief showing Aphrodite (F9) and a banqueter (F6) were found in a semi-circle in the

southern corner of room 3. Two other figurines were found in this same room, close to the entrance to room 4: a standing woman (F4) and the mould (F7) for the banqueter (F6). So the majority of the terracottas were found in rooms that were easily accessible from the street (rooms 1 and 3).

The front part of the seven-roomed House of Agathon consisted of areas 6 and 7, which together formed a large inner courtyard that could be entered from the street (fig. 3.22).[2] The living quarters behind this courtyard comprised a large central room (3) flanked by two smaller rooms (1–2 and 4–5) on either side. The statuettes of Helle and a standing woman were found in room 5, which was a storage area (F1, F5).

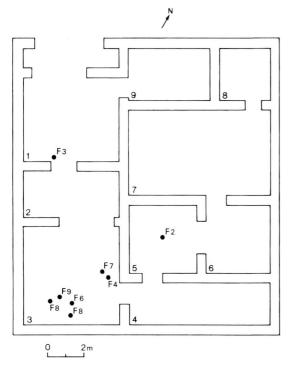

Fig. 3.21. Distribution of terracottas in the House of the Coroplast.

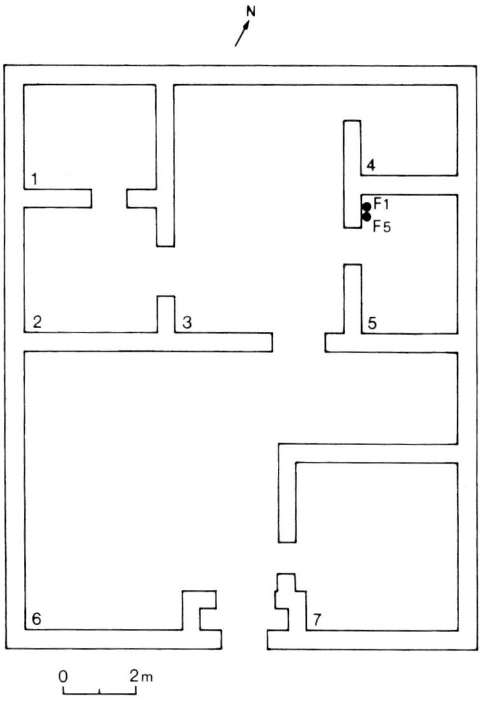

Fig. 3.22. Distribution of terracottas in the House of Agathon.

2.3. Manufacturing techniques

The hollow statuettes were made in moulds, which was common practice in the production of series of terracottas in Hellenistic times, as indeed also attested by the moulds from the House of the Coroplast.[3] Each mould or *matrix* was taken from a hand-modelled *patrix* (original), or from a previously moulded statuette. A uni-valve mould sufficed to make relief F9, but several partial moulds were used for the other terracottas, such as F1 (fig. 3.24). The head of the latter figurine was most probably made in two mould halves, separately from the body to which it was attached prior to firing.[4] The moulds for the head and those for the body may have been varied to produce different statuettes. The carefully detailed back of the head contrasts markedly with the woman's crude, plain back, which is a mere slab with a large vent-hole. F3 (whose head is now lost; fig. 3.26) has a similarly plain back; this is a common feature in Hellenistic terracottas. Although these backs at first sight seem to have been modelled by hand, it has been found that in this period moulds were used for even the plainest of rear sides[5], as also in the case of the statuette of Helle (fig. 3.28) and the representation of the banqueter (F6; fig. 3.29).

The rectangular vent-holes were made *ante cocturam* to ensure unhindered circulation of air during firing.[6] They are not a technical necessity because the hollow terracottas are open at the base and this alone

would have allowed sufficient evaporation of moisture and expansion of gases without any risk of the figurine cracking or exploding (Higgins, 1986: p. 66). The largely undetailed backs and the large vent-holes clearly indicate that the terracottas were meant to be viewed from the front only.

The two mould halves representing the sides of a horse (F8; figs 3.30 & 3.31) have marks showing how the two halves were to be joined to create a figurine. These marks, which are to be found on the moulds' outer rims, consist of small semi-globular protrusions to guide the coroplast's hand.

It is very likely that the statuettes were painted after firing, although only very few traces of pigments have survived. Pigments were originally painted on the white slip that was applied to F1 and F5; those on F1 are still in good condition. The slip concealed the reddish-yellow fabric and formed an even, neutral ground for the colours.

Some Hellenistic terracottas show signatures or other potters' marks, but it is questionable whether the *en creux* mark (?) KΣI observable in the interior of one of the figurines of a woman (F2; fig. 3.25) is indeed an inscription.

2.4. The colours of the fabrics

The colours and fabric of the terracottas from New Halos appear to imply local production (no chemical analyses have been performed). The Munsell Soil

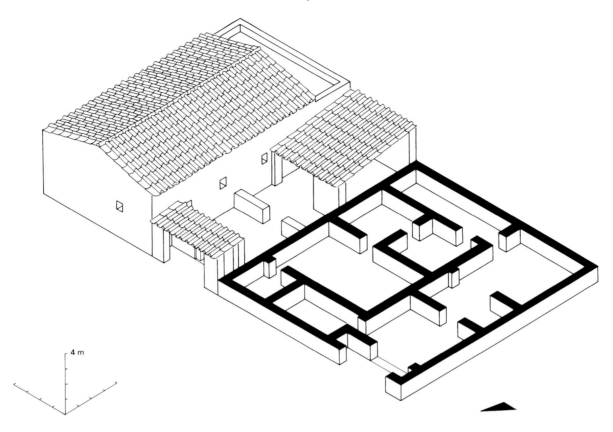

4 m

Fig. 3.23. Isometric reconstruction of the House of the Coroplast.

Color Charts make a distinction between pale and reddish terracottas. The terracottas found in the House of the Coroplast are all of the reddish shades, except F4. The colours Reddish Yellow (in practice an orange shade) and Light Red predominate. The colour of mould F4 is Pink to Reddish Yellow, *i.e.* between the two main terracotta shades. The majority of the statuettes from the Sepulchral Building in the upper town (26 specimens) also have reddish fabrics.[7]

The terracottas from the House of Agathon (F1, F5) are all of a very similar, slightly smudgy, soft fabric of varying shades that easily crumbles. The yellowish shades perhaps indicate slight burning. The colour of F1 is between Pale Yellow and Reddish Yellow. The fragments of F5 are Reddish Yellow and Light Red. Such variations in shades constituted no problems as the fabric was ultimately to be concealed by a white slip.

Leaving aside the possibility that the moulds from the House of the Coroplast were imported from elsewhere[8], the reddish shades suggest a local origin. The statuette of the banqueter and its mould, which are both Light Red to Red, were moreover made from the same type of paste, which may reasonably be assumed to have been acquired locally (F6–7).

Iron, which causes such red shades during firing in a kiln with an oxidising atmosphere, links the fabric

to the iron-containing reddish soil of New Halos. Some of the pottery found at Halos moreover shows the same red shades (Reinders, 1988: pp. 252–277). So we may conclude that the reddish fabric of most of the moulds and statuettes implies that terracotta was produced at New Halos.

2.5. The House of the Coroplast and the manufacture of terracottas

The discovery of moulds in the House of the Coroplast offered us an extraordinary opportunity to study the arrangement of rooms in what must have been a coroplast's house (fig. 3.23).[9] The house had nine rooms and comprised a large inner courtyard (areas 1–2), a room that was most probably a workshop (room 3), rooms 4 and 9 and the living quarters, rooms 5–8, in a separate part.

Areas 1–2 and room 3 in the western part of the house lay in line with the entrance. Rooms 4 and 9 were easily accessible from this part of the house, but were detached from the living quarters, which lay further away from the entrance. The two parts were connected via a door between rooms 4 and 5. When closed, this door will have closed room 4 off from the living quarters.

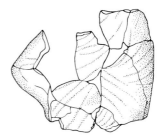

Fig. 3.25. Woman standing, fragment F2 (scale 1: 2).

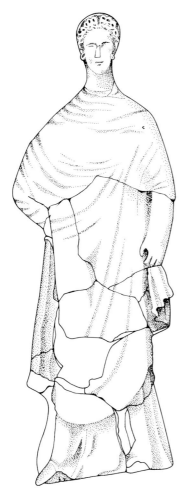

Fig. 3.24. Woman standing F1 (scale 1: 2).

The layout of a coroplast's workshop will have been very similar to that of a potter; both used clay as their raw material and some of the techniques they employed were also the same. A representation on an Attic *hydria* shows a roofed potter's workshop that opens onto an inner courtyard containing a kiln (Scheibler, 1983: p. 108). Much of the work will have been done out in the open, especially the preparation of the clay. Bright daylight was a prerequisite for the artisans to perform their tasks, but they had to be protected from direct sunlight. The uncovered areas 1–2 and the roofed room 3 that opened onto the courtyard of the House of the Coroplast will together have met these requirements.

Provisions for drying were also important, because the production volume depended on the required drying time and the amount of room available for the drying. Unfired figurines had to be dried to a leather-hard state under controlled, constant conditions, away from direct sunlight. The clay had to be allowed to dry slowly because otherwise there was a risk of the statuettes deforming due to irregular evaporation of

moisture. Irregular drying also involved a risk of cracking and fracturing as a result of shrinkage. The separately positioned room 9 had a roof, and may have contained the space needed to dry a series of terracottas and possibly also to store a supply of raw clay.

The final drying to a bone-dry condition could take place in the open air, in a courtyard or an uncovered room. This may have been done in the courtyard (areas 1 and 2) and possibly also in room 4. The finished products may have been put on display for sale in these same areas.

The next stage in the manufacture was the firing of the terracottas in a kiln. A large concentration of charcoal found in the northeastern corner of room 3 could represent evidence of a kiln (Reinders, 1988: p. 126), although this would mean that the kiln stood in a (partly?) roofed room. Another possibility is that several workshops collectively used a kiln situated elsewhere,[10] perhaps outside the town's perimeters, to save fuel and time. A kiln in a town house would of course have implied a great risk of fires.

It is quite well possible that terracottas were produced in the House of the Coroplast on a household basis, rather than by a workshop owner with labourers or slaves. Different kinds of people of different ages appear to have engaged in the production of pottery, including women (*cf.* fig. 3.24).[11] To summarise, the House of the Coroplast contained all the roofed rooms and open areas required for the efficient manufacture of terracottas in series.

2.6. The House of the Coroplast and terracotta production at other sites

There is documented archaeological evidence of other workshops in Greek towns that produced terracottas on a small scale.[12] Such workshops have been excavated at Abdera (Thráki),[13] Delos (two workshops)[14] and Eretria (Constantinou, 1957). Other sites that have yielded evidence for the small-scale manufacture of terracottas are Argos (Guggisberg, 1988: pp. 169, 208, 226), Árta[15] and Naukratis (Hogarth *et al.*, 1905: pp. 132–133). The evidence obtained at these sites tells us that coroplasts' workshops were situated at central locations in cities, where they were easily accessible from the towns' main roads (this was the case at Abdera and New Halos and with the two workshops on Delos). They were almost always situated in or very close to residential areas rather than industrial areas. They could evidently be easily combined with living quarters inside houses, as attested by House B at Abdera,[16] House B on Delos[17] and the House of the Coroplast at New Halos. They do not appear to have had any standard layout, except that the coroplast seems to have had access to a central courtyard that provided access to various comparatively small rooms (Abdera, New Halos). The number of rooms used for production appears to have varied from one single-storied, undivided room on Delos (room 80 on the 'Agora of the Italians') to several rooms as for example at Eretria (four rooms) and New Halos.

No indisputable evidence of a kiln was found in any of the aforementioned workshops,[18] but charcoal found among wasters at Eretria and the concentration of charcoal discovered at New Halos constitute possible indications of kilns.

Evidence of the kind found at some of the other terracotta-production sites, such as wasters or a water supply, was absent at the House of the Coroplast, though water may have been obtained from a cistern elsewhere in town. Neither were there any remains of stone basins for purifying and preparing the clay; perhaps basins made of perishable wood were used here. Also absent were blocks of raw clay, paint, pigments and tools, but it is conceivable that two knives found in room 4, near the entrance to room 3 (Reinders, 1988: p. 122, fig. 78; p. 259, Nos. 34.21–22), were used in the production process.

To summarise, although certain evidence is (still) lacking, the House of the Coroplast does have various elements in common with other investigated workshops. Additional, smaller-scale evidence of terracotta production is provided by the moulds, the evidence pointing to the use of local clay, the variety observable in the range of terracottas and a figurine found with its mould.

2.7. Local production at New Halos, conclusion

The presence of moulds, the arrangement of the rooms and open areas and certain elements in common with other sites at which terracotta is known to have been produced on a small scale indicate that terracotta was manufactured in the House of the Coroplast. The homogeneity of most of the terracotta fabrics also points to local manufacture. The reddish shades constitute a link with both the pottery excavated at New Halos and the local soil. The banqueter and its mould were moreover made from the same type of paste, making it very likely that the clay was obtained locally (F6–7).

The different areas in the house, roofed and unroofed, are in accordance with a coroplast's activities during the subsequent stages of his craft, including the possibility that he supplemented his income with the manufacture of other ceramic products such as pottery.

The differences observable between the layout of the House of the Coroplast and that of, say, the House of Agathon (Reinders, 1988: p. 116, fig. 71), which yielded only two terracottas and whose layout closely resembles that of the neighbouring House of the Ptolemaic Coins, may illustrate the differences between a house comprising a workshop (and possibly also a shop) and a private house, or a house in which craft activities were of lesser importance.

It is possible that the manufacture of terracottas at New Halos was initially based on imported moulds and terracottas, from which new figurines and moulds could easily be derived. This will have been less expensive than importing figurines on a large scale.[19] The statuette of Helle riding a ram (F5) also suggests local production, because it expresses a scene from a myth that is closely connected with Halos, as will be explained below.

2.8. Purpose of the terracottas

The terracottas from the House of the Coroplast were presumably not found in the context for which they were intended. Statuettes found in sanctuaries, burials and houses throughout the Greek world show that they were used for various purposes – sacred, funerary and domestic (Higgins, 1986: pp. 65, 119). Funerary use is attested at Halos in particular by the terracottas found in the Sepulchral Building.[20]

The figurines of a standing woman (F1) and Helle (F5) were found in one of the rooms of the House of Agathon; they may consequently have been used in a domestic context, perhaps in a private shrine. We may safely assume that they stood in some fixed place with their rear sides to the wall, considering their (largely) undetailed backs and the large vent-hole in F1, which, in our eyes at least, make it unlikely that they were used as merely decorative objects or toys. The fact that the majority of the terracottas were too fragile to have been handled frequently also makes it very unlikely that they were toys.

2.9. Subjects and types

The terracottas comprise free-standing statuettes and a relief. The majority portray solitary human figures. In F5 and F9 the human figure has been combined with an animal. They represent Helle and Aphrodite riding a ram and a billy-goat, respectively. The only representation of a single animal is that of a horse. The majority of the figures are female: F1–4 represent standing females, F5 and F9 riding females. One male

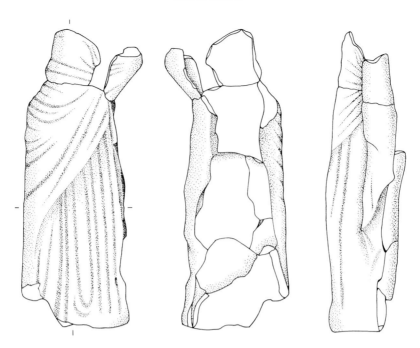

Fig. 3.26. Woman standing, fragment F3 (scale 1:2).

figure, a banqueter, is represented twice, by a statuette and its mould (F6–7).

Standing female figures
The elaborately draped standing women posing in an
elegant stance (F1–4; figs. 3.24–27) can be classified
as 'Tanagra figures', thus named after the Tanagra
cemetery (Boeotia) where such figurines were first
discovered in the 19th century. Three such figurines
were found in the House of the Coroplast (F2–4), one
in the House of Agathon (F1). They wear a long
chiton and a *himation* wrapped tightly around the
upper part of the body (the cloak is missing in the case
of F4). The clothing was originally probably painted
in bright colours, judging from pigments preserved
on terracottas from Tanagra, and the hair in red, apparently attesting to a contemporary custom of dy-
ing hair with henna (Higgins, 1986: pp. 119–120,
123).

As with many other Hellenistic terracottas of
women, it is difficult for us to distinguish between
goddesses and mortals. Mortal women, probably wor-
shippers, in a 2nd-century votive relief in Munich
closely resemble the Tanagra figures in many respects,[21] while a relief by Archelaos (*c.* 220–150 BC)
represents deities, personifications and mortals in the
same guise.[22] The statuettes were possibly identified
on the basis of their attributes, for instance as a muse
or a nymph. The absence of attributes does however
not necessarily imply that a portrayed figure is a
mortal, as is indicated by scenes painted on pottery
in which deities and demigods are portrayed in the
same ways as people of flesh and blood (Higgins,

1967: p. 98). An aperture in the left hand of F1 suggests that this figure held a separate attribute, perhaps
a fan, in her hand.

Some of the female figures and youths that were
placed in burials may represent the deceased, but, as
most of the figurines are female, it is unlikely that
they were all intended to be portrayals of the dead.
Perhaps the figurines formed part of the personal pos-
sessions of the deceased and had a decorative function. Series of identical terracottas placed in one and
the same grave were apparently specifically acquired
for the occasion (Higgins, 1986: p. 65).

The very popular Tanagra style of female statuettes is now thought to have originated in Athens, in
imitation of sculpture from around the middle of the
4th century BC.[23] This resulted in naturalistically
rendered figures with close attention to the folds in
the drapery, which conceal certain parts of the body
and accentuate others. Folds at the knees for example often indicate the posture, as is observable especially in F1. The latter figure is of the so-called
'muffled Tanagra' type, which was popular in early
Hellenistic times. Arms, hands and feet are frequently
hidden by the garments, as is demonstrated by F1–3.
But usually the contours of the body are clearly indicated.

The Attic terracottas were exported in large numbers and were copied in many other workshops, also
at New Halos, around the mid-3rd century. Stylistic
features of the first two of the three phases in the development of the Tanagra style are observable in F1–
3. These features, along with the frontal pose, point
to the early Hellenistic period, from the beginning

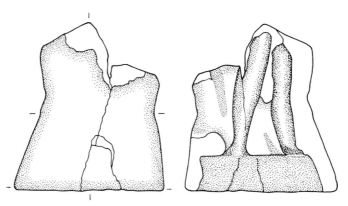

Fig. 3.27. Woman standing, fragment F4 (scale 1:2).

of the 3rd century to the end of the first quarter of the 3rd century.

In the first phase, the garments were loosely draped across the upper part of the body.[24] The *himation* has softly undulating, ridge-shaped folds and is wrapped round the body in a direction opposite to that of the stance, its diagonals sometimes strongly accentuated. The *chiton* conceals the leg bearing the weight of the body. The advanced knee of the other, flexed, leg adds depth to the representation. In a slightly later phase, the *himation* was wrapped tightly round the body, often hanging in zigzag folds down one side. The clothing's diagonal lines are accentuated, resulting in a strong contrast between the drapery and the female body beneath it.

The spool-shaped plinth in the fragmentary mould F4 was an old-fashioned element in the early Hellenistic period, by which time the plaque-shaped base had already been introduced as a technical innovation.[25] This mould must therefore have been taken from an earlier, 4th-century statuette. The two other terracottas of standing females (F1 and F3), and possibly also F2, should be reconstructed with plaque bases. The latter have disappeared because they were vulnerable, separately made plinths, whereas the spool-shaped base was incorporated in the mould.

Helle riding a ram
The statuette of Helle from room 5 of the House of Agathon represents a young girl sitting astride a strolling ram (F5; fig. 3.28). The myth of Helle and Phrixos concerns the two children of king Athamas, who repudiated their mother Nephele and married Ino, the daughter of king Kadmos. The envious stepmother threatened Phrixos and Helle with death by inventing a Delphian oracle demanding the sacrifice of Phrixos. In an attempt to save her children, Nephele sent a ram with a golden pelt which she ordered to fly to Kolchis. However, en route, Helle fell from the animal into the sea, thereafter called Hellespont (sea of Helle). Phrixos safely reached king Aiëtis at Kolchis, whom

he presented with the pelt of the ram that he had sacrificed to Zeus – the golden fleece.

Helle does not feature in an older version of the myth, according to which Athamas himself wanted to sacrafice his son on mount Laphystios (Pinsent, 1975: pp. 65 ff.). Zeus then sent a ram with a golden fleece to save Phrixos. This version was apparently connected with a custom in classical Halos according to which any oldest male member of the descendants of Athamas wishing to enter the *Leiton* (council house) was to be sacrificed to Zeus (Hunger, 1959: p. 334). Herodotos (VII 197) associated the custom of human sacrifice at (Classical) Halos with this myth. The myth seems to have been a cult legend related to the sanctuary of Zeus Laphystios in classical Halos (Reinders, 1988: pp. 159 ff.) and in the later version the human sacrifice was replaced by the ram.

Coins from Halos feature the bearded head of Zeus Laphystios on the obverse, and either Phrixos or Helle on the reverse, but never both together.[26] Helle is represented on the coins of Classical Halos, riding side-saddle on the ram whose horn she holds (Reinders, 1988: p. 236). The coins of the Hellenistic period depict Phrixos hanging alongside the ram while holding its horns or neck. The sea is cursorily indicated on some coins. Both representations resemble Melian terracotta reliefs portraying the girl sitting side-saddle on the ram and Phrixos hanging from its side, with the sea indicated by waves and, occasionally, fish.[27] The figures are (almost) adults, as can be inferred from their proportions in relation to the ram, Helle's mature upper body and the modelling of Phrixos' muscles. By contrast, Helle is depicted as a young girl in F5, as also in a Boeotian terracotta. However, in the latter statuette she is sitting side-saddle, as on the coins. Helle's posture is therefore quite unique; usually, female figures riding animals are depicted astride the animal (Amazons excepted). Perhaps her posture is explained by the fact that she is a child, although terracottas of children sitting astride animals usually seem to represent boys (Winter, 1903: pl. 305–306). Similarly, representations of Phrixos sitting astride an

Fig. 3.28. Helle riding a ram, fragment F5 (scale 1:2).

animal are also known from paintings (Helbig & Donner, 1868: 1251-1258) and gems.[28]

The general design of the terracotta of Helle does therefore not seem to have been derived from Melian or other reliefs, but is more in accordance with portrayals of young boys or Mercurius on a goat or ram (Winter, 1903: pl 305, 1–9; pl. 306, 5–7). This aspect is remarkable, as is the fact that the statuette of Helle is of Hellenistic date, whereas the coins of this period portray Phrixos. If the statuette's representation is true to the myth, it is conceivable that the white slip of which traces remain on the ram was painted with a golden or yellow shade to indicate the golden fleece.

Banqueter
The statuette and its mould of a banqueter with a *rhyton* and *patera* (F6–7) show a figure on a *kline* in the conventional attitude of half reclining, legs outstretched, trunk erect and head turned (fig. 3.29). Both were found in room 3 of the House of the Coroplast. The representation of this motif, known in the Greek world from the 7th century BC onwards, changed little throughout the ages and its conservatism strikes the eye (note the long beard and the style of the hair). A small table on which some standard attributes, such as drinking gear, stand is normally among the fixed elements. Changes in the perspective of the table, the turning of the upper part of the body and the increasingly naturalistically represented draperies reflect different phases in the development of this motif.

The subject, which has been extensively discussed by Dentzer (1971; 1982), was originally associated with Greek aristocracy. The banqueters were frequently provided with the symbols of their privileges defining their status: servants, horses, hounds, weapons and musical instruments. In addition, from the end of the sixth century onwards, members of the privileged, broader (town) class were portrayed as banqueters.

The aristocratic ideals were also applied to representations of deities and heroes, but terracottas of banqueters were apparently not bound to a particular god or cult because they have been found in sanctuaries of different deities. The type of terracotta represented by F6–7 may have been created as a mark of honour to a *heros*, for instance a mythological founder of a city. Many heroes had played important parts in political and social life, and founders of cities and important military leaders were frequently venerated as heroes after their deaths. Banqueters were often provided with attributes, for instance a *polos*, to mark them as heroes (Dentzer, 1982: p. 563).

In the Hellenistic period this formula was frequently applied to depictions of deceased individuals, identifying them with heroes. This custom may have been instigated by political interests because most of the Hellenistic towns were in the hands of powerful rulers. Such a ruler would often declare himself a *heros*, for instance the descendant of Alexander the Great, in order to legitimise his actions (Uhlenbrock & Thompson, 1990: p. 38). It is thought that putting an ordinary citizen on a par with the founder of a city, *i.e.* a *heros*, created a bond with that founder and helped to preserve a feeling of one's own identity (Dentzer, 1982: p. 565).

It is assumed that portrayals of a deceased individual show that person attending a funerary banquet. This may be a recollection of happy earthly occasions, with the family gathered around the deceased, or it could be the dead enjoying a banquet in the netherworld, with the banquet symbolising happy immortality. Another possibility is that the scenes represent a funerary rite, the sacrifice of a meal to the dead, or a recollection of the banqueting feast held by the relatives on the tomb during or after the funeral. In this context it may be added that a woman may have been seated at the foot of the *kline* that is now missing.[29]

Of the snakes, dogs and horses that are often represented in such scenes the latter animals are of interest to us here. According to Dentzer, the Thracian army ascribed a symbolic value to horses in connection with banqueting scenes. A horse in a banqueting scene is thought to indicate the rank of the deceased.[30] It is conceivable that this was also the case at New Halos, which was originally a military settlement, in particular as the mould of a horse (F8) was found near the relief of the banqueters. To summarise, the banqueter (F6–7) probably represents a *heros* or a deceased individual, possibly in the guise of a hero. The terracottas that were made from the mould may have been used for funerary or sacred purposes.

Horse
Two mould halves which when joined yielded a complete horse statuette were found in the House of the Coroplast (F8; figs 3.30 & 3.31). The animals made

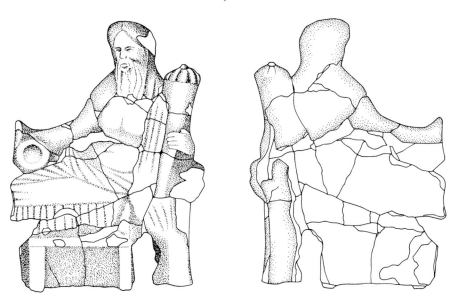

Fig. 3.29. Banqueter F6 (scale 1:2).

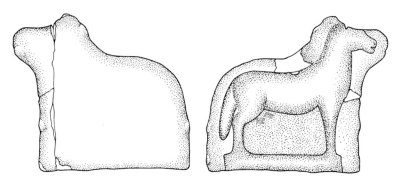

Fig. 3.30. Horse, half mould F8 (scale 1:2).

from this mould may have been used for various purposes. Offered in a sanctuary, they may have embodied a plea for a horse's health and fertility, or they may have been offered to please a specific deity. A horse in a burial possibly indicated the rank of a deceased (see above, in the section on the banqueter). The Demeter sanctuary at Proerna yielded statuettes of both a banqueter and a horse, but the first dated from the first half of the 5th century and the second from the first half of the 6th century.[31]

Relief showing Aphrodite riding a billy-goat
The fragmentary mould F9 from room 3 of the House of the Coroplast was intended for a disc-shaped relief representing Aphrodite. The semi-nude goddess sits side-saddle on a billy-goat that is jumping to the left as they ride through the air.[32] She is accompanied by a small goat. There was probably a second small goat originally, as can be seen in other depictions of this motif.

Billy-goats were sacrificed to this goddess; her epithet, *Epitragia,* derives from this custom. The fa-

mous Aphrodite Pandemos statue representing this motif by Skopas in Elis (350-320 BC; LIMC II, 1984: fig. 975) is now lost, but it probably influenced the many depictions of Aphrodite *Epitragia* in gems, terracottas, reliefs and other media. They always portray the goddess riding side-saddle and draped or semi-nude.

The stars, combined with the absence of indications of landscape in most of these depictions indicate the firmament through which Aphrodite rides at night. No stars are observable in F9; they may have been incised in the reliefs *ante cocturam*, or painted on after firing.[33] She is here a celestial goddess of light, derived from the cult of the Phoenician goddess Astarte, as also indicated by the crown with rays that is to be seen on her head in some depictions (LIMC II, 1984: fig. 964; in F9 the head is missing). The small goats are thought to symbolise Aix and Eriphoi, which together represented a constellation that served as a warning of approaching storms to ships' crews (Knigge, 1982: p. 154).

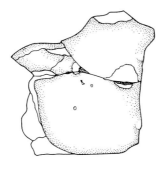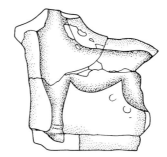

Fig. 3.31. Horse, half mould F8 (scale 1:2).

In other depictions of this motif two young men accompany the goddess; one preceding Aphrodite and holding a torch identified as Phosphoros, the morning star (Knigge, 1982: pp. 154 ff.). The man following the goddess is assumed to represent Hesperos, the evening star. The morning star apparently made way for the goddess of light while the evening star followed in her wake. Aphrodite thus reigned over both stars, personifying the star Aphrodite, our planet Venus. A ladder that is frequently to be seen in these depictions could symbolise the rising and setting of the stars.[34]

The planet (star) Aphrodite-Venus was associated with happiness and love, feelings that were also closely linked with the many ornaments and amulets showing this scene. The goddess usually holds the billy-goat's neck with one hand, while raising a veil from the side of her body with the other. This part of the scene is poorly preserved and unidentifiable in F9. The gesture may refer to her role as the patron deity of marriages, while the notoriously fertile billy-goat refers to her role as fertility goddess.

The representation of Aphrodite riding a goose is regarded as the counterpart of the Aphrodite Pandemos motif, the first symbolising spiritual love, the second physical love (LIMC II, 1984: figs. 941–945). These two separate aspects of the goddess were subjects of veneration in different sanctuaries (Knigge, 1982: pp. 164–166).

To summarise, Aphrodite Pandemos was worshipped as a planet (star) goddess associated with happiness and physical love. The fact that several reliefs could be produced from the mould suggests that the goddess was venerated in this guise somewhere in New Halos. The disc-shaped reliefs may have been suspended from a wall or the like. They are probably to be considered cheaper substitutes of reliefs made of costly materials such as silver like the (non-identical) silver medallion known from Athens (Kerameikos).

2.10. Conclusion

The terracottas from New Halos represent an important group of finds because we know exactly where they came to light in the houses in the lower town and because their find context dates them to the fairly narrow timespan from *c.* 300 to *c.* 265 BC. Most importantly, the moulds found in the House of the Coroplast indicate that this house contained a coroplast's workshop. This offered us a rare opportunity to study the arrangement of rooms in a house where terracotta was manufactured. The four moulds discovered here were used to produce reliefs and figurines representing four different motifs, including female figures of the Tanagra type. The three statuettes of standing women recovered from this house and the House of Agathon can be classified as early representatives of the Hellenistic Tanagra range.

There is a special link between Helle, represented by one of the figures, and the town of New Halos. This is evident from the coins depicting scenes from the myth of Helle and Phryxos that have been found in Halos. In classical Halos at least, this myth was related to the cult of Zeus. The mould for disc-shaped reliefs depicting Aphrodite show that this goddess was also worshipped at New Halos, especially in her role as goddess of physical love.

The representations of a banqueter prove that a traditional motif with early stylistic features derived from aristocratic iconography was still being reproduced at least three centuries after its introduction. These scenes may represent a deceased person portrayed as a *heros*, reflecting the secularised spirit of the time.

2.11. Notes

1. In the excavations carried out between 1978 and 1989. Of these terracottas, 37 figurines were found in the Sepulchral Building, a funerary structure containing a double grave (see Reinders, 1988: p. 137; Nos. 50.06–09, 51.07–08, 52.18–33, 53.07, 54.89–91, 54.93–95).
2. See Ch. 2: 3 of this volume for this house.
3. *Cf.* Higgins, 1967 (1): p. 137; Higgins, 1986: pp. 65–70, 119–

120; *Histoire et archeologie les dossiers* 81, 1984; Lunsingh Scheurleer, 1986: pp. 9–13; Nicholls, 1952; Uhlenbrock & Thompson, 1990.

4. *Cf.* Higgins, 1986: pp. 67, 69–70, figs 66–69.

5. Higgins, 1986: pp. 66–67, fig. 65a; p. 70, fig. 68; p. 119.

6. *Cf.* the rectangular vent-holes in Hellenistic Boeotian terracottas, including those from Tanagra (Higgins, 1986: pp. 66–68, 119; Lunsingh Scheurleer, 1986: p. 11).

7. See note 1.

8. For the transport of moulds see Higgins, 1967: p. 5; Laumonier, 1956: p. 26; Uhlenbrock & Thompson, 1990: pp. 16, 49.

9. For a survey of all the finds recovered from the House of the Coroplast see Reinders, 1988: pp. 254–263 and fig. 78.

10. *Cf.* Mollard-Besques, 1963: p. 15; Scheibler, 1983: p. 110.

11. Thompson, 1963b: p. 61; Scheibler, 1983: p. 111.

12. Mollard-Besques, 1963: p. 14; Higgins, 1967: p. 5; Uhlenbrock & Thompson, 1990: p. 15.

13. Graham, 1972: pp. 289, 295–296, 299; Lazarides, 1952: pp. 260 ff.; Lazarides, 1960: p. 49. pl. 20, pl. 27.

14. Chamonard, 1922–24: pp. 23, 72, 214, 221 ff., plates I–II; Lapalus, 1939: pp. 98 ff.; Laumonier, 1956: pp. 18–22.

15. Dakaris, 1967: p. 310, pl. 350, figs a–d; Daux, 1967: p. 681; Uhlenbrock & Thompson, 1990: p. 15.

16. The eleven rooms and areas of this house included a courtyard containing a cistern and the house's ground-plan corresponds to the plans of the smaller *prostas* and *oikos* houses at Priene.

17. This seven-roomed house (Insula VI, centre) has a medium-sized ground-plan in comparison with the other houses on this island. Its workshop, at the back of the building, was accessible only through the living quarters.

18. Laumonier, 1956: p. 19, refers to the possible presence of a kiln in House B on Delos.

19. *Cf.* Higgins, 1967: p. 5; Mollard-Besques, 1963: pp. 71 ff.; Thompson, 1963: p. 13.

20. See note 1.

21. Higgins, 1986: p. 132, fig. 157; Pollitt, 1986: pp. 196–197, fig. 210.

22. Pollitt, 1986: pp. 15–16, fig. 4; p. 295, for further literature.

23. *Cf.* Besques, 1984: p. 50; Thompson, 1952–1965; Higgins, 1986: pp. 118–119; Liepmann, 1975: p. 27; Uhlenbrock & Thompson, 1990: pp. 48 ff.

24. Thompson, 1963: pp. 28 ff.; Higgins, 1986: pp. 123–140; Uhlenbrock & Thompson, 1990: pp. 51 ff.; *cf.* figurines in Mollard-Besques, 1971–72; Winter, 1903: vol. 3, *jüngere Typen*.

25. Higgins, 1986: pp. 108–111, figs 128, 130–132; pp. 119, 124, 140–141, fig. 170; p. 150, fig. 183.

26. Moustaka, 1983: pp. 68–69; Reinders, 1988: pp. 164–165, 236–251.

27. Jacobsthal, 1931: figs 5, 31, 34, 69–70, 101–102.

28. Boardman & Wilkins, 1970: No. 785; Fossing, 1929: Nos. 894–895; Walters, 1926: No. 1134.

29. *Cf.* terracotta from Valtsa; *cf.* Leyenaar-Plaisier, 1979: p. 77 sub No. 153.

30. Dentzer, 1982: pp. 562 ff., and notes 30–32.

31. Daffa-Nikonanou, 1973: p. 126; p. 135, pl. 13, fig. 1, horse of the same type from the sanctuary at Ambetia.

32. For Aphrodite on a billy-goat see Knigge, 1982: esp. plates 31–33; LIMC II, 1984, Aphrodite, figs 947–976; Miller, 1979: pl. 22–23; Mitroploula, 1975.

33. *Cf. Venus te lijf*, pp. 110–111, No. 86, fig. 56, terracotta of Aphrodite on a billy-goat with painted stars.

34. Miller, 1979: p. 40, associates the ladder with a fertility ritual.

2.12. References

BESQUES, S., 1984. Les ateliers de Tanagra, production et diffusion aux IIIe et IVe s. *Histoire et Archeologie les Dossiers* 81, pp. 44–60.

BOARDMAN, J. & R.L. WILKINS, 1970. *Greek Gems and Finger-rings: Early Bronze Age to Late Classical*. Thames & Hudson, London.

CHAMONARD, J. 1922-24. *Le Quartier du Théâtre* (= Exploration archéologique de Délos 8). De Boccard, Paris.

CONSTANTINOU, J., 1957. Érétrie (part of: Chronique des fouilles et découvertes archéologiques en Gréce en 1956). *Bulletin de Correspondance Hellénique* 81, pp. 588–590.

DAFFA-NIKONANOU, A., 1973. *Thessalika iera Dimitros kai koroplastika anathimata*. Hetaireia Thessalikon Ereunon, Volos.

DAKARIS, S., 1967. Nomos Artis. *Archaiologikon Deltion* 19, Chronika 1964, pp. 310–311.

DAUX, G., 1967. Chronique des fouilles et découvertes archéologiques en Grèce en 1966. *Bulletin de Correspondance Hellénique* 91, pp. 623–890.

DENTZER, J.M., 1971. Aux origines de l'iconographie du banquet couché. *Revue Archéologique* 1971 (2), pp. 215–258.

DENTZER, J.-M., 1982. *Le Motif du Banquet Couché dans le Proche-Orient et le Monde Grec du VIIe au IVe Siècle avant J.C.* École français de Rome, Rome.

FOSSING, P., 1929. *Catalogue of the Antique Engraved Gems and Cameos*. Thorvaldsen Museum, Copenhagen.

GRAHAM, J.W., 1972. Notes on houses and housing districts at Abdera and Himera. *American Journal of Archaeology* 76, pp. 295–301.

GUGGISBERG, M., 1988. Terrakotten von Argos. Ein Fundkomplex aus dem Theater. *Bulletin de Correspondance Hellénique* 112, pp. 167–234.

HACKENS, T. & E. LÉVY, 1965. Trésor héllenistique trouvé à Délos en 1964: Les Byoux. *Bulletin de Correspondance Hellénique* 89, pp. 503–566.

HELBIG, W. & O. DONNER, 1868. *Die Wandgemälde der vom Vesuv Verschütteten Städte Campaniens*. Breitkopf und Härtel, Leipzig.

HIGGINS, R.A., 1967. *Greek Terracottas*. Methuen, London.

HIGGINS, R.A., 1986. *Tanagra and the Figurines*. Trefoil Books, London.

HOGARTH, D.G., H.L. LORIMER & C.G. EDGAR, 1905. Naukratis, 1903. *Journal of Hellenic Studies* 25, pp. 105–136.

HUNGER, H., 1959. *Lexikon der Griechischen und Römischen Mythologie*. Hollinek, Wien.

JACOBSTHAL, P., 1931. *Die Melischen Reliefs*. Berlin-Wilmersdorf.

KNIGGE, U., 1982. O Astir tis Aphroditis. *Mitteilungen des Deutschen Archäologischen Instituts, Athenische Abteilung* 97, pp. 153–170.

LAPALUS, E., 1939. *L'Agora des Italiens* (= Exploration archéologique de Délos 19). De Boccard, Paris.

LAUMONIER, A., 1956. *Les Figurines en Terre Cuite* (= Exploration archéologique de Délos 23). De Boccard, Paris.

LAZARIDES, D.I., 1952. Anaskafai kai ereunai en Abdirois. *Praktika*, pp. 260–273.

LAZARIDES, D.I., 1960. *Pilina eidolia Abdiron*. Athene.

LEYENAAR-PLAISIR, P.G., 1979. *Les Terres Cuites Grecques et Romaines, Catalogue de la Collection du Musée National des Antiquités à Leiden*. Rijksmuseum van Oudheden te Leiden, Leiden.

LIEPMANN, U., 1975. *Griechische Terrakotten, Bronzen, Skulp-*

turen (= Bildkataloge des Kestner-Museums Hannover 12). Kestner-Museum, Hannover.

LIMC II (*Lexicon Iconographicum Mythologiae Classicae II*), 1984. Artemis, Zürich-München.

LUNSINGH SCHEURLEER, R.A., 1986. *Grieken in het Klein, 100 Antieke Terracotta's*. Vereniging van Vrienden van het Allard Pierson Museum Amsterdam, Amsterdam.

MILLER, S., 1979. *Two Groups of Thessalian Gold* (= Classical Studies 18). University of California Press, Berkeley.

MITROPLOULA, E., 1975. *Aphrodite auf der Ziege*. Pyli Verlag, Athens.

MOLLARD-BESQUES, S., 1963. *Les Terres Cuites Grecques*. P.U.F., Paris.

MOLLARD-BESQUES, S., 1971-72. *Catalogue Raisonné des Figurines et Reliefs en Terre-Cuite Grecs, Étrusques et Romains, III, Époques Hellénistique et Romaine, Grèce et Asie Mineure, Louvre*. Éditions des Musées Nationaux, Paris.

MOUSTAKA, A., 1983. *Kulte und Mythen auf Thessalischen Münzen* (= Beiträge zur Archäologie 15). Konrad Triltsch, Würzburg.

NICHOLLS, R.V., 1952. Type, Group and Series: a reconsideration of some coroplastic fundamentals. *Annual of the British School at Athens* 47, pp. 217–226.

NINOU, K. (ed.), 1978. *Treasures of Ancient Macedonia*. Archaeological Museum of Thessalonike, Athens.

PINSENT, J., 1975. *Greek Mythology*. Hamlyn, London.

POLLITT, J.J., 1986. *Art in the Hellenistic Age*. Cambridge University Press, Cambridge.

REINDERS, H.R., 1988. *New Halos, a Hellenistic Town in Thessalia, Greece*. Hes, Utrecht.

SCHEIBLER, I., 1983. *Griechische Töpferkunst: Herstellung, Handel und Gebrauch der Antiken Tongefäße*. Beck, München.

THOMPSON, D.B., 1952. Three centuries of Hellenistic terracottas. *Hesperia* 21, pp. 116–164.

THOMPSON, D.B., 1954. Three centuries of Hellenistic terracottas. *Hesperia* 23, pp. 72–107.

THOMPSON, D.B., 1957. Three centuries of Hellenistic terracottas. *Hesperia* 26, pp. 108–128.

THOMPSON, D.B., 1959. Three centuries of Hellenistic terracottas. *Hesperia* 28, pp. 127–152.

THOMPSON, D.B., 1962. Three centuries of Hellenistic terracottas. *Hesperia* 31, pp. 244–262.

THOMPSON, D.B., 1963a. Three centuries of Hellenistic terracottas. *Hesperia* 32, pp. 276–292; 301–317.

THOMPSON, D.B., 1963b. *The Terracotta Figurines of the Hellenistic Period* (= Troy Supplementary monographs, 3). Princeton University Press, Princeton, N.J.

THOMPSON, D.B., 1965. Three centuries of Hellenistic terracottas. *Hesperia* 34, pp. 34–76.

UHLENBROCK, J.P. & D.B. THOMPSON, 1990. *The Coroplast's Art*. College Art Gallery, The College at New Paltz, State University of New York, New York.

Venus te Lijf. Liefde en Verleiding in de Oudheid. Allard Pierson Museum Amsterdam, Amsterdam 1985.

WALTERS, H.B., 1926. *Catalogue of the Engraved Gems and Cameos, Greek, Etruscan and Roman, in the British Museum*. London.

WILSON, L.M., 1930. Loomweights. In: D.M. Robinson, *Architecture and sculpture, houses and other buildings* (= Excavations at Olynthus 2). The Johns Hopkins Press, Baltimore, pp. 118–121.

WINTER, F., 1903. *Die Typen der Figürlichen Terrakotten* (= Die antiken Terrakotten III). Spemann, Berlin, *etc.*

3. LOOMWEIGHTS

YVETTE BURNIER AND STEVEN HIJMANS

A substantial number of loomweights were excavated at New Halos (Appendix 4). They were found in every house and in every room, in some cases in large quantities (*e.g.* in room 5 of the House of Agathon). Loomweights have also been found scattered throughout the fields of the former city area of New Halos.

Despite their abundance, loomweights have been extensively discussed in only a few site publications (Davidson, 1952; Deonna, 1938; Wilson, 1930) and although the warp-weighted loom, for which the loomweights were used, has been the object of several studies, only a small number of articles have focussed on loomweights specifically (Sheffer, 1981; Sheffer & Granger-Taylor, 1994; Vogelsang-Eastwood, 1990 with lit.). A comprehensive study is lacking.

3.1. Identification

A large number of pyramidal and discoid, and a smaller number of conical or trapezoid clay objects were found in New Halos. They all had one or two holes at the top. Although a variety of functions has been ascribed to such objects in the past century, ranging from identification-tags for cattle to votive offerings (*cf.* Deonna, 1938: pp. 151–152), it is now commonly agreed that their main function was that of loomweights for warp-weighted looms. This does not mean that all loomweights were of terracotta (see below), nor that all terracotta weights were used on the loom only. Incidental use of these objects for other purposes should not be ruled out, as they were easily available and cheap. Establishing the nature of such other, secondary functions is however a speculative enterprise, and we will here therefore restrict ourselves to their primary function as loomweights.[1]

3.2. The warp-weighted loom

The use of loomweights was limited to a single type of loom known as the warp-weighted loom. This loom originated in Europe in the Neolithic.[2] It consisted of two upright posts set at an angle against the wall, to whose top ends was fastened a horizontal cloth-beam. Halfway up the uprights were two forked brackets holding the heddle rod. Some distance below this the shed rod was fastened between the two uprights. The brackets for the heddle rod could be set at varying heights, depending on the circumstances, but the shed rod was in a fixed position. The warp was fastened to the cloth-beam and hung downwards, with the individual warp threads falling alternately over and behind the shed rod. Below the shed rod, the warp threads were bundled into a hank of perhaps about ten threads each and tied to a loomweight in order to keep the warp taut.[3] To ensure that the warp threads remained evenly spaced, the odd-numbered threads were individually attached to the heddle rod with loops of string. The weft was beaten upwards, so that the woven cloth eventually hung down from the cloth-beam. By rotating the cloth-beam the woven cloth could be rolled up at the top to ensure that the working height remained the same.

It is often stated that the warp-weighted loom was the most popular, or even the only type of loom in use in ancient Greece (*e.g.* Forbes, 1956: p. 200). It is illustrated on a small number of Greek vases (Carroll, 1987 [1965]: pp. 37–47), and appears to be the only type mentioned in literary sources (Crowfoot, 1936/1937). It is also the only type of loom to have left clear traces in the archaeological record in the form of its terracotta or lead weights. Concerning this assumed predominance of the warp-weighted loom a cautionary note may be in order, for contrary to the views of Forbes and others, it was certainly not the only type of loom in use. A small number of Attic red-figure vases illustrate simple textile frames or hand looms used for weaving or the *sprang* technique (Keuls, 1983: figs 14.25 & 14.31; Clark, 1983) and Carroll (1987 [1965]: pp. 57–58) assumes that the horizontal two-beamed loom, which had a long history of use in both Egypt and the Near East, was also known to the Greeks (*cf.* Ellis, 1976).

Neither the hand loom nor the horizontal two-beamed loom have left any traces in the archaeological record. Even the warp-weighted loom did not always leave traces, for there is evidence that stones and weights of mudbrick (Sheffer, 1981; Forbes, 1956: p. 200) and lead were used, besides terracotta loomweights. Stones used as weights are difficult to identify, mudbrick weights will generally have been too perishable to have survived, and lead weights may have been melted down and reused, thus leaving no evidence. To correctly assess the popularity of the warp-weighted loom *versus* other looms in ancient Greece we must therefore rely on literary and iconographic sources, and unfortunately precisely literary sources on Greek looms are very scarce (Crowfoot, 1937/1938: p. 38), as are illustrations in Greek art (though note that not only warp-weighted looms are illustrated, but also smaller hand-held looms). In other words, the almost complete silence of our literary and iconographic sources on warp-weighted looms makes it difficult to argue that no other looms were used by the Greeks. It is far more likely that the situation in Greece was not different from that in Palestine and Syria, where there is clear (albeit scant) evidence for the coexistence of various types of looms in the Iron Age and later times (Ellis, 1976: p. 77; *cf.* Forbes, 1956: p. 203).[4] Be that as it may, our evidence for

weaving in New Halos is limited to warp-weighted looms whose terracotta loomweights have survived.[5]

3.3. Types

Four types of loomweights have been found in Halos: pyramidal (68%), discoid (22%), trapezoid (8%) and conical (less than 2%) (Appendix 4; table 3.12, fig. 3.32). The pyramidal loomweights were clearly the most popular. They have a roughly square base with rounded corners. The top of the pyramid is usually rounded, rarely pointed. The sides are generally flat, sometimes slightly concave or convex. The pyramidal weights have a single hole just below the top.

The discoid loomweights were also fairly common, but it should be noted that over 75% of these weights were found in one house. Their shape in section ranges from rectangular with rounded edges to lenticular. The discs are not always perfect circles, as one side was sometimes flattened to enable them to stand. All the discoid loomweights have two holes set fairly close together. The position of these holes varies from just above the centre to close to the edge of the disc.

Though uncommon in New Halos, trapezoid loomweights were found in almost the same numbers as discoid loomweights in all the houses except the House of Agathon. These loomweights differ from the pyramidal weights not only in that they have a rectangular rather than square base, but also in that they have two holes just below the top instead of only one. Despite these differences, trapezoid loomweights are frequently described as pyramidal in the literature (*e.g.* Metzger, 1978: p. 86, No. 3).

Four conical loomweights were also found, in three houses. They have one perforation below the top, making them comparable with small pyramidal loomweights, from which they differ only in shape (a circular base rather than a square one).

The Halos loomweights are all made of terracotta, except for one discoid one, which is made of lead. The original ratio of terracotta and lead loomweights cannot be established on account of the possibility of lead loomweights having been recycled (*cf.* Robinson, 1941: p. 471). The composition of the clay of the loomweights is similar to that of the roof tiles of Halos, the temper generally being coarse to very coarse. There is a wide range of colours. We assume that all the loomweights were produced locally.

The existence of at least four different types of loomweights in New Halos is not unusual. Wilson (1930) identified no fewer than eight types at Olynthos, Deonna (1938: pp. 155–164) distinguished five types on Delos (all of lead except the discoid weights) and Metzger (1978: p. 86 & pl. 43) illustrates three types for Eretria. Generally one or two types predominate (Davidson, 1952), as is indeed also the case in New Halos, but there is no reason to assume that the four types attested in New Halos were not in use

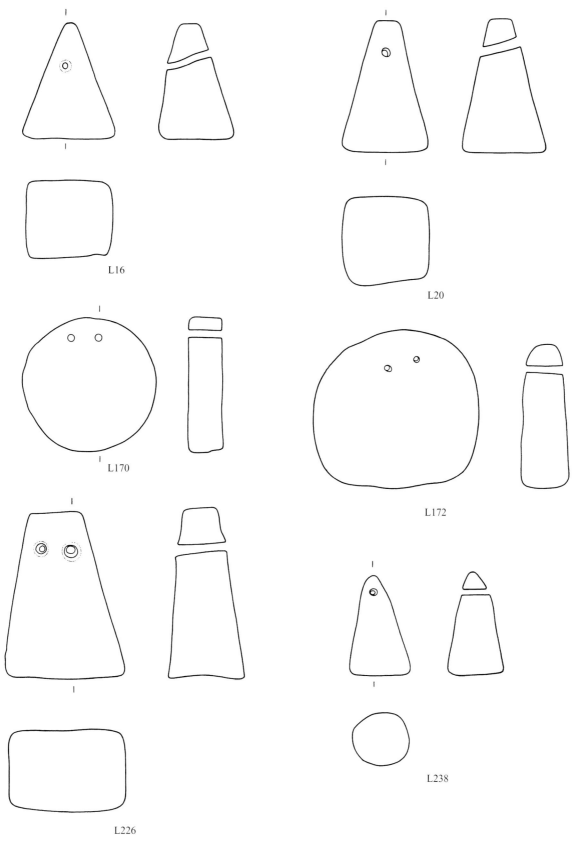

Fig. 3.32. Pyramidal, discoid, trapezoid and conical loomweights from the houses of New Halos (scale 1:2).

simultaneously. Terracotta loomweights were sturdy enough to have survived decades of use, and this virtually rules out the possibility that a shift in preference would be reflected in the archaeological record of the narrow timespan of only four decades for which New Halos was inhabited. This helps to explain why at least three of the four types were represented in all the houses.

3.4. Weight, decoration and use

The number of loomweights per house in Halos ranges from 11 to 139 (or from 11 to 37 if we ignore the House of Agathon). The average number of loomweights per house is about 21 (again excepting the House of Agathon). This number is rather low in comparison with that of Olynthos. If Bieber (1928: pp. 12–13) was indeed correct in assuming that about 65–70 loomweights will have been required to weave a woollen *peplos*, then none of the excavated houses in New Halos would appear to have contained sufficient weights for a normal loom except the House of Agathon, which yielded enough weights for two looms. On the other hand, modern reconstructions of warp-weighted looms are based on as few as 10 loomweights (*e.g.* Forbes, 1956: p. 202, fig. 31), and the number of loomweights needed could in fact vary considerably as it depended on the width of the loom, the type of material used for the weave and the number of warp threads per loomweight. Even so, the small number of loomweights per house may imply that only simple textiles were made in a household context in New Halos while larger and/or finer textiles were produced by (semi-)professional weavers elsewhere.

The loomweights themselves can in theory provide some indication of the types of cloth woven on the looms, because we may postulate a relationship between the weight of the loomweights and the nature and quality of the woven cloth. To ensure cloth of uniform quality, all the warp lines had to be equally taut, meaning that the employed loomweights had to be of roughly the same weight. The strength with which the weft could be beaten upwards (and thus the density of the cloth) also depended on the weight of the loomweights (Wilson, 1930: pp. 120–121). This suggests that lighter, smaller loomweights were used for fine cloth, and heavier ones for coarser fabrics. Evidence supporting this was found at Olynthos, where Wilson (1930: p. 121) recorded two distinct, clearly defined groups, one consisting of loomweights weighing about 105–120 g (3¾–4½ oz) and the other of weights of about 155–175 g (5½–6¼ oz).

The situation at New Halos is less clear-cut. There is a substantial degree of variation in the weights of the Halos loomweights, ranging from 35 g to 258 g. The loomweight of the latter weight, a large trapezoid specimen, is an exception, the normal range being 35–175 g. The discoid and pyramidal loomweights have weights covering this entire range, although the discoid ones are on average heavier than the pyramidal ones; the trapezoid weights are relatively heavy (94–172 g, plus the large loomweight of 258 g), while the four conical loomweights are among the lightest.

Even if we ignore the exceptionally large trapezoid loomweight, the weight of the lightest loomweights is hardly more than 20% of that of the heaviest ones. Despite this variation it proved impossible to use weight as a criterion for classifying the New Halos loomweights according to function. Over 80% of the loomweights fall within the far more restricted range of 80–120 g and the remaining 20% do not appear to show significant clusters within the lower (35–80 g) or higher (120–175 g) weight ranges, but are instead evenly distributed.

Marks or decorations bring us no closer to identifying different functional groups of loomweights, because only 32 of the loomweights bear a mark or some form of decoration. The most common mark (21 loomweights) is a deep cavity below the hole, which does not completely pierce the loomweight. Other impressed marks are round, bar-shaped or consist of three small cavities. Two loomweights, both from the House of Agathon, are marked with a letter (a *gamma* and a *tau* respectively) which was incised in the loomweight before it was fired.[6] One pyramidal loomweight shows a grid of two horizontal and two vertical lines on its base (also incised before firing) and one shows a simple incised line. Three loomweights, finally, were marked with a stamped decoration – two consist of a simple oval, the third of an oval framing a palmette-like motif.

It is difficult to determine what (if any) significance the marks may have had. It may be relevant to note that of the loomweights marked with a single deep cavity, the pyramidal ones weigh 50–70 g, the discoid ones 110–122 g and the trapezoid ones 142–157 g. It should however be borne in mind that less than 9% of the loomweights bear such a mark, and those loomweights moreover belong to three different groups.[7] All the other marks are restricted to one or two loomweights only, thus precluding analysis.

As neither the distribution of weight nor the presence of marks allows us to identify any separate groups of loomweights with specific functions we are left with the types themselves. If we ignore the shape, the most noticeable difference between individual loomweights is the fact that some have two holes and others only one. This difference is generally ignored (*e.g.* by Wilson, 1930 in her discussion of different functional groups of loomweights in Olynthos),[8] but there must have been a reason for it. The most obvious conclusion would appear to be that each hole was meant for one hank (or possibly one warp thread). Presumably two hanks per loomweight became nec-

Table 3.12. New Halos. Numbers and proportions of loomweights found in the six houses.

House No.	Pyramidal	Discoid	Trapezoid	Conical	Total
1	10	2	0	1	13
2	10	1	2	0	13
3	9	4	5	0	18
4	18	2	2	2	24
5	28	3	3	0	34
6	88	41	7	1	137
Total	163	53	19	4	239
	68%	22%	8%	2%	

essary if there was not enough space for one loom-weight per hank. This may have been the case if a fairly fine, closely spaced thread was used for the warp. The fact that the doubly pierced loomweights are on average heavier than the singly pierced ones supports this conclusion, as does the flat, discoid shape of the more common type of doubly pierced loomweight in Halos. Hanging next to each other, such discs will have occupied significantly less space than pyramidal or conical loomweights.

We therefore propose to identify two functional groups of loomweights in Halos: those with a single hole, used for fabrics with relatively thick, widely spaced warp threads, and those with two holes, used to weight two hanks of fine, closely spaced warp threads.

3.5. Conclusion

Too few houses have been excavated in New Halos to allow us to draw general conclusions concerning weaving practices in the town as a whole. What does emerge is that some weaving was done in all the excavated houses, and if our interpretation of the different functional groups of loomweights (singly and doubly pierced) is correct, then we may perhaps also conclude that different types of cloth were woven in each house. On the other hand, the small number of loomweights per house suggests that home weaving was limited in scope, and that part of the textile production took place in specialised weaving ateliers. It is tempting to link the House of Agathon to such specialised production, as it contained over six times the average number of loomweights, a disproportionate number of which were discoid. However, as the vast majority of the loomweights from this house were found in a storage area (room 5) they cannot be definitely associated with weaving activities inside the house.

3.6. Notes

1. Deonna (1938: p. 152) suggests that weights of these types were also used for fishing nets, but the majority of the weights he discusses are made of lead.
2. On the warp-weighted loom in general cf. Hoffmann, 1964; Wild, 1970: pp. 61–68; Engelhardt & Weichmann, 1990; Rast-Eicher, 1992. On Greek warp-weighted looms cf. Crowford, 1936/1937; Forbes, 1956: pp. 199–207; Carroll, 1987 [1965]: pp. 37–58; Carroll, 1983; Keuls, 1983; Carroll, 1988: pp. 22–24. Cf. Sheffer & Granger-Taylor, 1994, for an important recent discussion of ancient textiles and their production.
3. On the different ways in which a hank of warp threads could be tied to a loomweight cf. Carroll, 1983.
4. Although Carroll (1988: pp. 22–23) agrees with the general consensus that the warp-weighted loom was the chief loom of the Greek tradition, she does not address the one-sided nature of our evidence (cf. Rast-Eicher, 1992; Keuls, 1983).
5. There is no evidence in New Halos for the use of stone or mudbrick loomweights. One lead loomweight was found.
6. For letters on loomweights cf. Stillwell, 1952: p. 270; Davidson, 1952: p. 151, fig. 24.
7. We cannot be certain that the deep cavity is a mark at all. At Nemea loomweights have been found with sticks inserted through the holes which pierced the loomweight, and metal rings are also known to have been inserted through them. They were used to allow the hank to be coiled or tied around the loomweight in different ways (Carroll, 1983). In New Halos, wooden or metal pins may have been inserted in the deep cavity with a similar purpose in mind.
8. But cf. Vogelsang-Eastwood (1990) on crescent-shaped objects with two or more holes. She is probably right in rejecting the identification of those objects as loomweights, but certainly wrong in claiming (p. 99) that "... all man-made weights have one hole ... ".

3.7. References

BIEBER, M., 1971 [1928]. *Griechische Kleidung*. De Gruyter, Berlin.

CARROLL, D.L., 1983. Warping the Greek Loom: A second method. *American Journal of Archaeology* 87, pp. 96–98.

CARROLL, D.L., 1987 [1965]. *Patterned Textiles in Greek Art. A Study of their Designs in Relationship to Real Textiles and to Local and Period Styles*. Ann Arbor [= Dissertation Los Angeles, 1965].

CARROL, D.L., 1988. *Looms and Textiles of the Copts. First Millennium Egyptian Textiles in the Carl Austin Rietz Collection of the California Academy of Sciences*. California Academy of Sciences, San Francisco.

CLARK, L., 1983. Notes on small textile frames pictured on Greek vases. *American Journal of Archaeology* 87, pp. 91–96.

CHAVANE, M.-J. & S. AGUETTANT, 1975. Les petits objects. In: J. Pouilloux & G. Roux (eds), *Salamine de Chypre* (= Salamine de Chypre 6). De Boccard, Paris.

CROWFOOT, G.M., 1936/7. Of the warp-weighted loom. *Annual of the British School at Athens* 37, pp. 36–47.

DAVIDSON, G.R., 1952. *The Minor Objects* (= Corinth 12). American School of Classical studies at Athens, Princeton.

DEONNA, W., 1938. *Le Mobilier Délien* (= Exploration archéologique de Délos 18). De Boccard, Paris.

ELLIS, R.S., 1976. Mesopotamian crafts in modern and ancient times: Ancient Near Eastern weaving. *American Journal of Archaeology* 80, pp. 76–77.

ENGELHARDT, B. & E. WEICHMANN, 1990. Archäologische Bodenspuren eines Gewichtwebstuhls der frühen Eisenzeit mit einem Beitrag über Webversuche. In: M. Fansa, B. Renken & J. Döring (eds), *Experimentelle Archäologie in Deutschland* (= Archäologische Mitteilungen aus Nordwestdeutschland, Beiheft 4). Isensee, Oldenburg, pp. 424–426.

FORBES, R.J., 1956. *Studies in Ancient Technology*, Vol. 4. Brill, Leiden.

HOFFMANN, M., 1964. *The Warp-Weighted Loom* (= Studia Norvegica 14). Universitetsforlaget, Oslo.

KEULS, E.C., 1983. Attic vase-painting and the home textile industry. In: W.G. Moon (ed.), *Ancient Greek art and iconography*. The University of Wisconsin Press, Madison, pp. 209–230.

METZGER, I.B., 1978. Die Funde aus der Pyrai. In: A. Altherr-Charonin & C. Bérard (eds), *Eretria, fouilles et recherches* (= Eretria 6). Francke, Berne, pp. 81–88.

RAST-EICHLER, A., 1992. Die Entwicklung der Webstühle vom Neolithikum bis zum Mittelalter. *Helvetia Archaeologica* 23, pp. 56–70.

ROBINSON, D.M., 1941. *Excavations at Olynthos: Metal and Minor Miscellaneous Finds* (= Olynthus 10). The John Hopkins Press, Baltimore.

SHEFFER, A., 1981. The use of perforated clay balls on the warp-weighted loom. *Journal of the Tel Aviv University Institute of Archaeology* 8, pp. 81–83.

SHEFFER, A. & H. GRANGER-TAYLOR, 1994. Textiles. In: J. Aviram & G. Foerster (eds), *Masada 4, The Yigael Yadin excavations 1963–1965, final reports*. Israel Exploration Society, Jerusalem, pp. 149–282.

STILLWELL, A.N., 1952. *The potters' quarter, the terracottas* (= Corinth 15, Part 2). American School of Classical Studies at Athens, Princeton.

VOGELSANG-EASTWOOD, G., 1990. Crescent loomweights? *Oriens antiquus* 29, pp. 97–113.

WILD, J.P., 1970. *Textile Manufacture in the Northern Roman Provinces*. University Press, Cambridge.

WILSON, L.M., 1930. Loomweights. In: D.M. Robinson, *Excavations at Olynthus, architecture and sculpture, houses and other buildings* (= Olynthus 2). The John Hopkins Press, Baltimore, pp. 118–121.

4. THE METAL FINDS

STEVEN HIJMANS

4.1. Introduction

Metal finds such as iron tools, structural elements and the like have so far received little attention in the archaeology of ancient Greece. In site publications they are relegated to volumes or sections entitled 'minor finds' and treated as such.[1] General books on Greek construction, agriculture or daily life tend to devote little space to specific tools, hardware and other metal *instrumenta domestica*.[2]

In part this is due to the fact that especially the iron objects of ancient Greece are often so poorly preserved that it is difficult to determine their nature and function. Metal finds do not feature prominently in excavation reports either, simply because they are rare in comparison with other finds. Metals can be easily recycled and are relatively costly, and in antiquity the metal of worn artefacts was therefore saved and re-used rather than discarded. But the lack of interest is also partly due to the traditional focus of classical archaeology on the history of ancient art and (monumental) architecture. As a result, there are so few good publications focusing on Greek metal finds that even in comparatively recent publications (*e.g.* Waldbaum & Knox, 1983) most of the important parallels are British or German rather than Greek or mediterranean. Most metal artefacts may not be glamorous, but they played a significant role in all aspects of Greek daily life, and therefore deserve study. They illuminate domestic architecture and carpentry, and constitute major components of weapons, tools and farm equipment, as well as domestic utensils, ornaments and cosmetic articles. In short, metal artefacts cover a whole range of objects an average Greek could expect to encounter on any given day. It is hardly surprising, therefore, that the houses of New Halos yielded a significant number of metal artefacts. Their good state of preservation is an added boon.

In a volume such as this, it is obvious that the metal artefacts should receive full attention. The number and range of the metal artefacts found in New Halos is however too limited for their publication to significantly enhance our understanding of Greek metalwork in general. Too many categories of artefacts are altogether absent (many tools, domestic utensils, certain types of farm equipment, *etc.*) while many other potentially interesting artefacts are too poorly preserved to be adequately explained and interpreted. Had we been able to excavate more houses in New Halos before their destruction by ploughing, we would, of course, have had a far larger number of finds per category, and this in turn, I am confident, would have helped us identify and interpret many metal parts which now remain unrecognised through lack of parallels. As it is, we have opted for a straight-forward, basic catalogue which lists all the metal finds from the houses and attempts to identify the objects and/or their functions only in those cases in which they appear to be fairly clear.

Some metal finds from Halos are either complete or largely complete, missing only a handle or a shaft which was made of more perishable material (spear-heads, sickles, scissors, the strigil, various tools, *etc.*).[3] Most finds are however less easy to interpret, because they formed (a small) part of a larger structure or perishable object. Quite a large number of finds, fi-nally, are so fragmentary that their original shape and function cannot be determined.

The order in which the individual objects will be discussed below is based largely on this division: first all the complete, recognisable types, then the recog-nisable structural elements whose specific functions are less clear, and finally the unrecognisable frag-ments. Basic descriptions of the objects are given in the catalogue (Appendix 5). Wherever possible, the descriptions and measurements given are those of the object itself, excluding any corrosion. Even so, the measurements should not be regarded as completely accurate, reflecting as they do objects which can change in shape even from one week to the next by shedding thick layers of lime-scaled corrosion. All measurements are given in centimetres.

I have restricted myself to describing the types of metal objects and discussing their uses. The compo-sition of the metals and the employed production tech-niques will be discussed either in a subsequent volume or in a separate publication.

All but one (a silver snake) of the metal finds from New Halos are made of iron, lead or bronze, and they are all corroded. The lead finds have a thin, greyish-white patina and have lost most of their pliability, as a result of which some have cracked. The majority of the bronze objects are covered with a layer of lime-scale; if not, they have a green patina and show a few spots of bronze disease. Some of the thinner bronze objects are so corroded that they have become very brittle and readily fall to pieces (*e.g.* the bronze spoon M64). Other thin bronze objects show no deteriora-tion whatsoever. The iron finds are the most corroded, although the state of individual finds differs signifi-cantly: some are intact and have only a very thin, patina-like layer of corrosion, whereas others are cracked through and falling to pieces. The state of the finds is apparently not linked to their size or location, but presumably mainly to the composition of the iron. The limited resources of the New Halos project did not allow for immediate or large-scale preservation and restoration of the metal finds, but through the generous courtesy of our colleagues in the Ephorate of Vólos, the most important finds could later be sent to the Vólos museum to be cleaned and conserved.

The most significant of these objects are now on dis-play in the Archaeological Museum of Almirós; the others are in the museum's storage house of the New Halos excavation and survey project.

In preparation for the catalogue, all the metal ob-jects were individually described, and most were drawn on a scale of 1:1.[4] No identification of catego-ries and types was attempted until this work had been completed, to avoid the risk of overinterpretation. Therefore some objects (certain knives *e.g.*) are de-scribed simply as strips of iron in our archives. The objects were categorised primarily on the basis of the New Halos finds themselves. Subsequent study of par-allels and secondary literature led to limited adjust-ments, but no major changes.

4.2. Agricultural implements

The agricultural implements of Halos consist of sick-les, hoes, a spade and shears (fig. 3.33 and Appendix 5). To these we should probably add the ferrules dis-cussed under the heading of weapons, as some of those objects may have served to reinforce the wooden prongs of pitch-forks, weeding-sticks or com-parable wooden tools. Few publications deal exten-sively with Greek agricultural implements, partly on account of the scant remains: most were made of wood (*cf.* the ferrules), the rest primarily of iron, and they are therefore rarely found in good condition.[5] It is thus especially fortunate that we recovered such a significant series of implements at New Halos. Indeed, we have examples of all iron implements used for general agriculture barring a plough, as well as what appears to be evidence for wooden farm equipment in the form of ferrules. We also have millstones. No evidence was found for threshing sleds *(triboloi)* with their characteristic flintstones, but it is quite possible that they had not yet been invented by the early Hel-lenistic age, and that the threshing was done by ani-mals treading the grain.[6] So we may safely state that, barring the absence of ploughs, the range of farming implements found at New Halos is consistent with what an archaeologist would expect to find in a Hel-lenistic farming community.

The only direct evidence that the inhabitants of New Halos practised pastoralism as well as arable farming is provided by the four shear halves found in the houses.[7] Although it seems likely that their actual function within the houses was that of knives, they have been included here as we presume that they were originally used for shearing sheep.

Hoes
The two hoes (M1-2) found in New Halos do not dif-fer significantly from those used in the region today. Of the two, the *dikella* is a unique find. It is a dou-ble-sided hoe, not unlike a pick-axe, with a single hacking blade on one side, and a two-pronged one on

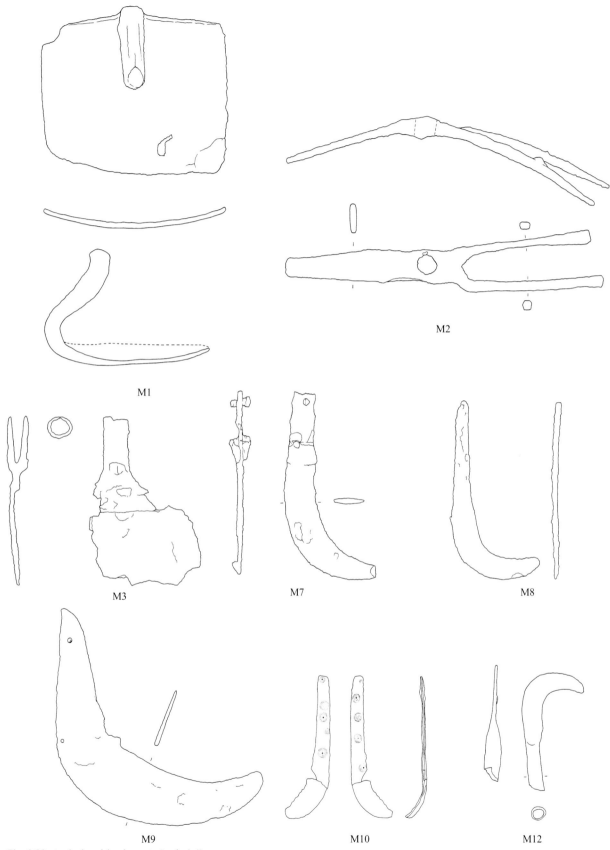

Fig. 3.33. Agricultural implements (scale 1:4)

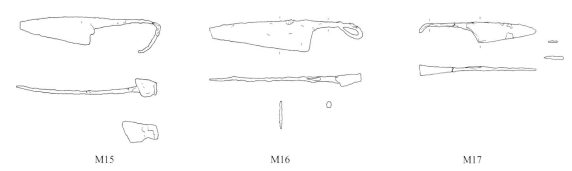

M15　　　　　M16　　　　　M17

Fig. 3.33. Agricultural implements (continued, scale 1:4).

the other (Isager & Skydsgaard, 1992: pp. 49–52, with refs). Although such hoes are mentioned in ancient sources, we believe that this is the first ancient Greek one to be published.[8]

Spade or shovel
Halos yielded one, incomplete shovel (M3). Although shovels are known from ancient written sources, this appears to be the first ancient Greek one to be published (Isager & Skydsgaard, 1992, p. 49; White, 1967: pp. 17–35). As it is incomplete, we cannot determine whether it was meant to be used primarily as a spade (for digging) or a shovel (for scooping up loose earth), but the latter seems more likely, as spades are generally more triangular.

Sickles and pruning knives
A wide range of sickles were found in New Halos, many of which are well-preserved (M4–13). Sickles are characterised by a curved blade (as opposed to the straight blades of knives), but apart from this basic common trait they vary greatly in size and shape. Most sickles found in New Halos are relatively small and have a broad, gently curving blade ending in a blunt, rounded point.

Sickles were used for pruning and harvesting, but it would be wrong to take the presence of sickles as necessarily indicative of arable farming or horticulture on a household basis. Gathering wild vegetables along the roadside was presumably as widespread in Greece in antiquity as it is today, so the presence of sickles need indicate no more than that members of the household occasionally engaged in this practice. Indeed, if we accept Boardman's (1971) suggestion, then some sickles may even have been used as strigils (Isager & Skydsgaard, 1992: pp. 52–53; White, 1967: pp. 71–103; Boardman, 1971: pp. 136–137; Jope, 1956: pp. 94–97).

Shears
The four shears found in New Halos are all of the same type (M14–17). Their most important function was probably the shearing of wool, although they may also have been cloth shears.[9] All the shears found in

M18

Fig. 3.34. Carpenter's tools (scale 1:4).

New Halos were broken at the spring – obviously the weakest point – and in all cases only one half was found. It seems likely, therefore, that the shear halves found in the houses were recycled and perhaps used secondarily as knives. The fact that in one instance the spring was actually bent back against the shank, forming a loop or eye (H23 5/m43), supports this suggestion.[10]

It is interesting to note that no shears were found in Olynthos (Robinson, 1941), nor does Raubitschek list any from Corinth (Raubitschek, 1988; Davidson, 1952; Stilwell, 1948). Precisely when the practice of shearing sheep was introduced into Greece is still a matter of debate (Daremberg & Saglio, 1896: 1241–1243, *s.v.* 'forfex', esp. figs. 3169–70; Gaitzsch, 1980: pp. 209–219, pls 69d, 74I; Ryder, 1983: p. 96 and pp. 696–700; White, 1967: pp. 119–120).[11]

4.3. Carpenter's tools

The carpenter's tools found at New Halos cannot be securely identified due to their poor state of preservation. The identification of the following objects as chisels and saws, respectively, should therefore be taken as tentative (fig. 3.34; Appendix 5).

Chisels
No object was found which can be securely identified as a chisel (M18–23). Some objects classified as nails may in fact have been chisels or engraving tools of some kind. They are characterised by thick, wedge-shaped shanks, usually rectangular in section, and the

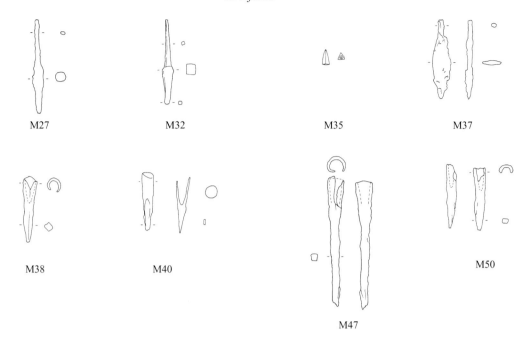

Fig. 3.35. Weapons (scale 1: 4).

lack of a clearly defined head. In those cases where they taper not to a point, but to a wedge-shaped end, their interpretation as a 'chisel' seems, on balance, preferable to that as a 'nail'. A second type of chisel consists of a shank which is pointed at both ends; one pointed end served to fasten the shank in its wooden handle, while the other was the tool point (Gaitzsch, 1980: pp. 148–174).

Saws
Two objects found in New Halos have been identified as small saws (M24–25). In both cases their exact size cannot be determined because intermediate pieces may be missing. The remaining parts are poorly preserved, with many of the teeth of the blade missing. It must be stressed, therefore, that in both cases the identification is tentative (Matthäus, 1984: pp. 150–154, with refs; Orlandos, 1966: pp. 33–38).

4.4. Weapons

Few weapons (fig. 3.35 and Appendix 5) were found in New Halos, despite the apparently military purpose of the city's foundation and the indisputably military nature of its strong fortifications. No remains of armour (helmets, body armour, *etc.*) or other forms of military equipment were found in the excavated houses, barring two spearheads (which are so poorly preserved that their identification is in fact only tentative). The arrowheads are quite likely to be evidence of hunting (both for food and in defence of flocks), although a military purpose cannot be excluded.

I have included in this section a category of objects that are usually classified as spear butts. I find this identification unlikely in the case of New Halos, because I do not see how one could explain the presence of 23 spear butts and yet only two spearheads in these houses. I therefore prefer to identify these objects as ferrules, *i.e.* iron points used to tip and strengthen the ends of wooden stakes, posts, shafts and implements. As I will argue below, these ferrules may have been parts of a range of different objects, including both agricultural implements and weapons. I have however included them here as I have been unable to determine which ferrules had which specific functions, and because comparable objects are generally discussed under the heading of weapons in other publications, too.

Arrowheads
The arrowheads found at New Halos belong to two types (M26–35). The common type, of iron, has a solid lozenge-shaped head tapering to a point at one end and a long, straight tang or shank at the other end which was inserted into the shaft of the arrow. In addition, one small bronze pyramidal arrowhead was found, with a hollow base into which the shaft was inserted. The small bronze arrowheads were probably used for small game and fowl; animal remains found at Halos include hare, and the total absence of bird remains could be attributable to the poor preservation of bone at this site. The larger arrowheads may have been used for predators and large game. Halos yielded remains of fox, badger, red deer and roe deer.[12]

Spearheads
The two spearheads listed in the catalogue are poorly preserved, and for this reason their identification is

only tentative (M36-37). Both are spearheads of the heavy hoplite spear, which was primarily a weapon of war. With type 1 the shaft of the spear was inserted into the hollow end of the spearhead, whereas the spearhead of type 2 ends in a long, thin shank or tang, which was inserted into the shaft. The two spearheads are both of iron.

Ferrules
All the ferrules found in New Halos are of the same type (M38–58); they consist of a tapering, solid point, ranging from quite sharp to very blunt, and a socket at the other end created by flattening that end during forging and folding the 'wings' round to form a tube. Sockets formed in this way often have a characteristic V-shaped opening on one side, resulting from the fact that the 'wings' did not fully close when bent round.

The function or functions of these objects is not immediately clear, and there is some confusion as to their exact identity. In our field-diaries, excavators have shown a consistent tendency to identify them as arrowheads.[13] This identification must be rejected. Although many of the ferrules are relatively small, they are too heavy for an arrow, and most of the shafts must moreover have had diameters of well over 1.5 cm for them to have fitted tightly into the hollow of the head. Such a diameter is inconceivable for arrow shafts.

We were long convinced that these were the heads of light hunting javelins of the kind that a shepherd might have carried with him to protect himself and his flocks for instance against wolves. There are however few published parallels, and most that have been published are of bronze. In addition, the bluntest ferrules (see above) will surely have been useless as weapons, unless we assume that the points of the specimens concerned broke off.

W.M. Flinders Petrie describes a number of similar bronze objects specifically as spear butts, "...varying from the tube to the ferrule".[14] Waldbaum and Knox (1983: No. 16) likewise, if tentatively, describe a Lydian one, also of bronze, as a spear butt, but they point out that it was originally identified as a spearhead.

A similar object of iron dating from Roman imperial times that was found at Fishbourne[15] has been identified as a ferrule, *i.e.* not necessarily a weapon at all, but simply a piece of iron used to strengthen and protect the end of a piece of wood. The Roman Brading Villa on the Isle of Wight also yielded a range of such objects, generally larger than those of New Halos, and varying more in type. The possibility that they were spearheads was rejected by Cleere (1958, pp. 66–67), who instead suggested that they were spikes which, mounted on sticks, were used for weeding, or the tips of wooden picks. Very similar objects

were until recently used in Britain to tip the points of wooden pitchforks (Manning, 1972: p. 188, No. 120).

Both the large numbers in which these objects were found in New Halos and their find contexts seem to preclude their identification as spear butts (one wonders where all the spearheads are). Some agricultural function would seem preferable, and more in keeping with the nature of the houses of New Halos. Even so, we feel that some (or even most) of these objects may very well have been used as tips of (hunting) javelins, although such a function is to be excluded in the case of the bluntest.

We therefore suggest the following possible functions for these ferrules (in descending order of likelihood): 1. tips of cattle prods, wooden pitchforks, pointed wooden hoes or other wooden farm implements; 2. points of light javelins used for throwing. In the case of the blunt ferrules this identification is to be rejected, unless we assume that the point has broken off; 3. spear butts (Marić, 1995: pl. 4,1–2 ('spear butts').

4.5. Domestic utensils (fig. 3.36 and appendix 5)

Knives
New Halos yielded few knives, and in none of the cases is the identification beyond doubt (M59–63). All knife blades are made of iron. The handle attachment varies, and is sometimes missing. The most common type is that of an iron prong or tang which was inserted into a wooden handle. See under shears (Ch. 3: 4.2) for a discussion of the reuse of the latter as knives after the spring had broken.

Ladle
One, very fragmentary bronze ladle was found in New Halos (M64). Its identification is certain as the shape was still clear during excavation, although much of the spoon subsequently crumbled to dust.

Pins and needles
Two bronze pins and the point of an iron pin or needle were found (M65–67). They are each of a different type. The pins listed in Appendix 5 did not have a primarily decorative function (although M66 has a carefully worked head). *Fibulas* and hairpins will be discussed below, under jewellery.

Strigil
New Halos yielded one small bronze strigil (M68; Raubitschek, 1998: 121–124; Robinson, 1941: pp. 172–180).

Weights
A few lead weights were found in New Halos (M69–72): a) a rectangular weight, flat-rectangular in section; b) a discoid weight, one side flat, one side convex;

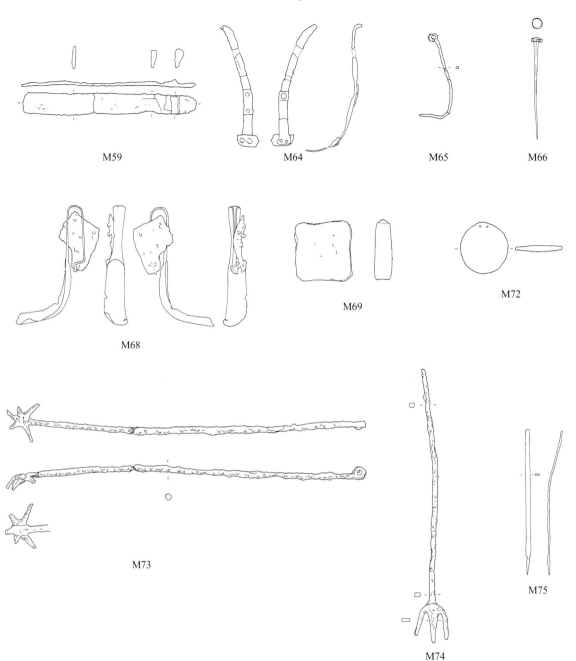

Fig. 3.36. Domestic utensils (scale 1:4).

c) a discoid loomweight, rectangular in section with two holes near the top.

'Meathook'

One iron object from New Halos, dubbed a 'magic wand' by its excavators, has parallels identified as meathooks (M73). It actually most closely resembles a modern Italian pasta spoon, being a pronged ladle or scoop with which pieces of meat, fish, large vegetables and the like could be fished out of pots of broth.

Fork

One long three-pronged fork was found in New Halos (M74).

Stylus

As Davidson (1952: p. 185) remarks, "The two essential requirements of the stylus – a point and a flat, blunt end for erasure – are fulfilled by many instruments", not all of them necessarily *styli*. Despite this *caveat* we feel certain that an object found in the House of Agathon (M75) is indeed a *stylus*, with a

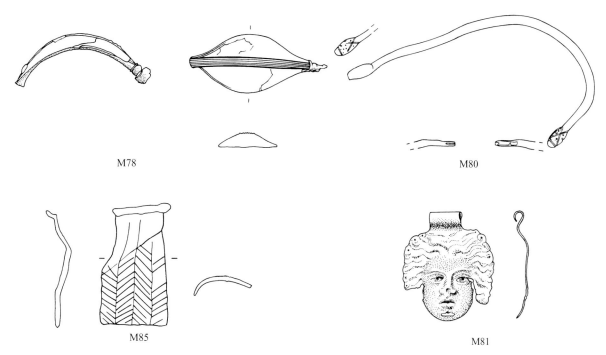

M78   M80

M85   M81

Fig. 3.37. Jewellery (scale 1:2).

well-defined point that was used for writing on wax tablets and a flat, blunt end for erasure. Whether the same holds for an object from the House of the *Amphorai* (M76) is less certain, as this object is broken, lacking a writing point.

4.6. Jewellery and cosmetic articles

Very little Hellenistic jewellery was found in the houses of New Halos (fig. 3.37 and Appendix 5). About half of the ornaments listed below derive from graves dating from the geometric period in the cemetery on top of which New Halos was built. There are also a few other finds which can be linked with cosmetics and toiletry.[16]

Hairpin, 'Glasinac' type
A bronze, ornamental (hair?)pin of the 'Glasinac' type was found in the House of the Coroplast (M77). Pins of this type have been found in former Yugoslavia as well as Greece (parallels are known from Olympia, Chauchitsa, Delphi, Pateli, Thermos (museum), and Kozani). The majority have been assigned earlier dates (Jacobsthal, 1956: p. 136, geometric period to 5th century BC; *cf.* Maier, 1956: pp. 69–70), but Philipp (1981: p. 98) has pointed out that they were still being used in Antigoneia (Albania) after the 3rd century BC, and I see no reason for an early date for this pin. The pin from the House of the Coroplast was lost in the earthquake of 1980. A similar hairpin, with a smaller end loop, was found in the Southeast Gate of New Halos.

Fibulae
Two *fibulae* were found (M78–79), both dating from the geometric period (mid-8th century BC), which means that they are stray finds from the cemetery from that period on top of which New Halos was built. The *fibulae* are of the same general type, which is typical of the region. Both lack the pin and catch, although the knob forming the transition from the bow to the pin and a small corner of the flat catch-plate survive on the *fibula* from the House of the Ptolemaic Coins.

Bronze bracelet
A bracelet from the House of the Geometric *Krater* (M80) is round in section, with both ends ending in flat, stylised snakes' heads decorated with a pattern of dots and lines. The bracelet is twisted out of shape. Snake-head bracelets were long popular in Greece (Philipp, 1981); without better parallels it is not possible to state with certainty whether this bracelet is Hellenistic or derives from the geometric-period burial, along with the *krater* which gave this house its name.

Bronze relief
A decorative bronze relief with a bent-over lip at the top was found in the House of the Ptolemaic Coins (M81). The relief shows the frontal face of a woman with long, wavy, Medusa-like hair.

Bronze inlay

A bronze inlay of a *pyxis*, originally rectangular in shape, came to light in the House of the Snakes (M82). Each side consists of a rectangle with two corner legs; the two short sides are of the same shape and dimensions; the front is of the same height as the short sides, but the back is lower, allowing space for the hinges of the *pyxis'* lost lid. Each side is decorated with two vertical lines defining the legs and two horizontal lines just below the middle. The *pyxis* itself (including the lid and bottom) was presumably made of thin wood.

4.7. Ritual objects

Snakes

Two small snakes were found in a stone vessel (*kadiskos*?) buried near the hearth of room 8 of the House of the Snakes, one made of iron (M83), the other of silver (M84). In view of the context in which they were found they have been listed as ritual objects here. They are thought to have formed part of an offering to Zeus Ktesios (Haagsma, forthcoming).

Miniature bed

A thin, flat, rectangular plaque of lead decorated with a chevron-pattern on one side was found in the House of the Geometric *Krater* (M85). It is presumably part of a miniature bed or couch whose four legs have disappeared. It closely resembles a miniature couch with all four legs intact which was found together with a miniature table (round, three-legged) in a child's grave in Eretria and is now in the Louvre. This bed is also of lead and closely matches ours in date, size, shape and chevron decoration.

4.8. Structural elements

The following section discusses metal objects which are recognisable *per se* and which formed part of some structure or object of perishable material that has since disappeared (fig. 3.38 and Appendix 5). In a few cases the finds' context and distribution provide indications of the nature of the structures concerned, *e.g.* in the case of decorative bosses of doors or chests. But the function of most of these metal remains can no longer be determined with certainty.

Nails

Many nails were found in Halos (M86–267). They range from tiny bronze pins to very large iron nails. Their common function was to fasten together two or more objects, generally made of wood. With modern nails this is achieved by hammering the nail through one object and into the next. This requires a strong nail of flawless iron, with a smooth, very straight shank. The quality of the thick, hand-forged nails of antiquity was not good enough for this kind

of use. Such nails were usually driven through a pre-drilled hole and went straight through both objects. The remainder of the shank was then hammered to one side, so that the nail in fact clinched the two objects together.[17] To avoid injury or damage from the exposed points, it was common practice to bend the points back first, and then hammer the shanks back in the same direction, so that the point was hammered into the wood. As a result, ancient nails such as those found in New Halos are typically shaped like an L or a rectangular J. In the catalogue we refer to such nails or shanks as 'clinched'.[18]

This aspect of the use of ancient nails has often been overlooked. Ginouvès & Martin (1985: p. 89) suggested that nails have always been used in the modern way, driven through wood without a pre-drilled hole, and many excavation reports typically present only a small selection of 'undamaged', *i.e.* straight nails, arranged typologically, without any mention of the possibility of clinching.[19] Cleere (1958: p. 56) mentions one clinched nail and suggests that in other cases the nail was 'too long' and its point broke off.[20]

Another aspect of traditional carpentry which is often overlooked by archaeologists is the fact that nails were by no means the only devices used to fasten together wooden elements. In fact many wooden objects, both small and large, were constructed with as few nails as possible, if any at all. In finer carpentry, pegging and doweling were usually the preferred methods of joining objects together, and various clamps were often used. Alternatives to nails used in construction include large clamps, lashing, *etc.* (Orlandos, 1966: pp. 45–49; Manning, 1976: pp. 143–153). So the fact that no small nails were found near the bronze inlay of a *pyxis* in the House of the Snakes (M82) does not necessarily imply that the inlay had become separated from the perishable material of the *pyxis* before being deposited, but could very well mean that no nails were used in the *pyxis'* construction. The inlay itself was definitely not fastened to the box with nails, as it contains no holes for any nails.[21] To avoid confusion or overly detailed descriptions, I have divided the nails into two basic groups: A. small nails and tacks; B. large nails. The small nails in principle all have lengths of at most 5.0 cm and the diameter of the shank is generally 0.4 cm or less. All the other nails I have classified as large. I believe that this subdivision reflects a real division between nails used for construction purposes (large) and nails used for fine carpentry and metalwork (small).

Cramps, clamps

Cramps and clamps of various types and shapes were found in New Halos (M268–278). In this section, cramps and clamps are understood to be all metal objects of a structural nature whose probable purpose

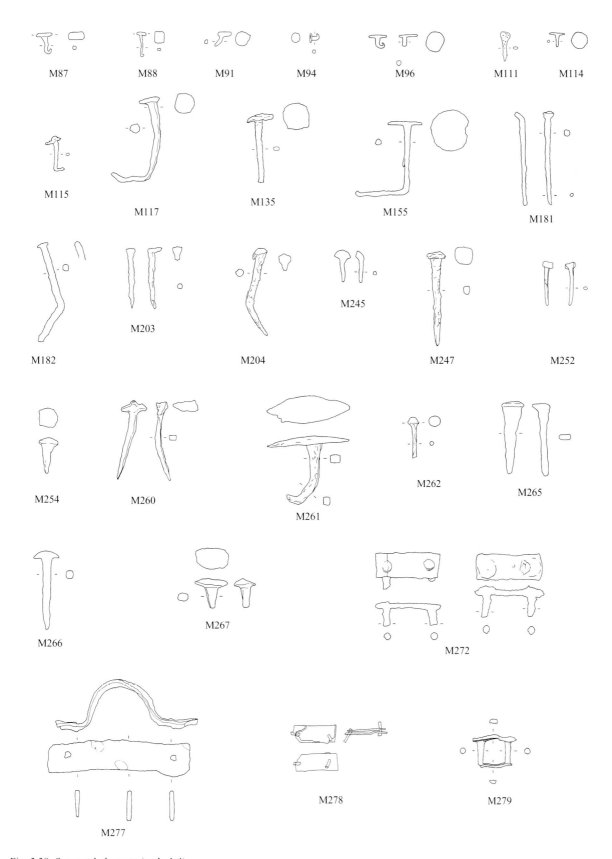

Fig. 3.38. Structural elements (scale 1:4).

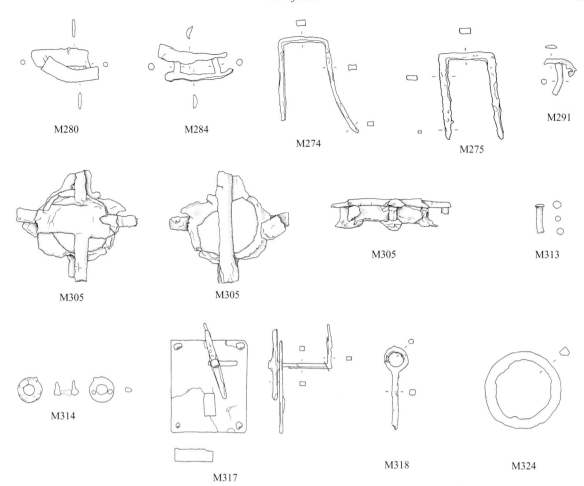

M280 M284 M274 M275 M291

M305 M305 M305 M313

M314 M317 M318 M324

Fig. 3.38. Structural elements (continued, scale 1:4).

was to hold together two objects. They differ from nails in that they were not somehow hammered through the two objects they were intended to connect, but fastened the objects together by metal bands or plates fastened to one of the objects at one end and to the other at the other end, or else fastened around the two objects. This broad definition clearly covers a wide range of types, which are listed separately in the catalogue.

Lead clamps

A very specific type of clamp found in New Halos is made of lead. Clamps of this type are in our excavation diaries generally referred to as 'pot repairs' (M279–307). It is certainly true that the practice of mending large, broken vessels with clamps of this type was widespread. Two lead strips were fastened across the fracture, one inside and one outside the vessel, and were connected to each other by two lead shanks inserted through a hole drilled through each of the two parts which were to be fitted together. We found a few pots which had been mended in this way

and which were obviously still in use at the time of the town's destruction. The majority of these clamps were however found in isolation, though sometimes still retaining a fragment of a sherd. When a pot was damaged beyond restoration it was apparently smashed and the lead clamps were removed and saved for remelting and reuse. In our opinion, the 'empty' lead clamps found in New Halos in association with clamps retaining the remains of a vessel should therefore be taken to constitute lead 'hoards', rather than clamps used to reinforce or repair objects made of perishable materials. The latter possibility can however not be altogether excluded, as lead clamps of this type are known to have been used elsewhere to repair stone and wooden objects, too.[22]

Fastenings

Various objects were found which can with a greater or lesser degree of certainty be identified as locks, keys, door handles and latches, keyholes, *etc*. These elements may have formed part of doors of houses, lids of chests, *etc*. (Briggs, 1956: pp. 415–416). Listed

in Appendix 5 are the objects which are likely to belong to this category on account of their shape or
context or known parallels. It will be clear that in
certain cases the identification can be only tentative.
Rings, fastened by two, three, or four prongs onto a
(wooden) surface, are regularly identified as keyhole
reinforcements (M314–316). Such elements were
placed over the hole for the key, and protected the
edges of the wood from damage. Whether this is the
only (or even the primary) function of such objects
is however doubtful. As Robinson observed, the distribution of these reinforcements in Olynthos (where
they are all of bronze) suggests that they were rarely
used for doors (unless there were internal doors with
locks), and cannot all have been used for keyholes in
chests and the like.[23] So the identification of these
objects as keyhole reinforcements is at best tentative,
and given here only for want of a better alternative.

A latch from the House of the Snakes consists of
a flat rectangular plate of iron with two rectangular
apertures, one horizontal and one vertical, and one
narrower than the other (M317). The door handle
passed through the narrow aperture; it consisted of a
long shank, clinched, and a cross-bar handle, both
rectangular in section. The aperture is such that the
shank could move a considerable distance from one
end to the other. The second, wider aperture is presumably the keyhole. Faint traces of the severely corroded nails which fastened the plate to the door were
observable in the four corners; a few minute fragments of the shanks of these nails were also found
(not described separately). A small rectangular sheet
of bronze (M598) was found in association with this
latch.

Pivot holes which held the metal (or metal-shod
wooden) rods on which the door swung were sometimes lined with metal. The Acropolis Gate of the
upper city yielded such objects *in situ,* of bronze
(Reinders, 1988: pp. 97–99, figs 60–61). Completely
different in shape, but apparently comparable in function, is the object (M319) from room 5 in the House
of Agathon (where it presumably lay as scrap metal)
which has been identified as a pivot hole lining
through comparison with a closely similar object
found at Isthmia (Raubitschek, 1998: pp. 133–134,
138, No. 486, pl. 75, with rcfs).

4.9. Parts of objects (fig. 3.39 and Appendix 5)

Lids
One complete lid of lead was found in New Halos,
and the fragments of another (M320 and 322). They
may have been the lids of *e.g. pithoi*. Completely different in size, shape and function is a small bronze
lid (M321) that was found in the House of the Snakes.

Rings
A variety of rings were found, of both iron and bronze,
in varying sizes (M323–337). These rings had no function *per se*, but formed part of larger objects. The potential range of uses is large. Iron rings may have been
fastened as handles to doors (of houses or cupboards),
lids (of chests), containers (wooden pails), *etc.*; the
handle may have been the ring itself, or a rope, chain
or swing-handle attached to it. Rings may also have
been attached to objects such as the ends of wooden
handles of tools and implements so that they could
be suspended from hooks; and rings may furthermore
have been used to connect leather straps to leather
bags, a bridle or rope to a harness, a gird to a pack-
saddle, *etc*. So it's not surprising that the former
function of individual rings can hardly ever be established.[24] One thing that is however certain is that none
of the circular rings listed here can be interpreted as
finger-rings[25], although it is not inconceivable that
some of the smaller bronze rings were parts of ornaments (for example elaborate earrings, necklaces or
bracelets). It should however be borne in mind that
even the smallest rings may have had the more mun-
dane function of a handle: *cf.* M321, a small bronze
lid of a *pyxis*, to which the bronze ring which served
as its handle is still attached; the diameter of that ring
is 0.5 cm.

Chain
New Halos yielded one object which may be the damaged link of a chain (M338). This identification is not
certain, however, and an alternative interpretation is
that it is a hook with an eye.

Rods
A small number of very long iron rods or pins were
found, whose function remains unclear (M339–340).
They cannot be classified as long nails because the
shanks have a uniform thickness and do not taper to
a point. One or two extremely long shanks without
heads (*e.g.* M446) probably belong to this category
but have been included under the heading of iron
shanks because they lack a head.

Hooks
Two hooks wcrc found in Ncw Halos (M341–345).
One is a very large (meat-)hook, the other is quite
small. No fish-hooks were found.[26]

Handles
Four metal handles were found in New Halos, but no
two are of the same type, so each has been described
separately (M346–349).

4.10. Metal fragments

The nature of a large number of metal finds could not
be determined, either because their state of preserva-

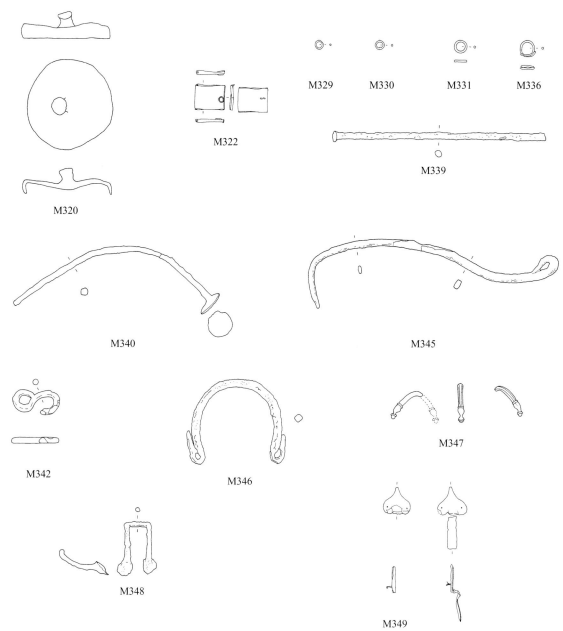

Fig. 3.39. Parts of objects (scale 1:4).

tion was too poor, or because they are fragments or elements of a larger whole, the nature of which now eludes us. Such objects are listed under a few general headings below, but, barring a few exceptions, no description of the individual objects is attempted (fig. 3.40 and Appendix 5).

Iron fittings, plating, reinforcements, structural elements

This category consists of iron plaques of a uniform thickness, with flat or curving, generally rectangular, shapes ranging from almost square to long and nar-

row, often with vertical shanks, holes for nails, *etc.* (M350–440). The potential range of functions is vast.

Iron shanks

All shanks without a head, and often also without a point, are listed in the catalogue (M441–565). Most are likely to be fragments of nails, but some of the shanks listed in the catalogue may of course have been (parts of) other objects; had it been found out of context, the door-handle M317, for example, would probably have been included in this section.

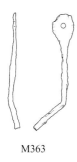
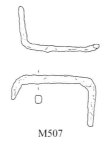
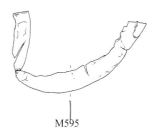

M363   M507   M595

Fig. 3.40. Metal fragments (scale 1:4).

Shapeless, unidentifiable and modern finds
Many iron 'artefacts' are listed in the catalogue under the headings of shapeless, unidentifiable and
modern finds (M566–607, M613–620 and M642–
648).

Wire
Various pieces of bronze wire were found, usually
bent and twisted, sometimes straight (M608–612). All
the wire is round in section.

Psimythion?
One short bar of lead (M621) – apparently a complete
object serving no logical function – may perhaps have
been used for producing 'white lead', or *psimythion*,
a cosmetic used to whiten skin.[27] The find context
however provides no proof of this, and I have therefore listed it here rather than under cosmetic articles.

Shapeless fragments of lead
Most shapeless fragments of lead come from room 5
of the House of Agathon and derive from objects
which melted in the fire that burned in the room at
the time of Halos' destruction (M622–641).

4.11. Notes

1. *Cf.* Fagerström (1988: p. 105): "Tools are rare in archaeologi-
    cal contexts, seldom adequately published, and almost never
    subject to serious study. Greece is no exception in this respect
    (...)".
2. As far as agriculture is concerned, Isager & Skydsgaard (1992)
    are the exception to this rule.
3. On wooden handles *cf.* Curwen, 1947.
4. I would like to thank Gerwin Abbingh, Annie Bakker, Gijsbert
    Boekschoten, Yvette Burnier, Jona Lendering, Margriet Haags-
    ma, Jan Jaap Hekman, Colette Kruyshaar, Douwtje van der
    Meulen, Bieneke Mulder, Gerald Wilmink and Christina
    Williamson for helping me at one time or another with the
    preparation of these descriptions and drawings.
5. On the lack of good publications *cf.* Isager & Skydsgaard, 1992:
    pp. 44–45; Amouretti, 1994: pp. 91–92.
6. Isager & Skydsgaard, 1992: p. 53 (with refs). Threshing floors
    would probably have been situated outside the city, and would, in
    any case, not have left any visible traces.

7. On the combination of arable farming and animal husbandry *cf.*
    Forbes, 1995.
8. For comparable Roman *asciae cf.* White, 1967: pp. 66–68, esp.
    fig. 44.
9. It is difficult to distinguish cloth shears and general-purpose
    scissors from sheep shears. The latter have pointed blades (as
    do the New Halos shears) rather than rounded ends, and are
    fairly large. The shears found at New Halos fall within the
    lower range of sizes recorded for ancient sheep shears
    (Gaitzsch, 1980; Ryder, 1983: pp. 695–700).
10. A variant of shears apparently still in use in Afghanistan has a
    wooden spring, but the shear halves from New Halos show no
    sign of this.
11. Patterson (1956: p. 193) believes that shears were not used until
    Roman times, and that the Greeks plucked wool.
12. On archery in general *cf.* Bergman *et al.,* 1988 (with refs).
13. They are also often identified as nails because the hollow end
    is frequently not noticed before cleaning, being filled with
    rust-coloured, firmly packed earth.
14. Flinders Petrie, 1917: p. 33, section 84; *cf.* esp. pl. XXXIX
    No. 191.
15. Cunliffe, 1971: pp. 134 ff., No. 53.
16. For examples of other artefacts associated with cosmetics, see *e.g.*
    the rare *unguentaria* (Ch. 3: 1.3; catalogue numbers P511–520).
17. This was still being done in the Middle Ages; *cf.* Thomson 1956:
    p. 394: "(...) The crude hand-made nails (...) could not be driven in
    with the careless abandon of the amateur carpenter of today – a
    hole had to be prepared first".
18. The term 'to clinch' may also be used for a rivet, *i.e.* an object
    used to bind together two metal elements (Ginouvès & Martin,
    1985: p. 65), but this use is secondary, and has for reasons of
    clarity been avoided in the catalogue.
19. Waldbaum & Knox, 1983: pp. 68–69; Manning, 1972: pp. 186–
    188; Dörner & Goell, 1963: *Taf.* 75 No. 20; Robinson, 1941:
    pp. 309–329, pls xci-xcv. Whether these nails were unused, or
    represent a parallel usage of nails without clinching, is diffi-
    cult to determine.
20. *Cf.* Chavane, 1975: p. 69, No. 191 (one clinched nail, presented
    without comment). Hurst, 1994: p. 304, does mention the de-
    liberate bending-over of the shafts, but without using the term
    clinched.
21. A clear example of misinterpretation resulting from a lack of
    understanding of traditional carpentry can be found in Hägg &
    Fossey (1980: p. 75), who discuss the meaning of a single
    bronze nail found in tomb 18 at Barbouna (for the tomb, *cf.*
    Hägg & Fossey, 1980: p. 72). They suggest that it served "... a

votive or other symbolic purpose" because it was the only nail in the tomb and because "... the bend in the nail is a neat right-angle and gives the impression of having been deliberately produced". In their view this means that the nail would have been "unusable in any case" and they add that a single nail "plainly does not of itself constitute evidence for the existence of a now disintegrated object such as a wooden box or the like". In fact, there is no reason to doubt that the bend is the result of clinching, for in New Halos the evidence for clinching includes even the tiniest bronze nails (*cf. e.g.* House of the Snakes, M349). Nor is there any reason why the bronze nail from the Barbouna tomb, with its nice, carefully made round head, should not have been part of a *pyxis* or some other perishable object constructed with the aid of doweling and/or pegging. The nail itself could have been used to fasten a decorative element, a handle ring or band (perhaps of leather?) or the like, which has disappeared because it, too, was made of perishable material.

22. Chavane, 1975: p. 18; she refers to Deshayes, 1963: pp. 95, 232 and pl. LXIX, for an example of a lead clamp used to repair a wooden object.

23. Robinson, 1941: pp. 259–260.

24. *Cf.* Davidson, 1952: p. 147.

25. Raubitschek, 1998: p. 61, tentatively classified the rings of this type which are of an 'appropriate size' as finger rings (*cf.* p. 168 app. E1 – but also her *caveat*, p. 61, note 81), while the rest, being too large or too small for finger rings, are described as 'rings for household vases' (p. 168, app. E2).

26. Evidence for fishing is rare (*cf.* Ch. 4: 4). A trident (unpublished) was found during excavations in the Southeast Gate. It is now on display in the Almirós museum.

27. I owe this suggestion to J.J. Hekman. *Psimythion* is produced by immersing either lead shavings or little blocks of lead in vinegar, which is subsequently allowed to evaporate, leaving behind a white powder. *Cf.* Forbes, 1955: pp. 39–40; Shear, 1936.

4.12. References

AMOURETTI, M.-C., 1994. l'Agriculture de la Grèce antique. Bilan des recherches de la dernière décennie. *Topoi* 4, pp. 69–94.

BERGMAN, C.A., E. McEWEN & R. MILLER, 1988. Experimental archery. *Antiquity* 62, pp. 658–671.

BOARDMAN, J., 1971. Sickles and Strigils. *Journal of Hellenic Studies* 91, pp. 136–137.

BRANIGAN, K., 1992. Metalwork and metallurgical debris. In: L. H. Sackett (ed.), *Knossos from Greek city to Roman colony. Excavations at the Unexplored Mansion II.* The British School of Archaeology at Athens, [S.l.], pp. 363–378.

BRIGGS, M.S., 1956. Building-construction. In: C. Singer, E. Jaffé, T.E. Williams & R. Raper (eds), *A history of technology,* Vol. 2. Clarendon Press, Oxford, pp. 397–448.

CARAPANOS, C., 1878. *Dodone et ses Ruines.* Hachette, Paris.

CHAVANE, M.J., 1975. *Salamine de Chypre, Tome 6: les Petits Objets.* Boccard, Paris.

CLEERE, H.F., 1958. Roman domestic Ironwork, as illustrated by the Brading, Isle of Wight, Villa. *Bulletin of the Institute of Archaeology 1,* pp. 55–74.

CUNLIFFE, B., 1971. *Excavations at Fishbourne 1961–1969,* Vol. 2. The finds. The Society of Antiquaries, London.

CURWEN, E.C., 1947. Implements and their wooden handles. *Antiquity 21,* pp. 155–158.

DAREMBERG, Ch. & E. SAGLIO, 1896. *Dictionnaire des Antiquités Grecques et Romaines,* Vol. 21. Hachette, Paris.

DAVIDSON, G.R., 1952. *The Minor Objects* (= Corinth 12). The American School of Classical Studies at Athens, Princeton.

DESHAYES, J., 1963. *La Nécropole de Ktima: Mission Jean Bérard, 1953–55.* Geuther, Paris.

DÖRNER, F.K., & GOELL, Th., 1963. *Arsameia am Nymphaios. Die Ausgrabungen im Hierothesion der Mithradates Kallinikos von 1953–1956* (= Istanbuler Forschungen 23). Mann, Berlin.

FAGERSTRÖM, K., 1988. *Greek Iron Age Architecture: Developments through Changing Times* (= Studies in Mediterranean Archaeology 81). Paul Åströms Förlag, Göteborg.

FORBES, H., 1995. The identification of pastoralist sites within the context of estate-based agriculture in ancient Greece: Beyond the 'Transhumance versus agro-pastoralism' debate. *Journal of the British School at Athens* 90, pp. 325–338.

FORBES, R.J., 1955. *Studies in Ancient Technology,* Vol. 3. Brill, Leiden.

GAITZSCH, W., 1980. *Eiserne Römische Werkzeuge* (= BAR International Series 78). British Archaeological Reports, Oxford.

GINOUVÈS, R. & R. MARTIN, 1985. *Dictionnaire Méthodique de l'Architecture Grecque et Romaine, Tome 1: Matériaux, Techniques de Construction, Techniques et Formes du Décor* (= Collection de l'école Française de Rome 84). École française d'Athènes, [Athènes].

HÄGG, I. & J. FOSSEY, 1980. *Excavations in the Barbouna Area at Asine, 1973-1977. The Hellenistic Necropolis and Later Structures on the Middle Slopes* (= Boreas 4.4). Universitetsbibliotheket, Uppsala.

HENIG, M., 1994. Objects mainly of metal, bone and stone. In: H.R. Hurst (ed.), *Excavations at Carthago.* Oxford University Press, Oxford, pp. 261–281.

HOEPFNER, W., 1983. *Arsameia am Nymphaios, 2. Der Hierothesion des Königs Mithradates I. Kallinikos von Kommagene nach den Ausgrabungen von 1963 bis 1967* (= Istanbuler Forschungen 33). Wasmuth, Tübingen.

HURST, H.R., 1994. *Excavations at Carthage. The British Mission, vol. 2.1. The Circular Harbour, North Side. The Site and Finds other than Pottery.* Oxford University Press, Oxford.

ISAGER, S. & J.E. SKYDSGAARD, 1992. *Ancient Greek Agriculture. An Introduction.* Routledge, London & New York.

JACOBSTHAL, P., 1956. *Greek Pins and their Connexions with Europe and Asia* (= Oxford Monographs on Classical Archaeology 3). Clarendon Press, Oxford.

JOPE, E.M., 1956. Agricultural implements. In: C. Singer, E. Jaffé, T.E. Williams & R. Raper (eds), *A history of technology,* Vol. 2. Clarendon Press, Oxford, pp. 81–102.

MANNING, W.H., 1972. The Iron Objects. In: S. Frere (ed.), *Verulamium excavations,* Vol. 1. Thames and Hudson, Oxford, pp. 163–195.

MANNING, W.H., 1976. Blacksmithing. In: D. Strong & D. Brown (eds), *Roman crafts.* Duckworth, London.

MARIĆ, Z., 1978. Depo Pronaden U Ilirskom Gradu Daors. *Glasnik Zemaljskog Muzeja Bosne I Hercegovine U Sarajevu NS 33,* pp. 23–67.

MARIĆ, Z., 1995. Die hellenistische Stadt oberhalb Ošanii bei Stolač (Ostherzegowina). *Bericht der Römisch-Germanischen Kommission* 76, pp. 31–72.

MATTHÄUS, H., 1984. Untersuchungen zu Geräte- und Werkzeugformen aus der Umgebung von Pompei. *Bericht der Römisch-Germanischen Kommission,* pp. 73–158.

MAIER, F., 1956. Zu einigen bosnisch-herzegowinischen Bronzen in Griechenland. *Germania* 34, pp. 63–75.

ORLANDOS, A., 1966. *Les Matériaux de Construction et la Technique Architecturale des Anciens Grecs, Tome 1* (= École Française d'Athènes, Travaux et Mémoires 16). Boccard, Paris.

PATTERSON, S., 1956. Spinning & weaving. In: C. Singer, E. Jaffé, T.E. Williams & R. Raper (eds), *A history of technology*, Vol. 2. Clarendon Press, Oxford, pp. 191–220.

PETRIE, W.M. FLINDERS, 1917. *Tools and Weapons Illustrated by the Egyptian Collection in University College, London*. University Press, London.

PHILIPP, H., 1981. *Bronzeschmuck aus Olympia* (= Olympische Forschungen 12). Walter de Gruyter & Co., Berlin.

RAUBITSCHEK, I.K., 1998. *The Metal Objects, 1952–1989* (= Isthmia 7). The American School of Classical Studies at Athens, Princeton.

REINDERS, H.R., 1988. *New Halos. A Hellenistic town in Thessalia, Greece*. Hes, Utrecht.

ROBINSON, D.M., 1941. *Excavations at Olynthus: Metal and Minor Miscellaneous Finds* (= Olynthus 10). The John Hopkins Press, Baltimore.

RYDER, M.L., 1983. *Sheep and Man*. Duckworth, London.

SAHM, B. 1994. Sonstige Metallfunde. In: Th. Kalpaksis, A. Furtwängler & A. Schnapp (eds), *Eleftherna 2. 2*. Ekdosis Panepistimiou Kritis, Rethymno, pp. 116–120.

SHEAR, T.L., 1936. Psymithion. In: E. Capps (ed.), *Classical studies presented to Edwards Capps on his seventieth birthday*. Princeton University Press, Princeton, pp. 314–317.

SINGER, C., E. JAFFÉ, T.E. WILLIAMS & R. RAPER (eds), 1956. *A history of technology*, Vol. 2. Clarendon Press, Oxford.

STILLWELL, A.N., 1948. Metal and glass objects. In: A.N. Stillwell & J.L. Benson (eds), *The potters' quarter* (= Corinth 15, part 1). American School of Classical Studies at Athens, Princeton, pp. 114–131.

THOMSON, R.H.G., 1956. The medieval artisan. In: C. Singer, E. Jaffé, T.E. Williams & R. Raper (eds), *A history of technology*, Vol. 2. Clarendon Press, Oxford, pp. 382–396.

WALDBAUM, J.C. & R. KNOX, 1983. *Metalwork from Sardis: the Finds through 1974*. Harvard University Press, Cambridge, Mass.

WELLS, B. (ed.), 1993. *Agriculture in Ancient Greece. Proceedings of the Seventh International Symposium at the Swedish Institute at Athens, 16–17 May 1990* (= Acta Instituti Atheniensis Regni Sueciae 42). Åström, Stockholm.

WIEGAND, Th. & H. SCHRADER, 1904. *Priene, Ergebnisse der Ausgrabungen und Untersuchungen in den Jahren 1895–1898*. Reimer, Berlin.

WHITE, K.D., 1967. *Agricultural Implements of the Roman World*. Cambridge University Press, London, *etc.*

WHITE, K.D., 1975. *Farm Equipment of the Roman World*. Cambridge University Press, Cambridge.

5. COINS

H. REINDER REINDERS

Coins were found in all six of the houses excavated between 1978 and 1993 (fig. 3.41). Although the majority of the coins are very small, on average 20 to 36 coins were found in each house thanks to the volunteers' careful excavation work; no metal detectors were used.

Due to the presumed earthquake that caused the sudden end of the occupation of New Halos and the site-formation process, the artefacts reflect daily life at the time of the disaster (Haagsma & Reinders, 1991). The coins from this 'closed context' give a fair account of the bronze emissions that were in circulation around 265 BC in this part of Thessaly (appendix 6). Other sites have yielded larger quantities of coins, but the information they provide on mint circulation is generally less accurate due to a long period of occupation, a complex stratigraphy or an uncertain context.

In this contribution on the site finds of New Halos we will focus on the autonomous bronze coinage of the cities of Achaia Phthiotis at the beginning of the 3rd century BC and the evidence provided by the coins concerning the dates of occupation of the city of New Halos. Contacts between New Halos and other cities will be discussed, as will the place of these bronze emissions in the mint circulation of the 3rd century BC.

5.1. Silver and bronze denominations

The cities of Halos and Melitaia in Achaia Phthiotis had struck bronze coins in the 4th century until 344 BC, when Thessaly came under Macedonian control (Rogers, 1932). At the beginning of the 3rd century BC, a number of cities in Achaia Phthiotis – Ekkara, Phthiotic Thebai, Peuma, Larisa Kremaste and New Halos – started to strike coins again (fig. 3.42). No silver coins from this period are known from these cities, with exception of a silver emission of Phthiotic Thebai. The cities obviously relied on foreign silver emissions for external trade and exchange. Bronze coins from New Halos are relatively rare; no bronze coins have been found in excavations at other sites. Evidence concerning the provenance of coins from Halos in private and public collections is either non-existent, lacunary or doubtful. So, for the time being we have the impression that the bronze coins of New Halos were intended mainly for internal use and for contacts with neighbouring cities.

Only seven out of a total of 150 Hellenistic coins found during the excavation of the six houses are of silver. They are three hemidrachms from Lokris, two silver coins from Thespiai and one silver coin from Thebai and Corinth each (Appendix 6). Other sites have likewise yielded only few silver coins; at Kassope, for instance, only 36 coins out of a total of 1170 were of silver (Dakaris, 1989: pp. 61–62). At New Halos we must however take into account the possibility that citizens took precious objects and silver coins with them when they left their houses, alarmed by the foreshocks of the impending disaster.

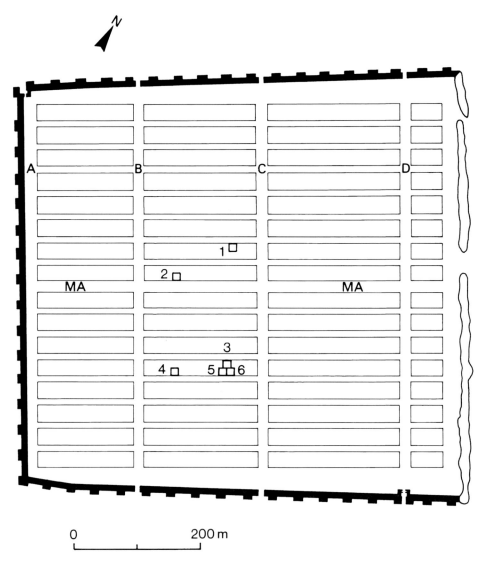

Fig. 3.41. Plan of the lower town of New Halos, showing the location of the excavated houses.

After the Macedonian conquest of central Greece in 344 BC, local coins remained in use in many regions, even long after their presumed date of emission. Moreover, some cities like Larisa in Pelasgiotis continued to strike coins throughout the late 4th and early 3rd centuries, until the large bronze emissions of Antigonos II Gonatas came into circulation (Furtwängler, 1990: pp. 118–121). In the third century BC many cities again started to strike their own coins. In Demetrias, the coins of the Macedonian kingdom however dominated over the coins struck by the city itself: 48 against 8% (Furtwängler, 1990: p. 110). The site finds from Eretria show the opposite; the coins of the town itself, the Euboian league and neighbouring towns outweigh the Macedonian emissions in a ratio of 46 to 20% (Furtwängler, 1990: p. 107).

The city of Chalkis continued to strike silver emissions in the 3rd century BC, though with inter-

vals. Picard (1979: p. 349) pointed out that bronze coins were initially only supplementary to silver emissions for minor expenses. In the course of the 3rd century, the role of bronze coinage however became more important: large bronze denominations gradually replaced the silver coins for internal use and interregional commercial contacts. In Chalkis this development started relatively late in the 3rd century BC. Picard's fourth category, dated 245–196 BC, comprises bronze coins only. Unlike in Euboia, in Aigyptos silver coinage intended for internal use had been replaced by large bronze denominations already during the reigns of Ptolemaios I and Ptolemaios II.

Coins struck by the Hellenistic kingdoms have been found in many excavated sites; at Demetrias and Eretria they represented 48 and 20%, respectively, of the total number of coins recovered (Furtwängler,

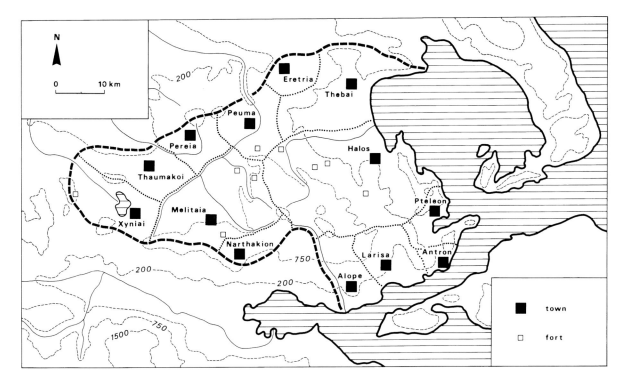

Fig. 3.42. Cities in Achaia Phthiotis.

1990: pp. 107–110). In New Halos, the great number of large bronze coins of Ptolemaios II came as a surprise. The coins had not been recognised as such before cleaning. Their weight outweighs that of the large bronze coins of the Macedonian kingdom: an average weight of 15 g in the case of the Ptolemaian coins as against 6 g in the case of the Kassandros coins showing a naked youth on a horse on the reverse.

Coins of Ptolemaios II are known from several excavations, for instance from Kourion on Kypros, Mirgissa in Aigyptos and Koróni in Attikí. Of the 400 coins found in excavations at Eretria, 21 were of Ptolemaios I and II. The Ptolemaian coins of Koróni, which represented 75% of the total number of coins found at this site, are linked to the presence of a Ptolemaic army camp during the Chremonidian war (268/7–262/1 BC). In my opinion the Ptolemaian coins of New Halos have nothing to do with any military action or cultural contacts of the kind suggested for Demetrias and Eretria (Furtwängler, 1990: p. 106).

Large denominations struck by Antigonos II Gonatas and the Aetolian league were not encountered in the houses of New Halos. Such coins were found in large quantities in towns like Demetrias and Eretria, but also in the cemetery of Medeon in Phokis. In Eretria 53 out of a total of 400 coins were struck by Antigonas Gonatas, and of the total of 99 coins found in Medeon 17 were struck by Antigonos and 18 by the Aetolian league (Hackens, 1976: pp. 123–25). The town of New Halos was obviously destroyed before

Fig. 3.43. Coin of Antigonos II Gonatas (Appendix 6, C21).

these emissions were in wide circulation, for only two coins of Antigonos Gonatas (fig. 3.43) have been found and none of the Aetolian league.

The majority of the coins found at New Halos are however small bronze denominations, which tell us something about the city's contacts with other cities. Cities and districts all over Greece are represented. Coins from towns in Thessaly and neighbouring districts have been found: Peuma, Larisa Kremaste, Malis, Lokris, Histiaia and Chalkis. But the finds also include odd specimens, such as coins from Korkyra, Sykiona and Orthagoria; they may have been lost by travellers. The provenance of the coins found in Halos is indicated in figure 3.44 and Appendix 6.

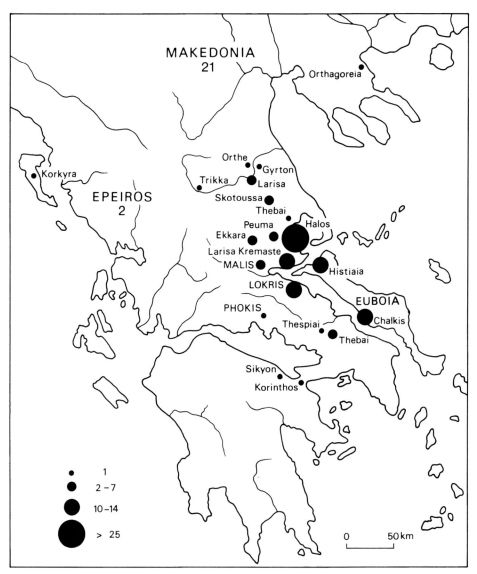

Fig. 3.44. Provenance of the coins found in the six excavated houses of New Halos.

5.2. Autonomous coinage in Achaia Phthiotis

The coins found at Halos reveal considerable mint activity in some of the towns of Achaia Phthiotis in the first quarter of the 3rd century BC: Ekkara, Thebai, Larisa, Halos and Peuma. Barring those of Ekkara, the coins of all these towns bear the monogram XA. Rogers (1932: p. 11) associated these coins with Demetrios Poliorketes' military campaign in central Greece in 302 BC. Furtwängler (1990: pp. 223–224; note 104) argued that XA represents the sign of a mintmaster of a mint union of these towns, rather than a league of the four towns initiated by Demetrios Poliorketes, as I presumed (Reinders, 1988). Characteristic of the early-3rd-century-BC coins of these towns are their good quality, the regular die positions and perhaps also some stylistic resemblance.

It is hard to say in which period this mint union was active. Several dies are known of the small denomination, series 7, of the coins of New Halos (Reinders, 1988: Appendix 6). These coins are all of good quality and have regular die positions. There are also many small coins, of series 8–20, of considerably poorer quality, a die position of approx. 7, and a wider range of weights than the coins of series 7. The obverse of the majority of these coins is vaguely struck. Coins of both series 7 (fig. 3.45) and series 8–20 were found at New Halos. They were hence in circulation at the time when the town was destroyed. There is no evidence suggesting that the city was still striking coins at that time. Since the coins of series 8–20 do not bear the monogram XA we assume that they date from the time when the mint union had ceased its activity. For the time being we have dated the coins

Fig. 3.45. Coin of New Halos, series 7.5 (Appendix 6, C60).

bearing XA to 302–288 BC. The coins without this monogram may have been struck during the interregnum, 288–277 BC, before Antigonos II Gonatas assumed power, or during a later period when Thessaly again came under Macedonian control and the large bronze emissions of Antigonos came into general use. The role of bronze coinage became more important for internal use during the 3rd century BC. The large bronze coins of the Ptolemies have been found at New Halos, Eretria and other cities. Increasing use of larger bronze coins is also reflected by the coins struck by the towns of Achaia Phthiotis and other towns in Thessaly. The towns of Achaia Phthiotis struck bronze denominations with average weights of approx. 2 and 6 g. The latter coins are comparable with the issues of Kassandros, which have an average weight of 6 g, and the bronze emissions of Antigonos II Gonatas and the Aetolian league later in the 3rd century BC. They even outweigh the bronze coins issued during the reign of Demetrios Poliorketes, when large silver coins were struck, too.

5.3. Dating evidence

The presumed dates of the coinage of the coins found during the excavation of the six houses in New Halos are presented in Appendix 6. As can be seen in the appendix, no coins from the second half of the third century BC were found. The coins of Ptolemaios II tell us when the occupation of New Halos came to an end. Svoronos (1904–1908) and Cox (1959) assumed that the letters between the eagle's legs indicate Ptolemaios' regnal year. Mørkholm (1991: p. 101) is however of the opinion that the letters represent some kind of issue numbering. If the letters indeed indicate the years of the reign of Ptolemaios II "this would spread a quite homogeneous coinage over a period of twenty years". If they are some kind of issue numbering, the series may have been produced "over a period of between five and ten years".

So far, thirteen coins of Ptolemaios II have been found in New Halos, seven of which are too badly worn to allow us to distinguish the letters concerned, while the others bear the letters O (3), T (2) and Φ (2). If the letters represent regnal years, the letter Φ gives a *terminus post quem* of 265/4 BC for the end of the occupation in New Halos. The Ptolemaic coins are worn to varying extents. Those bearing T and Φ

are the least worn, so the letters at least provide a chronological sequence. McCredie associates comparable coins found in Attica with the Chremonidean war (268/7–262/1), when Ptolemaios II supported Athens. So, whether the letters represent regnal years or some kind of issue numbering, they were definitely in use in the 260s (Mørkholm, 1991: p. 105). No coins of the later bronze emissions of Ptolemaios II, which according to Mørkholm coincided with the introduction of the Arsinoe gold *c.* 261/0 BC, have been found in New Halos. The absence of coins of Antigonos Gonatas and the Aetolian league also seems to confirm a date shortly after 265/4 BC for the end of the occupation of New Halos.

If our hypothesis that the occupation of New Halos came to an end on account of an earthquake is correct, then the coins found in this city give a good impression of the coins that were in circulation in Achaia Phthiotis around that time.

The coins of New Halos confirm the dates of the coins of the neighbouring towns of Peuma, Phthiotic Thebai and Larisa Kremaste. Rogers dated these coins to 302–286 BC and linked the mint activity to the influence of Demetrios Poliorketes, who favoured the autonomous position of Greek cities in his campaigns against Kassandros. He may have continued this policy during his reign. Other mints were active in Thessaly besides those of the four towns of Achaia Phthiotis: Larisa, Skotoussa, Orthe and Ekkara. The latter two towns have not been localised with certainty; Orthe was a town in Thessaliotis and Ekkara lay in Achaia Phthiotis. The state of preservation of the coins from these two towns suggests that they date from the first quarter of the 3rd century BC, and not the 4th century BC as assumed by Rogers (1932).

The coins' state of preservation also provides some clues to the presumed dates of the coinage of the other coins found at Halos. The silver coins from Corinth and Thespiai are fairly worn, suggesting that their presumed date, between 400 and 344 BC, is correct. The same holds for the two silver hemidrachms from Lokris, which have been dated to 369–338 BC.

Picard studied the coins struck by Chalkis (1979). Coins of his first and second categories are represented among the site finds of New Halos; these coins Picard dated to 318–290 and 274–252 BC, respectively. The coins found in New Halos seem to confirm these dates.

Coins of the Euboian league showing a standing cow on the obverse and a bunch of grapes on the reverse were according to Wallace (1956: pp. 130–131; Pl. XV, 10–16) struck towards the middle of the 2nd century BC. Excavations in Eretria showed that a 3rd-century date is more likely (Bloesch, 1965: p. 288). Picard (1979: pp. 170–171) listed these coins as issues 22–26, for which he gave a date between 253/2 and 245 BC. One coin of the Euboian league, presumably of Picard's emission 24, bearing EYBO l., was found in New Halos. Picard (1979: pp. 272–274) associates emissions 22–26 with the period in which the cities of Euboia acknowledged Alexandros of Corinth, a grandson of Demetrios Poliorketes, as king (253/2–245 BC). But the coin's poor state of preservation and the fact that the town of New Halos no longer existed in that period suggest an earlier date for emissions 22–26. The coins of another Euboian town, Histiaia, are generally dated to 338–300 BC. The extent of wear of the coins found at New Halos seems to confirm this date.

It has long been a matter of discussion whether the Macedonian coins bearing the monogram ΔHMHTPH on the boss of a Macedonian shield are to be ascribed to Demetrios Poliorketes (306–283 BC) or to Demetrios II (306–229 BC). Head (1977: p. 232) dated these coins to the latter's reign, contradicting Newell (1927: pp. 118–120). The *Sylloge Nummorum Graecorum* (SNG Copenhagen 1224–1228) agrees with Head. Three coins bearing the monogram ΔHMHTPH were found in New Halos and also three coins of Demetrios Poliorketes showing a male head with a Corinthian helmet on the obverse and a prow on the reverse. The fact that no coins from the second half of the 3rd century have been found in the houses of New Halos implies that the coins bearing the monogram ΔHMHTPH are to be ascribed to Demetrios Poliorketes.

5.4. Phrixos, Protesilaos and Thetis

Legends relating to the respective cities are indicated on the coins from New Halos, Thebai and Larisa. The story of Phrixos, who is represented clinging to a ram on the 3rd-century coins of New Halos, relates to the legendary founder of Halos, Athamas, whose children Phrixos and Helle fled to Kolchis on the ram with the golden fleece (Reinders, 1988: pp. 164–166; *cf.* Ch. 3: 2 for a terracotta relief showing Helle riding a ram). Helle seated sideways on the ram is depicted on the 4th-century coins of Halos. The representation of Phrixos in the reverse dies of the 3rd-century coins of New Halos (fig. 3.45) is different from that in, for instance, ceramic reliefs. According to Moustaka (1983: p. 68), this is evidence of the "*Eigenständigkeit des Stempelschneiders der seine Bildtypen schuf, ohne an Vorbilder streng gebunden zu sein*".

The coins from the city of Phthiotic Thebai show Protesilaos leaping ashore from the prow of a galley on the obverse (Moustaka, 1983: p. 64). Protesilaos came to Troy with forty ships carrying people from Phylake, Pyrasos, Antron and Pteleos, and was the first to be killed in the Trojan War. He himself came from Phylake, a town which is generally assumed to have lain in the territory of 3rd-century Phthiotic Thebai (Stählin, 1928: p. 173).

Thetis, Achilles' mother, is depicted on the coins of Larisa in Phthiotis. She is seated on a hippocamp and holds a shield in her left hand. Depictions of Achilles are not to be found on coins of the towns of Achaia Phthiotis, although his homeland was in the Óthris region.

Contrary to those of New Halos, Thebai and Larisa, the coins of the city of Peuma show no legend on the reverse, but simply an enlarged version of the monogram XA. A young male (?) head adorns the obverse of the coins of Peuma. The link with either Thetis or Achilles suggested by Heyman seems implausible (Moustaka, 1983: p. 68). Did Peuma lack a legendary prehistory like the other towns; a new town without a predecessor?

The bronze emissions of New Halos, Thebai, Larisa and Peuma are linked not only by the presence of the monogram XA, but also by the choice of legends relating to the cities' history, the coins from Peuma being an exception.

5.5. Copper production in Achaia Phthiotis

The production costs of bronze coins were considerable. A high temperature of about 1300°C was necessary to extract copper from ore and to melt the copper to produce the bronze flans. A bronze coin was of little value in comparison with a silver coin of the same weight and the agio for exchanging bronze currency for silver was unfavourable. There was another factor influencing the cost of production: the bronze of the flans was much harder than silver, so a larger number of dies was required for large emissions as the dies soon became worn. In this respect it is not surprising that in the early 3rd century BC cities refrained from producing their own bronze emissions and instead relied on the bronze emissions of the Hellenistic kingdoms and, as in the case of Demetrias, cities like Larisa. But how are we to explain the bronze emissions of New Halos, Phthiotic Thebai, Peuma and Larisa Kremaste from around this time? As we have seen in the preceding sections, these four cities in Achaia Phthiotis struck bronze coins of good quality with fixed die positions. Those of New Halos and Larisa have denominations of 2 and 6 g. These coins bear the monogram XA of a mintmaster (?), and all except those from Peuma show mythological scenes on the reverse. The foundation of the new towns of Halos and Peuma and the considerable size of the towns of Phthiotic Thebai and Larisa Kremaste imply a period of prosperity in Achaia Phthiotis, in which the population increased and there was consequently a need for small change.

Considering the high production costs of bronze coins there is only one factor that might explain the bronze emissions of the four towns: the occurrence of copper ore in Achaia Phthiotis, especially in the Óthris mountains. The geological maps of 'Almyros' and 'Anavra' show a few present-day copper mines. More convincing is the evidence obtained by Papastamatiki & Dimitriou (1987), who published a survey of the distribution of slags – *scoriae* – in Achaia Phthiotis and the production of copper in Pelasyía. Copper slags had already been found at several sites before the Geological Institute at Athens started an investigation to determine their distribution. This survey yielded evidence on slags at the following sites: Pelasyía, Archáni, Perivóli, Stírfaka, Limogárdi, Anávra, Spartiá and Áyii Theódori. The total volume of the slags was estimated to be 150,000 m³, about 100,000 m³ of which were found near Pelasyía.

The village of Pelasyía lies close to the site of the ancient town of Larisa Kremaste. Larisa is also called Larisa Kremaste or Pelasgian Argos in the Iliad. Artefacts found among the slags indicate that copper was produced at this site in the 4th/3rd century BC. The slags were found to contain metallic ore, indicating that they were produced in smelting for the purpose of extracting copper at a temperature of between 1100–1400°C. It is tempting to link the production of copper near Larisa with the minting activities of Larisa, Halos, Thebai and Peuma in Hellenistic times. According to Papageorgiou *et al.* (1990) the "production of material was primarily for coinage production, undoubtedly not simply for local use".

5.6. Contacts between New Halos and other cities

It is generally accepted that the "area of circulation of bronze coinages was restricted within the narrow territory of the city itself and perhaps a few neighbouring city states" (Mørkholm, 1991: p. 6). The coins found in the six excavated houses however show that this is not correct, at least not as far as the region of Achaia Phthiotis in the first quarter of the 3rd century BC is concerned. Bronze coins from neighbouring cities, Lokris and the cities of Euboia circulated in New Halos. They inform us about the districts and cities with which New Halos had contacts. Picard (1979: p. 320) already mentioned contacts between the towns of Euboia and Thessaly. A collection found at Almirós, between Thebai and New Halos, included 15 coins from Chalkis, 25 from Histiaia and 2 coins struck by the Euboian league. And 8 of the 27 coins found at Gorítsa came from Chalkis and Histiaia. With one doubtful exception, all the Euboian coins found in Thessaly came from sites close to the coast. The finds from Halos confirm that there were contacts between the coastal cities of Thessaly and the cities of Euboia: of the total of 150 coins, 14 came from Chalkis, 10 from Histiaia and one is a coin struck by the Euboian league.

Furtwängler observed differences in the contacts of the towns of Demetrias, New Halos and Eretria. Contrary to findings in Demetrias, in Halos coins of the city itself and of neighbouring cities in Achaia Phthiotis prevail (figs 3.44 and 3.46). Coins from Malis, Lokris, Histiaia and Chalkis point to contacts

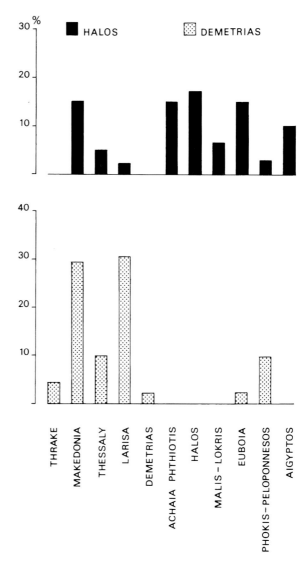

Fig. 3.46. Comparison of the provenance of the coins found at Demetrias and New Halos.

in a southerly direction. Coins from Larisa in Pelasgiotis, which were frequently encountered at Demetrias, were found in only small numbers at New Halos. Demetrias relied strongly on the bronze coinage of the Macedonian kingdom, which is not surprising considering it was a Macedonian capital and army camp. The four cities of Achaia Phthiotis were on the other hand autonomous, at least in the first quarter of the 3rd century BC, and New Halos continued to mint its own coins as it had done before the Macedonian conquest by Philippos II in 344 BC. The finds of Eretria are comparable with those of Halos: a predominance of coins struck by the city itself and of coins from neighbouring cities and districts.

5.7. Conclusion

The coins found in the excavations of the houses at New Halos come from a closed context. The letters on the reverse of the coins of Ptolemaios II representing the king's regnal year provide an accurate date for the end of the occupation of New Halos: shortly after 265/4 BC. Coins from the cities of New Halos, Peuma, Phthiotic Thebai and Larisa Kremaste bearing the monogram XA reveal the activity of a mint union in Achaia Phthiotis that struck autonomous coins of good quality in the first quarter of the third century BC. The finds also provide a fairly good impression of the coins that were in circulation around that time, and of the contacts between the city of Halos and cities in the neighbourhood, on Euboia and in Lokris.

5.8. References

BLOESCH, H.J., 1965. Coins of the Euboian League. *Archaioloyikon Deltion* 20, p. 288.

COX, D.H., 1959. *Coins from the Excavations at Curium* (= Numismatic Notes and Monographs 145). The American Numismatic Society, New York.

FURTWÄNGLER, A.E., 1990. *Demetrias, eine Makedonische Gründung im Netz Hellenistischer Handels- und Geldpolitik.* Habilitationsschrift, Heidelberg.

GAEBLER, H., 1906-1935. *Die Antiken Münzen von Makedonia und Paionia I–II.* Reimer, Berlin.

HACKENS, T., 1976. Les monnaies de la nécropole. In: C. Vatin (ed.), *Médéon de Phocide V, Tombes Hellénistiques, objets de métal, monnaies.* De Boccard, Paris, pp. 123–129.

HAAGSMA, M.J. & R. REINDERS, 1991. Verwoesting door een aardbeving. *Tijdschrift voor Mediterrane Archeologie* 8, pp. 16–25.

HEAD, B.V., 1977. *Historia Nummorum, a Manual of Greek Numismatics.* Spink, London.

METCALF, D.M., 1983. *Coinage of the Crusades and the Latin East in the Ashmolean Museum Oxford.* Royal Numismatic Society and Society for the Study of the Crusaders and the Latin East, London.

MØRKHOLM, O., 1991. *Early Hellenistic Coinage, from the Accession of Alexander to the Peace of Apamea (336–188 B.C.).* Cambridge University Press, Cambridge.

MOUSTAKA, A., 1983. *Kulte und Mythen auf Thessalischen Münzen.* Triltsch, Würzburg.

NEWELL, E.T., 1927. *The Coinages of Demetrius Poliorcetes.* Oxford University Press, London.

PAPASTAMATAKI, A., D. DEMETRIOU & B. ORPHANOS, 1994. Mining and metallurgical activities in Pelasgia. The production of copper in antiquity. In: R. Misdraki-Kapou (ed.), *La Thessalie. Quinze années de recherches archéologiques, 1979–1900. Bilans et perspectives, Tome A.* Athina, pp. 243–248.

PAPASTAMATAKI, A. & D. DIMITRIOU, 1987. *The production of Copper in Pelsagia* (= Praktika tis Akadimias Athinon 62). Athens.

PICARD, O., 1979. *Chalcis et la Confédération Eubéenne. Étude de Numismatique et d'Histoire (IVe–Ier siècle).* De Boccard, Paris.

REINDERS, H.R., 1988. *New Halos, a Hellenistic town in Thessalía (Greece).* Hes, Utrecht.

ROGERS, E., 1932. *The Copper Coinage of Thessaly.* Spink and Son, London.

STÄHLIN, F., 1928. *Das Hellenische Thessalien.* Engelhorn, Stuttgart.

SVORONOS, N., 1904-1908. *Ta Nomismata tou Kratous ton Ptolemaion.* Athinai.

WALLACE, W.P., 1956. *The Euboian League and its Coinage* (= Numismatic notes and monographs No. 134). The American Numismatic Society, New York.

Fig. A. The location of the city of New Halos. The lower town is situated in the yellow cornfields at the transition point between the foothills of Mount Óthris and a swamp at the back of Soúrpi bay. View from the east.

Fig. B. Map of Magnesia by Gazi (1814). Unlike the cartographers of the 18th century, Gazi located Halos correctly in relation to the villages of Almirós, Plátanos and Soúrpi.

Fig. C. House of the Coroplast. View of the foundations from the north.

Fig. D. House of the Broken *Amphorai*. View of the foundations from the northeast.

Fig. E. Lower town of New Halos during the excavation of the House of Agathon in 1987. View from the upper town.

Fig. F. Foundations of the House of Agathon. View from the northwest.

Fig. G. Lower town of New Halos during the excavation of the House of Agathon (detail)

Fig. H. Christos Katrantzonis from Plátanos, the owner of the plot of land where the House of Agathon was situated.

Fig. I. Excavation of the courtyard of the House of the Ptolemaic Coins in 1989. View from the northeast.

Fig. J. House of Agathon and House of the Ptolemaic coins. View of the foundations of the living quarters from the east.

Fig. K. Excavation of the House of the Snakes in 1993. View from the west.

Fig. L. Broken pottery. House of the Ptolemaic Coins.

Fig. M. Terracotta figurine of a banqueter with a *rhyton* and *patera*, found in the House of the Coroplast.

Fig. N. Coin from the city of New Halos (enlarged). Phryxos clinging to the ram with the golden fleece is depicted on the obverse. The letters ΑΛΕΩΝ appear below the ram and the monogram ΧΑ at upper left.

Fig. O. Coin from the city of Larisa Kremaste (enlarged). Thetis seated on a hippocamp carrying a shield to her son Achilles is depicted on the obverse. The letters ΛΑΡ appear below the shield and the monogram ΧΑ on the shield.

Fig. P. Prosília. Summer camp of Sarakatsan families in the Óthris mountains. Contacts with pastoralists and analysis of the animal remains recovered in the excavations of the houses in New Halos induced a discussion of the possibility of transhumance in Hellenistic times.

CHAPTER 4. ENVIRONMENT, VEGETATION AND ANIMAL REMAINS

1. INTRODUCTION

H. REINDER REINDERS

In the first volume on the Hellenistic city of New Halos (Reinders, 1988) the location of the city itself and its environment were the main subjects of investigation. The city was strategically located on a small stretch of land between the foothills of the Óthris mountains and a backswamp bordering the Pagasitikós gulf. At the start of the investigation, before small-scale excavations were initiated, a hypothesis was put forward that New Halos was a port, like its precursor 'Old' Halos, the city which was besieged and destroyed by a Macedonian army under Parmenion in 346 BC. I was of the opinion that the occupation of New Halos declined and came to an end in Late Hellenistic or Roman times, after the inlet of the Pagasitikós gulf, which I had assumed had been used as an anchorage ground for New Halos, had silted up.

In order to check this hypothesis, André van Marrewijk and Paulien de Roever conducted 200 borings in the backswamp between the eastern wall of New Halos and the Pagasitikós gulf. The cores revealed the sequence of the different phases in the silting process, but yielded no evidence on the date of the start of sedimentation. Bottema (1988; 1994) subsequently investigated the palynological contents of a core obtained from the backswamp in order to reconstruct the vegetational history of the Almirós plain. ^{14}C dates obtained for this core and analysis of mollusc shell assemblages in other cores by Van Straaten (1988) led to the conclusion that my hypothesis was wrong: around 3,000 BP a salt marsh already existed in this area, and in Hellenistic times there was no inlet with open water or any other anchorage ground close to the city wall. The lower course of the river Amphrysos intersecting the backswamp may however have been navigable up to a landing place close to the wall (Ch. 1: 2).

Our environmental research was not confined to the Hellenistic period and the site and immediate surroundings of New Halos alone, but also extended to the preceding and subsequent periods and the city's territory (Reinders, 1988). In order to determine the area's vegetational history, Bottema (1994) studied cores from the two Zerélia lakes in the Almirós plain. Contacts with pastoralists in the Almirós-Soúrpi plain and the Óthris mountains and analysis of the faunal remains recovered in the excavations of the houses in New Halos induced a discussion of the possibility of transhumance in Hellenistic times (Reinders, 1994; 1996; Reinders & Prummel, 1998). This chapter pre- sents the results of the analysis of the animal remains by W. Prummel and of a study of the vegetational history of the Óthris mountains conducted by H. Woldring.

1.1. References

BOTTEMA, S., 1988. A reconstruction of the Halos environment on the basis of palynological information. In: H.R. Reinders, *New Halos, a Hellenistic town in Thessalía, Greece*. Hes, Utrecht, pp. 216-226.

BOTTEMA, S., 1994. The prehistoric environment of Greece: A Review of the palynological record. In: P.N. Kardulias (ed.), *Beyond the site, regional studies in the Aegean area*. University Press of America, New York/London, pp. 45–68.

REINDERS, H.R., 1988. *New Halos, a Hellenistic town in Thessalía, Greece*. Hes, Utrecht.

REINDERS, H.R., 1994. *Schaarse Bronnen*. Inaugural lecture, University of Groningen.

REINDERS, H.R., 1996. *Veehouders in Thessalië. De Overgang naar een Sedentair Bestaan*. Thessalika Erga, Groningen.

STRAATEN, L.M.J.U. VAN, 1988. Mollusc shell assemblages in core samples from ancient Halos (Greece). In: H.R. Reinders, *New Halos, a Hellenistic town in Thessalía, Greece*. Hes, Utrecht, pp. 227–235.

REINDERS, H.R. & W. PRUMMEL, 1998. Transhumance in Hellenistic Thessaly. *EA* 3, pp. 81–95.

2. FOREST VEGETATION AND HUMAN IMPACT IN THE ÓTHRIS MOUNTAINS

HENK WOLDRING

2.1. Introduction

This section discusses the results of a botanical study of forest plant communities in the Óthris mountains (fig. 4.1), southern Thessaly, Greece, carried out in June 1990. Since only little specified phytogeographical and phytosociological information was hitherto available on this area, it seemed worthwhile to include the newly obtained information in this volume.

The research involved inventorying the arboreal species in 68 vegetation plots (fig. 4.2; eight were unusable), including *lianas*, such as *Hedera, Lonicera, Smilax* and *Clematis*. Other taxa like *Cistus, Coridothymus capitatus, Hypericum empetrifolium, Asparagus acutifolius, Anthyllis hermanniae* and *Genista acanthoclada* are typical components of the so-called 'phrygana'. These species were also included, since they indicate a certain stage of degradation of the vegetation (Ch. 4: 2.5).

The second aim of the study was to examine the impact of man and his livestock on the vegetation in

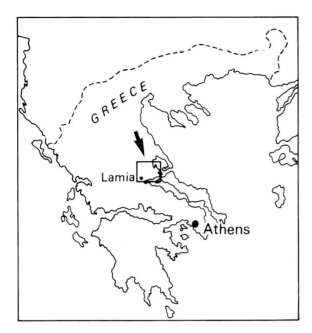

Fig. 4.1. Map of Greece showing the location of the Óthris mountains.

a region that was fairly inaccessible until recent times. In the Óthris, the impact of man must be largely attributable to transhumance, a farming system involving the movement of a farmer and his flock into the mountains for the warm season (Ch. 4: 2.4). A prerequisite for mountain farming is the presence of summer pasture. In this context attention was given to the extent of pasture in the mountains and to features which might provide evidence on the natural altitude of the tree line in the Óthris.

2.2. Methods

The sample plots (releves) were where possible set out along diagonal courses intersecting in the Óthris. The plots range in altitude from sea level to 1500 m. Each plot measures approximately 50×50 m. The frequency of the species in each releve, the altitudes and density of the vegetation cover are presented in table 4.1. Species recorded in only one or two releves are listed in table 4.2. Figure 4.2 shows the distribution of the phytosociological vegetation zones in the Óthris mountains as concluded from the botanical inventory.

With respect to the phytosociological classification the author refers to *Vegetation Südosteuropas* (Horvat *et al.*, 1974), which contains a comprehensive discussion of Mediterranean plant communities, and to the *Carte de la végétation potentielle de la région méditerranéenne, feuille No. 1: Méditerranée orientale* (Quézel & Barbéro, 1985).

2.3. Geographical situation and climate

The Óthris mountains are located between 22°21'/22°51'E and 38°53'/39°3'N and cover an area of roughly 900 km². In the south, hilly country borders the Maliakós Kólpos (Maliakós gulf). The basin of Dhaukli with lake Xínias borders the mountains to the west. The Thessalian plain extending in a northwesterly direction begins near Melitaia. In the northeast, the mountains gradually descend to the alluvial plain of Almirós, an extensive agricultural area (fig. 4.2).

The main ridges of the mountain chains run roughly SW–NE, while the central part is west-east oriented. Another ridge runs from the eastern central part and is NW–SE oriented. This ridge includes the 1726-m-high Gioúzi, the highest summit in the Óthris mountains. A north-south cross-section is saddle-shaped, *i.e.* with a fairly flat centre and steep sloping sides (fig. 4.3). Rivers run almost radially in all directions from the summits (Stählin, 1924).

The area under study has a climate with dry, hot summers and mild, relatively wet winters. A general feature of Mediterranean climates is a period of summer drought, whose duration shortens with increasing altitude. In the lowermost zone, the Eu(thermo)-Mediterranean, this period lasts about six months and in the mountains two to three months.

The Óthris mountains are situated in the lee of the Píndhos mountains. As a consequence, average rainfall is low, 500–600 mm/year. By contrast, the Ionian west coast catches 1100 mm/year. Westerly winds prevail and bring rain mainly outside the summer season, *e.g.* July: 10–25 mm, December: 100 mm. The Thessalian plain is one of the most arid regions in Europe.

With regard to plant life, two main types of substrate play a role: calcareous rocks, *e.g.* limestone, and non-calcareous rocks such as serpentine and slate. It is a general rule in phytosociology that calcicole plants, *i.e.* lime-demanding species, thrive on calcareous substrates, whereas calcifugous species occur on soils poor in lime, *i.e.* siliceous or decalcified substrates. But in the Mediterranean climate, the depth of the soil is also an important factor with respect to plant growth, since water supply is more constant in deep soils (Quézel, 1981). Limestone dominates in the southern and eastern parts of the Óthris area. The so-called 'blue chalk' near Kokkotí is of Triassic origin. In the west and north, principal substrates are of serpentine and slate.

2.4. Transhumance

Transhumance is a form of pastoralism that is particularly common in the Balkans, Turkey and the Mediterranean. According to the literature, transhumance proper involves the movement of a farmer and his livestock into the mountains for the summer period

Table 4.1. Numbers 1–5 indicate the frequency of the occurrence of each species. 1 = rare; 2 = occasional; 3 = frequent; 4 = abundant; 5 = dominant; loc = local.

Releve No.	1	2	3	4	5	6	7	8	9	10	11	12	13	14	16	17	18	19	20	21	22	23	24
Altitude (m)	70	280	320	470	600	600	430	200	220	300	380	580	700	1050	1200	1225	1225	1230	1275	1300	1250	1140	1105
Cover (%)	60	-	-	80	-	-	100	70	50	70	60	30	80	70	20	30	50	<10	35	<10	±30	70	40
Pistacia lentiscus	2	-	-	-	-	-	-	-	-	-	-	-	-	-	-	-	-	-	-	-	-	-	-
Olea europaea (cult.)	2	1	1	-	-	-	-	2	2	-	-	-	-	-	-	-	-	-	-	-	-	-	-
Anthyllis hermanniae	-	-	-	-	-	-	-	-	-	2	-	-	-	-	-	-	-	-	-	-	-	-	-
Cistus incanus	-	-	2	2	3	1	-	-	-	2	-	-	-	-	-	-	-	-	-	-	-	-	-
Hypericum empetrifolium	-	-	1	2	-	-	-	-	-	-	-	-	-	-	-	-	-	-	-	-	-	-	-
Calicotome villosa	-	-	-	-	-	-	3	-	-	-	-	-	-	-	-	-	-	-	-	-	-	-	-
Lembotropis nigricans	-	-	-	-	-	1	-	-	-	-	-	-	-	-	-	-	-	-	-	-	-	-	-
Phillyrea latifolia	1	3	3	3	3	3	-	3	3	4	4	3	2	1	-	-	-	-	-	-	-	-	-
Pistacia terebinthus	-	1	2	2	-	-	-	1	1	1	1	-	1	-	-	-	-	-	-	-	-	-	-
Paliurus spina-christi	1	-	1	-	-	-	2	-	1	-	-	-	-	-	-	-	-	-	-	-	-	-	-
Spartium junceum	1	-	-	1	1	-	-	-	-	-	2	-	-	-	-	-	-	-	-	-	-	-	-
Arbutus unedo	-	-	-	5	-	3	-	-	-	4	-	-	-	-	-	-	-	-	-	-	-	-	-
Quercus ilex	-	-	-	-	-	3	-	-	-	-	-	-	-	-	-	-	-	-	-	-	-	-	-
Quercus coccifera	5	5	4	5	3	-	-	5	5	4	5	5	5	4	-	-	4	4	5	-	2	5	1
Smilax aspera	-	-	-	-	-	-	1	-	-	-	-	1	-	-	-	-	-	-	-	-	-	-	-
Lonicera etrusca	-	-	-	-	-	-	2	-	-	-	-	-	-	1	-	-	-	-	1	-	-	-	-
Lonicera implexa	-	-	-	-	-	-	-	-	-	1	-	-	-	-	-	-	1	-	-	-	-	-	-
Asparagus acutifolius	-	-	-	-	-	-	-	-	-	-	1	-	1	-	-	-	-	-	-	-	-	-	-
Crataegus schraderana	-	-	-	-	-	-	-	-	-	-	-	-	1	-	-	-	-	-	-	-	-	-	-
Pyracantha coccinea	-	-	-	-	-	-	-	-	-	-	-	-	1	1	-	-	-	-	-	-	-	-	2
Rosa sempervirens	-	-	-	-	-	1	2	-	-	-	-	-	-	-	-	-	-	-	-	-	-	-	-
Ruscus aculeatus	-	4	2	-	-	-	2	-	-	-	-	-	-	-	-	-	-	-	-	-	-	-	-
Osyris alba	-	-	-	-	-	-	-	-	1	-	-	-	-	-	-	-	-	-	-	-	-	-	-
Erica arborea	-	-	-	2	-	-	-	-	-	-	-	-	-	-	-	-	-	-	-	-	-	-	-
Ostrya carpinifolia	-	-	-	-	-	-	-	-	-	-	-	-	-	-	-	-	-	-	-	-	-	-	-
Cotinus coggygria	-	-	1	3	2	-	-	-	-	-	-	-	-	-	-	-	-	-	-	-	-	-	-
Cercis siliquastrum	-	-	2	-	-	-	3	-	-	3	-	-	-	-	-	-	-	-	-	-	-	-	-
Colutea arborescens	-	-	-	-	-	-	-	-	-	1	1	-	1	-	-	-	-	-	-	-	-	-	-
Fraxinus ornus	-	-	-	-	-	2	3	-	-	-	-	1	-	1	-	1	-	-	-	-	-	-	-
Acer monspessulanum	-	-	-	-	-	1	-	-	-	-	-	-	-	-	-	-	1	-	1	-	-	-	-
Pyrus amygdaliformis	-	-	-	-	-	-	-	-	1	1	-	1	-	-	-	-	1	-	1	-	-	-	-
Quercus brachyphylla	-	-	-	-	-	-	-	-	-	-	-	-	-	-	-	-	1	-	1	-	-	-	-
Quercus pubescens	-	-	-	-	-	-	-	-	-	-	-	-	-	2	-	-	3	-	2	-	-	-	-
Quercus cerris	-	-	-	-	-	-	-	-	-	-	-	-	-	-	-	-	-	-	-	-	-	1	-
Quercus frainetto	-	-	-	-	4	5	5	-	-	-	-	-	-	-	-	-	-	-	-	-	-	-	5
Abies cephalonica	-	-	-	-	-	-	-	-	-	-	-	-	-	1	-	5	3	3	-	-	-	-	-
Prunus cocomilia	-	-	-	-	-	-	-	-	-	-	-	-	-	-	2	1	2	2	2	4	3	2	2
Crataegus laciniata	-	-	-	-	-	-	-	-	-	-	-	-	1	2	2	-	-	2	2	4	2	1	2
Ilex aquifolium	-	-	-	-	-	-	-	-	-	-	-	1	-	-	-	-	1	1	1	1	-	-	-

Table 4.1. (continued).

Releve No.	1	2	3	4	5	6	7	8	9	10	11	12	13	14	16	17	18	19	20	21	22	23	24
Altitude (m)	70	280	320	470	600	600	430	200	220	300	380	580	700	1050	1200	1225	1225	1230	1275	1300	1250	1140	1105
Cover (%)	60	-	-	80	-	-	100	70	50	70	60	30	80	70	20	30	50	<10	35	<10	±30	70	40
Hedera helix	-	-	-	-	-	-	-	-	-	-	-	-	-	-	-	-	1	-	-	-	-	-	1
Clematis vitalba	-	-	-	-	-	-	-	-	-	-	-	-	-	-	-	-	-	-	-	-	-	-	-
Daphne oleoides	-	-	-	-	-	-	-	-	-	-	-	-	-	-	2	2	-	-	-	-	-	-	-
Juniperus oxycedrus	-	2	-	1	3	3	3	-	-	1	-	2	-	2	5	2	4	3	2	3	1	2	4
Prunus spinosa	-	-	-	-	1	-	-	-	-	-	1	-	-	-	-	-	-	-	-	-	-	-	-
Crataegus spec.	-	-	-	-	-	-	-	-	-	-	-	1	-	1	1	1	1	-	-	4	1	-	1
Rosa spec.	-	-	-	-	-	1	2	-	-	-	-	-	-	-	-	-	1	4	2	4	-	-	-
Rubus spec.	-	-	-	-	-	2	1	-	-	-	-	-	-	-	-	-	-	-	1	-	-	-	-

Releve no.	25	26	27	28	29	30	31	32	33	34	35	36	37	38	39	40	41	42	43	44	45	46	47
Altitude (m)	960	920	820	820	600	500	380	50	120	420	500	520	700	860	950	1150	1200	1200	1500	220	300	550	800
Cover (%)	85	80	75	90	70	60	70	100	40	±30	75	65	40	50	30	80	40	50	15	80	60	80	90
Pistacia lentiscus	-	-	-	-	-	-	-	4	3	2	2	2	1	-	-	-	-	-	-	3	4	4	-
Olea europaea	-	-	-	-	-	-	-	3	3	3	2	2	1	-	-	-	-	-	-	3	4	-	-
Anthyllis hermanniae	-	-	-	-	-	-	-	2	-	-	1	-	-	-	-	-	-	-	-	-	1	-	-
Cistus incanus	1	1	-	-	-	1	-	1	1	2	3	3	2	4	3	-	-	-	-	3	2	-	2
Hypericum empetrifolium	-	-	-	-	-	-	-	1	-	3	-	-	-	-	-	-	-	-	-	-	1	-	-
Calicotome villosa	-	-	-	-	-	1	-	1	2	-	-	-	-	-	-	-	-	-	-	-	-	1	-
Lembotropis nigricans	1	-	-	-	-	-	-	1	-	3	1	-	-	-	-	-	2	-	-	2	-	-	-
Phillyrea latifolia	-	2	3	-	2	3	4	3	3	3	3	4	1	1	-	-	3	-	-	3	2	4	3
Pistacia terebinthus	-	-	-	-	-	2	1	-	-	1	2	3	-	-	-	-	2	-	-	2	1	1	-
Paliurus spina-christi	-	-	-	-	-	-	-	3	-	-	-	1	-	-	-	-	-	-	-	-	-	-	-
Spartium junceum	-	-	-	3	3	3	3	2	-	3	3	4	4	-	2	-	-	-	-	1	-	2	-
Arbutus unedo	-	-	-	-	-	-	-	1	-	-	4	1	-	-	-	-	-	-	-	-	-	-	-
Quercus ilex	-	-	4	-	5	5	4	3	3	3	3	3	5	5	5	4	3	5	-	4	5	5	5
Quercus coccifera	-	-	-	-	2	1	-	3	1	-	3	1	1	1	1	1	1	1	1	-	-	-	-
Smilax aspera	-	-	-	-	-	-	-	3	1	1	1	1	-	1	1	1	1	1	1	-	-	-	-
Lonicera etrusca	-	-	-	-	-	-	-	-	-	-	-	-	-	-	-	-	1	-	-	-	-	-	-
Lonicera implexa	-	-	-	-	-	-	-	-	1	1	1	-	-	-	-	-	-	-	-	-	-	-	-
Asparagus acutifolius	-	-	-	-	-	-	-	-	1	1	1	-	-	-	-	-	-	-	-	3	1	1	-
Crataegus schraderana	-	-	-	-	1	-	-	2	-	-	-	-	-	-	-	-	-	-	-	-	-	-	-
Pyracantha coccinea	1	-	1	-	2	1	-	-	-	-	-	-	-	-	-	-	-	-	-	-	-	-	-
Rosa sempervirens	-	-	-	-	-	-	-	2	-	-	-	-	-	-	-	-	-	-	-	-	-	-	-
Ruscus aculeatus	1	-	-	-	-	-	-	-	-	-	-	-	-	-	-	-	-	-	-	-	-	-	-
Osyris alba	-	-	-	-	-	-	-	-	-	1	-	-	-	-	-	-	-	-	-	-	-	-	-
Erica arborea	-	-	-	-	-	-	-	-	-	3	3	-	-	-	-	-	-	-	-	-	-	-	-
Ostrya carpinifolia	-	-	-	-	-	-	-	-	-	-	1	-	-	-	-	-	-	2	-	-	-	-	-
Cotinus coggygria	-	-	-	-	-	-	-	-	-	-	-	-	-	-	-	-	-	-	-	-	-	-	-

Table 4.1. (continued).

Releve no.	25	26	27	28	29	30	31	32	33	34	35	36	37	38	39	40	41	42	43	44	45	46	47
Altitude (m)	960	920	820	820	600	500	380	50	120	420	500	520	700	860	950	1150	1200	1200	1500	220	300	550	800
Cover (%)	85	80	75	90	70	60	70	100	40	±30	75	65	40	50	30	80	40	50	15	80	60	80	90
Cercis siliquastrum	-	-	-	-	-	-	-	1	-	-	-	-	-	-	-	-	-	-	-	-	-	-	-
Colutea arborescens	-	-	-	-	-	-	-	2	-	-	-	-	-	-	-	-	-	-	-	-	-	-	-
Fraxinus ornus	1	-	-	-	-	-	-	-	-	-	-	-	-	-	-	-	-	-	-	-	-	-	-
Acer monspessulanum	-	-	-	-	-	-	-	-	-	-	-	-	-	-	-	-	-	-	-	-	-	-	-
Pyrus amygdaliformis	-	1	-	2	-	1	2	2	-	1	-	1	1	-	2	1	-	-	-	-	-	2	2
Quercus brachyphylla	-	-	3	-	3	-	-	-	-	-	-	3	-	-	-	-	-	-	-	-	-	-	-
Quercus pubescens	1	2	3	2	3	1	1	2	-	1	-	2	-	2	-	1	-	-	-	-	-	-	1
Quercus cerris	-	-	-	-	-	-	-	-	-	-	-	-	-	-	-	-	-	-	-	-	-	-	2
Quercus frainetto	5	2	-	3	4	-	-	-	-	-	-	-	-	-	-	-	-	-	-	-	-	-	3
Abies cephalonica	-	-	-	-	-	-	-	-	-	-	-	-	-	1	3	5	-	-	-	-	-	-	-
Prunus cocomilia	-	-	-	-	-	-	-	-	-	-	-	-	-	-	-	-	-	-	3	-	-	-	-
Crataegus laciniata	2	-	2	2	-	-	-	-	-	-	-	-	-	-	-	-	-	-	1	-	-	-	-
Ilex aquifolium	-	-	-	-	-	-	-	-	-	-	-	-	-	-	-	4	3	-	1	-	-	-	-
Hedera helix	-	-	-	1	-	-	-	-	-	-	-	-	-	-	-	-	-	-	-	-	-	-	-
Clematis vitalba	-	-	-	-	-	-	-	-	-	-	-	-	-	-	1	-	-	-	-	-	-	-	-
Daphne oleoides	-	-	-	-	-	-	-	-	-	-	-	-	-	-	-	1	3	2	4	-	-	-	-
Juniperus oxycedrus	3	4	2	-	1	-	-	-	-	1	1	1	2	2	3	3	3	3	3	-	-	2	2
Prunus spinosa	-	-	-	2	-	-	-	-	-	-	-	-	-	-	1	-	-	-	-	1	1	-	-
Crataegus spec.	-	-	-	-	-	-	1	-	-	1	-	1	-	-	-	-	1	-	-	-	-	-	-
Rosa spec.	1	-	-	-	-	-	-	-	-	-	-	-	-	-	1	2	1	-	1	-	-	-	-
Rubus spec.	1	-	-	2	2	-	-	1	-	-	-	2	-	-	-	2	-	-	-	-	-	-	-

Releve no.	48	49	50	51	52	53	54	55	56	57	58	59	60	61	62	64	71	72	73	74	75	76
Altitude (m)	900	1000	950	1120	900	1120	1150	1290	1200	1200	1200	1150	1140	1050	1150	600	700	1020	640	750	1040	580
Cover (%)	95	25	50	15	60	80	35	15	50	50	50	60	30	90			100	10–90	60	90	85	90
Pistacia lentiscus	-	-	-	-	-	-	-	-	-	-	-	-	-	-	-	-	-	-	-	-	-	-
Olea europaea	-	-	-	-	-	-	-	-	-	-	-	-	-	-	-	-	-	-	-	-	-	-
Anthyllis hermanniae	-	-	-	-	-	-	-	-	-	-	-	-	-	-	-	-	-	-	-	-	-	-
Cistus incanus	-	-	-	-	1	-	2	-	-	-	-	-	-	-	-	-	1	-	1	-	1	2
Hypericum empetrifolium	-	-	-	-	-	-	-	-	-	-	-	-	-	-	-	-	-	1	1	-	-	2
Calicotome villosa	-	-	-	-	-	-	-	-	-	-	-	-	-	-	-	-	-	-	-	-	-	-
Lembotropis nigricans	-	-	-	-	-	-	-	-	-	-	-	-	-	-	-	-	-	-	-	-	-	-
Phillyrea latifolia	2	-	-	-	-	-	-	-	-	-	-	-	-	-	-	-	4	-	-	-	-	-
Pistacia terebinthus	-	-	-	-	-	-	-	-	-	-	-	-	-	-	-	-	-	1	-	-	-	-
Paliurus spina-christi	-	-	-	-	-	-	-	-	-	-	-	-	-	-	-	-	-	-	-	-	-	-
Spartium junceum	-	-	-	-	-	-	-	-	-	-	-	-	-	-	-	-	-	-	1	-	-	-
Arbutus unedo	-	-	-	-	-	-	-	-	-	-	-	-	-	-	-	-	-	-	-	-	-	1
Quercus ilex	-	-	-	-	-	-	-	-	-	-	-	-	-	-	-	-	-	-	-	-	-	5

Table 4.1. (continued).

Releve no.	48	49	50	51	52	53	54	55	56	57	58	59	60	61	62	64	71	72	73	74	75	76
Altitude (m)	900	1000	950	1120	900	1120	1150	1290	1200	1200	1200	1150	1140	1050	1150	600	700	1020	640	750	1040	580
Cover (%)	95	25	50	15	60	80	35	15	50	50	50	60	30	90	-	-	100	10-90	60	90	85	90
Quercus coccifera	3	4	5	2	4	4	3	4	4	4	3	-	-	1	-	-	3	5	3	-	-	2
Smilax aspera	-	-	-	-	-	-	-	-	-	-	-	-	-	-	-	-	1	-	-	-	-	-
Lonicera etrusca	-	-	1	-	-	-	-	-	1	1	1	-	-	-	-	-	-	-	1	-	-	-
Lonicera implexa	-	-	-	-	-	-	-	-	-	-	-	-	-	-	-	-	-	-	-	-	-	1
Asparagus acutifolius	1	-	-	-	-	-	-	-	-	-	-	-	-	-	-	-	-	-	-	-	-	-
Crataegus schraderana	1	-	-	-	-	-	-	-	-	-	-	-	-	-	-	-	-	-	-	-	-	-
Pyracantha coccinea	-	-	-	-	-	1	-	-	-	-	1	-	-	-	-	-	-	-	-	-	-	-
Rosa sempervirens	-	-	-	-	-	-	-	-	-	-	-	-	-	-	-	-	-	-	-	-	-	1
Ruscus aculeatus	1	-	-	-	-	-	-	-	-	-	-	-	-	-	-	-	-	-	-	-	-	-
Osyris alba	-	-	-	-	-	-	-	-	-	-	-	-	-	-	-	-	-	-	-	-	-	-
Erica arborea	-	-	-	-	-	-	-	-	-	-	-	-	-	-	-	-	-	-	-	-	-	-
Ostrya carpinifolia	-	-	-	-	-	-	-	-	-	-	-	-	-	-	-	-	-	1	-	-	-	-
Cotinus coggygria	-	-	-	2	-	-	1	-	-	-	-	-	-	-	-	-	-	-	-	-	-	-
Cercis siliquastrum	-	-	-	-	-	-	-	-	-	-	-	-	-	-	-	-	-	-	-	-	-	-
Colutea arborescens	-	-	-	-	-	-	-	-	-	-	-	-	-	-	-	-	-	-	1	-	-	-
Fraxinus ornus	-	-	-	-	-	-	-	-	-	-	-	-	-	-	-	-	-	1	-	-	-	-
Acer monspessulanum	1	-	1	-	-	-	-	-	-	-	-	-	-	-	-	-	-	1	-	-	-	-
Pyrus amygdaliformis	2	-	-	2	-	-	1	-	-	-	-	-	-	-	-	-	-	1	-	-	-	-
Quercus brachyphylla	2	-	-	-	-	-	-	-	cf.2	-	-	-	-	-	-	-	-	-	-	-	-	-
Quercus pubescens	-	-	-	-	-	-	-	-	4	-	2	-	-	1	-	2	4	-	2	-	-	1
Quercus cerris	-	1	1	-	-	5	1	-	-	-	-	-	-	-	-	-	-	-	-	-	-	-
Quercus frainetto	4	5	2	5	5	4	4	-	2	2	5	5	-	4	1	5	4	1	5	5	5	2
Abies cephalonica	-	-	-	-	-	-	-	-	-	2	5	5	3	4	4	-	-	-	-	-	-	-
Prunus cocomilia	-	-	2	-	-	1	1	-	-	1	2	1	3	1	4	-	-	3	-	-	-	-
Crataegus laciniata	-	-	-	-	-	-	-	1	-	1	1	-	1	-	2	-	-	-	-	-	-	-
Ilex aquifolium	-	-	-	-	-	-	-	-	-	1	1	4	1	1	1	-	-	-	-	-	-	-
Hedera helix	-	-	-	-	-	-	-	-	-	-	-	-	-	-	-	-	-	1	-	-	-	-
Clematis vitalba	-	-	1	-	-	-	-	-	-	-	1	-	2	-	-	-	-	-	-	-	-	loc.
Daphne oleoides	-	-	-	-	-	-	-	-	-	-	-	2	2	-	-	-	-	-	-	-	-	-
Juniperus oxycedrus	4	4	4	4	3	2	2	4	4	4	4	-	4	1	4	2	2	2	3	1	1	-
Prunus spinosa	2	-	1	1	-	1	-	2	-	1	-	-	1	-	-	-	1	-	-	-	-	-
Crataegus spec.	-	-	-	-	-	-	1	-	-	-	-	-	-	-	-	-	-	-	-	-	-	-
Rosa spec.	1	-	1	1	-	-	-	2	-	2	1	-	-	1	1	-	1	-	1	-	1	-
Rubus spec.	-	-	2	4	-	-	-	-	-	-	-	-	-	-	-	-	-	-	-	-	1	2

Table 4.2. List of species not included in table 4.1. 1 = rare; 2 = occasional; 3 = frequent.

Releve No.

3	*Ficus carica*	1
6	*Carpinus orientalis*	2
7	*Clematis flammula*	1
11	*Prunus cerasifera*	1
14	*Quercus macrolepis* (*aegilops*)	3
	Amygdalus webbii	1
	Prunus mahaleb	1
22	*Ulmus glabra*	locally dominant
	Malus sylvestris	1
24	*Viscum album*	3
26	*Viscum album*	1
32	*Myrtus communis*	1
	Rosa sempervirens	2
	Rhamnus alaternus	1
33	*Vitex agnus-castus*	1
	Juniperus phoenicea	1
	Coronilla emerus	1
39	*Juglans* (cult.)	1
40	*Ephedra fragilis* ssp. *campylopoda*	1
45	*Coridothymus capitatus*	1
57	*Acer campestre*	1
60	*Berberis vulgaris*	1
63	*Taxus baccata*	2
	Tilia sp.	1
	Sorbus graeca	1
69	*Rhamnus lycioides*	1
	Clematis flammula	1
71	*Ligustrum vulgare*	1
	Crataegus heldreichii	1
72	*Rhamnus graeca*	3
	Crataegus heldreichii	1
76	*Platanus orientalis*	locally dominant

(Reinders & Prummel, 1998). The absence of winter fodder and the climatic hazards of the winter season do not allow domestic herbivores to remain at high altitudes throughout the year. The winter residence is usually in the adjacent valleys. Some farmers in the Óthris region still practise this form of pastoralism.

A prerequisite for transhumance is the presence of summer pasture in the mountains. Such pasture by nature occurs near and above the timberline, *i.e.* in the subalpine and alpine zones. In Greece these zones are to be found above 1800 m in the Pelopónnisos and above 2000/2200 m in the northern Píndhos mountains. It is indicated in Ch. 4: 2.5, under zone IV, that the Óthris mountains would on account of their elevation be forested by nature and a herbaceous cover would occur only under exceptional, local conditions. The open woodland with large stretches of herbaceous vegetation which currently exists above 1300–1400 m is certainly the result of a period of mountain pastoralism. Another consequence of the interference of man besides the clearance of forest is that certain anthropogenic plant communities spread at the expense of those occurring naturally. Nitrate-demand-

ing species, plants resistant to or favoured by trampling, grazing and burning, expand whereas taxa susceptible to disturbance decline or even disappear.

Palynological analysis of marsh sediment from Mavrikopoúla at an altitude of 1200 m in the central Óthris has shown that the shift from forest to open woodland and extensive herb vegetation that has occurred in that area is not a recent development, but the result of millennia of human interference. It is assumed that the practice of transhumance as a regular element of the farming economy developed during the period of occupation of New Halos. The system gained progressively more importance until, around 1000 BP, the pasturing of livestock ultimately became the principal aim for farmers to move into the mountains. The history of the mountain vegetation in the last 3000 years will be discussed in Ch. 4: 3.

2.5. Vegetation zones

Zone I: Eu-Mediterranean Oleo-Lentiscetum aegeicum vegetation. In the south-facing foothills, up to about 600 m, we find the Eu-Mediterranean *Oleo-Lentiscetum aegeicum*, an association within the Oleo-Ceratonion. As its name implies, this unit is typical of the Aegean. According to Quézel & Barbéro (1985), pre-forest plant communities in this zone indicating regression of the initial forest vegetation belong to the Pistacio-Rhamnion. The scrubland of the Oleo-Ceratonion, which is generally described as *maquis* or *matorral,* may develop in areas with an average annual rainfall of less than 600 mm and an average temperature of 17–18°C. Plant communities of this type are at present widespread in the coastal regions of Greece, including the southern part of the Óthris.

According to Mayer & Aksoy (1986), the *Oleo-Lentiscetum aegeicum* should be regarded as a degradation phase of the Oleo-Ceratonion. By nature, carob (*Ceratonia siliqua*) grows only on the narrow strips of land enclosing the southern part of the Pagasitikós gulf. The absence of the heat-demanding and frost-intolerant (fatal frost around -11°C) carob in the southern Óthris will be attributable to climatic conditions. Its occurrence along the Pagasitikós gulf is one of the northernmost in its area of distribution. As the area of the Maliakós gulf is quite small, frosts may occasionally occur, and frost is fatal for this species.

Besides the widely cultivated olive (*Olea europaea*), mastic tree (*Pistacia lentiscus*) is characteristic of the vegetation of the southern foothills. In addition, kermes oak (*Quercus coccifera*), mock privet (*Phillyrea latifolia*) and strawberry tree (*Arbutus unedo*) are components of the natural vegetation. The general occurrence of kermes oak, mock privet, strawberry tree and rockrose (*Cistus incanus*) could be the result of occasional fires or even intentional burning, since these species are pyrophytes, which means that

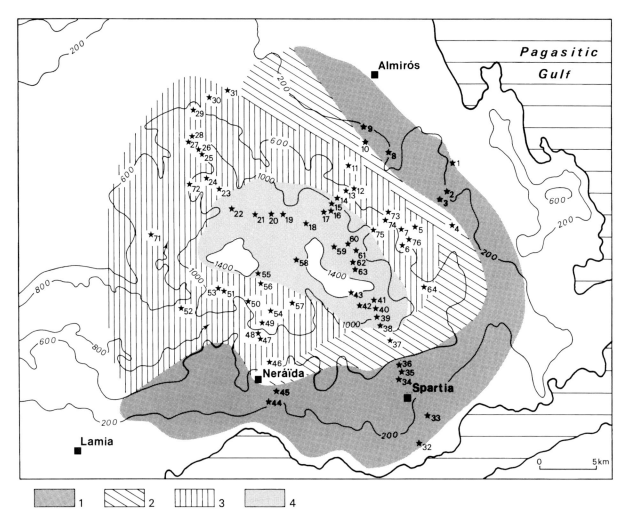

Fig. 4.2. Map of the area showing the releve numbers (1–76) and the different vegetation zones. Legend: 1. Eu-Mediterranean *Oleo-Lentiscetum aegeicum* vegetations; 2. Eu-Mediterranean *Quercion-ilicis* vegetations; 3. Deciduous oak forests (*Quercion frainetto-cerris*); 4. The Oro-Mediterranean vegetation zone (*Abietion cephalonicae*).

they are able to regenerate rapidly from subterranean buds, which enhances their competitiveness. It is almost tradition for shepherds to deliberately burn kermes oak *maquis*, to thus obtain young shoots as fodder for their livestock.

The absence of Aleppo pine *(Pinus halepensis)*, a typical species of the Eu-Mediterranean in the south of Greece, could be due to fire. This pine occupies largely the same habitat as the pyrophytes mentioned above. The trees themselves are not resistant to fire, which makes regeneration dependent on the germination of seeds. The rapid regrowth of the pyrophytes may retard the development of the tiny seedlings and thus prevent the re-establishment of the pine forest (Le Houérou, 1981). On the other hand, the pollen diagram of Mavrikopoúla clearly indicates the absence of pines in the Óthris mountains in the last 3000 years. Extensive stands of Aleppo pine occur on the nearby

Isle of Euboia (Évvia), up to altitudes of 1000 m. Horvat *et al.* (1974: p. 91) indicated the frost-intolerance of this pine species. The aforementioned occasional occurrence of frost can be considered the factor preventing the establishment of Aleppo pine in the Óthris region.

Further degradation of the *maquis* by fire or grazing leads to phrygana, a type of vegetation composed of low, thorny, cushion-shaped plants. Releve No. 1 represents such a vegetation. Climate and edaphic factors restrict the occurrence of phrygana in the Óthris. This vegetation has its optimum on siliceous soils, and such soils are not common in the southern part of this mountain range. Several abandoned olive groves were observed in the vicinity of releve No. 1 (table 4.1). The failure of olive cultivation must probably be ascribed to the siliceous soils which occur here.

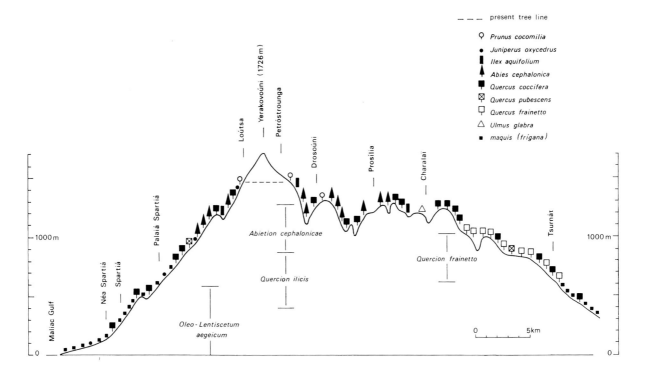

Fig. 4.3. North-south cross-section of the Óthris mountains, showing the altitudinal distribution of phyto-sociological units and the main arboreal components (after Reinders, 1994).

Anthyllis hermanniae is a typical phrygana species. Though accidentally poorly represented in the tables, this species locally dominates the degraded vegetation. *Genista acanthoclada* and *Cistus* ssp. occur scattered throughout the *Oleo-Lentiscetum aegeicum*, where plant growth is seriously disturbed.

Zone II: Eu-Mediterranean Quercion-ilicis vegetation. Due to increasing altitude, temperatures are lower further inland (average temperature 14–15°C) and the summer drought is less restrictive because of the increasing rainfall. Such a climate prevails in the southern and eastern foothills, above 600 m. According to Horvat *et al.* (1974), sclerophyllous taxa of the *Quercion ilicis* dominate here on account of the climate. Nowadays, this type of vegetation extends up to 900–1000 m. The transition from the *Oleo-Lentiscetum aegeicum* to *Quercion ilicis* is imperceptible. In the southern hills mastic tree disappears between 500 and 600 m, but a considerable number of species are represented in both vegetation units. This is partly due to anthropogenic disturbance, but it also indicates the small variation in temperature and precipitation. The regional variant is an association of holm oak (*Quercus ilex*) and eastern strawberry tree (*Arbutus andrachne*), the *Andrachno-Quercetum ilicis*. According to Quézel & Barbéro (1985), pre-forest vegetation belongs to the Pistacio-Rhamnion, like the vegetation of

the *Oleo-Lentiscetum aegeicum.*

Curiously, the characteristic species were virtually not encountered. Holm oak is represented in only two vegetation records (releve Nos. 6 and 76). The species is however over-represented, because it has not been recorded elsewhere. Eastern strawberry tree is entirely absent. Deciduous species, such as manna ash (*Fraxinus ornus*) and oriental hornbeam (*Carpinus orientalis*), which usually occur in holm-oak forests, are also poorly represented. Holm oak and manna ash play a principal role in the forest vegetation of the Greek west coast, where annual rainfall is twice as high as in this part of the Óthris. The extreme rareness of holm oak is probably attributable to the lower amounts of precipitation on the Óthris massif.

The relative dryness favours drought-resistant species such as the 'weedy' kermes oak. In areas where the initial forest has regressed, the kermes oak also invades the aforementioned other vegetation units. The species occurs at altitudes ranging from 0 to 1275 m, which demonstrates its ecological tolerance. In Greece the reduction of holm-oak forest leads to the expansion of *maquis* vegetation led by kermes oak. Burning is not likely to have been a principal factor responsible for the expansion of *coccifera* oaks in the *Quercion ilicis*. Though not known as a pyrophyte, holm oak is to some extent resistant to fire. In a former pine forest containing large numbers of *ilex* oaks in the understorey near Salerne (Var, France), which burned down about seven years ago, almost all the pines were killed. The above-ground vegetation of holm oaks was also destroyed, but, unlike the pines, these specimens readily regenerated from subterranean parts and dominated the new vegetation after seven years. Le Houérou (1981) also noted a high resistance to fire in holm oak. Even the above-ground parts may survive an average fire. These examples indicate that fire is not the limiting factor for holm oak in the Óthris mountains.

Besides kermes oak, strawberry tree, mock privet and terebinth (*Pistacia terebinthus*) are commonly present in the *Quercion ilicis*. They usually occur as shrubs, but occasionally kermes oak and mock privet attain the arborescent stage. In the secluded domain of the Xeniás monastery (releve No. 3) and in the similarly sacred precinct of the church near Kofí (releve No. 11) specimens of kermes oak and mock privet measured up to about 10 m. Other species characteristic of the Quercion ilicis are *Clematis flammula*, *Rosa sempervirens* and *Lonicera etrusca*. Only the latter is fairly common. *Cistus* (mostly *C. incanus*) and *Asparagus acutifolius* are the most common species of the herbaceous stratum. The *Quercion ilicis* plant community on the eastern and southern sides of the mountains directly adjoins the Oro-Mediterranean conifer forest. This implies that the traditional Sub-Mediterranean zone, the transition from a Mediterranean vegetation to an inland vegetation of a more continental climate, is not clearly represented in these parts.

According to Quézel (1981), disturbance of primary forest may lead to an upward shift of thermophile, sclerophyllous species. It is tempting to ascribe the absence of a well-developed Sub-Mediterranean zone to destruction of the forest. According to Horvat *et al.* (1974), however, the Ostryo-Carpinion, the phytosociological unit in the Sub-Mediterranean climate, is favoured by destruction of primary forest. The almost complete absence of the Ostryo-Carpinion is therefore most likely to have its ground in the geographical situation: southern exposure and limited rainfall in this area apparently restrict the development of the Sub-Mediterranean vegetation zone.

Zone III: Deciduous oak forests (Quercion frainetto-cerris). Sclerophyllous vegetation dominates in the eastern part of the mountains and deciduous oak forests encircle them on the southwest and northern sides roughly between the 600 and 1200 m contour lines and decreasing in altitude to the north. Here the influence of the Mediterranean is tempered to such an extent as to result in a more continental climate, which is manifested especially in more severe and longer periods of frost. Owing to the windward position of these parts, the average rainfall is 600 to 800 mm/year (Horvat *et al.*, 1974).

Quézel & Barbéro (1985) identified the *Quercion frainetto-cerris*, an alliance of Hungarian oak (*Q. frainetto*) and Turkey oak (*Q. cerris*), on non-calcareous substrates, and locally on calcareous substrates the Ostryo-Carpinion, in which the downy oak (*Q. pubescens*) plays a significant part. Horvat *et al.* (1974) are of the opinion that by nature the Ostryo-Carpinion prevails.

Thermophile sclerophyllous shrubs are present in places in the undergrowth and in forest verges. Where the canopy is not closed, brushwood of kermes oak and prickly juniper (*Juniperus oxycedrus*) develops, which demonstrates that the climate is not strictly continental. Hungarian oak, which is the most common oak species in the lowlands and hills of the Balkans, ranges from 400 to 1100 m in the north (releve Nos. 5–7 and 24–29) and from 800 to 1200 (1290) m in the southwest (releve Nos. 47–57). Its presence at increasing altitudes towards the southwest indicates the species' intolerance for long periods of summer drought. Hungarian oak and Turkey oak prefer non-calcareous soils. In the northern Óthris, where slate is common at the surface, the author recorded extensive coppices of Hungarian oak (releve Nos. 24 and 25), which may be remnants of the immense oak forests described by Stählin (1924) when he travelled through the northern Óthris at the beginning of the twentieth century. Big stumps of felled trees support this conjecture. A unique mixed stand of Hungarian oak and Greek fir is to be found in the domain of a monastery near Dhrampala (releve No. 61). The mixed occurrence and uniform height of about 25 m of both species suggest that the trees were planted. Nevertheless, this stand of trees shows the vegetational potential of both species.

Vast, almost monospecific oak forests occur in the southwestern part of the mountain range, in the area of Drístela (releve Nos. 51–54). The extensiveness of the forests must be ascribed to the slight variation in altitude in this area. Hungarian oak is the most common species, locally attended by downy oak and Turkey oak. Downy oak dominates on calcareous substrates in the area between Vourliá and Vourlítsa (releve No. 56) and is found scattered elsewhere, especially on the northern flanks. Though at home in the Sub-Mediterranean, this drought-resistant species is ecologically fairly indifferent and easily invades other vegetation zones, even that of the Eu-Mediterranean *maquis*. The expansion of the species is favoured by disturbance of the primary vegetation. An abrupt change from downy oak to Hungarian oak forest is to be found halfway between releve Nos. 56 and 57. At releve No. 51, tree stumps of felled oaks measured up to 1 m in diameter. Judging from living trees at this location, the stumps must derive from Hungarian oaks.

Large-scale fellings are carried out in the surroundings of Drístela. Felled trees measured 20–30 cm in diameter and were aged 50–80 years. Except for these fellings, the forests in this area give the impression of being fairly undisturbed. Different deciduous oak species dominate the arborescent vegetation by nature.

Another deciduous oak species represented in a few places is *Quercus brachyphylla* (releve Nos. 36, 47 and 48). In the mountains of the Pelopónnisos this thermophilous species appears together with Hungarian oak at altitudes around 1000 m (Horvat *et al.*, 1974). The species takes the place of sessile oak (*Q. petraea*) in the warmer regions of Greece.

Zone IV: The Oro-Mediterranean vegetation zone (Abietion cephalonicae) and the tree line. By nature, summer drought-resistant conifer forest covers most mountains in the Mediterranean climate zone. In the mountain zone, the summer drought lasts two to three months. *Abies* species play a dominant role in the Óthris mountains above 900 m altitude. The Greek fir (*Abies cephalonica*) is the only fir species represented in the tables. Macedonian fir (*A. borisii-regis*) could not be identified with certainty. However, according to Quézel & Barbéro (1985) and Horvat *et al.* (1974), Macedonian fir plays a prominent role in the region; this slightly continental species may in fact be more common in the western part of the mountains, which the present author did not visit.

Abies cephalonica dominates the central mountains between 900 and 1400 m altitude. In dense stands there is no undergrowth due to the lack of daylight. Up to 1200–1300 m, kermes oak is a frequent component at clearances. A prominent species in the fir zone is *Daphne oleoides*, an evergreen dwarf shrub. It has a high incidence in sunlit places within the forest and appears to be associated with the Greek fir at first sight. However, *Daphne oleoides* is also common, and thicket-forming, in open situations beyond the fir forest. According to Browicz (1983), the species indeed occurs in montane forest, but is particularly common in the subalpine and alpine zones.

Two other species which frequently occur in the Greek fir zone are holly (*Ilex aquifolium*) and prickly juniper. Holly is well-known from the Sub-Atlantic climate zone in Europe, but is also common throughout the Mediterranean basin. Its occurrence in the Mediterranean mountains demonstrates a certain tolerance for dry conditions during the growing season. Because of its leathery and prickly foliage, holly seems to be avoided by herbivores, which in some areas leads to pure open stands of this species, as near Skopiá (1220 m).

Selective grazing also seems to favour prickly juniper. This species is distributed in all vegetation zones, but it is particularly abundant in the highest parts, above 1300 m. It usually occurs as a low shrub, but in the mountain area this conifer occasionally attains the arborescent stage (3–4 m). Juniper species usually indicate regression of the primary forest. In the fir zone they spread where the fir forest is cleared. Considering the size and age of the specimens, the mountain zone must be a favourable habitat. The remoteness of the locations may be another factor favouring the survival of such forms.

The effects of overgrazing are observable in the composition of the vegetation of a rock formation at the foot of the Pilioúras (1200 m). The steep rock contains a cave with running water, which is most probably the underground course of the Mégas Lákkos (see Ch. 4: 2.6). On top of the rock, beyond the reach of grazers, grow a number of woody species which are not, or virtually not, found elsewhere. Because they have not been included in the tables, these species will be enumerated here: hop hornbeam, a full-grown unique specimen of a kind not found elsewhere; Greek rowan (*Sorbus graeca*); a maple species, probably *Acer heldreichii*, and a lime species, probably *Tilia platyphyllos* (the latter species could be identified only with binoculars because of their inaccessible location). Several low, broad shrubs of yew (*Taxus baccata*) occur on the slopes opposite the rock formation. The foliage of this conifer contains a toxic substance, lethal to livestock. Lime and Greek rowan were not encountered elsewhere. One wonders whether species may have disappeared from the region as a result of the voracity of livestock.

Nowadays, Greek fir ranges from 900 to 1300 (1400) m in altitude and is consequently found in the Oro-Mediterranean vegetation belt. Above the fir zone is a vegetation containing patches of light-demanding species, such as prickly juniper, hawthorn (*Crataegus* spp.), Greek plum (*Prunus cocomilia*), sloe (*P. spinosa*), *Rosa* spp., bracken (*Pteridium*) and *Daphne oleoides*. These species are also common in open locations in the Greek fir zone and expand where the forest has been cleared. Judging from the degraded plant growth found above 1300 m, the present upper limit of the fir forest is not a natural one. There are indeed other indications suggesting that, without human interference, the belt of Greek fir would have extended to higher altitudes than it does today.

On the Isle of Kefalloniá, Knapp (1965) distinguished two subzones in the *Abies cephalonica* belt, which here ranges from 800 to 1620 m. In the lower subzone kermes oak accompanies Greek fir, whereas the former is absent in the upper subzone. In the Óthris, kermes oak attends the Greek fir practically to its upper limit. Ecological conditions apparently preclude the presence of kermes oak at this altitude, and the degraded zone above 1300 m may be considered the equivalent of the upper fir zone mentioned by Knapp (1965).

Does the open woodland above the fir zone indicate the proximity of the natural tree line? A feature of the tree line in the Mediterranean mountains is the presence of subalpine woody vegetation, which often forms the transition to the treeless alpine zone. In the

Mediterranean mountains this is a typical vegetation, indicated by Polunin (1980) as the hedgehog zone. The woody vegetation found above the fir zone differs greatly from this type of vegetation in character, and the subalpine zone is consequently not developed in the Óthris mountains.

In the Mediterranean climate zone summer-drought-resistant conifer forest usually occurs up to the tree line and in the Óthris *Abies cephalonica* (and possibly also *A. borisii-regis*) would probably constitute the natural tree line. An indication of its natural potential altitude is the range of Greek fir forest on the Parnassós mountain, about 50 km to the south of the Óthris. According to Polunin (1980), Greek fir here constitutes practically monospecific stands up to the tree line at 1900 m altitude. Daubenmire (cited in Crawford, 1989) has calculated a reduction in the tree line of 110 m for each degree of increase in latitude. Converted, this would result in a reduction of about 50 m and a (hypothetical) tree line at about 1850 m altitude in the Óthris. Local ecological conditions may cause variations in the tree line, but the indicated altitude suggests that the upper part of the mountains would be forested by nature.

This conclusion implies an almost complete absence of natural grasslands and the necessary clearance of forest in order to create them. The pollen diagram of Mavrikopoúla (section 4.3) demonstrates that this process did not start recently; the present grasslands are the outcome of at least three millennia of human interference.

2.6. The Greek plum

As a widespread and characteristic species of the montane zone, the Greek plum (*Prunus cocomilia*) deserves some attention. This species is indeed so dominant that a mountain chain in the eastern part of the Óthris, the 'Koromiliés', took its name from this plum: *cocomilia/coromilia* is the Greek word for plum. The small gnarled trees or shrubs attain a maximum height of 5 m with relatively thick trunks. Most specimens have large spines and glabrous young twigs. The yellow to reddish fruits are hardly stalked. Though much sought after by the local people, they are not considered very palatable by modern standards. Juvenile shrubs are not easily distinguished from the sloe (*Prunus spinosa*). A 55-year-old tree, with a stem diameter of only 20 cm, in the yard of a local house illustrates the species' slow rate of growth.

The taxonomy of the spinous variety is somewhat problematic. The *Flora Europaea* (Tutin *et al.*, 1964) does not mention a spiny variety. Davis (1965–1988) mentions two varieties:
- var. *cocomilia*; glabrous twigs and leaves, branches spineless. Distribution: Italy, Greece, Turkey;
- var. *puberula*; twigs and leaves sparsely pubescent, branches with spines. Endemic in Turkey.

The morphology of the Óthris specimens is intermediate, spiny with glabrous leaves and twigs. Apparently there is a high degree of morphological variation within this species. *Cocomilia* specimens collected near Burdur in southwest Turkey have a dense, grey pubescence on the twigs and leaves. Davis' description of the Turkish species matches these specimens.

The profusion of thorns to some extent safeguards the Greek plum against the voracity of animals, and doubtless favours its presence over species more susceptible to grazing. Sheep, and possibly goat, too, are surprisingly known to feed on considerable quantities of the fruits, evidently without incurring injury. In the yard of a *stáni* at Loútsa the author found large quantities of fruit stones in the sheep dung. The distances the animals cover daily must contribute greatly to the dispersal of the fruit stones and could be a reason for the abundance of *cocomilia* specimens in the mountains.

In many places grazing has led to the development of 'forest steppe' landscapes. In the Mégas Lákkos (which means 'large hole'), a depression at an altitude of 1400 m, the Greek plum is the only arborescent species to be found. The trees are located along the edges and on some higher 'islands'. Bracken and grasses make up the ground-flora. The depression has no river-outlet, but drains underground, via an outlet which is easily obstructed. The typical pattern and thin dispersal of trees and the extensive ground-flora could result from occasional inundation and not primarily from grazing, as is the case elsewhere in the mountains.

The species is found at altitudes of 950–1500 m. In the belt of the Greek fir it occurs mostly in open places and forest verges. Above the conifer belt, the plum constitutes a significant part of the woody vegetation, insofar as woody species are still present. The species was recorded even at 1500 m altitude (relevé No. 43), the highest point in the examined plots. According to Polunin (1980), Greek plum takes on a dwarfed and procumbent habit in the subalpine zone, in Greece generally above 1700 m. Such adaptations also apply to other woody species at this altitude. The changes in habit reflect conditions different from those in the lower zones. Specimens from different altitudes are morphologically identical, which also indicates that the subalpine zone is not represented in the Óthris.

2.7. Conclusion

The inventory of the woodlands of the Óthris revealed the following phytosociological units:

Zone I: Eu-Mediterranean *Oleo-Lentiscetum aegeicum*, 0–600 m, seaside *maquis*; principally in the foothills along the Maliakós gulf;
Zone II: Eu-Mediterranean *Andrachno-Quercetum*

ilicis, (400)600–1000 m, woodlands with predominantly sclerophyllous (leathery-leaved) plant communities on the eastern and southern slopes;

Zone III: Sub-Mediterranean *Quercion frainetto-cerris*, (400)600–1200 m, montane deciduous oak forests in the west and north;

Zone IV: Oro-Mediterranean *Abietion cephalonicae*, 900–1300/1400 m, mountainous conifer forests, concentrated mainly in the central Óthris.

The exploitation of the woodlands in this part of the Greek mainland has led to a reduction of primary, natural forests and the expansion of plant communities resistant to factors such as burning and felling by man and browsing by domestic herbivores.

Besides extensive olive cultures in the thermo-Mediterranean, secondary plant communities have arisen consisting of *maquis* and locally of phrygana. Up to 1000 m leathery-leafed scrubland dominates the eastern and southern slopes. The present distribution of kermes oak in the hills and mountains up to 1200 m must be largely attributable to the small amounts of precipitation in combination with grazing and burning of the vegetation.

The restricted occurrence of secondary plant communities in the northern and western parts of the mountains indicates a low degree of disturbance of the deciduous oak forests. Relatively high precipitation and limited grazing probably allow rapid regeneration of the oak species.

The reduction of *Abies cephalonica* forest has in many places resulted in open scrubland, often consisting of prickly juniper and members of the *Rosaceae* family (*Crataegus, Prunus, Rosa* and *Rubus*). This vegetation is also abundantly represented above the present upper limit of the Greek fir forest.

The absence of subalpine vegetation, the growing habits of the woody species in the upper mountains and the presence of kermes oak in the upper part of the Greek fir zone indicate that the natural tree line is not present in the Óthris. The altitude of Greek fir stands on the Parnassós mountain suggests that the (hypothetical) altitude of the natural tree line would lie above 1800 m. Without human interference the upper mountain zone would probably have consisted of conifer forest dominated by Greek silver fir.

2.8. Acknowledgements

Professor S. Bottema critically read the manuscript, Mrs G. Entjes-Nieborg processed the manuscript and Mr J.H. Zwier provided the illustrations.

2.9. References

BROWICZ, K., 1983. *Chorology of Trees and Shrubs in South-West Asia and Adjacent Regions.* Polish Scientific Publishers, Warszawa-Poznan.

CRAWFORD, R.M.M., 1989. *Studies in Plant Survival. Ecological Case Histories of Plant Adaptations to Adversity* (= Studies in Ecology 11). Blackwell, Oxford.

DAVIS, P.H. (ed.), 1965-1988. *Flora of Turkey and the East Aegean Islands.* 10 vols. Edinburgh University Press, Edinburgh.

HORVAT, I., V. GLAVA & H. ELLENBERG, 1974. *Vegetation Südosteuropas* (= Geobotanica Selecta Bd. IV). Gustav Fischer Verlag, Stuttgart.

KNAPP, R., 1965. *Die Vegetation von Kephallinia, Griechenland.* Königstein.

LE HOUÉROU, H.N., 1981. Impact of man and his animals on Mediterranean vegetation. In: F. di Castri, D.W. Goodall & R.L. Specht (eds), *Mediterranean-type shrublands* (= Ecosystems of the World 11). Elsevier, Amsterdam, Oxford, New York, pp. 479–522.

MAYER, H. & H. AKSOY, 1986. *Wälder der Türkei.* Gustav Fischer, Stuttgart.

POLUNIN, O., 1980. *Flowers of Greece and the Balkans. A Field Guide.* Oxford University Press, Oxford, New York, Toronto, Melbourne.

QUÉZEL, P., 1981. Floristic composition and phytosociological structure of sclerophyllous matorral around the Mediterranean. In: F. di Castri, D.W. Goodall & R.L. Specht (eds), *Mediterranean-type shrublands* (= Ecosystems of the World 11). Elsevier, Amsterdam, Oxford, New York, pp. 107–122.

QUÉZEL, P. & M. BARBÉRO, 1985. *Carte de la Végétation Potentielle de la Région Méditerranéenne. Feuille No. 1: Méditerranée Orientale.* CNRS, Paris.

REINDERS, H.R. & W. PRUMMEL, 1998. Transhumance in Hellenistic Thessaly. *Environmental Archaeology* 3, pp. 81–95.

STÄHLIN, F., 1924. *Das Hellenische Thessalien. Landeskundliche und geschichtliche Beschreibung Thessaliens in der hellenistischen und römischen Zeit.* Reprint: Verlag Adolf M. Hakkert, Amsterdam, 1967.

TUTIN, T.G., V.H. HEYWOOD, N.A. BURGES, D.H. VALENTINE, S.M. WALTERS & D.A. WEBB (eds), 1964. *Flora Europaea.* 5 Vols. Cambridge University Press, Cambridge.

3. LATE HOLOCENE VEGETATIONAL HISTORY OF THE ÓTHRIS MOUNTAINS

HENK WOLDRING

3.1. Introduction

Archaeological surveys have shown that the Thessalian plains were occupied in the Neolithic already, if not earlier. There was a marked increase in the number of inhabited sites in the Middle Bronze Age, but only few Late Bronze Age sites are known. The only evidence for occupation in the Iron Age consists of barrows (*tumuli*) (personal communication, H.R. Reinders).

The archaeological excavation of the remains of the Hellenistic city of New Halos (Reinders, 1988) triggered research into the city's prehistoric environment and man's impact on it. A reconstruction of the palaeoenvironment of the Halos area (Bottema, 1988) demonstrated deforestation of the coastal lowlands in the Younger Holocene. Whether the vegetation at higher elevations, *e.g.* in the nearby Óthris mountains, was affected at the same time and to the same extent has so far remained uncertain due to a lack of palaeoenvironmental data.

Unlike for the plains, we have fairly little archaeological evidence for the mountains. The greater part of the little evidence that is available is moreover of a relatively late date. Fortifications dating from the Hellenistic and Ottoman periods are the earliest known evidence of man's presence in the Óthris.

The aim of the vegetational study whose results are presented in this chapter was to determine the development of the mountain vegetation in relation to man's activities by means of palynological analysis of a sediment core taken from a marsh in the central Óthris.

3.2. The coring of Mavrikopoúla

A sediment core was taken from the Mavrikopoúla marsh in the central Óthris mountains (co-ordinates 22°35'E, 39°0'N; fig. 4.4). The coring (depth 7.41 m) was performed by S. Bottema and H.R. Reinders in 1993. The spring-fed marsh is situated at 1200 m altitude. The basin measures approx. 40×80 m. The lithology is as shown in table 4.3. Four radiocarbon dates have been obtained (table 4.4).

In the Almirós plain, sediment cores (Zerélia II and Zerélia 8) were taken from two small lakes (fig. 4.5: 3). Only one radiocarbon date (Zerélia II: 1440± 90, GrN-12051) was available at the time of the publication of the results of their analysis. As their age and geographical location make these cores of interest for the present study, new radiocarbon dates are presented in table 4.5.

For information on the geology and climate of southern Thessaly the reader is referred to Reinders (1988), Bottema (1988) and Woldring (this volume).

3.3. The pollen assemblage zones of the Mavrikopoúla core

Two zones have been distinguished in the Mavrikopoúla pollen record (fig. 4.4). Zone 1 (spectra 1–14; 725–435 cm) shows high AP (arboreal pollen) values, which gradually decrease from 80% to less than 50% in the upper part. *Quercus robur*-type pollen and *Abies* pollen make up most of the AP pollen sum. Gramineae and Cyperaceae are important NAP (non-arboreal pollen) components. Zone 1, which spans about 2000 years (*c.* 3000–1000 BP), has been divided into two subzones.

Subzone 1a (spectra 1–6, 725–590 cm) initially shows high deciduous oak and sclerophyllous oak values. This is followed by a rapid increase in *Abies* at the expense of both oak taxa. The decline of the oak taxa is followed by a temporary increase in *Pteridium* and *Urtica dioica* values. Deciduous oak values again increase towards the end of this subzone. The maximum *Abies* values (spectrum 5) are accompanied by an increase in *Hedera* of about 15%.

Subzone 1b (spectra 7–14; 590–435 cm) shows a further reduction in AP values, but deciduous oaks on the whole show the highest values. *Salix* and *Ilex* show a slight increase in the lower half of this subzone. *Abies* values are considerably lower than in the previous subzone. The steady increase in NAP values is mainly attributable to an increase in Gramineae. *Pteridium*, Umbelliferae and *Peplis* show appreciable values in two or more spectra.

Zone 2 (spectra 15–36; 435–0 cm) spans the past 1000 years. AP values have decreased to between 10 and 40%. Grasses largely make up the NAP pollen sum. Several taxa that are considered to be anthropogenic make an appearance or increase, such as *Hordeum/Triticum* species, *Plantago lanceolata*, *Rumex acetosa* and *Urtica dioica* species. Zone 2 has also been divided into two subzones:

Subzone 2a (spectra 15–27; 435–225 cm). Low AP values result from a drastic decrease in deciduous oak values and a reduction in *Abies* values to almost zero. Gramineae and Cyperaceae make up most of the NAP values. Significant peaks of *Juncus/Luzula*-type pollen and *Dryopteris* and *Salix* pollen are observable in this zone.

Subzone 2b (spectra 28–36; 225–0 cm). The AP values are slightly higher than those of subzone 2a. Deciduous oak pollen almost disappears, but *Quercus calliprinos*-type pollen shows a gradual increase to about 10% in the upper spectrum. *Abies* values of about 5% in the upper spectrum indicate the return of this conifer in recent times. This part of the diagram shows a low, but uninterrupted *Juniperus* curve. Caryophyllaceae, Compositae and *Plantago*-type pollen further increase, whereas the average Gramineae values are slightly lower than in the previous subzone. A steep decline of Gramineae and an increase in Cyperaceae, *Ranunculus repens*-type pollen and *Galium*-type pollen occurred in subrecent times.

3.4. Comparison and correlation of the Mavrikopoúla record with other pollen diagrams from Greece

Extensive palynological research has been carried out in Greece in the past three decades. To inform the reader about local differences in the development of the vegetational history, various pollen diagrams of the Mediterranean climate zone will be briefly reviewed below and, where relevant, compared with the Mavrikopoúla record. The pollen diagram of Pertoúli

Table 4.3. Lithography of the coring of Mavrikopoúla.

0–100 cm	organic, peaty matter
100–300 cm	silt or clay mixed with organic remains; leaves and stems of seed plants were incorporated (possibly alive) during the deposition
300–550 cm	clay and silt mixed with varying amounts of carbonaceous or peaty matter
550–675 cm	dark grey clay, wood remains, concretions
675–715 cm	dark grey to black clay, wood remains
715–725 cm	peat
725–731 cm	black humic clay, concretions
730–735 cm	loose organic remains
735–741 cm	probably gravelly matter (lost)
741 cm	bedrock; end of coring

Trikálon has been included in this review because the site concerned lies in the mountain zone and it is interesting to compare its pollen record with the identified macro-remains. The locations of the coring sites are indicated in figure 4.5.

Two pollen diagrams come from locations in the plains bordering the Óthris. Halos (Bottema, 1988) is situated in the coastal plain between the Óthris mountains and the Pagasitikós gulf (elevation 0 m above sea level). The pollen diagram of this site spans seven millennia of vegetational history. Deciduous oak values are around 30% from 7000 to 5000 BP and decline in the fifth and fourth millennia, reaching values of between 5 and 10% from 3000 BP onwards. Since deciduous oaks dominate the arboreal pollen set, the AP curve shows a similar development. The values of the Halos diagram suggest an oak park landscape in the mid-Holocene. Such a forest will most probably have been the consequence of the long periods of summer drought characterising the Thessalian climate.

Zone 3 of the Halos diagram, which spans the past three millennia ([14C] date at the base 2890±70 BP) bears a close resemblance to the Mavrikopoúla diagram in that the AP values in both diagrams show a drastic reduction after 3000 BP, with the difference that this reduction occurs slightly later, and less drastically initially, in the Mavrikopoúla diagram. The deciduous oak values of the Mavrikopoúla diagram correspond to those of Halos zone 3 from AD 1000 onwards. Halos zone 3 also shows a drastic reduction in sclerophyllous oaks and in *Ostrya/Carpinus orientalis* and *Abies*, the latter species ultimately becoming rare. *Olea* shows a conspicuous decline towards the present. The

Mavrikopoúla diagram also shows higher olive values in the Roman period (spectra 3–10) than in later times.

The strong decrease in arboreal pollen observable in the Halos and Mavrikopoúla diagrams shortly after 3000 BP points to a severe reduction in arborescent components in the plains and the mountains. The diagrams of Mikro Zerélia, a small lake in the Almirós plain (elevation 150 m above sea level), however show no evidence of early deforestation. The diagram of Zerélia II ([14C] date at the base 1440±90 BP) shows strongly varying values (10–70%) of deciduous oaks until around the 12th century AD, when the values definitively drop to 5% or less. This may correspond to the collapse of oak forest that is assumed to have occurred in the Mavrikopoúla area around AD 1000. The regular intervals observable in the deciduous oak curve between *c.* AD 500 and 1200 suggest felling of the forest approximately once every 200 years. The Zerélia II diagram spans subzone 1b and zone 2 of the Mavrikopoúla diagram. Radiocarbon dates obtained for Zerélia II and Zerélia 8 were given in table 4.5.

Litóchoro Katerínis (Athanasiadis, 1975) is a coastal marsh (elevation 25 m above sea level) southwest of Thessaloníki. The AP sum includes alder (*Alnus*). The dominating pollen values suggest that this species grew in stands in the swamps. The AP values of the Litóchoro diagram are a little lower after 3000 BP than before this date, which is mainly due to a gradual decrease in deciduous and sclerophyllous oaks. *Abies* vanishes from the pollen record after 2000 BP. *Olea, Ilex, Sanguisorba minor* and *Rumex* are rare throughout, whereas *Plantago* (*s.l.*) and Gramineae show a steady increase in the past three millennia.

The diagram of Kopaïs in Boeotia (Viotía) (Turner & Greig, 1974) reveals a decline of deciduous and sclerophyllous oaks from *c.* 5000 BP, the beginning of zone K7 (base radiocarbon-dated to 5205±120 BP). In the Kopaïs diagram, *Abies* values are insignificant in the Younger Holocene. Grass values increase in this period.

Two diagrams from the southern Peloponese, from Kiládha (Bottema, 1990) and lake Lérna (Jahns, 1993), show a sharp increase in Mediterranean pollen types, *e.g. Phillyrea, Olea, Pistacia* and *Erica*, around 3000 BP (Kiládha data inferred from other cores taken in this area). The total AP values of the Kiládha diagram increase from 30 to 90% around this time, whereas those of the diagram of lake Lérna, which lies only a short distance from Kiládha, de-

Table 4.4. Radiocarbon dates of organic fractions from the coring of Mavrikopoúla.

GrN-22372	420–422.5 cm	1000±100 BP	cal. 1Σ: AD 900–1170	2Σ: AD 810–1250
GrN-22373	422.5–425 cm	1030±100 BP	cal. 1Σ: AD 880–1160	2Σ: AD 700–1250
GrN-25490	595–600 cm	2160±50 BP	cal. 1Σ: 356–98 BC	2Σ: 368–46 BC
GrN-20636	695–700 cm	2940±60 BP	cal. 1Σ: 1260–1060 BC	2Σ: 1380–990 BC

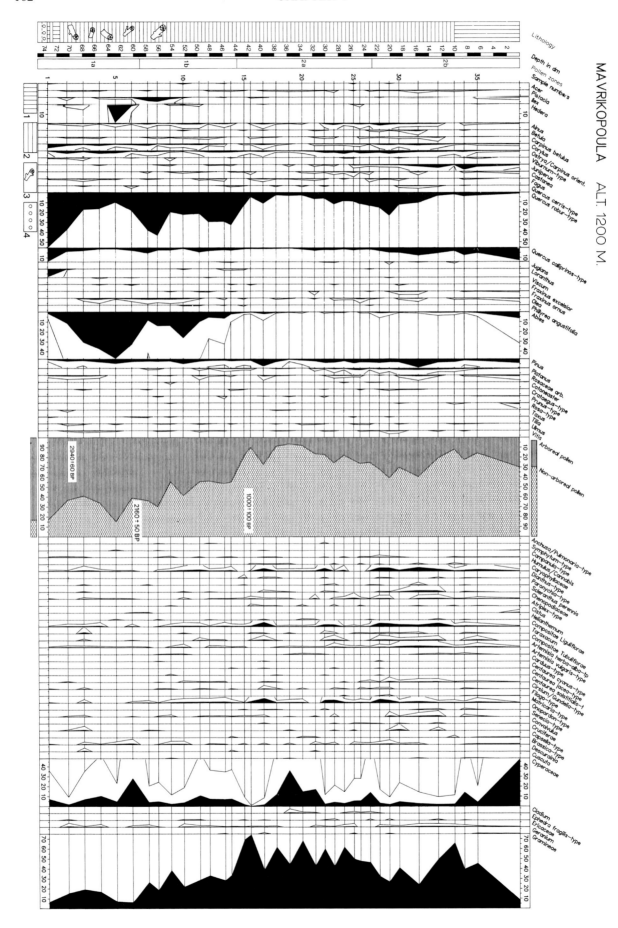

Fig. 4.4. The pollendiagram of Mavrikopoúla.

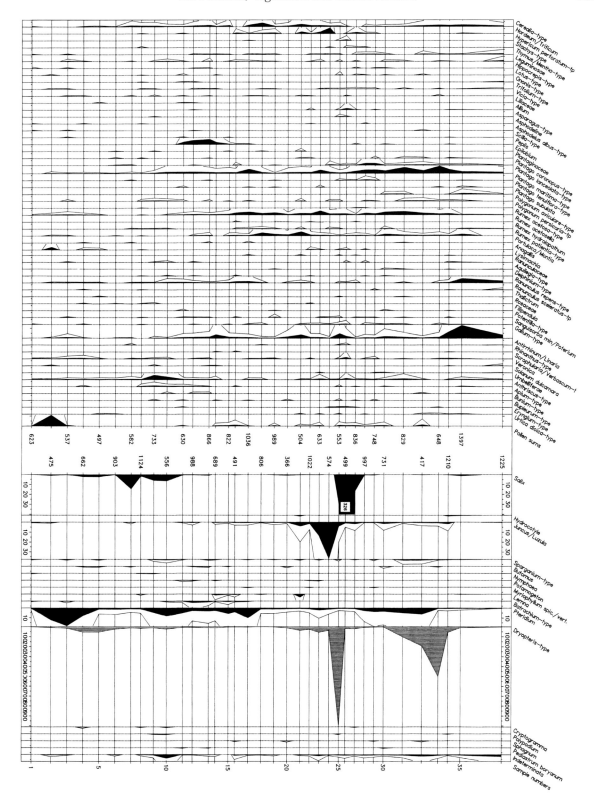

Fig. 4.4. (continued).

Fig. 4.5. Map of Greece, showing the core locations discussed in Ch. 4: 3.2: 1. Mavrikopoúla; 2. Halos; 3. Zerélia; 4. Litóchoro; 5. Kopaïs; 6. Kiládha; 7. Lérna; 8. Trichonís; 9. Pertoúli.

Table 4.5. Radiocarbon dates of Zerélia 8 and Zerélia II.

Zerélia 8	GrN-23187	1340–1347 cm	460±60 BP	organic fraction
	GrN-23236	1340–1347 cm	7310±80 BP	calcareous fraction
Zerélia II	GrN-23188	670–679 cm	710±40 BP	organic fraction
	GrN-23401	670–679 cm	750±80 BP	macroremains
	GrN-23237	670–679 cm	5880±180 BP	calcareous fraction
	GrN-23266	670–679 cm	770±110 BP	1 cm organic fraction

crease. In both diagrams the deciduous oak values show a decrease after 3000 BP.

Lake Trichonís (Bottema, 1982) is situated in the Mediterranean climate zone on the west coast of mainland Greece (elevation 20 m above sea level). Higher deciduous oak and lower sclerophyllous oak values were measured in the deposit above a layer of volcanic ash, which may derive from the Santorini eruption that took place *c.* 3200 BP. A decline in deciduous oak is observable around 2000 BP, accompanied by a conspicuous increase in *Platanus*. Total AP values remain more or less constant throughout the Trichonís diagram.

Pertoúli Trikálon (Athanasiadis, 1975) is the only mountain site discussed here. It lies in central mainland Greece at an altitude of 1275 m. *Abies* is the dominant tree at this site. Useful additional information supplementing the palynological data was obtained in an analysis of macro-remains, which revealed the occurrence of wood, seeds and fir stomata throughout the sediment. This proves that high fir pollen values indicate stands of this species close to the coring site.

The Pertoúli Trikálon diagram shows a slow but steady decline of AP values from 3000/3500 BP. A stronger decrease is observable halfway zone Z1–Z3.

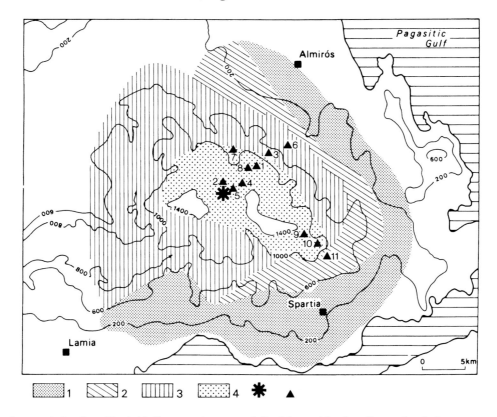

Fig. 4.6. Surface sample locations, Nos 1–11 (for vegetation zones cf. fig. 4.2; star: Mavrikopoúla core location).

This point may correspond to the collapse of AP values in the Mavrikopoúla area around AD 1000, with the distinction that fir caused the decline of AP at Pertoúli Trikálon. The steep rise of AP values in recent times is attributable to the return of fir. The data relating to the younger period are uncertain as ^{14}C dates indicate sediment inversion in the top metre.

The palynological evidence of Greece shows a general reduction in AP values in the Younger Holocene, which in most cases relates to the decline of deciduous oak. A final reduction in deciduous oak forest took place around 3000 BP. At the same time 'new' arborescent species such as *Fraxinus ornus, Olea, Platanus, Juglans* and *Castanea* spread, along with indicator herbs such as *Hordeum/Triticum, Plantago lanceolata, Rumex acetosa* and *Sanguisorba minor*. The expansion of these groups and the decline of deciduous oak forest is generally associated with greater human impact, agriculture and pastoralism.

3.5. Reconstruction of the Óthris mountain vegetation

Zone 1 (3000–1000 BP). Around 3000 BP, when sedimentation started in Mavrikopoúla, oak taxa dominated the forest (fig. 4.4). Wood remains point to the occurrence of oak specimens close to the marsh. The pollen evidence suggests a predominance of deciduous and sclerophyllous oaks. Two sclerophyllous oak species are to be considered with respect to the Óthris, namely kermes oak *(Quercus coccifera)* and holm oak *(Q. ilex)*. Holm oak is rare in the Óthris today. The 1990 inventory revealed only one natural occurrence of this species. Its almost complete absence has been ascribed to the low amounts of rainfall in this area (see section Ch. 4: 2.5, zone II). However, according to Reinders (personal communication), holm oak does occur here and there at low altitudes in the eastern outskirts of the Óthris. In the Óthris holm oak does not exceed the 800 m contour, whereas kermes oak is widespread up to 1300 m altitude.

The pollen of these oak species can usually be distinguished from that of deciduous oak species. *Quercus robur*-type pollen may be assumed to derive from Hungarian oak *(Q. frainetto)* and/or downy oak *(Q. pubescens)*. Most pollen of Turkey oak *(Q. cerris)*, a species which at present occurs locally in the western mountains, can be distinguished from that of *Q. robur*-type pollen by its slightly larger size, thinner exine and less clearly discernible scabrae (Woldring & Bottema, manuscript). Pollen of *cerris* oak was only occasionally encountered in the samples.

The oak forest that existed at the time represented by spectrum 1 must have been studded with the parasitic *Loranthus*, a member of the mistletoe family (Loranthaceae). *Loranthus* prefers deciduous oak species as hosts. The high values of deciduous oaks and the insect-pollinated *Loranthus* imply that Hungarian or downy oak dominated around the Mavrikopoúla basin around 3000 BP. The high proportion of sclerophyllous oaks in this spectrum implies that there were sufficient suitable habitats for this species, in spite of the predominance of the aforementioned species. The deciduous oak forest may have been dotted with kermes oak stands, or this species may have grown as scrub in the understorey.

In the course of subzone 1a, *Abies* forest expanded in the surroundings of Mavrikopoúla, apparently at the expense of deciduous and sclerophyllous oaks. Wood remains of both fir and oak were found in the lower half metre of the sediment (fir at 725, 695–690 and 570 cm, oak at 695–690 cm depth). This suggests that there were stands of trees of these taxa close to the present centre of the basin. The embedded wood remains are not necessarily contemporary with the sediment sequence: branches may have become stuck in the mud and trunks may have become mired in the sediment. Radiocarbon analyses often reveal substantial differences in age between stratigraphically correlated wood remains and sediments. This implies that the stratigraphic order in which wood remains occur in a sediment is of limited value. The succession of oak and fir indicated by the pollen evidence is thought to approach the actual development most closely.

Spectrum 5 shows that ivy (*Hedera*) expanded when deciduous oak had declined to minimum values. The ivy values are remarkable since this species is insect-pollinated. The pollen of insect-pollinated species is usually not, or only poorly dispersed, and such species are consequently mostly represented in small proportions in pollen samples. This must mean that ivy was abundantly present at the time of spectrum 5. The species flowers in particular at the top of (solitary) trees, preferably deciduous oaks. This suggests that there were open tree stands in the areas where the oak forest was not replaced by fir. Openings or thin tree stands are also indicated by bracken (*Pteridium*). This fern sporulates more readily in open locations than in dense shadow.

Wood remains indicate the local development of willow (*Salix*) shrub towards the end of subzone 1a, in spite of low pollen values. Fir and oak, which had grown near the coring site initially, had apparently been replaced by willow. This points to a gradual rise of the water table and extension of the marsh, preventing the nearby growth of dry-land taxa. The increase in *Urtica dioica*-type pollen in spectrum 2 probably also represents a local development. The quite sudden appearance of this pollen and its high values suggest disturbance of the marsh. One of the habitats of *Urtica dioica* proper, the likely producer of the pollen, is areas with a strongly fluctuating water level, such as marginal areas between dry land and water or marsh. Occasional pollen of water lily (*Nymphaea*), crowfoot species (*Batrachium*-type), pondweed (*Pota-*

mogeton) and duckweed (*Lemna*) indicates the presence of open water in the basin.

The slight decrease in AP values in subzone 1a is followed by a more pronounced decrease in subzone 1b, but deciduous oaks nevertheless regained territory in this period. The beginning of subzone 1b was found to be approximately 2000 BP by extrapolating the ^{14}C dates. The woodlands which existed from the Roman period to AD 1000 were dominated by Hungarian oak and possibly downy oak. At the beginning of subzone 1b holly (*Ilex aquifolium*) was present in the understorey or at open locations. The renewed expansion of bracken also suggests stretches devoid of trees, or at least an open tree canopy.

Wood remains and increasing pollen values indicate the local expansion of *Salix* shrub in subzone 1b. Willow shrub may have developed on exposed sands as a pioneer plant. The occurrence of such habitats can also be inferred from the presence of water purslane (*Peplis portula*). This annual colonises sands or loam which are inundated during the winter and emerge from the water for the growing season. The pollen of *Peplis* differs from that of other Lythraceae in its smaller size, different length/width ratio and characteristic pore. Although it has been included under NAP, this plant tends to spread on bare, moist soils, so we may assume that it grew in the marsh. The small amounts of pollen of *Batrachium* and *Lemna* imply the presence of stretches of permanent water, besides temporarily inundated areas.

Zone 2 (1000–0 BP). A drastic thinning of the forest occurred at the beginning of zone 2, approximately 1000 BP. Arboreal species such as fir (*Abies*) and kermes oak did not grow in the direct vicinity of the basin for several centuries. Deciduous oaks were also severely reduced, but they partly made up for lost ground towards the end of subzone 2a, together with *Ostrya* or *Carpinus orientalis*. The low AP values suggest open, apparently disturbed woodland and such areas were suitable habitats for the members of the *Ostrya/Carpinus orientalis* group.

Although the herb vegetation became more diverse during subzone 2a, this stratum is totally dominated by grasses, including taxa producing pollen with grains larger than 40 μm. The observable expansion in the range of pollen types, which came to include pollen of species such as *Rumex acetosa* and *R. patientia*, *Plantago lanceolata*, *Urtica dioica*, Caryophyllaceae and Compositae, indicates a shrinking tree cover, disturbance and grazing. The peaking of *Juncus, Dryopteris* and *Salix*, respectively, towards the end of subzone 2a probably reflects gradually drier conditions in the marsh. The convincing pollen evidence for the local occurrence of willow is supported by the occurrence of wood remains at 245 cm depth. Cyperaceae, though included in the pollen sum, may also have formed part of the marsh vegetation.

Generally speaking, the AP sums of subzones 2a

and 2b do not differ significantly, though some major shifts in plant communities occurred at the beginning of subzone 2b. The last deciduous oaks disappeared from the surroundings of Mavrikopoúla and *Ostrya carpinifolia/Carpinus orientalis* also declined. They were apparently replaced by kermes oak and, to a lesser extent, by a juniper species, probably prickly juniper (*Juniperus oxycedrus*), which is currently one of the leading arboreal species at high altitudes. The occurrence of *Crataegus, Prunus, Rosa* and other woody Rosaceae, all taxa with poor pollen dispersal, indicates a development towards a very open landscape in which large stretches of herbal vegetation were interspersed with isolated shrubs or impoverished scrubland.

The NAP taxa present a similar picture. Several taxa which appear or increase in subzone 2a expand further in subzone 2b. The same groups or taxa indicating disturbance in subzone 2a are present in even larger numbers throughout the greater part of subzone 2b. Part of the deforested area was colonised by bracken. Most of the taxa had by this time declined considerably or even disappeared altogether. Species with *Ranunculus repens*-type and *Galium*-type pollen seem to have replaced the 'anthropogenic' taxa. *Ranunculus repens*-type and *Galium*-type pollen may however also derive from the marsh vegetation and may consequently be overrepresented.

The quantity of *Dryopteris*-type spores also suggests a local origin. During the period represented by subzone 2b, ferns that produced spores of this type spread again for a considerable time in the basin. The same holds for Cyperaceae, several species of which will have found suitable habitats on the peaty, acid substrate. The increasing amounts of Cyperaceae pollen observed in the samples from the uppermost layers probably derive from sedges (*Carex* spp.) or spike-rushes (*Eleocharis* spp.) growing in the marsh.

3.6. Relation between the present vegetation, modern pollen precipitation and the Mavrikopoúla diagram

During the fieldwork, surface samples were collected in the mountains (fig. 4.6). These samples provided information on the pollen production and dispersal of the palynologically identified individual taxa. Pollen records of surface samples are also useful in comparing and interpreting palynological results of pollen diagrams. It is assumed that pollen conservation in moss cushions, the medium in which the pollen rain is optimally trapped, is restricted to a few decades.

Most samples were collected in the *Abies cephalonica* zone. Sites 3, 6 and 11 lie in the *Quercus frainetto* zone, but scattered Greek fir specimens were also found at sampling sites 3 and 6. Forest or (degraded) woodland was observed at all the sites. Two surface samples (2 and 5) were collected in the im-

mediate vicinity of the marsh (within 100 m distance). Local vegetation: *Juncus* species dominate the marsh vegetation. The outer zone consists of *Juncus effusus*(-type) plants, the central part of the marsh vegetation is made up of *Juncus conglomeratus* and *Dryopteris*-type ferns. Other recorded marsh plants include *Epilobium* sp., *Veronica* sp., *Carex* cf. *echinata*, *Carex* cf. *vulpina*, *Juncus articulatus*, *Holcus*, *Prunella*, cf. *Cynosurus*, *Galium* cf. *palustre*.

Vegetation at the sampling sites (Bottema & Reinders: fieldnotes 1993):
1. Grazed area with *Pteridium* and open herb vegetation. *Juniperus, Prunus cocomilia*. At 100 m Greek fir forest.
2. Mavrikopoúla area. Woodland with *Prunus cocomilia* and *Pteridium*. At higher elevations *Abies cephalonica* and *Ilex aquifolium*.
3. Degraded forest with scattered *Abies* and *Quercus frainetto*, browsed *Juniperus* and *Quercus coccifera*. Herbs: *Onopordon*, Labiatae, Gramineae.
4. In *Abies* forest.
5. Mavrikopoúla area. Woodland with *Prunus cocomilia* and *Ilex aquifolium*, herbs and grasses.
6. Well-developed *Quercus coccifera* forest with some *Juniperus oxycedrus*, *Abies* and a single *Phillyrea latifolia*.
7. Grazed area, altitude 1100 m. Deciduous oaks, *Quercus coccifera*, *Hedera helix*, *Crataegus* sp., *Acer* sp.
8. Dry valley, altitude 1100 m. *Quercus* sp., *Abies cephalonica*, *Hedera helix*, *Crataegus* sp., *Acer* sp.
9. South slope of Pilioúras, alt. 1200–1300 m. Predominating *Abies cephalonica*; also *Ilex aquifolium*, *Juniperus oxycedrus*, *Quercus coccifera*.
10. North slope of Paloúki, altitude 1100–1200 m. Predominating *Abies cephalonica*; also *Ilex aquifolium*, *Juniperus oxycedrus*, *Quercus coccifera*.
11. South slope of Visaloúda, altitude 800–900 m. Dense *Quercus frainetto* forest.

On the whole, the pollen values of the present-day samples (fig. 4.7) agree fairly well with the values recorded in the Mavrikopoúla diagram for the last thousand years. The AP values, whose share is significantly higher in the present-day pollen record, present some problem. This difference is mainly caused by kermes oak, although this species was not observed at four sites (Nos 1, 2, 4 and 5). The values point to abundant pollen production and dispersal. The share of kermes oak in the pollen rain is especially high where specimens attain the tree stage, as at site No. 6. Comparison of the present-day record with the Mavrikopoúla diagram strongly suggests that the expansion of kermes oak is a phenomenon of quite recent date.

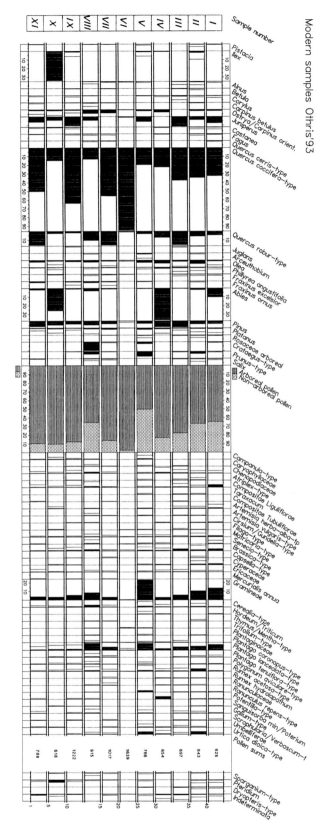

Fig. 4.7. Pollen spectra of surface samples Nos 1–11 (I–XI).

As already indicated by Athanasiadis (1975), *Abies* values are only high in a nearby fir forest. Apparently, the pollen of this species is not transported over long distances. This also holds for the insect-pollinated holly, *Ilex aquifolium*. The presence of insect-pollinated taxa in the pollen record implicates that the producers formed part of the vegetation at the sampling site. The pollen values agree quite well with the occurrence of this species at four of the sampling sites (Nos 2, 5, 9 and 10). Holly must have been dominant at site No. 10.

Olive (*Olea europaea*) is widely cultivated in the southern foothills of the Óthris below 400 m. Although *Olea* is not found in the mountains, its pollen was encountered in all the surface samples. This points to excellent pollen dispersal of the species. Fair amounts of olive pollen must have been transported uphill by thermal air flows. The same must hold for terebinth (*Pistacia terebinthus*) and mastic tree (*Pistacia lentiscus*), which are not represented above 700 masl. Also plane (*Platanus*) shows a good pollen dispersal. It is a local tree that preferably grows alongside stream valleys and is therefore not encountered in the vegetation of the sampling locations.

The AP values of the present-day samples compare well with those of the uppermost sample of the Mavrikopoúla diagram. When the probably local Cyperaceae are excluded from the pollen sum, the AP percentages of spectrum 26 are between 50 and 60%. The fairly high AP values in the surface samples and spectrum 26 in the Mavrikopoula diagram, relative to the values in the rest of zone 2, suggest some recovery of forest in recent times, maybe as a consequence of reduced farming activities in the mountain zone.

Nevertheless, evidence of grazing is abundantly provided by the appreciable values of ribwort plantain and sorrel species. The present grazing rates may favour these taxa, even though part of the pollen, like that of *Olea*, may come from lower elevations. The limited share of grasses in the present-day samples, relative to the palynological evidence of the last millennium, indeed points to a declining number of foraging animals.

Another anthropogenic pollen type present in all the modern samples is *Urtica dioica*. Plants which produce this type of pollen have not been observed in the mountain vegetation, which could suggest that most of this pollen also comes from lower elevations, such as the vegetation in the olive zone.

3.7. Mountain vegetation and human impact

Changes in pollen compositions are often the consequence of changes in climate factors, but several Near Eastern and Mediterranean pollen diagrams, especially from the 4th millennium onwards, provide evidence of changes caused by human interference. Such interference may include forest felling, burning of the vegetation, pasturing of livestock and crop cultivation. When these activities are practised on an a sizeable scale, they may result in changes in pollen values, the disappearance of plant taxa and the appearance of 'new' ones.

The Mavrikopoúla pollen diagram also provides ample evidence of the presence of man in the mountains. Subzone 1a, which spans approximately the 3rd millennium BP, differs from the other subzones in that tree pollen dominates the pollen record. The appearance of ribwort plantain and the expansion of *Pteridium* however point to grazing, if still on a modest scale. The spread, probably local, of stinging nettle (*Urtia dioica*-type) observable in spectrum 2 likewise suggests disturbance in the marsh proper. The appearance or expansion of these taxa reflects the presence of man in the Mavrikopoúla area. No, or only very little evidence of human impact seems to be observable at the beginning of the diagram, in spectrum 1. High AP values and the absence of anthropogenic indicators suggest a quite undisturbed, oak-dominated mountain forest at the time of spectrum 1.

The diagram shows a change from oak forest to conifer forest in subzone 1a. Radiocarbon dates indicate that the occupation of New Halos coincided approximately with spectrum 5. The decrease in the deciduous oak values to approx. 10% that can be inferred from this spectrum implies a substantial decline of the oak forest. By the beginning of the Christian era the deciduous oak forest had largely recovered. Do these developments reflect forest succession, shifts induced by climate changes or human interference?

Surface samples indicate a sharp decrease in fir pollen only a few hundred metres away from the fir forest (Bottema, 1974), but the values of fir pollen in surface samples taken in or close to the fir forest also rarely exceed 10%. We may therefore safely assume that the *Abies* values recorded in subzone 1a represent a dense fir forest that took up most of the dry land up to the edge of the marsh. High fir pollen values, mostly more than 20%, were found in the area of Pertoúli Trikálon, a peat bog in central Greece (altitude 1139 m; Athanasiadis, 1975). As at Mavrikopoúla, macro-remains (wood, stomata, seeds, needles) from the site confirm that fir grew close to the coring site.

The Mavrikopoúla fir values should however be interpreted with due care. Firstly, there may have been a dense fir forest around the marsh, but other plant taxa (*e.g. Pteridium*, woody Rosaceae and *Ostrya/Carpinus orientalis*) indicate that more open conditions were also to be found in this area. Secondly, a natural shift from deciduous oak forest to more summer-drought-resistant conifer forest could in the Mediterranean point to increased aridity, at least during the growing season. But other eastern Mediterranean pollen diagrams show no signs of any increasing aridity during this timespan. If anything, the palynological

evidence suggests more humid conditions. The shift in forest plant communities during subzone 1a must therefore represent a local event rather than a general expansion of fir forest in the Óthris mountains.

What conditions may have caused an oak forest to give way to a conifer forest? Thoreau (1993) devoted a life-long project to this question in the forests near Concord, Massachusetts, New England, in northeast USA. Though the main focus of his research was the dispersal of seeds, he also comprehensively described forest succession as a related theme. In forest succession, the dominating forest tree is almost always replaced by a different one after some calamity such as a fire, windfall or felling. Such changes should indeed occur, as the absorption of nutrients for several decades depletes the soil, making a healthy growth of a second generation of the same species unlikely. This rule holds in particular where dense forest is concerned. In open forest, the integration of other species is more gradual.

In Massachusetts much of the forest is dominated by pitch pine (*Pinus rigida*), white pine (*P. strobus*) and several broad-leaved oak species. According to Thoreau, white pine succeeds pitch pine, oaks succeed white pine and pitch pine, but, conversely, white pine also succeeds oak.

The drastic decline in oak around Mavrikopoúla observable from spectrum 1 onwards suggests a calamity. Could the forest have been set on fire or were the trees felled for some reason? Besides *Abies, Pteridium* also responds to a decline of oak forest. This fern spreads under several conditions. Repeated dosage of ammonia, grazing, fire and felling favour the species. Once established, it has little to fear; few plants are able to grow in the acid substrate of decaying leaves, and grazing herbivores, wild and domestic, avoid the poisonous leaves (Weeda *et al.*, 1985–1994). The presence of bracken suggests an open area. It retreated only at the approach of the fir forest. It would seem that the fir forest did not immediately displace the oak forest, but that the oaks just vanished, leaving an open situation for a fairly long period. Apparently (part of) the oak forest was erased by some disaster around 3000 BP.

Fire is not likely to have been the cause of the oak decline. The pollen samples contain no evidence of burning in the form of charcoal particles. Grazing is likewise assumed to be an unlikely cause. Long-term grazing would probably have led to an expansion of kermes oak, but it would also have prevented the establishment of the fir forest which dominates most of subzone 1a. It has been ascertained that livestock roaming in forests are partial to seedlings and young shoots of fir (Horvat *et al.*, 1974, pp. 544–545).

Other possible causes of the retreat of oak are pathogens, periods of severe drought, windfall and the felling of trees by humans. No pathogens capable of causing massive destruction of oaks are known in

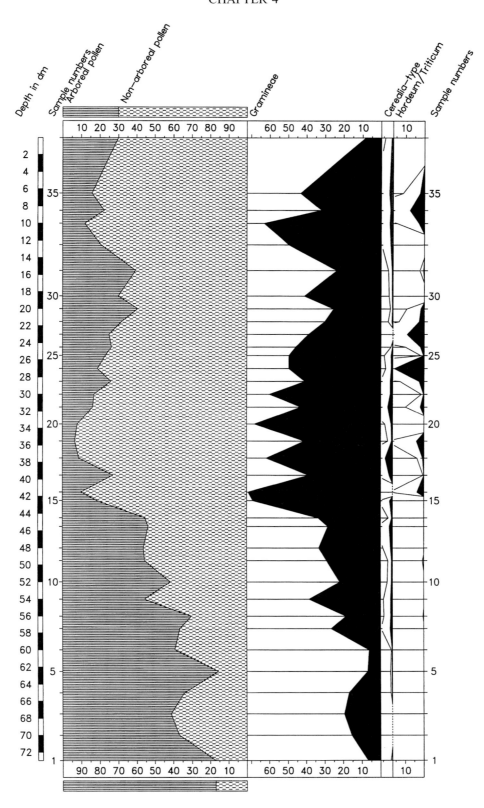

Fig. 4.8. Total values of Hordeum/Triticum-type, including pollen <40 μm (see Ch. 4: 3.8).

Europe, and deep-rooting trees are not very susceptible to windfall. It is true that greater drought can lead to an expansion of summer-drought-resistant fir forest (see section 4.2), but it can also favour the development of sclerophyllous oak.

Major shifts in vegetation are observable in pollen diagrams from the Near East and Greece from around 3000 BP. The diagram of Halos, which lay in the coastal plain between the Óthris mountains and the Pagasitikós gulf, shows a reduction in deciduous oak values around this time from approx. 30% to less than 10% (Bottema, 1988). Kermes oak, hop hornbeam/oriental hornbeam and fir also show lower values after *c.* 2900 BP. Bottema ascribes the decline of deciduous oaks in the Halos diagram to human interference.

From the calibrated radiocarbon dates (table 4.4) it can be concluded that the decline of oak forest took place around the transition from the Late Bronze Age to the Iron Age. This is known to have been a turbulent time of warfare, political instability and shifts in population (Trojan War, collapse of the Hittite Empire, the Sea People). It could be that humans sought shelter in the mountains for some time, but it is more likely that ore smelting, for which tremendous amounts of oak wood are required, was responsible for the decline of the oak forests. Considerable amounts of wood may moreover have been needed for activities associated with warfare, such as the construction of seaworthy vessels. At the time of subzone 1a, the coastal plains and lower hills were largely depleted of wood (Bottema, 1988). This fact not only demonstrates an increase in the population of southeast Thessaly, but also a great lack of wood and/or grazing terrain since the Neolithic. This lack of wood in the plains will ultimately have led to the exploitation of the mountain forests.

Halos is first mentioned in the works of Herodot (480 BC); the town was destroyed by a Macedonian army in 346 BC (Reinders, 1988). New Halos was occupied 302–265 BC. Extrapolation of the Mavrikopoúla radiocarbon dates suggests that the heydays of New Halos coincided with the minimum oak values around spectrum 5. It may be assumed that the exploitation of the mountain forest and the transport and possible processing of timber contributed to the town's success.

The slight decrease in AP values in subzone 1a is followed by a more pronounced decrease in subzone 1b (*c.* AD 0–1000). By this time, the oak forest had largely recovered, but probably not to the density reflected in spectrum 1. Some of the pollen types indicating human impact in zone 2 are already represented by uninterrupted curves in subzone 1b, *e.g.* Caryophyllaceae, Compositae Liguliflorae, *Plantago lanceolata* and *Rumex acetosa*. This proves that man was present and that his impact on the environment was intensifying.

Of the NAP taxa, grasses show the highest increase. Grasses may have grown in the marsh proper, but the total pollen picture points to the spread of terrestrial grass species in the thinned forest or deforested areas. Being tolerant to grazing, grasses may expand under these conditions. This also holds for bracken, which spreads under ceaseless browsing of the plant cover while competing plants disappear. The presence of ribwort plantain also suggests grazing during subzone 1a.

Subzone 1b provides ample evidence that humans moved into the mountains at the beginning of the Christian era. Whether they did so in the context of transhumance cannot be inferred from the pollen diagram. It is however tremendously hazardous to keep herbivores at such elevations over the winter. Due to the mountain topography, the creation of a forage stock large enough to cover the winter season is an immense, if not impossible task. The livestock would therefore usually have to rely on dead plant cover and twigs of shrubs. And although herbivores may forage on such matter, occasional snow or ice, which would make the forage inaccessible, are hazards which farmers would not venture. Even in recent times, unexpected early snowfall has made victims among people and animals (Reinders, 1994).

The collapse of forest plant communities at the beginning of zone 2, *c.* AD 1000, coincides with a further and quite drastic increase in a range of anthropogenic species. There can be no doubt that human impact triggered the shifts in plant communities. The grassland acreage will have been even larger, given the fact that grasses are among the commonest plants around Mavrikopoúla.

The sharp decline of deciduous oaks and the disappearance of fir at the beginning of subzone 2a show that large-scale felling led to a substantial contraction of the mountain forest. Grazing and browsing in the clearings prevented the regeneration of forest and occasioned the development of grassland rich in herbs. The steady increase in ribwort plantain (*Plantago lanceolata*-type) and sorrel (*Rumex acetosa*-type) in zone 2 suggests a gradual expansion of the grassland area. These species spread optimally under a moderate grazing regime, since they cannot compete in a dense grass cover or under heavy grazing. The continuously high values of members of Caryophyllaceae and Compositae in the first half of subzone 1b could even point to relatively high grazing pressure. Various taxa with *Matricaria*-type pollen (*e.g. Achillea* spp., *Anthemis* spp., *Matricaria* spp.) are secondary choice of grazing herds.

The evidence presented above implies the herding of large flocks and/or herds in the mountains. As mentioned above, year-round herding of domestic animals in the mountains will have involved almost insuperable problems. The combination of these facts suggests that transhumance played a role in the subsistence economy of southern Thessaly, at least after AD 1000.

3.8. Corn growing

Some attention must be paid to the Cerealia and *Hordeum/Triticum*-type pollen in this section of the diagram. Cerealia-type pollen differs from other grass pollen only in the size of the grains (>40 μm). The same holds for *Hordeum/Triticum*-type pollen, which however has larger pores and clearly defined, thicker annuli. These features are usually combined with a peculiar exine structure and shape of the pollen grain. A fairly large proportion of the pollen grains smaller than 40 μm were found to show all the features of pollen of the *Hordeum/Triticum* type.

Measurements by Beug (1961) of a range of pollen of grass species mounted in glycerine jelly prove that part of the pollen of *Hordeum* and *Triticum* species remains below the 40 μm standard. Veenman (1997) measured the differences between pollen of *Hordeum* and *Triticum* mounted in glycerine jelly and silicone oil, the traditional mounting medium at the Groningen Institute of Archaeology. Conversion of the values obtained by Beug shows that even more pollen grains of this type would range below 40 μm in the silicone oil medium. Figure 4.8 gives the pollen ratios of Mavrikopoúla after the removal of the assumed *Hordeum/Triticum* types smaller than 40 μm from among the grasses and their inclusion with the actually determined *Hordeum/Triticum*-type pollen. *Hordeum/Triticum*-type pollen <40 μm shows the highest values in subzone 2a.

Besides the pollen of cultivated and wild barley and wheat species, *Hordeum/Triticum*-type pollen also includes that of *Aegilops*, a genus related to *Triticum*. Several wild *Hordeum* (4) and *Aegilops* (9) species producing pollen of this type belong to the flora of Greece; most can grow at altitudes in the mountain zone.

Leaving aside the possible expansion of one of the natural species, it is known that cereals were widely cultivated in the mountains in the past. The fairly indisputable occurrences of pollen of the *Hordeum/Triticum* type (predominantly in subzone 2a) suggest the introduction of a new species in this mountainous area. Unfortunately, the diagram shows no further evidence of corn growing, such as pollen of accompanying arable weeds. This suggests that either wild *Hordeum* or *Aegilops* spread in the vicinity of the marsh. The long-awned inflorescences characteristic of these genera may provide some protection against grazing; they are often found untouched in grazed areas. So the increase in *Hordeum/Triticum*-type pollen in zone 2a could imply grazing, but also cereal cultivation.

3.9. Local developments

The picture that emerges from the marsh vegetation reveals possible local disturbance (*Juncus*), pollution (*Juncus, Lemna*) or eutrophication by nitrate/phosphate (*Lemna, Urtica*). Several rushes expand where the water level frequently changes (*e.g. Juncus effusus*), in trampled and dense substrates (*e.g. Juncus squarrosus, J. bufonius, J. compressus*) and in areas that suffers eutrophication (*Juncus effusus, J. conglomeratus*) and desiccation (*Juncus effusus*) (Weeda et al., 1985–1994). Dynamic local conditions can be inferred from the rapid succession of *Juncus/Luzula, Dryopteris* and *Salix*. The plant remains incorporated in the sediment suggest that increased deposition of silt and silicates triggered these changing conditions. The abrupt spread and succession of local taxa and the disappearance of water plants suggest gradually drier conditions or a variable (ground)water level.

The changes in the water balance are largely attributable to increased sedimentation in the past thousand years. This period saw the deposition of approximately four metres of sediment, a marked increase in comparison with the three metres of sediment that were laid down in the preceding two millennia. These figures clearly indicate increased erosion as a consequence of deforestation and the devastation of mountain biotas since AD 1000.

The small quantities of *Scleranthus, Polygonum aviculare, Artemisia vulgaris, Centaurea solstitialis, Onopordon* and *Eryngium* pollen found in subzone 2 point to a trodden, heavily grazed open vegetation. Most of these species have modest pollen dispersal, which would imply that the pollen derived from plants growing in the vicinity of the marsh. The number of cattle and/or sheep and goats in the mountains must have been considerable, given the acreage of pasture that was available there for several centuries. We may assume that the farmers' livestock was strongly dependent on the marsh for drinking water, since most rivers in the Óthris mountains run dry in summer. All herbivores other than goats have to be watered almost daily, which would imply the frequent concentration of herds and flocks near the marsh. In such areas the surface of the soil will soon have become devoid of vegetation and only the aforementioned species that are fairly resistant to trampling and are ignored by grazing animals will have been able to survive or even benefit from such conditions.

No sediment was observed between 60 and 5 cm beneath the surface. The 55-cm gap revealed seemingly abrupt changes. A marsh vegetation differing from the vegetations represented at lower levels, which grew mostly on clay or silt, may have developed on the peaty surface. *Ranunculus repens*- and *Galium*-type pollen and especially Cyperaceae pollen includes the pollen of species common in marshy habitats. The values of Cyperaceae, *Ranunculus re-*

pens-type pollen and *Galium*-type pollen in the surface samples taken at locations outside the marsh indeed indicate that the increasing amounts of pollen of these groups in the uppermost spectra were produced locally.

3.10. Transhumance

The evidence of human impact in the Mavrikopoúla diagram differs somewhat from that generally found in contemporary pollen diagrams from the Mediterranean climate zone. The difference concerns especially the values of cultivated trees such as *Olea* and *Juglans* and trees and shrubs whose expansion is usually associated with man's exploitation of the environment, such as *Pistacia, Platanus* and *Castanea*. Olive grows in the thermo-Mediterranean climate zone (maximum altitude 600 m). The same holds for the most common *Pistacia* species in this area, *P. terebinthus* and *P. lentiscus*, which do not grow above the 700-m contour line. These species may be assumed to be represented by the highest pollen values at sites in the thermo-Mediterranean zone. *Platanus orientalis* is not restricted to the Eu-Mediterranean climate zone, but this species is not found above 1000 m altitude. In Greece, *Juglans* and *Castanea* are to be found up to altitudes of approx. 1500 m, but the 1990 inventory (Ch. 4: 2) showed that both species are now rare in the Óthris mountains. Only *Juglans* was recorded once (table 4.2). The Mavrikopoúla record indicates that these species were indeed rare in the past three millennia. The most likely reason for their (almost complete) absence is the comparatively small amount of precipitation.

The evidence presented above shows that climate factors such as temperature and precipitation influence the spread of certain arborescent species, which explains the low pollen values in the Mavrikopoúla diagram.

The palynological data imply that if cereal cultivation was at all practised, it took place on a small scale only. This is not surprising, considering the total absence of any archaeological evidence of permanent farming settlements. This, and the hazards of the winter season, would also preclude any subsistence strategy based on year-round farming in the mountains.

The scanty evidence available on arboriculture and crop cultivation contrasts markedly with the abundant indications of pastoralism. Grasslands started to expand during the occupation of New Halos. Grazing indicators like ribwort plantain and sorrel are represented in almost all the spectra. Along with the man-induced shifts in the composition of the forest, these species represent almost three millennia of human impact in the mountainous zone.

The Mavrikopoúla diagram shows that the extent of mountain pasture was limited at the beginning of the 3rd millennium BP. The progressively larger expanses of grassland made it more and more attractive to bring livestock into the mountains during the growing season. The advantages of such seasonal moves were twofold. The incorporation of the mountains in the farming system implied a substantial extension of the grazing area, and the absence of the animals in spring and summer facilitated farming activities like crop cultivation near the settlements.

It may be assumed that the exploitation of the mountain environment eventually led to the development of a farming system which included the mountain pastures. Such a farming economy, usually referred to as transhumance, seems to have become firmly established here by 1000 BP. This economy did in all probability not arise at once, but evolved gradually, following man's felling activities in the mountain zone. These activities enforced humans to be in the mountains for weeks or months at a time. To ensure a sufficient supply of food, they will have taken live animals along with them. We may assume that the exploitation of the mountains shifted more and more towards pasturing during subzone 1b, when ever more clearances were made in the forest and the grazing areas expanded. This development may have begun already in the upper part of subzone 1a, so during the occupation of New Halos or slightly later. From this time onwards the AP values show a steady decline. The regular growth of the population in this period probably triggered the shift towards transhumance. Around 1000 BP, by which time most of the forest had been cleared, pasturing will have become the main aim for people to move into the mountains for the summer period. The steady growth of the population from Hellenistic times onwards assumedly triggered the development of this economy, which has persisted to the present day.

3.11. Final remarks

Pollen diagrams from the eastern Mediterranean show major shifts in vegetation towards the end of the 4th millennium BP. These shifts include palynological indications of human influence, crop cultivation and pasture (van Zeist *et al.*, 1975; Bottema & Woldring, 1984(1986)). The pollen records from these times differ from those of previous periods in the appearance of pollen of 'new' species such as walnut *(Juglans)*, sweet chestnut *(Castanea sativa)*, olive *(Olea)*, manna ash *(Fraxinus ornus)* and plane *(Platanus orientalis)* and the spread of ribwort plantain *(Plantago lanceolata)*, sorrel *(Rumex)*, burnet *(Sanguisorba minor)* and knotgrass *(Polygonum aviculare)*. The period around 3000 BP concerned includes the transition from the Late Bronze Age to the Iron Age. The radical palynological changes coincide with the fall of the Hittite Empire, political instability, warfare and population shifts in the eastern Mediterranean.

It is in this turbulent period that the Mavrikopoúla marsh developed. Is this a coincidence or is the development of the marsh perhaps attributable to an intentional diversion of an existing water course? There is some evidence showing that forest grew in this area before the marsh developed here. A purely humic layer containing twigs and other organic remains at a depth of 735–730 cm probably represents the top layer of the original forest soil. The absence of pollen in samples from this depth could indeed imply that a low water table and occasional desiccation prevented its preservation.

Remains of oak and fir wood were frequently encountered in the bottom metre of the sediment (see Ch. 4: 3.5). Oak and fir do not tolerate permanently wet conditions, which suggests that the rising water table killed the trees growing near the coring location. The presence of oak and fir at this location means that no marsh or open water existed here before the sedimentation process began, *c.* 3000 BP.

The pollen record provides convincing evidence of human impact (animal grazing) from Roman times onwards (the beginning of subzone 1b), but the shifts in forest plant communities observable before the Roman period do not seem to have been climate-induced either. These shifts, along with the occurrence of stinging nettle in the lowermost spectra, similarly suggest human interference, such as felling of the oak forest for example to obtain wood for use as fuel in ore smelting or for building ships. The event that led to a reduction in oak forest created pastureland whose acreage gradually expanded, as can be inferred from the continuous decrease in AP values. Deciduous oaks regenerated to some extent in subzone 1b, which spans approximately the period from Roman times to AD 1000. The recovery of oak in this period coincides extraordinarily well with the decline of agriculture from 200 BC to AD 900 postulated by Athanasiadis (1975) on the basis of the pollen diagram of Litochóro.

Human impact is most evident from AD 1000 onwards. The landscape that then developed probably resembled the present-day mountain landscape around Mavrikopoúla. Any attempts at cereal growing will have been altogether abandoned in the 15th century or so, when the main goal of man's presence in the mountains seems to have been the herding of livestock (high values of ribwort plantain and sorrel).

The palynological changes observable in the youngest time phase point to overgrazing (reduction in grazing indicators, increase in juniper and kermes oak), but the subrecent expansion of fir suggests diminished grazing pressure. The spread of fir may actually have been a local event (pollen of fir is poorly dispersed), except that the uppermost part of the Zerélia diagram shows a similar picture, suggesting that fir expanded not only around Mavrikopoúla, but possibly over large parts of the northern flanks of the

Óthris mountains, too. Fir pollen is conspicuously absent from the Halos diagram. The slight increase in AP values to approx. 30% observable in the uppermost spectrum (No. 36) and the even higher AP values found in the surface samples confirm diminished exploitation of the mountain zone by grazing.

Possible causes of the recent decrease in grazing pressure are population movements after the retreat of the Turks from Thessaly (1881) and the decline of the transhumance tradition in the past decades (Reinders, 1996).

3.12. Acknowledgements

Thanks are due to Professor S. Bottema for having critically read and discussed the text of this chapter. The wood fragments were identified by Mrs J.N. Bottema-Mac Gillavry. Mrs G. Entjes-Nieborg processed the manuscript.

3.13. Note

Taxa not included in the diagram of Mavrikopoúla.

Spectrum 2: *Sorbus*-type 0.2, *Cyclamen* 0.2, *Phyllitis* 0.2;

spectrum 3: *Cerinthe*-type 0.2, *Cheilanthes* 0.2, *Ophioglossum* 0.6;

spectrum 4: *Sorbus*-type 0.1, *Spiraea*-type 0.1, *Aellenia*-type 0.1;

spectrum 6: *Papaver* 0.3, *Typha latifolia* 0.1, *Botrychium* 0.1;

spectrum 7: *Sambucus nigra*-type 0.2;

spectrum 8: *Euphorbia* 0.1, *Secale* 0.3, Polygonaceae 0.1, *Polygonum convolvulus*-type 0.1;

spectrum 9: *Malus*-type 0.1, *Iris* 0.1;

spectrum 10: *Caltha* 0.5, *Alisma* 0.2;

spectrum 11: *Ixiolirion* 0.2;

spectrum 12: *Picea* 0.1, *Marrubium* 0.1, *Valerianella* 0.1, *Cheilanthes* 0.1, *Isoetes* 0.1;

spectrum 13: *Picea* 0.2, *Mercurialis annua* 0.1;

spectrum 14: *Sambucus nigra*-type 0.1, *Aesculus* 0.1, *Papaver* 0.1, *Valeriana* 0.1;

spectrum 15: *Malva* 0.1;

spectrum 16: *Salvinia* 0.2;

spectrum 17: *Valeriana* 0.1, *Polygonum amphibium* 0.1, *Alisma* 0.2, *Anthoceros punctatus*-type 0.1;

spectrum 18: *Nepeta* 0.2, *Plantago media*-type 0.1;

spectrum 19: *Frangula* 0.1, *Daphne* 0.1, *Herniaria*-type 0.1, *Genista*-type 0.1, *Plantago media*-type 0.1, *Salvinia* 0.1;

spectrum 21: *Populus* 0.2, *Reseda* 0.2, *Equisetum* 0.4;

spectrum 22: *Scorzonera*-type 0.1, *Phyllitis* 0.1;

spectrum 23: *Cotinus* 0.2, *Euphorbia* 0.2, *Rubus* 0.2;

spectrum 25: *Lathyrus*-type 0.4, *Littorella*-type 0.2;

spectrum 26: *Rhamnus* 0.2, *Bidens*-type 0.4, *Xanthium* 0.2, *Mercurialis annua* 0.2, *Alhagi* 0.4;

spectrum 27: *Aesculus* 0.1;

spectrum 28: *Bidens*-type 0.2, *Cynocrambe* 0.1, *Scabiosa columbaria*-type 0.1, *Littorella*-type 0.1, *Helleborus* 0.1, *Nigella* 0.2, *Ranunculus arvensis*-type 0.1;

spectrum 29: *Scorzonera*-type 0.1, *Scabiosa columbaria*-type 0.1, *Ephedra distachya*-type 0.1, *Teucrium* 0.1, *Plantago major*-type 0.1, *Armeria/Limonium* 0.1, *Lycopodium* 0.1;

spectrum 30: *Heliotropium*-type 0.1, *Onobrychis*-type 0.1, *Equisetum* 0.1;

spectrum 32: *Armeria/Limonium* 0.2;

spectrum 33: *Salsola* 0.2, *Xanthium* 0.2, *Prunella*-type 0.2;

spectrum 34: *Heliotropium*-type 0.1, *Datisca* 0.1, *Sideritis* 0.1, *Teucrium* 0.1, *Valerianella* 0.1, *Epipactis* 0.1;

spectrum 35: *Arceuthobium* 0.1, *Celtis* 0.1, *Jasione*-type 0.2, *Herniaria*-type 0.1, *Linum* 0.1;

spectrum 36: *Plumbago* 0.1.

3.14. References

ATHANASIADIS, A., 1975. Zur postglazialen Vegetationsentwicklung von Litochóro Katerínis und Pertoúli Trikálon (Griechenland). *Flora* 164, pp. 99–132.

BEUG, H.-J., 1961. *Leitfaden der Pollenbestimmung für Mitteleuropa und die Angrenzenden Gebiete.* Gustav Fischer Verlag, Stuttgart.

BOTTEMA, S., 1974. *Late Quaternary Vegetation History of Northwestern Greece.* Thesis, Groningen.

BOTTEMA, S., 1982. Palynological investigations in Greece with special reference to pollen as an indicator of human activity. *Palaeohistoria* 24, pp. 257–289.

BOTTEMA, S., 1988. A reconstruction of the Halos environment on the basis of palynological information. In: H.R. Reinders, *New Halos. A Hellenistic town in Thessalia, Greece.* Hes, Utrecht, pp. 216–226.

BOTTEMA, S., 1990. Holocene environment of the Southern Argolid: A pollen core from Kiladha Bay. In: T.J. Wilkinson & S.T. Duhon (eds), *Franchthi Paralia. The sediments, stratigraphy, and offshore investigations. Excavations at Franchthi Cave, Greece.* Fasc. 6. Indiana University Press, Bloomington & Indianapolis, pp. 117–138.

BOTTEMA, S. & H. WOLDRING, 1984 (1986). Late Quaternary vegetation and climate of southwestern Turkey. Part II. *Palaeohistoria* 26, pp. 123–149.

JAHNS, S., 1993. On the Holocene vegetation history of the Argive Plain (Peloponnese, Greece). *Vegetation History and Archaeobotany* 2, pp. 187–203.

HORVAT, I., V. GLAVA & H. ELLENBERG, 1974. *Vegetation Südosteuropas* (= Geobotanica Selecta Bd. IV). Gustav Fischer Verlag, Stuttgart.

REINDERS, H.R., 1988. *New Halos, a Hellenistic Town in Thessalía, Greece.* Hes, Utrecht.

REINDERS, H.R., 1994. *Schaarse Bronnen.* Inaugural lecture, University of Groningen.

REINDERS, H.R., 1996. *Veehouders in Thessalië.* Thessalia Erga, Groningen.

THOREAU, H.D. 1993. *Faith in a Seed.* Island Press, Washington DC.

TURNER, J. & J.R.A. GREIG, 1975. Some Holocene pollen diagrams from Greece. *Review of Palaeobotany and Palynology* 20, pp. 171–204.

VEENMAN, F., 1997. Landgebruik in de Agro Pontino (1900–1100 BC). Een archeologisch en paleobotanisch onderzoek. Undergraduate thesis, Groningen.

WEEDA, E.J., R. WESTRA, Ch. WESTRA & T. WESTRA, 1985-1994. *Nederlandse Ecologische Flora, Wilde Planten en hun Relaties.* Deel I, II and V. Hilversum/Haarlem.

WOLDRING, H., this volume. Forest vegetation and human impact in the Óthris mountains (Ch. 4: 2).

WOLDRING, H. & S. BOTTEMA, manuscript. Late Glacial and Holocene vegetation history of Central Anatolia: the palynological record of Eski Acıgöl.

ZEIST, W. VAN, H. WOLDRING & D. STAPERT, 1975. Late Quaternary vegetation and climate of southwestern Turkey. *Palaeohistoria* 17, pp. 53–143.

4. ANIMAL HUSBANDRY AND MOLLUSC GATHERING

WIETSKE PRUMMEL

4.1. *Introduction*

This section discusses the animal remains found in six houses dating from 302-265 BC in Hellenistic Halos, Achaia Phthiotis, Thessaly. The excavation was carried out by the HALOS team of the University of Groningen, the Netherlands, between 1978 and 1993, under the supervision of H. Reinder Reinders. Among the members of the team were students of Groningen and other universities, young archaeologists and volunteers. The town of New Halos was situated at the narrow transition of the Almirós and Soúrpi plains. To the west of the town were the northeastern foothills of the Óthris mountains (highest summit 1726 m). About 1.5 km to the east of the town was the shore of Soúrpi Bay, which forms part of the Pagasitikós gulf. Between the town and Soúrpi Bay was a salt marsh, which was protected from the sea by a coastal barrier in Hellenistic times (Reinders, 1988: fig. 20) (fig. 4.9).

The town of New Halos consisted of two parts, a rectangular lower town and a triangular upper town. The lower town was the residential part of the town. This part may have contained an *agora*. Otherwise the lower town consisted entirely of houses. The foundations of 16 buildings which were presumably not residential houses were found in the upper town. This part of the town may have contained some of the town's public buildings (Reinders, 1988: p. 208).

The lower town measured about 720×625 m. It was divided into 64 housing blocks by a southwest northeast oriented main avenue, four avenues (A, B, C and D) extending at right angles to the main avenue and fourteen streets running parallel to the main avenue, seven on either side (fig. 4.10). Each housing block consisted of two rows of 15-m-long houses. No space was left between the houses or between the houses and the street. Each house was entered from the adjacent street. The number of houses in each block varied. The blocks between avenues A and B, B and C and C and D consisted of about 26 houses each. The blocks between avenue D and the east wall comprised much smaller numbers of houses. The total number of houses in the town is estimated to have been about 1440 (Reinders, 1988: p. 113). The houses in the individual blocks have been numbered from west to east, those of the north row first, followed by those of the south row.

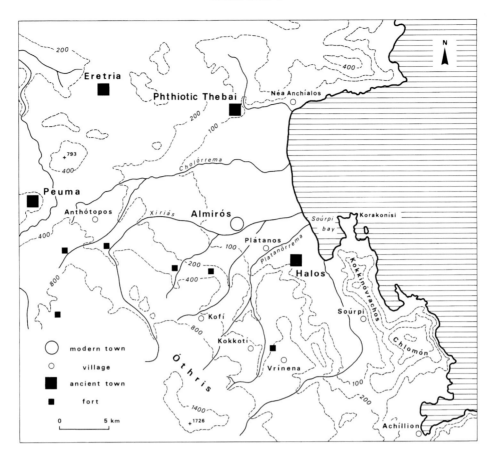

Fig. 4.9. The Almirós and Soúrpi plains along the Pagasitikós gulf, showing ancient sites and modern towns and villages.

The six excavated houses of New Halos lie between avenues B and C. Two lie to the north of the main avenue, *i.e.* the House of the Coroplast (house 1) and the House of the Geometric *Krater* (house 2) (fig. 4.10: 1 and 2), the other four to the south of it: the House of the Snakes (house 3), the House of the *Amphorai* (house 4), the House of the Ptolemaic Coins (house 5) and the House of Agathon (house 6) (fig. 4.10: 3-6). Houses 5 and 6, which measured 12.5 m along the street, were slightly smaller than houses 1, 2, 3 and 4, which were 15 m wide. The entrances of houses 1 and 3 were oriented towards the north, those of houses 2, 4, 5 and 6 towards the south (fig. 4.10).

Most houses had a main living room (*oikos*), several small rooms adjoining the main living room (in some houses two on either side of the main living room), a few other rooms or a corridor and a courtyard (Reinders, 1988; Reinders, 1989; Haagsma, 1991; Dijkstra, 1994). This pattern is clearly observable in houses 1, 5 and 6 and less clearly in houses 2, 3 and 4. House 2 was only partly excavated, as was the courtyard of house 5. There was no room for any byres or stables inside or between the houses. The

areas between the enceinte and the housing blocks, which are 20–35 m wide and 720 m long along the north and south walls and 15 m wide and 625 m long along the west and east walls (Reinders, 1988: fig. 15), will have been the only places where livestock may have been kept inside the town. The enceinte's defensive function however makes it unlikely that flocks of animals were kept in these areas on a regular basis, or that flocks passed through the gates daily to leave the town for grazing.

The six houses were not rebuilt at any time during the town's occupation. So the surface was never levelled with secondary refuse such as animal remains for the construction of a new house. Secondary refuse may however have been incorporated in fresh mud floors in the houses (Rowley-Conwy, 1994), for instance in small-scale renovation activities. The layout of some of the houses was presumably changed at some point. The floors of the closely built houses of New Halos were kept as clean as possible (compare Rowley-Conwy, 1994). Waste may have been stored in one of the rooms or the courtyard for a while, but it will have been removed from the houses fairly quickly, presumably to be deposited on arable fields

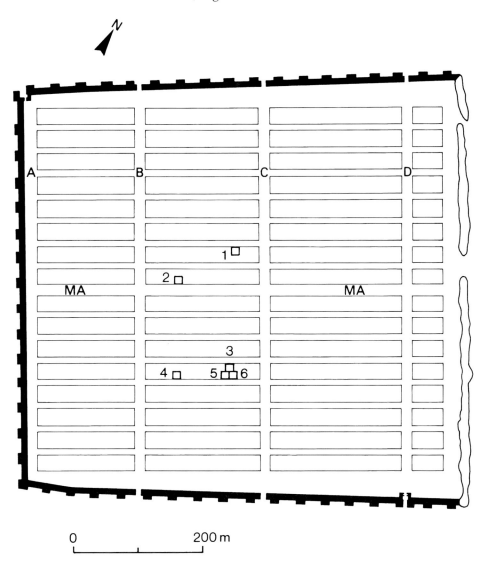

Fig. 4.10. Plan of the lower town of New Halos, showing the locations of the six excavated houses: 1. House of the Coroplast (house 1); 2. House of the Geometric *Krater* (house 2); 3. House of the Snakes (house 3); 4. House of the *Amphorai* (house 4); 5. House of the Ptolemaic Coins (house 5); 6. House of Agathon (house 6) MA: Main Avenue (MA), A–D: Avenues A, B, C and D.

outside the town to fertilise the soil (Hodkinson, 1988: p. 49). The town's short occupation period and the absence of large-scale levelling for new houses will be the main causes of the small numbers of animal remains found in the six houses (table 4.6, a total of 1424 animal remains). Six houses, a *katagogion*, a *stoa*, an *agora*, four streets and a canal in the town of Kassope in Ípiros, which were built within a span of about 340 years from some time in the 4th century until 30 BC, yielded more than 45,300 animal remains (Friedl, 1984; Boessneck, 1986). The rather poor preservation conditions in New Halos may also account for some loss of animal remains.

The town of New Halos was abandoned in or soon after 265 BC, possibly because of an earthquake (see

section 5.2). The inhabitants had enough time to leave their houses. No human skeletons were found. Some of the animal remains found in the houses may represent the food, bone implements and most recent waste present in the houses when the earthquake started. This would imply that the composition of the animal remains of the six houses is partly accidental. The remains of individual meals and the use of individual bone implements may thus be reflected by the animal remains. Waste from former meals and lost bone implements will have been incorporated in the mud floors of the houses and the courtyards during the approximately 37 years of the houses' use. Since short-term accumulation of animal remains can hardly be distinguished from long-term accumulation, oppor-

Table 4.6. Animal remains from six houses in Hellenistic New Halos (Thessaly, Greece). NR = Numbers of remains; %idf = Proportion of the identified remains; %dom = Proportion of the remains from domestic mammals; %udf = Proportion among the unidentified mammalian remains.

House	1: House of the Coroplast 15*15			2: House of the Geometric Krater 15*15			3: House of the Snakes 15*15			4: House of the Amphorai 15*15			5: House of the Ptolemaic Coins 12.5*15			6: House of Agathon 12.5*15			Total		
Width * depth in m	NR	%idf	%dom	NR	%idf	%dom	NR	%idf	%dom	NR	%idf	%dom	NR	%idf	%dom	NR	%idf	%dom	NR	%idf	%dom
Domestic mammals																					
Equids (*Equus* sp.)	2	2.3	3.7	1	2.8	14.3	8	3.4	14.0	-	-	-	2	2.0	7.1	-	-	-	13	1.5	4.0
Dog (*Canis familiaris*)	-	-	-	-	-	-	4	1.7	7.0	-	-	-	-	-	-	-	-	-	4	.5	1.2
Pig (*Sus domesticus*)	4	4.6	7.4	-	-	-	5	2.1	8.8	4	2.1	3.3	2	2.0	7.1	1	.4	1.8	16	1.8	5.0
Cattle (*Bos taurus*)	23	26.4	42.6	2	5.6	28.6	12	5.0	21.1	23	12.0	19.2	5	5.0	17.9	9	4.0	16.1	74	8.4	23.0
Sheep (*Ovis aries*)	2	2.3	3.7	3	8.3	42.9	1	.4	1.8	7	3.7	5.8	2	2.0	7.1	19	8.4	33.9	34	3.9	10.6
Goat (*Capra hircus*)	1	1.1	1.9	-	-	-	-	-	-	1	.5	.8	1	1.0	3.6	3	1.3	5.4	6	.7	1.9
Sheep/goat (*Ovis/Capra*)	22	25.3	40.7	1	2.8	14.3	27	11.3	47.4	85	44.5	70.8	16	15.8	57.1	24	10.7	42.9	175	19.9	54.3
Total	54	62.1	100.0	7	19.4	100.0	57	23.9	100.0	120	62.8	100.0	28	27.7	100.0	56	24.9	100.0	322	36.7	100.0
Wild mammals																					
Brown hare (*Lepus europaeus*)	1	1.1		-	-		-	-		-	-		-	-		-	-		1	.1	
Vole (*Microtus* sp.)	-	-		-	-		-	-		2	1.0		-	-		-	-		2	.2	
Fox (*Vulpes vulpes*)	-	-		-	-		1	.4		-	-		-	-		-	-		1	.1	
Badger (*Meles meles*)	-	-		-	-		-	-		1	.5		-	-		-	-		1	.1	
Red deer (*Cervus elaphus*)	-	-		-	-		3	1.3		4	2.1		1	1.0		2	.9		10	1.1	
Roe deer (*Capreolus capreolus*)	-	-		-	-		-	-		-	-		-	-		2	.9		2	.2	
Cf. ivory	1	1.1		-	-		-	-		-	-		-	-		-	-		1	.1	
Total	2	2.3		-	-		4	1.7		7	3.7		1	1.0		4	1.8		18	2.1	
Reptiles																					
Tortoise (*Testudo* sp.)	-	-		-	-		-	-		13	6.8		-	-		-	-		13	1.5	
Fish																					
Species unknown	1	1.1		-	-		-	-		-	-		-	-		-	-		1	.1	
Marine gastropoda																					
Patella caerulea	2	2.3		-	-		33	13.9		1	.5		3	3.0		2	.9		41	4.7	
Gibbula divaricata	-	-		-	-		-	-		-	-		-	-		1	.4		1	.1	
Gibbula albida	-	-		-	-		1	.4		-	-		-	-		1	.4		2	.2	
Cerithium vulgatum	-	-		2	5.6		3	1.3		-	-		3	3.0		2	.9		10	1.1	
Luria lurida	-	-		1	2.8		-	-		-	-		-	-		-	-		1	.1	
Tonna galea	1	1.1		-	-		-	-		-	-		-	-		-	-		1	.1	
Bolinus brandaris	-	-		1	2.8		5	2.1		-	-		2	2.0		-	-		8	.9	

Table 4.6. (continued).

House	1: House of the Coroplast 15*15			2: House of the Geometric *Krater* 15*15			3: House of the Snakes 15*15			4: House of the *Amphorai* 15*15			5: House of the Ptolemaic Coins 12.5*15			6: House of Agathon 12.5*15			Total		
Width * depth in m	NR	%idf	%dom	NR	%idf	%dom	NR	%idf	%dom	NR	%idf	%dom	NR	%idf	%dom	NR	%idf	%dom	NR	%idf	%dom
Hexaplex trunculus	2	2.3					13	5.5		1		.5	2	2.0		1		.4	19	2.2	
Buccinulum corneum										1		.5							1	.1	
Conus ventricosus				1	2.8					1		.5							2	.2	
Unidentified gastropod							1	.4		1		.5	1	1.0		1		.4	4	.5	
Marine bivalvia																					
Arca noae	1	1.1					4	1.7					1	1.0		32	14.2		38	4.3	
Glycymeris insubrica													2	2.0		2	.9		4	.5	
Mytilus galloprovincialis	1	1.1					1	.4											1	.1	
Pinna nobilis													2	2.0					3	.3	
Chlamys cf. glabra																2	.9		2	.2	
Spondylus gaederopus	5	5.7					1	.4					1	1.0		2	.9		9	1.0	
Ostrea edulis	2	2.3					46	19.3					1	1.0		1	.4		50	5.7	
Psoedochama gryphina													1	1.0					1	.1	
Acanthocardia tuberculata							14	5.9		1		.5	6	5.9		1		.4	22	2.5	
Cerastoderma glaucum	15	17.2		20	55.6		30	12.6		36	18.8		31	30.7		56	24.9		188	21.4	
Mactra stultorum							3	1.3											3	.3	
Donacilla cornea							1	.4					3	3.0					4	.5	
Callista chione													1	1.0					1	.1	
Tapes decussatus							1	.4								30	13.3		31	3.5	
Venus verrucosa	1	1.1		4	11.1		4	1.7		7	3.7		6	5.9		28	12.4		50	5.7	
Unidentified bivalve							1	.4		1		.5	1	1.0					3	.3	
Unident. (marine) mollusc	1	1.1					2	.8								1		.4	4	.5	
Total marine Mollusca	30	34.5		29	80.6		165	69.3		50	26.2		67	66.3		163	72.4		504	57.4	
Terrestrial gastropoda																					
Helix figulina							6	2.5		1		.5	1	1.0		2	.9		10	1.1	
Lindholmia lens							2	.8											2	.2	
Helicella sp.							4	1.7					4	4.0					8	.9	
Total terrestrial Gastropoda							12	5.0		1		.5	5	5.0		2		.9	20	2.3	
Total identified	87	100.0		36	100.0		238	100.0		191	100.0		101	100.0		225	100.0		878	100.0	
Proportion identified		65.9			90.0			59.1			41.6			85.6			82.7			61.7	

Table 4.6. (continued).

House	1: House of the Coroplast 15*15		2: House of the Geometric *Krater* 15*15		3: House of the Snakes 15*15		4: House of the *Amphorai* 15*15		5: House of the Ptolemaic Coins 12.5*15		6: House of Agathon 12.5*15		Total	
Width * depth in m	NR	%udf	NR	%udf	NR	%udf	NR	%udf	NR	%udf	NR	%udf	NR	%udf
Unidentified mammalian remains														
Of cattle/horse size	5	11.1	2	50.0	62	37.6	37	13.8	6	35.3	8	17.0	120	22.0
Of sheep/goat/pig size	25	55.6	2	50.0	43	26.1	228	85.1	10	58.8	39	83.0	347	63.6
Of unknown size	15	33.3	-	-	60	36.4	3	1.1	1	5.9	-	-	79	14.5
Total unidentified	45	100.0	4	100.0	165	100.0	268	100.0	17	100.0	47	100.0	546	100.0
Proportion unidentified		34.1		10.0		40.9		58.4		14.4		17.3		38.3
Total	132		40		403		459		118		272		1424	
Human, *Homo sapiens*	1		-		-		-		-		-		1	

Table 4.7. Mammalian remains from six houses in Hellenistic New Halos (Thessaly, Greece). Bone weights. G = Weight in g; %idf = Proportion of the identified remains; %dom = Proportion of the remains from domestic mammals; %udf = Proportion of the unidentified mammalian remains.

House	1: House of the Coroplast 15*15			2: House of the Geometric *Krater* 15*15			3: House of the Snakes 15*15			4: House of the *Amphorai* 15*15			5: House of the Ptolemaic Coins 12.5*15			6: House of Agathon 12.5*15			Total		
Width * depth in m	G	%idf	%dom	G	%idf	%dom	G	%idf	%dom	G	%idf	%dom	G	%idf	%dom	G	%idf	%dom	G	%idf	%dom
Domestic mammals																					
Equids (*Equus* sp.)	45	7.5	7.5	26	26.8	26.8	433	61.2	66.2	-	-	-	72	36.2	43.1	-	-	-	576	18.8	20.3
Dog (*Canis familiaris*)	-	-	-	-	-	-	10	1.4	1.5	-	-	-	-	-	-	-	-	-	10	.3	.4
Pig (*Sus domesticus*)	22	3.6	3.7	-	-	-	22	3.1	3.4	13	1.5	1.6	6	3.0	3.6	5	.9	1.0	68	2.2	2.4
Cattle (*Bos taurus*)	464	76.8	77.3	52	53.6	53.6	87	12.3	13.3	435	48.6	52.5	30	15.1	18.0	284	51.2	58.2	1352	44.2	47.7
Sheep (*Ovis aries*)	5	.8	.8	12	12.4	12.4	6	.8	.9	75	8.4	9.0	9	4.5	5.4	114	20.5	23.4	221	7.2	7.8
Goat (*Capra hircus*)	13	2.2	2.2	-	-	-	-	-	-	6	.7	.7	24	12.1	14.4	21	3.8	4.3	64	2.1	2.3
Sheep/goat (*Ovis/Capra*)	51	8.4	8.5	7	7.2	7.2	96	13.6	14.7	300	33.5	36.2	26	13.1	15.6	64	11.5	13.1	544	17.8	19.2
Total	600	99.3	100.0	97	100.0	100.0	654	92.5	100.0	829	92.6	100.0	167	83.9	100.0	488	87.9	100.0	2835	92.7	100.0
Wild mammals																					
Brown hare (*Lepus europaeus*)	2	.3		-	-		-	-		-	-		-	-		-	-		2	.1	
Vole (*Microtus* sp.)	-	-		-	-		-	-		0	.0		-	-		-	-		0	.0	
Fox (*Vulpes vulpes*)	-	-		-	-		1	.1		-	-		-	-		-	-		1	.0	
Badger (*Meles meles*)	-	-		-	-		-	-		0	.0		-	-		-	-		0	.0	
Red deer (*Cervus elaphus*)	-	-		-	-		-	-		66	7.4		32	16.1		16	2.9		166	5.4	
Roe deer (*Capreolus capreolus*)	-	-		-	-		52	7.4		-	-		-	-		51	9.2		51	1.7	
Cf. ivory	2	.3		-	-		-	-		-	-		-	-		-	-		2	.1	
Total	4	.7		-	-		53	7.5		66	7.4		32	16.1		67	12.1		222	7.3	
Total identified	604	100.0		97	100.0		707	100.0		895	100.0		199	100.0		555	100.0		3057	100.0	
Proportion identified		94.8			87.4			89.5			74.7			86.9			91.6			85.6	
Unidentified mammalian remains		% udf			% udf			% udf			% udf			% udf			% udf			% udf	
Of cattle/horse size	13	39.4		12	85.7		64	77.1		127	41.9		20	66.7		24	47.1		260	50.6	
Of sheep/goat/pig size	13	39.4		2	14.3		11	13.3		171	56.4		8	26.7		27	52.9		232	45.1	
Of unknown size	7	21.2		-	-		8	9.6		5	1.7		2	6.7		-	-		22	4.3	
Total unidentified	33	100.0		14	100.0		83	100.0		303	100.0		30	100.0		51	100.0		514	100.0	
Proportion unidentified		5.2			12.6			10.5			25.3			13.1			8.4			14.4	
Total	637			111			790			1198			229			606			3571		
Human, *Homo sapiens*	10			-			-			-			-			-			10		

tunities for distinguishing individual short-term activities or long-term characteristics of diet and bone tools are limited.

The animal remains will be discussed in relation to the individual houses to reveal possible differences in the use of the houses and in the inhabitants' social status, for instance associated with the house's size or its situation within the town. The distribution of the animal remains over the individual rooms and the courtyards has also been analysed to obtain clues to the use of the rooms. The animal remains of Hellenistic Halos have moreover been compared with those from contemporary sites in eastern Mediterranean countries (Boessneck, 1973; Friedl, 1984; Boessneck, 1986; Uerpmann et al., 1992; Ruscillo, 1993; Nobis, 1994; Vila, 1994).

4.2. Material and methods

The excavations and the recovery of the animal remains were carried out by hand. Some wet sieving was executed in 1979 and 1993. The wet sieving yielded a very small number of tiny unidentifiable bone fragments and some recent terrestrial gastropod shells, but no additional archaeological evidence. We may conclude that the vertebrate and mollusc remains recovered by hand are representative of the remains preserved in the soil. The recovery of animal remains was possibly less precise in houses 1, 2 and 5 than in the other houses (pers. comm. H.R. Reinders).

The animal remains are stored in the *apothiki* of the Museum of Almirós (Magnísia, Greece), where they were studied in 1993 and 1994. A reference collection of recent vertebrate skeletons was not available during identification. Therefore, some animal remains may have remained unidentified which might have been identified in the presence of a reference collection (table 4.6). A reference collection of marine mollusc shells was gathered on the beaches of Amaliápolis and Almirós, a short distance from the site (table 4.22). Terrestrial gastropods were collected in the surroundings of the site. This reference material was identified with the aid of the zoological collections of the University of Amsterdam. Their curator, R. Moolenbeek, assisted with the identifications.

The animal remains are of a pinkish red colour due to the red colour of the substrate. The vertebrate remains are poorly preserved. Many bones broke before and after deposition and during excavation and post-excavation handling. The mollusc shells are well preserved, although many shells broke either while they lay on the floor, during their time in the soil or during excavation and post-excavation handling.

The animal remains were counted (table 4.6), weighed (table 4.7) and measured (tables 4.20 and 4.21). The age and sex of each animal were established where possible. All information available on burning, gnawing, pathological alterations, processing

into bone artefacts or the use of parts of animals was recorded.

4.3. Results

In total, 1424 animal remains were collected in the six excavated houses: 900 bones and bone fragments and 524 mollusc shells or shell fragments. A total of 354 vertebrate remains could be identified. They account for 39.3% of all the vertebrate remains. The other 546 vertebrate remains, corresponding to 60.7% of the total, could not be identified to species. Small numbers of mollusc shell fragments could not be identified to species. The proportion of unidentified mammal remains, 38.3% of all the animal remains, is rather high (table 4.6).

Domestic and wild mammals account for 94.7% and 5.3% of the identified mammalian remains from all six houses (table 4.6: NR = 340), or 92.7% and 7.3% of the identified mammalian bone weight (table 4.7: G = 3057 g). The weights of the bone fragments of domestic mammals from these six houses range from 1.6 to 13.9 g, with an average of 8.8 g (tables 4.6 and 4.7).

The almost complete absence of fish remains (a single vertebra) and total absence of bird remains are striking. Poor preservation may have resulted in underrepresentation of these groups. Fish and bird remains are however also very rare among the more than 45,300 animal remains found in the town of Kassope in Ípiros (approx. 0.1 and 0.9%, respectively) (Friedl, 1984: Tab. 1; Boessneck, 1986: Tab. a), among the remains from the Hellenistic-Roman layers of Troy (0.5 and 1.8%, respectively) (Uerpmann et al., 1992: Tab. 2a) and among the 3rd/2nd-century-BC remains from Messene on the Pelopónnisos (Nobis, 1994).

In total, 524 mollusc shells and shell fragments were found. They account for 59.7% of the identified animal remains (table 4.6). Their weight, 2326 g, accounts for 43.2% of the total weight of the identified animal remains (5383 g). Because of the totally different build of mammals and molluscs, their numbers of remains and weights cannot be easily compared. A gastropod may leave behind only one shell, a bivalve two valves, but a sheep about 150 bones. A complete sheep bone represents on average 100 g of meat (an entire sheep about 15 kg of meat). Each gastropod or bivalve shell represents only a few grams of food. Consequently, molluscs will have been of less importance in the subsistence system than their numbers and proportions (table 4.6) suggest. A naturally perforated and slightly worn shell of Venus verrucosa suggests that empty shells, i.e. from dead molluscs, were also brought into the town. No water-worn shells (Reese, 1982: p. 140; Reese, 1992) were found in the six houses of New Halos. Shells of recently died molluscs collected on beaches are however rarely water-worn, and are difficult to distinguish from shells that are the refuse of mollusc meals. The large numbers of shells and shell fragments, their presence in each house (table 4.6) and the fact that rooms that yielded large quantities of domestic mammal remains generally also contained large quantities of mollusc shells (tables 4.8 and 4.9) show that molluscs were regularly consumed in New Halos. The fact that many of the identified mollusc species (table 4.6) are also highly appreciated nowadays is another argument for assuming that the mollusc shells found in the six houses are mainly the remains of mollusc meals.

Remains of domestic mammals account for 36.7% of the identified animal remains from all six houses in terms of numbers (table 4.6). The proportions of the other groups of animals are: wild mammals 2.1%, reptiles 1.5%, fish 0.1% and molluscs 59.7%. Two groups of houses were distinguished in a cluster analysis of the proportions of remains of domestic mammals, wild mammals, reptiles combined with fish, and molluscs among the identified remains. The remains of domestic mammals among the identified remains account for more than 60% of the remains from houses 1 and 4. The proportion of (fragments of) mollusc shells found in these two houses is less than 35% (cluster 1). The proportions of wild mammals and reptiles/fish among the identified remains from these houses are relatively high (2.3 and 3.7%, and 1.1 and 6.8%, respectively). The remains of domestic mammals among the identified remains from houses 2, 3, 5 and 6 (the first of which was however a very small sample) account for less than 30%. The mollusc remains from these four houses account for 70% (cluster 2). The proportions of wild mammal remains found in the houses of cluster 2 vary between 0 and 1.8%. No remains of reptiles or fish were found in these houses (fig. 4.11, table 4.6). The clustering does not correlate with the difference in care paid to the recovery of the remains (see above).

That the high proportion of domestic mammal (and other vertebrate) remains in the houses of cluster 1 is not attributable to higher identification rates for vertebrate (mammalian) remains in these houses is demonstrated by the identical clusters based on the proportions of mollusc remains, domestic and unidentified mammalian remains combined, wild mammals and reptile/fish remains (fig. 4.12). There are no obvious reasons to assume that conditions favoured the preservation of vertebrate remains in houses 1 and 4 and that of mollusc shells in houses 2, 3, 5 and 6. This means that houses 1 and 4 contained mostly vertebrate remains and houses 2, 3, 5 and 6 mostly mollusc shells or shell fragments.

These differences may be attributable to long-term differences in consumption habits of the inhabitants of the houses of clusters 1 and 2, the inhabitants of the cluster 1 houses having consumed relatively large proportions of meat dishes and few mollusc meals and those of the cluster 2 houses the opposite. However,

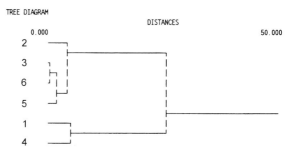

Fig. 4.11. Hellenistic New Halos. Cluster analysis of the six houses on the basis of the proportions of numbers of remains (NR) of domestic mammals, wild mammals, reptiles+fish and molluscs, single linkage method (nearest neighbour).

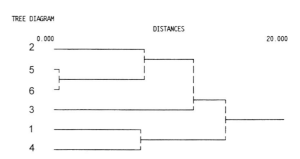

Fig. 4.12. Hellenistic New Halos. Cluster analysis of the six houses on the basis of the proportions of numbers of remains (NR) of domestic mammals+unidentified mammalian remains, wild mammals, reptiles+fish and molluscs, single linkage method (nearest neighbour).

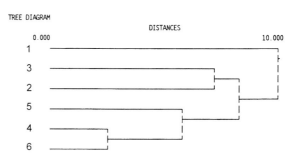

Fig. 4.13. Hellenistic New Halos. Cluster analysis of the six houses on the basis of the proportions of numbers of remains (NR) of equids, dogs, pigs, cattle and sheep and goats, single linkage method (nearest neighbour).

a mollusc meal results in a huge quantity of waste, *i.e.* shells, whereas a meat dish with the same amount of energy and protein yields only a few bones. So the difference in the proportions of meat dishes and mollusc meals between the houses of clusters 1 and 2 may be exaggerated.

Another explanation for the difference in the proportions of mollusc remains between the houses of clusters 1 and 2 could be that the inhabitants of the houses of cluster 2 consumed one or more mollusc meals shortly before they left their houses. The waste

Table 4.8. Six houses in Hellenistic New Halos. Numbers of animal remains in each room of each of the six houses. Room adjoin. m.l.r. = a small room adjoining the main living room (m.l.r.).

House 1: House of the Coroplast	Room/area										
	1 Court-yard	2 Court-yard	3 Room	4 Room	5 Room adjoin. m.l.r.	6 Room adjoin. m.l.r.	7 Main living room	8 Room adjoin. m.l.r.	9 Room adjoin. m.l.r.	?	TOTAL
DOMESTIC MAMMALS											
Equids (*Equus* sp.)	-	-	-	2	-	-	-	-	-	-	2
Pig (*Sus domesticus*)	1	-	1	1	-	-	-	-	-	1	4
Cattle (*Bos taurus*)	1	1	2	9	1	-	5	-	2	2	23
Sheep (*Ovis aries*)	-	-	-	1	-	-	1	-	-	-	2
Goat (*Capra hircus*)	-	-	-	1	-	-	-	-	-	-	1
Sheep/goat (*Ovis/Capra*)	1	1	4	14	-	-	1	-	-	1	22
Total domestic mammals	3	2	7	28	1	-	7	-	2	4	54
WILD MAMMALS											
Hare (*Lepus europaeus*)	-	-	-	-	-	-	1	-	-	-	1
cf. ivory	-	-	-	-	-	-	1	-	-	-	1
TOTAL MAMMALS	3	2	7	28	1	-	9	-	2	4	56
FISH	-	-	1	-	-	-	-	-	-	-	1
Marine molluscs	1	5	5	5	5	-	7	-	-	2	30
TOTAL MOLLUSCS	1	5	5	5	5	-	7	-	-	2	30
Unidentified mammalian remains	1	-	16	20	-	1	6	-	1	-	45
TOTAL	5	7	29	53	6	1	22	-	3	6	132
%	*4*	*5*	*22*	*40*	*5*	*1*	*17*	*-*	*2*	*5*	*100*
Man (*Homo sapiens*)	1	-	-	-	-	-	-	-	-	-	1

House 2: House of the Geometric *Krater* (partly excavated)	Room/area					
	1 Room adjoin. m.l.r.	2 Room adjoin. m.l.r.	3 Main living room	4 Court-yard?	5 Room	TOTAL
DOMESTIC MAMMALS						
Equids (*Equus* sp.)	-	-	-	-	1	1
Cattle (*Bos taurus*)	-	2	-	-	-	2
Sheep (*Ovis aries*)	-	3	-	-	-	3
Sheep/goat (*Ovis/Capra*)	-	1	-	-	-	1
Total domestic mammals	-	6	-	-	1	7
TOTAL MAMMALS	-	6	-	-	1	7
Marine molluscs	9	3	9	5	3	29
TOTAL MOLLUSCS	9	3	9	5	3	29
Unidentified mammalian remains	1	-	3	-	-	4
TOTAL	10	9	12	5	4	40

Table 4.8. (continued).

House 3: House of the Snakes	Room/area									
	1 Room	2 Room	3 Room	4 Court- yard	5 Room adjoin. m.l.r.	6 Room adjoin. m.l.r.	7+9 Room: corri- dor	8+11 Main living room	TOTAL	
DOMESTIC MAMMALS										
Equids (*Equus* sp.)	-	-	-	7	1	-	-	-	8	
Dog (*Canis familiaris*)	2	-	-	2	-	-	-	-	4	
Pig (*Sus domesticus*)	3	-	-	1	1	-	-	-	5	
Cattle (*Bos taurus*)	-	-	-	9	2	-	-	1	12	
Sheep (*Ovis aries*)	-	-	-	1	-	-	-	-	1	
Sheep/goat (*Ovis/Capra*)	2	-	-	11	10	-	1	3	27	
Total domestic mammals	7	-	-	31	14	-	1	4	57	
WILD MAMMALS										
Fox (*Vulpes vulpes*)	-	-	-	1	-	-	-	-	1	
Red deer (*Cervus elaphus*)	-	-	-	3	-	-	-	-	3	
TOTAL MAMMALS	7	-	-	35	14	-	1	4	61	
Marine molluscs	18	2	3	103	4	12	8	15	165	
Terrestrial molluscs	-	-	-	10	-	-	-	2	12	
TOTAL MOLLUSCS	18	2	3	113	4	12	8	17	177	
Unidentified mammalian remains	8	-	-	136	19	1	-	1	165	
TOTAL	33	2	3	284	37	13	9	22	403	
%	*8*	*1*	*1*	*70*	*9*	*3*	*2*	*5*	*100*	

House 4: House of the *Amphorai*	Room/area										
	1 Room	2 Room	3 Room	4 Room	4a Room	5 Room	6 Room	7 Open area	8 Court- yard	?	TOTAL
DOMESTIC MAMMALS											
Pig (*Sus domesticus*)	2	1	-	-	-	-	1	-	-	-	4
Cattle (*Bos taurus*)	-	2	5	-	-	4	9	2	-	1	23
Sheep (*Ovis aries*)	-	1	1	1	-	3	-	1	-	-	7
Goat (*Capra hircus*)	-	1	-	-	-	-	-	-	-	-	1
Sheep/goat (*Ovis/Capra*)	1	5	6	2	-	21	44	5	-	1	85
Total domestic mammals	3	10	12	3	-	28	54	8	-	2	120
WILD MAMMALS											
Vole (*Microtus* sp.)	-	2	-	-	-	-	-	-	-	-	2
Badger (*Meles meles*)	-	-	-	-	-	-	1	-	-	-	1
Red deer (*Cervus elaphus*)	-	-	2	-	-	-	1	-	-	1	4
TOTAL MAMMALS	3	12	14	3	-	28	56	8	-	3	127
Tortoise (*Testudo* sp.)	-	-	-	-	-	-	13	-	-	-	13
Marine molluscs	13	5	4	1	-	12	13	2	-	-	50

Table 4.8. (continued).

House 4 (continued). House of the *Amphorai*	Room/area										
	1 Room	2 Room	3 Room	4 Room	4a Room	5 Room	6 Room	7 Open area	8 Court-yard	?	TOTAL
Terrestrial molluscs	-	-	-	-	-	-	1	-	-	-	1
TOTAL MOLLUSCS	13	5	4	1	-	12	14	2	-	-	51
Unidentified mammalian remains	2	21	8	4	-	70	151	12	-	-	268
TOTAL	18	38	26	8	-	110	234	22	-	3	459
%	*4*	*8*	*6*	*2*	*-*	*24*	*51*	*5*	*-*	*1*	*100*

House 5: House of the Ptolemaic Coins	Room/area							
	1 Room adjoin. m.l.r.	2 Room adjoin. m.l.r.	3 Main living room	4 Room adjoin. m.l.r.	5 Room adjoin. m.l.r.	6 Court-yard *	7 Room	TOTAL
DOMESTIC MAMMALS								
Equids (*Equus* sp.)	-	-	1	-	1	-	-	2
Pig (*Sus domesticus*)	-	-	1	-	1	-	-	2
Cattle (*Bos taurus*)	-	1	-	2	-	1	1	5
Sheep (*Ovis aries*)	-	-	1	-	1	-	-	2
Goat (*Capra hircus*)	-	-	1	-	-	-	-	1
Sheep/goat (*Ovis/Capra*)	-	1	9	-	2	3	1	16
Total domestic mammals	-	2	13	2	5	4	2	28
WILD MAMMALS								
Red deer (*Cervus elaphus*)	-	-	1	-	-	-	-	1
TOTAL MAMMALS	-	2	14	2	5	4	2	29
Marine molluscs	8	7	33	7	8	1	3	67
Terrestrial molluscs	-	-	5	-	-	-	-	5
TOTAL MOLLUSCS	8	7	38	7	8	1	3	72
Unidentified mammalian remains	1	2	10	-	2	1	1	17
TOTAL	9	11	62	9	15	6	6	118
%	*8*	*9*	*52*	*8*	*13*	*5*	*5*	*100*

* incompletely excavated

Table 4.8. (continued).

House 6: House of Agathon	Room/area							
	1 Room adjoin. m.l.r.	2 Room adjoin. m.l.r.	3 Main living room	4 Room adjoin. m.l.r.	5 Room adjoin. m.l.r.	6 Court-yard	7 Room	TOTAL
DOMESTIC MAMMALS								
Pig (*Sus domesticus*)	-	-	-	-	-	1	-	1
Cattle (*Bos taurus*)	-	-	-	-	4	2	3	9
Sheep (*Ovis aries*)	-	-	2	-	16	1	-	19
Goat (*Capra hircus*)	-	1	-	-	2	-	-	3
Sheep/goat (*Ovis/Capra*)	1	3	3	1	5	5	6	24
Total domestic mammals	1	4	5	1	27	9	9	56
WILD MAMMALS								
Red deer (*Cervus elaphus*)	-	-	-	-	-	2	-	2
Roe deer (*Capreolus capreolus*)	-	-	-	-	2	-	-	2
TOTAL MAMMALS	1	4	5	1	29	11	9	60
Marine molluscs	12	2	10	-	95	34	10	163
Terrestrial molluscs	-	-	-	-	2	-	-	2
TOTAL MOLLUSCS	12	2	10	-	97	34	10	165
Unidentified mammalian remains	-	1	1	-	1	12	32	47
TOTAL	13	7	16	1	127	57	51	272
%	5	2	6	0	47	21	19	100

of the meals will then still have been in the houses at the time when they were abandoned. The inhabitants of the houses of cluster 1 may have consumed their last mollusc meal in the house a longer time ago. The refuse of those meals will then have been largely removed from the houses concerned by the time the people left. If molluscs were consumed on a seasonal basis, for instance only in January or February as on present-day Lésvos (Ruscillo, 1993: p. 205), this could mean that the houses of cluster 2 were abandoned in the season of mollusc consumption and the houses of cluster 1 in another season. Another factor making it difficult to interpret the difference between clusters 1 and 2 is that houses 5 and 6 were only partly excavated.

The proportions of remains of domestic species are as follows, in the order of their numbers (NR): sheep and goat (with sheep/goat: 66.8%), cattle (23.0%), pig (5.0%), equids (4.0%) and dog (1.2%). Sheep and cattle were represented in each of the six houses, pig in five houses, goat and equids each in

four houses and dog in one house (table 4.6). Cluster analysis of the proportions of the remains of the different domestic mammals led to another distinction (fig. 4.13): houses 4, 5 and 6 are characterised by high proportions of small ruminants (sheep and goat: more than 65% of NR) (cluster 1) whereas houses 1, 2 and 3 are characterised by less than 60% remains of small ruminants (cluster 2). This clustering is partly attributable to the many knucklebones (astragali) of sheep and goat that were found in houses 5 and 6 (table 4.16) and may have very little to do with the proportions of mammal meat consumed.

The animal remains were unequally distributed over the rooms of the six excavated houses, some rooms containing more animal remains than others (tables 4.8 and 4.9). Rooms with large numbers of animal remains are regarded as rooms in or near which food was prepared and consumed, or where waste of cooking and meals was stored before being discarded. In the completely excavated houses with a clear arrangement of rooms, *i.e.* houses 1, 3 and 6,

Table 4.9. Six houses in Hellenistic New Halos, grouped after the clustering of remains of domestic mammals and molluscs among the identified animal remains. Main rooms in which domestic mammal remains, unidentified mammal remains, wild mammal remains, mollusc remains and all remains (Total) were found, in sequence of decreasing numbers of remains. Abbreviations: Main liv. room = main living room; Room adj. main liv. room = small room adjoining main living room.

Cluster 1: domestic mammalian remains >60%; mollusc remains <35%

House 1: House of the Coroplast
Domestic mammals	room 4; room 3 + main liv. room
Unident. mammals	room 4; room 3; main liv. room
Wild mammals	main liv. room
Molluscs	main liv. room; room 3 + room 4 + room adj. main liv. room + courtyard
Total	room 4; room 3; main liv. room

House 4: House of the Amphorai
Domestic mammals	room 6; room 5; room 3; room 2
Unident. mammals	room 6; room 5; room 2; room 7 (open area)
Wild mammals	room 6; room 2; room 3
Molluscs	room 6; room 1; room 5
Total	room 6; room 5; room 2; room 3

Cluster 2: domestic mammalian remains <30%; mollusc remains >70%

House 3: House of the Snakes
Domestic mammals	courtyard; room adj. main liv. room; room 1
Unident. mammals	courtyard; room adj. main liv. room; room 1
Wild mammals	courtyard
Molluscs	courtyard; room 1; main liv. room; room adj. main liv. room
Total	courtyard; room adj. main liv. room; room 1; main liv. room

House 6: House of Agathon
Domestic mammals	room adj. main liv. room; courtyard + room 7; main liv. room
Unident. mammals	room 7; courtyard
Wild mammals	room adj. main liv. room; courtyard
Molluscs	room adj. main liv. room; courtyard; room adj. main liv. room; main liv. room; room 7
Total	room adj. main liv. room; courtyard; room 7; main liv. room

House 5: House of the Ptolemaic Coins (NB: courtyard incompletely excavated)
Domestic mammals	main liv. room; room adj. main liv. room; courtyard
Unident. mammals	main liv. room
Wild mammals	main liv. room
Molluscs	main liv. room; four rooms adj. main liv. room
Total	main liv. room; three rooms adj. main liv. room

House 2: House of the Geometric Krater (NB: partly excavated)
Domestic mammals	room adj. main liv. room
Unident. mammals	main liv. room
Molluscs	main liv. room + room adj. main liv. room; probable courtyard
Total	main liv. room; two rooms adj. main liv. room

large concentrations of animal remains were found in the main living room, one or more of the small rooms adjoining the main living room and the courtyard (but not in the courtyard of house 1). This means that most meat and molluscs were stored, prepared and consumed in the main living room and the adjoining rooms and in the courtyard. Part of the waste remained in these rooms and the courtyard, where it accumulated in corners and was incorporated in floors.

The remains of domestic mammals and the unidentified mammalian remains were equally distributed over the houses: the room with the largest number of remains of domestic mammals is also the room with the largest number of unidentified mammalian remains (tables 4.8 and 4.9). This will be due to the fact that the unidentified remains are mainly bone fragments of the same domestic mammals as the identified remains. Some wild mammal remains may be included among the unidentified remains.

The mollusc remains and the mammalian remains were unequally distributed over the rooms of the six houses (tables 4.8 and 4.9). In most houses, mollusc remains were most numerous in the main living room and the adjacent rooms. The mammalian remains, however, were most numerous in the courtyards and in other rooms than in the main living room or adjacent rooms (tables 4.8 and 4.9). This difference is possibly connected with the different ways in which mammalian meat and molluscs are prepared. Mammalian meat is detached from the bones partly during the preparation of a meal and partly during the meal itself. Molluscs, either cooked or uncooked, are always lifted from their shells during a meal. The distribution of bones and shells suggests that most food was prepared and consumed in the main living room and the small adjacent rooms. The kitchen refuse of the New Halos houses, *i.e.* bones, was systematically removed from the main living room and adjacent rooms to the courtyard or other rooms, whereas part of the waste of the meals, *i.e.* a few bones or in the case of a mollusc meal a large quantity of shells, was left on the floors of the rooms in which the food had been eaten and thus stood a chance of ending up in the corners and floor of the main living room and the adjacent small rooms.

4.4. Domestic mammals

Equids: All parts of the body are represented by remains (table 4.10). An astragalus found in house 3 was identified as the bone of an ass, *Equus asinus*, an ulna from house 1 as probably the bone of a horse, *Equus caballus* (table 4.20). The other equid remains come from ass or horse, or possibly from their cross-breeds, mule and hinny. Dental and epiphyseal age data indicate that about one fifth of the equids were slaughtered when they were less than 12–15 months old and about four fifths when they were more than 12–24 months old, or even more than 42 or 48 months old.[1] The size of the ass astragalus from house 3 (table 4.20) falls within the range of the ass astragali from Kassope (Friedl, 1984: Tab. 35, *e.g.* GH 41.5–47.0, n = 5) and the Hellenistic deposits of Messene (Nobis, 1994: p. 309, GH 44.5–50, n = 3). The horse ulna from house 1 is from a horse of moderate size.

Table 4.10. Six houses in Hellenistic New Halos, remains of equids, *Equus* sp. Distribution over the skeleton. NR = Number of remains. Unburned and burned material combined.

House	1		2		3		5		Total	
	NR	%	NR	%	NR	%	NR	%	NR	%
Maxilla	-	-	-	-	1	13	2	100	3	23.1
Costae	-	-	-	-	1	13	-	-	1	7.7
Humerus	-	-	-	-	1	13	-	-	1	7.7
Ulna	1	50	-	-	-	-	-	-	1	7.7
Femur	-	-	-	-	1	13	-	-	1	7.7
Tibia	-	-	-	-	1	13	-	-	1	7.7
Astragalus	-	-	1	100	2	25	-	-	3	23.1
Calcaneus	-	-	-	-	1	13	-	-	1	7.7
Metatarsus III	1	50	-	-	-	-	-	-	1	7.7
Total	2		1		8		2		13	100.0

Table 4.11. Six houses in Hellenistic New Halos, remains of pig, *Sus domesticus*. Distribution over the skeleton. NR = Number of remains. Unburned and burned material combined.

House	1		3		4		5		6		Total	
	NR	%	NR	%	NR	%	NR	%	NR	%	NR	%
Mandibula	1	25	2	40	1	25	-	-	-	-	4	25.0
Maxilla/mandibula	-	-	-	-	-	-	1	50	-	-	1	6.3
Humerus	1	25	-	-	-	-	1	50	-	-	2	12.5
Ulna	-	-	-	-	1	25	-	-	-	-	1	6.3
Metacarpus IV	1	25	-	-	-	-	-	-	-	-	1	6.3
Pelvis	-	-	2	40	-	-	-	-	-	-	2	12.5
Femur	1	25	1	20	-	-	-	-	-	-	2	12.5
Tibia	-	-	-	-	1	25	-	-	-	-	1	6.3
Calcaneus	-	-	-	-	-	-	-	-	1	100	1	6.3
Phalanx 2	-	-	-	-	1	25	-	-	-	-	1	6.3
Total	4		5		4		2		1		16	100.0

Dog: Four dog bones, the incisor and premolar part of a left mandible with canine and premolars, a loose lower left canine and fragments of a left radius and a left ulna were found in house 3, the lower canine and radius in room 1, the mandible and ulna in the courtyard. The mandible and the loose canine are from two dogs. The mandible is from a subadult dog: the premolars were only slightly worn. The canine is from a dog of more than 6 months old. The measurements of the mandible (table 4.20), which was possibly not full-grown, fall within the range of those of the dog mandibles of Kassope (Friedl, 1984: Tab. 38). The dog was of moderate size.

Pig: All parts of the body are represented by the pig remains found in New Halos (table 4.11). Mandibles are slightly overrepresented, probably because mandible parts and teeth are easily identified. Dental age data show that some pigs were killed when they were between 12 and 16 months old, others when they were between 16 and 24 months old.[2] The epiphyseal age data confirm the young age at which most pigs were killed.[3] The few measurements available (table 4.20) indicate that the Halos pigs fell within the size range of the Kassope pigs (Friedl, 1984: Tab. 30). They were smaller than contemporary pigs in Hungary and Romania (Friedl, 1984: pp. 95–96). A fragment of a left humerus found in house 1 (room unknown, inventory number X 208; Reinders, 1988: Catalogue 30.08) had been gnawed by a dog.

Cattle: All parts of the body are represented by remains (table 4.12). Skull elements, including upper and lower jaws, are rare. These elements usually disintegrate into many fragments, of which the teeth tend to survive fairly well and are easily identified. The low frequency of skull fragments probably means that joints of cattle meat, without the skull, were brought into the houses rather than complete animals. This could mean that the cattle were killed and butchered outside the housing blocks, or even sacrificed at a central place, and that the meat was sold to or divided among the citizens. The astragali, knucklebones, a total of ten, nine from the right side and one from the left[4], are overrepresented among the cattle remains (13.5%, table 4.12). In house 6 they accounted for 44% of the cattle remains.

About 20% of the cattle were killed when they were between 1 and 1½ years old, about 13% when they were between 1½ and 2–2½ years old and two thirds when they were more than 2–2½ years old (table 4.13, epiphyseal age data). The adult age of most cattle is confirmed by a mandible with erupting M_3 from an animal that was 25–28 months old and an upper jaw from a cow that was more than 19–24 months old. About the same proportions were found for the 4th–1st-century-BC remains from Kassope, where about one third of the cattle were killed as young animals and about two thirds as adult animals (Friedl, 1984: pp. 25–27).

Table 4.12. Six houses in Hellenistic New Halos, remains of cattle, *Bos taurus*. Distribution over the skeleton. NR = Number of remains. Unburned and burned material combined.

House	1		2		3		4		5		6		Total	
	NR	%	NR	%	NR	%	NR	%	NR	%	NR	%	NR	%
Cranium	-	-	-	-	-	-	-	-	1	20	-	-	1	1.4
Maxilla	-	-	-	-	1	8	-	-	-	-	-	-	1	1.4
Mandibula	-	-	-	-	1	8	1	4	-	-	-	-	2	2.7
Hyoid	-	-	-	-	-	-	1	4	-	-	-	-	1	1.4
Vertebrae	1	4	-	-	-	-	-	-	-	-	-	-	1	1.4
Costae	2	9	-	-	1	8	2	9	1	20	-	-	6	8.1
Scapula	-	-	-	-	1	8	3	13	-	-	-	-	4	5.4
Humerus	3	13	-	-	-	-	3	13	-	-	1	11	7	9.5
Radius	3	13	-	-	1	8	-	-	-	-	-	-	4	5.4
Ulna	-	-	-	-	1	8	-	-	-	-	-	-	1	1.4
Carpus	4	17	-	-	1	8	-	-	-	-	-	-	5	6.8
Metacarpus III/IV	1	4	-	-	-	-	1	4	-	-	1	11	3	4.1
Pelvis	1	4	-	-	-	-	-	-	-	-	-	-	1	1.4
Femur	1	4	-	-	1	8	3	13	1	20	-	-	6	8.1
Tibia	3	13	-	-	1	8	1	4	1	20	1	11	7	9.5
Astragalus	3	13	1	50	-	-	2	9	-	-	4	44	10	13.5
Metatarsus III/IV	-	-	-	-	-	-	5	22	1	20	-	-	6	8.1
Metapodium	-	-	-	-	1	8	-	-	-	-	-	-	1	1.4
Phalanx 1	-	-	-	-	2	17	-	-	-	-	1	11	3	4.1
Phalanx 2	1	4	1	50	-	-	-	-	-	-	1	11	3	4.1
Phalanx 3	-	-	-	-	-	-	1	4	-	-	-	-	1	1.4
Total	23		2		12		23		5		9		74	100.0

The cattle of New Halos (table 4.20) were on average smaller than those of 4th–1st-century-BC (Hellenistic–Roman) Kassope in Ípiros. Estimates of the height of the withers of these cattle are 117, 133 and 134 cm (Friedl, 1984: pp. 33–36 and Tab. 8–11). The larger size of the Kassope cattle could be attributable to the greater amount of precipitation, that in Ípiros being twice to four times as high as that in Thessaly, implying that feeding conditions for cattle were much better in Ípiros than in Thessaly. We have no information on the size of the 3rd-century-BC cattle of Kassope as the Kassope cattle measurements were not categorised according to period. The average size of the Kassope cattle might have been greater due to improvements in cattle breeding introduced during the Roman period, the last part of the occupation period of Kassope. The few measurements available for cattle from the Hellenistic Kabiren sanctuary (Kabirion) in Thebai in Boiotia (Viotía) show that these cattle were of an intermediate size between those of New Halos and Kassope (Boessneck, 1973: p. 13). The cattle of Hellenistic Messene were slightly larger than those of New Halos (Nobis, 1994: pp. 305–306).

The astragalus from house 2, room 2 (inventory number H21 19), one of the rooms adjacent to the main living room, was centrally perforated from the dorsal to the plantar plane. The diameter of the hole is 7 mm. Wear resulting from handling was visible at the distal end, on the lateral condylus and to a lesser extent on the medial condylus. The astragalus from room 9 in house 1, which may have been a shop (inventory number X 53), was worn, presumably also as a result of handling. The cattle astragali from room 2 (inventory number HA 166/12) and room 3 (inventory number HA 173/02) of house 4 and those from room 5 of house 6 were charred. No other cattle bones in house 4 and only one other cattle bone in house 6 (a second phalanx found in room 7, inventory number HA 166/12) were charred. No less than 60% of the cattle astragali from the six houses was charred. This proportion is very high, especially when we consider that only 12.2% of all the cattle bones from the six houses was charred. No signs of use were observable on the charred cattle astragali. The astragalus found in the courtyard of house 1 (inventory number X 667) was not perforated, worn or charred.

The cattle astragali probably had a special meaning in New Halos and were for that reason handled, burned and kept in the houses. A calcined cattle astragalus was found in the Kabirion in Thebai (Boessneck, 1973: p. 10; no exact date given). Two cattle astragali from Thebai are perforated, one from the Roman period, which is perforated in a latero-medial direction, the other from an unidentified period, which is perforated in a dorso-plantar direction like the New Halos astragalus. A cattle astragalus with a dorso-plantar perforation was found in 2nd/1st century BC Eleutherna on Kríti (Vila, 1994). The diaphysis of a metacarpus or metatarsus from the living room of house 3 was severely worn.

Sheep and goat: Of the 215 remains of sheep or goat (table 4.6), 34 could be identified as deriving from sheep (table 4.14) and 6 as deriving from goats (table 4.15). The remaining bones, mostly skull and spinal elements and phalanges, were identified as sheep/goat (table 4.16). All parts of the body are represented by the remains, though in unequal proportions. Skull elements, especially maxillae and mandi-

Table 4.13. Six houses in Hellenistic New Halos, epiphyseal age data of cattle, *Bos taurus*. The %fused for each age group represents the proportion of cattle killed at an older age; the %unfused for each age class represents the proportion of cattle killed at a younger age; the interval% is the proportion of cattle killed between the former and this age class.

Age in months	Element	Fused		Unfused		Total	Interval
		N	%	N	%	N	%
7–10	Acetabulum	1	100	-	-	1	-
12–20	Humerus distal	1		-			
	Radius proximal	1		1			
	Phalanx 2 proximal	2		-			
	Sum	4	80	1	20	5	20
20–24	Phalanx 1 proximal	1	100	-	-	1	-
24–30	Tibia distal	1		1		2	
	Metatarsus distal	1		-			
	Sum	2	67	1	33	3	13
Total		8	80	2	20	10	

NB: a vertebra with an unfused epiphysis was found as well.

Table 4.14. Six houses in Hellenistic New Halos, remains of sheep, *Ovis aries*. Distribution over the skeleton. NR = number of remains. Unburned and burned material combined.

House	1		2		3		4		5		6		Total	
	NR	%	NR	%	NR	%	NR	%	NR	%	NR	%	NR	%
Humerus	-	-	-	-	-	-	1	14	-	-	-	-	1	2.9
Metacarpus III/IV	-	-	-	-	-	-	1	14	-	-	1	5	2	5.9
Tibia	-	-	-	-	1	100	1	14	-	-	1	5	3	8.8
Astragalus	1	50	3	100	-	-	2	29	2	100	16	84	24	70.6
Calcaneus	1	50	-	-	-	-	1	14	-	-	-	-	2	5.9
Metatarsus III/IV	-	-	-	-	-	-	1	14	-	-	1	5	2	5.9
Total	2		3		1		7		2		19		34	100.0

Table 4.15. Six houses in Hellenistic New Halos, remains of goat, *Capra hircus*. Distribution over the skeleton. NR = number of remains. Unburned and burned material combined.

House	1		4		5		6		Total	
	NR	%	NR	%	NR	%	NR	%	NR	%
Horncore	-	-	-	-	1	100	-	-	1	16.7
Humerus	1	100	-	-	-	-	-	-	1	16.7
Radius	-	-	-	-	-	-	1	33	1	16.7
Astragalus	-	-	1	100	-	-	2	67	3	50.0
Total	1		1		1		3		6	100.0

bles, are well represented, partly because loose teeth and tooth fragments are easily recognised. Metapodials and phalanges, which do not hold any meat, are underrepresented (table 4.16). Sheep and goat astragali (16%, table 4.16) are overrepresented to an even greater extent than cattle astragali. The largest number of astragali were found in house 6 (tables 4.14–4.16). Unlike with the cattle astragali, both sides of the sheep and goat astragali are equally represented (sheep: 12 left, 12 right; goat: 2 left, 1 right; sheep/goat: 3 left, 2 right, 3 from unidentified side). Leaving aside the astragali, which had a special meaning (see below), the scarcity of lower limb parts of the skeleton suggests that all or most of the sheep and goats were killed elsewhere, just like the cattle.

Few sheep and goats were killed as lambs or kids, *i.e.* in their first year. A slightly larger number were killed in their second year. The majority of the sheep and goats were killed when they were more than 2 years old or even more than 3½ years old (tables 4.17–4.18). Two foetal femora found in the courtyard of house 3, presumably from the same foetus, indicate the killing, death or abortion of a pregnant ewe or mother-goat. Some sheep were killed as lambs, others as subadults or adults. Goats were killed when they were more than 3–4 months old.[5] The dental and epiphyseal age data of both species together suggest that 20–25% of the sheep and goats were killed in their

first or second year and 75–80% when they were more than two years or more than 3½ years old.[6]

The sheep and goat remains from Kassope (all periods) show a kill-off pattern that is almost identical to that of New Halos: 30% of the animals were killed when they were less than 3½ years old, 70% when they were more than 3½ years old. The fact that most sheep and goats were killed at an advanced age in these Hellenistic towns indicates that the main aim of sheep and goat husbandry was the production of milk and/or wool. Sheep and goats were slaughtered only when they were no longer capable of producing offspring, enough milk for human consumption or enough good-quality wool. Their meat was then consumed. The proportion of young sheep and goats among the Hellenistic sacrifices made at the Kabirion near Thebai was much greater than at New Halos and Kassope (Boessneck, 1973: Tab. 10: 19% infantiles, 35% juveniles, 45% subadults and adults). This could be attributable to a preference for young sheep and goats as sacrificial animals at this particular sanctuary.

The horn core from an adult billy goat that was found in the living room of house 5 (room 3, inventory number HA 54) shows that some billy goats at least reached an adult age. A cut mark on the horn core shows that it was cut from the animal or its skin. It may represent waste of a horn processor or a tan-

Table 4.16. Six houses in Hellenistic New Halos, remains of sheep, *Ovis aries*, goat, *Capra hircus*, and sheep/goat, *Ovis/Capra*. Distribution over the skeleton. NR = number of remains. Unburned and burned material combined.

House	1 NR	1 %	2 NR	2 %	3 NR	3 %	4 NR	4 %	5 NR	5 %	6 NR	6 %	Total NR	Total %
Cranium	-	-	-	-	1	4	-	-	-	-	-	-	1	.5
Horncore	-	-	-	-	-	-	-	-	1	5	-	-	1	.5
Maxilla	2	8	1	25	10	36	13	14	3	16	9	20	38	17.7
Mandibula	-	-	-	-	8	29	20	22	-	-	1	2	29	13.5
Maxilla/mandibula	-	-	-	-	1	4	2	2	2	11	-	-	5	2.3
Epistropheus	7	28	-	-	-	-	-	-	-	-	-	-	7	3.3
Vert. cervicales	-	-	-	-	-	-	1	1	-	-	-	-	1	.5
Vert. thoracalis	-	-	-	-	-	-	-	-	1	5	-	-	1	.5
Vert. lumbales	2	8	-	-	-	-	-	-	-	-	-	-	2	.9
Vertebrae	-	-	-	-	1	4	-	-	-	-	-	-	1	.5
Costae	2	8	-	-	-	-	8	9	6	32	1	2	17	7.9
Scapula	2	8	-	-	-	-	1	1	-	-	-	-	3	1.4
Humerus	1	4	-	-	3	11	3	3	1	5	2	4	10	4.7
Radius	-	-	-	-	-	-	7	8	1	5	3	7	11	5.1
Ulna	-	-	-	-	-	-	1	1	-	-	-	-	1	.5
Carpus	1	4	-	-	-	-	-	-	-	-	-	-	1	.5
Metacarpus III/IV	-	-	-	-	-	-	4	4	-	-	1	2	5	2.3
Pelvis	-	-	-	-	-	-	2	2	-	-	-	-	2	.9
Femur	1	4	-	-	2	7	7	8	1	5	3	7	14	6.5
Tibia	2	8	-	-	2	7	8	9	1	5	2	4	15	7.0
Astragalus	2	8	3	75	-	-	5	5	2	11	23	50	35	16.3
Calcaneus	1	4	-	-	-	-	1	1	-	-	-	-	2	.9
Metatarsus III/IV	1	4	-	-	-	-	6	6	-	-	1	2	8	3.7
Phalanx 1	1	4	-	-	-	-	3	3	-	-	-	-	4	1.9
Phalanx 3	-	-	-	-	-	-	1	1	-	-	-	-	1	.5
TOTAL	25		4		28		93		19		46		215	100.0

Table 4.17. Six houses in Hellenistic New Halos, combined dental age data of sheep, *Ovis aries*, and goat, *Capra hircus*. N = number of jaws or teeth.

Approximate age in months	N	%
1–2	-	-
3	-	-
4–8	-	-
9	1	2
10–17	1	2
<18–24	1	2
18–24	2	5
>24	5	13
>24, distinctly worn teeth	22	55
>24, heavily worn teeth	7	18
>24, worn down teeth	1	2
Total	40	100

Imprecise dental age data of two adult sheep or goats (more than 24 months old).

ner. We have no information on the sex of the sheep or of the other goats.

Due to the high degree of fragmentation, no complete long bones are available for estimating the height of the withers of the sheep and goats of New Halos (table 4.20). The measurements of the New Halos sheep fall within the ranges of those of the Hellenistic town of Kassope in Ípiros (Friedl, 1984: pp. 59–61 and Tab. 19-21), Hellenistic Messene in the western Pelepónnisos (Nobis, 1994: pp. 307–308) and Thebai in Viotía (Boessneck, 1973: Tab. 11–12).[7] The Kassope, Messene, Thebai and New Halos sheep were small. The Kassope sheep had withers heights of between 55 and 67 cm, with an average of about 60 cm.

The measurements of the New Halos goats (table 4.20) fall within the size range of the Kassope goats, which had withers heights of between 68 and 72 cm (Friedl, 1984: pp. 63–64 and Tab. 22–24). The goats whose remains were found at the Hellenistic Kabirion site near Thebai were of the same size (Boessneck, 1973: Tab. 11–12).

Besides being overrepresented, the sheep and goat

Table 4.18. Six houses in Hellenistic New Halos, combined epiphyseal age data of sheep, *Ovis aries*, and goat, *Capra hircus*. The %fused for each age represents the proportion of sheep/goat killed at an older age; the %unfused for each age class represents the proportion of sheep/goat killed at a younger age; the interval% is the proportion of sheep/goat killed between the former and this age class.

Age in months	Element	Fused	%	Unfused	%	Total	Interval
3–4	Humerus distal	2		-			
	Radius proximal	2		-			
	Sum	4	100	-	-	4	-
5–10	Phalanx 1 proximal	3	100	-	-	3	-
15–24	Tibia distal	3	75	1	25	4	25
36–42	Femur proximal	1		-			
	Calcaneus proximal	1		-			
	Sum	2	100	-	-	2	
Total		12	92	1	8	13	

NB: a lumbar vertebra with unfused epiphyses was found as well.

Imprecise age data: 2 femora of a foetal sheep/goat (presumably of one individual), a femur of an infantile sheep/goat (birth – *c.* 3 months), and an astragalus of a juvenile sheep (*c.* 3–12 months).

Table 4.19. Six houses in Hellenistic New Halos, remains of red deer, *Cervus elaphus*. Distribution over the skeleton. NR = number of remains. Unburned and burned material combined.

House	3		4		5		6		Total	
	NR	%	NR	%	NR	%	NR	%	NR	%
Antler	-	-	3	75	1	100	-	-	4	40.0
Humerus	-	-	1	25	-	-	-	-	1	10.0
Tibia	1	33	-	-	-	-	-	-	1	10.0
Astragalus	1	33	-	-	-	-	-	-	1	10.0
Phalanx 1	-	-	-	-	-	-	1	50	1	10.0
Phalanx 2	1	33	-	-	-	-	1	50	2	20.0
Total	3		4		1		2		10	100.0

astragali show some other peculiarities. Three sheep astragali from house 4 (room 2, inventory number HA 126/30), house 5 (the main living room, inventory number 80 (of a juvenile sheep)) and house 5 (room 5, inventory number HA 46) are badly worn and were thus definitely used as knucklebones. Two sheep astragali found in house 2 (both in room 2, inventory numbers H21 27 and 30) are fairly worn, suggesting that they served the same purpose.

Eighteen of the 24 sheep astragali are charred. They were found in room 2 (inventory number HA 126/30; also worn) and room 4 (inventory number HA 109/15) of house 4, in the main living room of house 5 (inventory number HA 80; also worn) and in room 5 of house 6 (15 of the 16 astragali found in this room). All three goat astragali are charred. They were found in room 2 of house 4 (inventory number HA 166/9) and in room 5 of house 6 (two astragali). Four of the eight sheep/goat astragali, all from room 5 of

house 6, are charred. So in total, 25 of the 35 sheep, goat and sheep/goat astragali (71%) are charred. Only three other sheep or goat bones are charred: a femur and a metatarsus found in room 3 of house 4 and a phalanx 1 from room 4 of the same house. The total number of charred bone fragments of sheep, goat and sheep/goat, 28, accounts for 13% of all the sheep and goat remains. A total of 207 charred (138) and calcined (69) sheep, goat and sheep/goat astragali were found in the Kabirion in Thebai (21% of all the astragali of these species) (Boessneck, 1973: p. 10) (no details are given on the exact dates of the astragali).

4.5. Wild mammals

Brown hare, Lepus europaeus: The brown hare is represented by a single bone fragment (inventory number X 339), the distal end of a left humerus of an adult

Table 4.20. Six houses in Hellenistic New Halos. Bone measurements of adult mammalian bones in mm, after the system of Von den Driesch (1976). br = burned, us = used.

Abbreviations of the measurements:

Antler: Cf burr: circumference of the burr; DCf burr: distal circumference of the burr (antler-base); Cf pedi: circumference of the pedicel; Dia pedi: maximum diameter of the pedicel.

Maxilla/mandible: L/BP4: length/width of fourth premolar; L/BM1: length/width of first molar; L/BM2: length/width of second molar; L/BM3: length/width of third molar.

Mandible: H-Dia: height of the diastema; HfP_3: height of mandible in front of third premolar; HfM_1: height of mandible in front of first molar; HhM_3: height of mandible behind third molar; $LP_{1\,or\,2}\text{-}P_4$: lenth of the premolar row.

Epistropheus: BFcr: maximum width of the cranial articular surface.

Humerus, radius, ulna, metacarpus, femur, tibia, astragalus, metatarsus, phalanx 1 and 2: GL: maximum length; GLpe (phalanx 1): maximum length of abaxial half; Bp: maximum width of the proximal end; BFp (radius): width of the proximal articular surface; BPC (ulna): maximum width of the proximal articular surface; Dp: maximum diameter of the proximal end; DC (femur): diameter of caput femoris; SD: minimum width of the diaphysis; Bd: maximum width of the distal end; BT (humerus): width of the trochlea; Dd: maximum diameter of the distal end.

Astragalus: GLl: maximum length of the lateral half; GLm: maximum length of the medial half; Dl: maximum depth of the lateral half; Dm: maximum depth of the medial half; GH: maximum height, GB: maximum width, BFd: width of distal articular surface).

Calcaneus: GL: maximum length; GB: maximum width.

Phalanx 3: DLS: diagonal length of sole; Ld: length of dorsal surface; MBS: width in the middle of the sole.

Dog, *Canis familiaris*

Mandible

House	3
Room	4
Inv.-No.	HA 267
HvP_3	16.4
$LP_2\text{-}P_4$	30.2
$LP_1\text{-}P_4$	34.4

Equids, *Equus* sp.

Ulna, probably of horse, *Equus caballus*

House	1
Room	4
Inv.No.	X 725
BPC	(44.2)

Astragalus, probably of ass, *Equus asinus*

House	2
Room	5
Inv.No.	H21 9
GH	43.5
GB	46.1
BFd	38.1

Pig, *Sus domesticus*

Mandible

House	3
Room	5
Inv.No.	HA 238
LM_3	31.8
BM_3	18.0

Humerus

House	1
Room	-
Inv.No.	X 204 C
SD	13.9

Phalanx 2 of toe 2 or 5

House	4
Room	7
Inv.No.	HA 116
Bp	11.7
SD	8.6

Table 4.20. (continued).

Cattle, *Bos taurus*

Humerus

House	1
Room	4
Inv.No.	X 725
BT	73.7

Radius

House	1
Room	4
Inv.No.	X 725
BFp	73.3

Astragalus

House	1	1	1	2	4	4	6	6
Room	1	9	9	2	2	3	5	5
Inv.No.	X 667	X 53	X 68 A	X 19	HA 166	HA 173		
GLl	69.4	(75.0)	63.7	.	68.5	58.5	70.6	66.0
GLm	64.5	68.1	60.0	.	62.6	35.8	66.0	62.4
Dl	39.6	.	35.4	.	38.7	.	38.8	37.3
Dm	.	.	34.5	(38.2)	40.6	.	40.0	.
Bd	42.9	.	40.4	.	49.5	.	47.8	43.7
		us		us	br	br	br	br
				holed				

Metatarsus

House	4
Room	5
Inv.No.	HA 127
Bd	56.4
Dd	31.8

Phalanx 2

House	1	2	6
Room	5	2	7
Inv.No.	X 155	X 32	X 24
GL	38.0	35.6	.
Bp	27.3	28.4	30.5
SD	21.8	23.1	25.3
Bd	22.4	(24.6)	.
			br

Phalanx 3

House	4
Room	3
Inv.No.	HA 130
DLS	46.7
Ld	42.4
MBS	12.2

Sheep, *Ovis aries*

Humerus

House	4
Room	3
Inv.No.	HA 130
BT	31.7

Metacarpus

House	4	6
Room	5	3
Inv.No.	HA 127	
Bp	23.7	.
SD	.	12.4

Tibia

House	3	4	6
Room	4	7	6
Inv.No.	HA 267	HA 131	
SD	.	16.0	15.0
Bd	24.6	28.3	.

Table 4.20. (continued).

Astragalus

House	1	2	2	4	4	5	5	6	6	6	6	6	6	6	6	6	6	6	6	6	6	6
Room	4	2	2	2	4	3	5	5	5	5	5	5	5	5	5	5	5	5	5	5	5	5
Inv.No.	X 725	H21 27/30	H21 27/30	HA 126	HA 1092	HA 80	HA 46															
GLl	.	29.9	28.4	29.1	29.7	.	30.3	30.7	33.2	31.9	(28.8)	26.5	27.7	29.5	31.0	28.3	29.2	26.5	26.6	26.8	.	25.7
GLm	28.4	28.4	26.1	26.8	(27.6)	.	28.5	28.4	.	30.1	27.1	25.8	26.2	28.6	.	.	.	.	.	25.7	24.2	25.1
Dl	.	16.1	.	15.5	15.2	12.4	15.9	17.5	16.9	17.5	16.5	15.5	15.1	15.9	.	.	.	.	14.0	14.5	15.3	13.1
Dm	16.7	16.2	.	.	.	12.5	16.6	17.5	.	18.0	16.5	.	.	17.7	.	.	.	.	.	15.0	.	.
Bd	.	18.7	19.3	18.9	18.8	13.3	19.1	20.8	.	20.1	18.2	17.0	18.4	19.2	.	17.6	.	18.3	18.2	17.2	17.6	16.1
		us?		us+br	br	us+br	us	br	br	br	br	br	br	br	br	br	br	br	br	br	br	br

Calcaneus

House	4
Room	5
Inv.No.	HA 127
GL	61.4
GB	20.4

Metatarsus

House	4	6
Room	5	3
Inv.No.	HA 153	
Bp	19.9	20.5
Dp	.	19.1
SD	.	14.5

Goat, *Capra hircus*

Humerus

House	1
Room	4
Inv.No.	X 725
Bd	32.7
BT	30.9

Radius

House	6
Room	2
Bp	31.4
BFp	30.0

Astragalus

House	4	6	6
Room	2	5	5
Inv.No.	HA 166		
GLl	.	30.5	.
GLm	28.9	28.6	29.5
Dl	.	15.4	.
Dm	.	16.2	.
Bd	(19.0)	18.1	.
	br	br	br

Sheep/goat, *Ovis aries/Capra hircus*

Maxilla

House	1	1	2	3	3	3	3	3	3	4	4	6	6	6
Room	4	4	2	1	5	5	5	8	4	6	9	2	3	7
Inv.No.	X 725	X 725	H21 47	HA 256	HA 224	HA 224	HA 224	HA 216	HA 246	HA 178	HA 157			H23 24
LM^1	.	.	.	11.6	14.2	15.7	.	.	14.2	13.5	.	.	.	10.8
BM^1	.	.	.	11.4	10.1	9.5	.	.	10.8	12.5	.	.	.	11.5
LM^2	.	17.5	.	.	15.9	18.4	16.1	14.6	.	.	18.4	.	16.7	14.0
BM^2	.	11.1	.	.	9.3	12.0	10.7	9.9	.	.	11.0	.	9.1	12.2
LM^3	17.8	.	16.4	.	.	.	15.4	.	.	.	.	16.0	.	.
BM^3	10.5	.	10.4	.	.	.	9.6	.	.	.	.	10.7	.	.

Table 4.20. (continued).

Mandible

House	3	3	4	4	4	4	4	4	4	4	4	4	6
Room	5	5	5	5	5	5	5	5	5	6	6	6	2
Inv.No.	HA 224	HA 224	HA 117	HA 127	HA 127	HA 127	HA 127	HA 127	HA 127	HA 115	HA 150	HA 178	
H-Dia	.	.	.	.	.	.	.	12.9	.	.	.	.	
HfM$_1$	.	.	.	.	.	.	.	20.3	.	.	.	.	.
LM$_1$	.	13.2	.	11.9	.	13.8	.	11.3	.	14.6	.	11.0	.
BM$_1$	.	7.4	.	7.9	.	7.8	.	7.6	.	9.6	.	7.5	.
LM$_2$	16.4	.	.	15.6	.	.	16.0	14.2	.	.	.	.	16.9
BM$_2$	7.3	.	.	9.2	.	.	8.3	8.1	.	.	.	.	7.9
LM$_3$	.	.	.	23.4	.	.	.	.	.	.	23.0	.	.
BM$_3$	.	.	8.5	.	8.5	8.4	7.9	.	9.0	.	9.3	.	.

Epistropheus

House	1
Room	4
Inv.No.	X 719
BFcr	42.3

Humerus

House	6
Room	7
Inv.No.	H23 42
SD	15.0

Radius

House	5	6
Room	3	6
Inv.No.	HA 53	
SD	15.6	15.4

Tibia

House	4	6
Room	6	2
Inv.No.	HA 137	
SD	13.6	.
Bd	.	28.5
Dd	.	19.9
		sheep?

Astragalus

House	1
Room	3
Inv.No.	X 615
GLl	25.9
Dl	14.1

Metatarsus

House	4
Room	5
Inv.No.	HA 127
Bp	21.0
	goat?

Phalanx 1

House	1
Room	2
Inv.No.	X 629D
GLpe	37.0
Bp	12.9
SD	10.1
Bd	12.3

Brown hare, *Lepus europaeus*

Humerus

House	1
Room	7
Inv.No.	X 339
Bd	12.4

Table 4.20. (continued).

Roe deer, *Capreolus capreolus*

Antler
House	6
Room	5
Cf burr	(85)
DCf burr	(61)
Cf pedi	(60)
Dia pedi	19.1

Red deer, *Cervus elaphus*

Antler
House	5
Room	3
Inv.No.	HA 51
Cf burr	(140)
DCf burr	(115)

Tibia
House	3
Room	4
Inv.No.	HA 230
Bd	45.0
Dd	34.3

Astragalus
House	3
Room	4
Inv.No.	HA 269
GLm	49.6
Dl	27.6

Phalanx 1
House	6
Room	6
GLpe	57.6
SD	19.1
Bd	21.4

Phalanx 2
House	3	6
Room	4	6
Inv.No.	HA 261	
GL	(40.5)	40.6
Bp	.	20.2
SD	.	15.5
Bd	.	16.7

Red fox, *Vulpes vulpes*

Mandible
House	3
Room	4
Inv.No.	HA 267
LP_4	6.5
BP_4	3.0
LM_1	10.8
BM_1	4.2

Badger, *Meles meles*

Radius
House	4
Room	6
Inv.No.	HA 178
Bp	11.6

individual. It was found in the main living room of house 1 (Reinders, 1988: Catalogue 37.43). The distal width (Bd) of the humerus (table 4.20) suggests a brown hare of medium size. Four humeri of brown hares from Kassope have distal widths of between 11.5 and 12.5 mm (Friedl, 1984).

Brown hares live in open steppe areas and farmland (arable) landscapes. They are also found in swamps and small woods, shelter belts and hedgerows (Corbet & Harris, 1991: pp. 154–161). The Almirós and Soúrpi plains, with their cultivated fields and woodlets, will have afforded good living conditions for the brown hare. The animal is still to be found in this part of Greece today.

Unidentified vole, Microtus sp.: A left and a right mandible from presumably the same individual of an unidentified vole species were found in room 2 of house 4 (inventory number HA 146/08). *Microtus* species occurring in the eastern part of Thessaly at present are *M. guentheri*, *M. thomasi* and *M. nivalis* and possibly *M. majori*.

M. guentheri has an eastern Mediterranean distribution area including the Almirós and Soúrpi plains. It is a species of dry, open landscapes (wheat fields, meadows, *etc.*). In Europe it is a lowland species (in Lebanon it is to be found up to altitudes of 2100 m) (Niethammer, 1982a). *M. thomasi* lives in open, especially pastured landscapes, from sea level to altitudes of 1700 m in the southwestern part of the Balkan (Niethammer, 1982b). *M. nivalis*, a mountainous species which is usually found at altitudes above 1000 m, is in Greece restricted to Mount Ólimbos, Mount Íti (province of Phthiotis, *i.e.* near New Halos), Mount Dírfis on the island of Euboia (Évvia) and Potamoi in the province of Dráma in Macedonia (Krapp, 1982: pp. 268-269). Unless the mandibles were brought to the site from Mount Íti, an identification as *M. nivalis* can be excluded. *M. majori* is also a mountainous species, which is thought to live or have lived on Mount Ólimbos (Storch, 1982: p. 460). In conclusion, the mandibles are most probably of either *M. guentheri* or *M. thomasi*.

These species dig extensive tunnel and storage systems often extending to depths of 20–45 cm beneath the surface (Niethammer, 1982a; 1982b). Such depths were reached in the excavation of the houses of New Halos. The vole jaws are therefore probably of (sub)recent intrusive origin.

Fox, Vulpes vulpes: The cheek tooth part of the left mandible of a (sub)adult fox was found in the courtyard of house 3 (inventory number HA 267/30). The animal was of a small size (table 4.20). The fox bones from Kassope are also small (Friedl, 1984: p. 169).

Foxes are very adaptable animals. They presumably lived in the plains as well as in the Óthris mountains. The waste of the town itself may have attracted the animal concerned, as human food remains may constitute part of the fox's diet (Wandeler & Lüps, 1993: p. 165). Foxes are regularly encountered in the site's surroundings nowadays. Foxes may have been hunted for their skins. The mandible could be the remnant of a skin that was brought to the town with the skull attached. As in Kassope, there is no evidence for the consumption of fox at New Halos (Friedl, 1984: p. 169).

Badger, Meles meles: The proximal half of a radius of a badger was found in room 6 of house 4 (inventory number HA 178). Badger was likewise represented by one fragment at Kassope (Friedl, 1984: p. 173).

Badgers live in both flat and mountainous areas, mostly in low densities. They live in wooded as well as open habitats, of which they prefer the first. The plains, the riverbed, the marsh along the coast and the Óthris mountains will all have afforded suitable habitats for badgers. The species is still to be found in Greece today. Badgers build extensive setts, systems of tunnels and chambers with widths of up to 80 m and depths of several metres, in which they remain during the daytime and the winter season. Setts are built mostly in woods, at the edges of woods, in hedgerows and copses, and rarely in open fields. They show a preference for slopes (Corbet & Harris, 1991: p. 419; Lüps & Wandeler, 1993: pp. 876–906). Badgers and foxes may share the same sett or den (Lüps & Wandeler, 1993: p. 896).

The easiest way of catching a badger, which is usually active outside its sett only during the night and at dusk, is by digging it out of its sett after a small dog has located it (Criel *et al.*, 1997). Inhabitants of New Halos will have caught badgers for their pelts and their meat.

Red deer, Cervus elaphus: Red deer was represented in houses 3, 4, 5 and 6 by a total of ten bone fragments (tables 4.6, 4.7 and 4.19). Antler was well represented with four fragments. Whether these fragments derive from shed antlers or from hunted red deer could not be established. All the antlers were fully developed. Shed antlers can be collected in late winter/early spring, just after they have been shed. Male red deer with fully developed antlers can be hunted from July/ August to February/March/April (Bützler, 1986: pp. 119–121).

The three phalanges are proximally fused, indicating that they derive from hunted red deer that were more than 3 years old. The tibia is distally fused and comes from an animal that was more than 4 years old. This would suggest that only adult red deer were hunted. The New Halos red deer were of the same small stature as those whose remains were found at Kassope (Friedl, 1984: pp. 138–155), Thebai (Boessneck, 1973: p. 22) and Messene (Nobis, 1994: p. 312).

The three fragments of antler found in room 3 of house 4 (inventory numbers HA 148/5, HA 173 and HA 173/1) were charred. A charred cattle astragalus was found in the same room. The right red deer astragalus from the courtyard of house 3 (inventory number HA 269/6) was neither worn nor charred; it was presumably not used as a knucklebone (as was the case at Thebes: Boessneck, 1973: pp. 9–10).

In Greece, red deer are nowadays only found in the Píndhos mountains and in some mountainous areas in Macedonia and Thráki (Bützler, 1986: pp. 114–115). The species originally occupied large parts of the mainland and the peninsulas of Greece. Red deer is a highly adaptable species, but has a preference for open woodland and wooded river valleys (floodplains), especially in flat terrains. In most European countries such terrains have been converted into agricultural, urban and industrial areas. Red deer is now restricted to quiet open terrains, even without much cover, to coniferous woodlands on poor soils and, for instance in Greece, to mountainous areas (Bützler, 1986: p. 125). Red deer will have lived in the Almirós and Soúrpi plains, the river valleys in these plains and the Óthris mountains in Hellenistic times.

Roe deer, Capreolus capreolus: Roe deer was represented by only two fully developed antlers in room 5 of house 6. One is an almost complete right antler of a hunted roe deer (not a shed antler). The diameter of the pedicel (19.1 mm, table 4.20) and the limited number of pearls on the burr show that the buck was about 4–5 years old (data for central European roe deer: diameter of pedicel 18–20 mm; Habermehl, 1985: p. 46). The other, more fragmented antler of unknown side shows more and larger pearls on the burr. This suggests that it belonged to an older buck than the almost complete antler. Whether this was a shed antler or an antler from a hunted roe deer buck could not be established. Both antlers are charred and partly calcined. They show black, grey and white stains caused by fire. 21 charred sheep and goat astragali were found in the same room. The roe deer antlers and the sheep and goat astragali may have been burned in the same fire.

The adult male roe deer bears fully developed antlers from April to October, November or December, when the antlers are shed (Von Lehmann & Sägesser, 1986: p. 246). This means that animals with fully developed antlers were obtainable from about April to December.

The present distribution area of roe deer in Greece is restricted to the northern part of the country: the Píndhos mountains, Mount Ólimbos and the Rodópi mountains (Von Lehmann & Sägesser, 1986: pp. 241–243). Roe deer is a highly adaptable species, which lives in coastal plains and in mountainous areas up to the tree line. It has a preference for the edges of woods (Von Lehmann & Sägesser, 1986: pp. 254–255). It was more widely distributed in Greece originally. The

last roe deer of the Pelopónnisos were seen and shot in the oak forest Kapellis in Ílis *c.* AD 1898. In 1922, roe deer were shot in the forest of Melissouryía on Chalkidikí. At an unknown date, probably around the beginning of the 20th century, a roe deer buck was shot on Mount Ólimbos (Niethammer, 1942–1949). Roe deer were to be found in the Almirós and Soúrpi plains, including the coastal parts, and the Óthris mountains in Hellenistic times.

Cf. Proboscidea, elephants: A bead of a dental, originally white, now yellowish material, presumably ivory, was found in the main living room of house 1 (inventory number X 322; Ch. 4: 4.11).

4.6. Reptiles

Tortoise, Testudo sp.: Twelve carapace fragments and one fragment of the pelvis of a terrestrial tortoise of the genus *Testudo*, with a total weight of 13 g, were found in room 6 of house 4.[8] One of the carapace fragments is from an abdominal carapace. The type of carapace, dorsal or abdominal, from which the other fragments derive could not be established. The thirteen tortoise fragments may come from one individual. Three *Testudo* species occur in the surroundings of Halos nowadays: *T. graeca*, *T. hermanni* and *T. marginata* (Engelmann *et al.*, 1986: pp. 196–200). Which species is, or possibly are, represented at New Halos could not be established.

Three carapace fragments are charred.[9] One of them is charred only on the inside of the (abdominal) carapace. This means that it was burned after the animal had been butchered, or after the carapace had in some other way become fragmented. The only other charred bone fragment found in room 6 of house 4 is a vertebral fragment of a sheep/goat/pig-sized mammal (inventory number HA 115/30). It was found close to the charred carapace fragment HA 115/15. See Ch. 4: 4.12 on the possible cause of the charring.

Nine carapace fragments of at least two terrestrial tortoises, *Testudo* sp., were found in the Hellenistic layers of the Kabirion sanctuary near Thebai (Boessneck, 1973: p. 26). A total of 37 carapace and extremity fragments of terrestrial tortoises, presumably *T. hermanni*, were found in Kassope (no information on period). *Mauremys caspica*, which lives in fresh water, and *Caretta caretta*, the common sea turtle, were represented at Kassope in small numbers (Friedl, 1984: pp. 189–190).

4.7. Fish

A worn vertebra of a medium-sized fish was found in room 3 of house 1 (inventory number X 666, weight 2 g) (Reinders, 1988: Catalogue 33.86). The height of the corpus is about 15.2 mm, its length about 9.5 mm. The species could not be identified due to the absence of a reference collection. A total of 61 fish remains from at least six species of marine fish were found in Kassope. The fish species represented in the 3rd-century-BC layers are not specified (Friedl, 1984: pp. 191–196; Boessneck, 1986). No fish remains were found at the Kabirion near Thebai (Boessneck, 1973). Five unidentified fish remains were found among the 1681 vertebrate remains recovered from the Hellenistic-Roman layers at Troy (Quadrat K13, Complex A) (Uerpmann et al., 1992). The proportion of fish remains at Troy is only slightly larger than that at New Halos: a single fish bone among 900 vertebrate remains. The only fish remains encountered in the 3rd–2nd-century layers at Messene is a vertebra of *Thunnus thynnus* (Nobis, 1994: p. 303).

4.8. Marine gastropoda

Patella caerulea, limpet (Greek: petalída)[10]: The limpet is represented by 41 shells and shell fragments, with a total weight of 18 g. Limpet remains were found in all the houses except house 2. Most of them, 33, were found in house 3. The measurable shells are between 22.5 and 36.4 mm long (n = 9) (table 4.21). Only adult specimens of moderate size were collected. Limpets may have lengths of between 20 and 66 mm (Poppe & Goto, 1991: p. 69). Empty shells were found in large numbers on the Amaliápolis and Almirós beaches in 1993/94 (table 4.22). The species is at present frequently found along almost all Greek coasts (Delamotte & Vardala-Theodorou, 1994: p. 204), but seems to be rare in the gulf of Thessaloníki (Sakellaríou, 1957: pp. 193–194). Limpets live on rocky shores, from the intertidal to water of a few m depth (Poppe & Goto, 1991: p. 69; Delamotte & Vardala-Theodorou, 1994, p. 204). Limpets are popular food in Mediterranean countries; they are eaten cooked (Davidson, 1981: p. 190).

Gibbula divaricata: A single shell of *Gibbula divaricata* with a weight of 4 g was found in room 7 of house 6. The shell derives from a rather large specimen (width 21.4 mm; *cf.* Poppe & Goto, 1991: p. 79: 12-20 mm in width (diameter)) (table 4.21). *Gibbula divaricata* lives on rocky coasts, from the intertidal to water of a few m depth (Poppe & Goto, 1991: p. 79). Shells of the species were found in large quantities on the Amaliápolis and Almirós beaches in 1993/94 (table 4.22). Delamotte & Vardala-Theodorou (1994: p. 207) report the occurrence of the species in central Greek waters. The species was not reported by Sakellaríou (1957) for the gulf of Thessaloníki. No *Gibbula* species is mentioned as Mediterranean seafood by Davidson (1981) and Bini (1965). A slightly larger relative of the *Gibbula* species, *Monodonta turbinata* (Greek: *tróchos*), is regularly consumed in Mediterranean countries after being boiled (Davidson,

1981: p. 191). This species and the related species *Monodonta articulata* are nowadays to be found in large numbers along the coast of the Pagasitikós gulf (table 4.22).

Gibbula albida: A shell of this species was found in the courtyard of house 3 and another in the courtyard of house 6. Their total weight is 5 g. Their widths, 19.0 and 19.2 mm (table 4.21), indicate that the animals were medium-sized (Poppe & Goto 1991: p. 78: 10–24 mm). *Gibbula albida* lives on all kinds of substrates, in water with depths of up to 20 m. It is common in the eastern Mediterranean (Poppe & Goto, 1991: p. 78). Shells of the species are to be found in small numbers on the two beaches of the Pagasitikós gulf today (table 4.22). Delamotte & Vardala-Theodorou (1994: p. 207) report the occurrence of the species in the western Greek waters. It was not found in the gulf of Thessaloníki (Sakellaríou, 1957).

Cerithium vulgatum, horn shell, needle shell (Greek: kerátios): A total of ten shells of this species were found in the main living room and the possible courtyard of house 2, in room 5, the corridor and the courtyard of house 3, in the main living room of house 5, and in the courtyard of house 6. The total weight of the shells is 36 g. The lengths of the shells vary between 26.7 and 49.7 mm (n = 4) (table 4.21). The shells derive from small to medium-sized animals. Poppe & Goto (1991: p. 112) state that adult shells of this species can attain lengths of up to 66 mm. The species lives on sandy and muddy substrates, in shallow water (Poppe & Goto, 1991: p. 112). Horn shells are to be found in large numbers on the Amaliápolis and Almirós beaches today (table 4.22). Delamotte & Vardala-Theodorou (1994: pp. 214-215) report the occurrence of the species in central Greek waters. Sakellaríou (1957: pp. 199-200) found the species in the gulf of Thessaloníki. It is regularly consumed after being boiled (Davidson, 1981: p. 194).

Luria lurida, lurid cowrie: A lurid cowrie with a weight of 8 g was found in room 2 adjacent to the main living room of house 2. The shell is from a medium-sized animal (table 4.21) (*cf.* Poppe & Goto, 1991: p. 124: 28–45 (63) mm in height). The lurid cowrie lives under stones and rocks near sponges, often in sandy biotopes, in water with depths of 1 to 60 m (Poppe & Goto, 1991: p. 124). Shells of the species are rarely found on the beaches of the Pagasitikós gulf today (table 4.22). Sakellaríou (1957: p. 207) found the species in small numbers in the gulf of Thessaloníki. Delamotte & Vardala-Theodorou (1994: p. 220) report the occurrence of the species in central Greek waters.

Fig. 4.14. Hellenistic New Halos. Shell of *Tonna galea* from the main living room of the House of the Snakes (house 3) from two points of view, inventory number HA 248/5.

Tonna galea, cask shell: A complete cask shell with a weight of about 250 g was found in the main living room of house 3 (inventory numbers HA 248/5 and HA 247/9) (fig. 4.14). Its height, approx. 190 mm, indicates that it derives from a medium-sized adult individual (*cf.* Poppe & Goto, 1991: p. 128: (150)180–200(250) mm in height). The species lives on all kinds of substrates, in water with depths of 20–80 m (Poppe & Goto, 1991: p. 128). No shell fragments of this species were found on the Amaliápolis and Almirós beaches in 1993/94 (table 4.22). Beach finds of this species are rare since the species lives in deep water.

Sakellaríou (1957: pp. 208–209) found the species in small numbers in the gulf of Thessaloníki. Delamotte & Vardala-Theodorou (1994: p. 222) report the occurrence of the species in southern, central and western Greek waters.

Bolinus brandaris (Greek: porphýra): Eight shells of this species, with a total weight of 63 g, were found in the main living room of house 2, in room 1, room 6 adjacent to the main living room and in the courtyard of house 3 and in the main living room and the adjacent room 4 of house 5 (table 4.6). Only one of

Table 4.21. Six houses in Hellenistic New Halos. Measurements of mollusc shells. (Min = minimum value; Max = maximum value; x = mean; n = number of measurements. Bivalvia: L = left hand valve; R = right hand valve.)

MARINE GASTROPODA

Patella caerulea, limpet (Greek: petalída)

House	1	3	3	3	3	3	3	3	3	3	3
Room	2	1	9	11	4	4	4	4	4	4	4
Inv.No.	X 629B	HA 205/99	HA 228/9	HA 247/65	HA 253α/67	HA 253α/81	HA 253/24	HA 253α/53	HA 253/10	HA 262α/23	HA 274/3
Length	36.4	32.7	.	22.5	26.7	25.3	24.5	29.8	.	24.7	34.2
Width	29.6	27.2	20.8	18.9	20.7	20.6	.	22.6	27.1	18.0	27.3
Height	.	7.2	.	.	6.5	.	7.4	6.4	9.7	5.7	6.1

Key of species:

1   *Gibbula divaricata*
2   *Gibbula albida*
3   *Cerithium vulgatum*, horn shell, needle shell (Greek: kerátios)
4   *Luria lurida*, lurid cowrie
5   *Tonna galea*, cask shell
6   *Bolinus brandaris* (Greek: porphýra)
7   *Hexaplex trunculus*, (rock) murex (Greek: porphýra)
8   *Buccinulum corneum*
9   *Conus ventricosus*, cone shell.

Species	1	2	2	3	3	3	3	3	3	3	3	4	5	6
House	6	3	6	2	2	3	3	3	5	6	6	2	3	3
Room	7	4	6	3	4	5	9	4	3	6	6	2	11	4
Inv.No.	X 56	H21 267/20	-	-	HA 17	HA 224/17	HA 220/7	HA 267/31	HA 81	HA 6	-	-	HA 248/5	H23 246/7
Height	19.5	.	.	42.4	43.3	.	.	26.7	.	.	49.7	35.7	190.0	.
Width	21.4	19.0	19.2	14.6	13.6	15.1	14.0	9.3	17.0	15.0	16.8	21.9	149.0	49.0

Species	7	7	7	7	8	9	9
House	3	3	3	3	4	2	4
Room	1	11	4	4	2	2	5
Inv.No.	HA 256/2	HA 232/36	HA 266/9	HA 267/38	HA 146/3	-	HA 127/35
Height	24.5	53.0	13.6	57.1	46.0	34.8	26.0
Width	18.4	34.9	7.4	43.0	21.0	18.4	14.7

MARINE BIVALVIA

Acanthocardia tuberculata, red-nosed cockle (Greek: methýstra, kydóni) (equivalve species)

House	3			
	min	max	x̄	n
Length	35.5	47.5	41.5	5
Height	26.7	50.2	39.4	5
½ Depth	11.1	21.1	16.6	8

Cerastoderma glaucum, lagoon cockle (Greek: methýstra, kydóni) (equivalve species)

House	1				2				3			
	min	max	x̄	n	min	max	x̄	n	min	max	x̄	n
Length	16.1	26.5	21.9	4	22.0	28.9	26.3	3	-	-	-	-
Height	14.6	25.7	21.2	5	18.0	31.0	24.9	8	20.4	36.2	29.7	4
½ Depth	6.4	11.3	9.4	7	7.6	13.2	10.7	14	8.7	20.2	14.4	5

House	4				5				6			
	min	max	x̄	n	min	max	x̄	n	min	max	x̄	n
Length	21.5	47.8	28.8	8	12.2	35.2	25.4	12	16.3	48.8	26.7	22
Height	20.4	40.3	27.4	11	19.6	33.7	25.1	11	15.4	39.2	23.7	29
½ Depth	7.3	16.6	12.0	12	6.6	18.2	11.9	20	6.4	20.6	10.5	32

Table 4.21. (continued).

Total for the six houses

	min	max	x̄	n
Length	12.2	48.8	26.3	49
Height	14.6	40.3	24.9	68
½ Depth	6.4	20.6	11.2	90

Ostrea edulis, oyster (Greek: strídi) (inequivalve)
Left hand valves

House	1				3				Total			
	min	max	x̄	n	min	max	x̄	n	min	max	x̄	n
Length	.	.	.	0	24.5	50.4	41.4	7	24.5	50.4	41.4	7
Height	.	.	63.7	1	34.2	75.0	55.9	10	34.2	75.0	56.6	11

Right hand valves

House	1				3				Total			
	min	max	x̄	n	min	max	x̄	n	min	max	x̄	n
Length	.	.	24.5	1	21.4	48.7	32.2	3	21.4	48.7	30.3	4
Height	.	.	27.0	1	29.8	71.6	49.9	6	27.0	71.6	46.6	7

NB: the L and R valves of house 1 are presumably paired.

Venus verrucosa, warty venus (Greek: kýdoni) (equivalve)

House	1				2				3			
	min	max	x̄	n	min	max	x̄	n	min	max	x̄	n
Length	.	.	44.4	1	.	.	.	0	35.3	40.5	37.9	2
Height	.	.	42.7	1	36.7	39.3	38.0	2	31.1	39.7	35.7	3
½ Depth	.	.	16.7	1	13.6	14.5	14.1	3	12.1	15.3	14.1	3

House	4				5				6			
	min	max	x̄	n	min	max	x̄	n	min	max	x̄	n
Length	37.9	45.4	42.1	4	35.6	43.8	39.7	2	37.1	44.9	41.2	8
Height	34.6	44.1	38.2	4	33.7	39.2	36.7	4	34.1	42.1	36.8	6
½ Depth	11.1	17.8	14.4	4	13.2	15.1	13.9	4	11.3	17.4	14.2	11

Total for the six houses

	min	max	x̄	n
Length	35.3	45.4	41.0	17
Height	31.1	44.1	37.3	20
½ Depth	11.1	17.8	14.3	26

Key of species
1   *Arca noae*, Noah's ark or turkey wing (Greek: kalógnomi)
2   *Glycymeris insubrica* (Greek: melokídono)
3   *Chlamys* cf. *glabra*
4   *Spondylus gaederopus*, spiny or thorny oyster (Greek: spóndylos)
5   *Psoedochama gryphina*
6   *Mactra stultorum*, through shell
7   *Donacilla cornea*, wedge shell (Greek: kochíli)
8   *Callista chione*, smooth venus (Greek: ghialisterí achiváda)
9   *Tapes decussatus*, carpet-shell (Greek: chávaro)

Species	1	1	1	1	1	1	1	2	2	2	3
House	1	3	3	3	3	6	6	5	6	6	6
Room	-	5	11	11	11	3	6	1	5	6	5
Inv.No.	X	HA	HA	HA	HA			H21			
	203	210/19	232/31	232/34	247/51	-	-	14	-	-	-
Valve	.	L	L	L	L	L	L	R	.	.	.
Length	.	52.4	59.3	.	53.9	59.4	.	49.9	.	.	41.1
Height	.	24.1	21.6	.	28.5	27.2	22.2	50.0	56.9	43.0	40.5
½ Depth	14.4	14.0	13.6	16.1	14.4	13.8	14.6	19.5	21.3	15.1	8.6

Table 4.21. (continued).

Species	4	4	4	4	5	6	6	7	7	7	7	8
House	1	1	1	6	5	3	3	3	5	5	5	5
Room	3	4	7	5	2	3	1	3	3	3	3	2
Inv.No.	X 656B	X 731A	X 310	-	HA 44	HA 202/6	HA 202/18	HA 193/12	HA 39	HA 81	HA 81	HA 53
Valve	R	L	R	R	(R)	.	L	L	L	R	R	.
Length	.	71.4	.	98.2	20.6	43.5	49.3	16.9	15.6	15.8	17.9	57.4
Height	.	70.8	.	.	23.5	.	.	10.4	10.4	10.7	.	44.8
½ Depth	24.0	23.6	34.0	39.6	10.3	.	.	.	3.1	3.3	3.6	15.2

Species	9	9	9	9
House	6	6	6	6
Room	5	5	5	5
Inv.No.	-	-	-	-
Valve	L	L	R	R
Height	18.1	23.8	.	.
½ Depth	6.2	7.6	7.8	6.7

TERRESTRIAL GASTROPODA

Key of species
1   *Lindholmia lens*
2   *Helicella* sp.

Species	1	2	2	2	2	2
House	3	3	5	5	5	5
Room	4	4	3	3	3	3
Inv.No.	HA 253/39	HA 271/8	HA 80	HA 81	HA 86	HA 86
Height	.	6.8	10.3	8.7	9.4	9.4
Width	11.0	12.2	.	13.6	16.3	.

them was measurable. Its width shows that it derives from a large animal (table 4.21) (*cf.* Poppe & Goto, 1991: p. 134: (50)70–90 mm in length, which is about 1.8 times its width). The species lives on sand and sandy/muddy substrates, in water with depths of 1-200 m (Poppe & Goto, 1991: p. 134). Small numbers of the shells were found on the Amaliápolis and Almirós beaches in 1993/94 (table 4.22). Delamotte & Vardala-Theodorou (1994: p. 228) found the species in central and southern Greek waters, Sakellaríou (1957: pp. 202–203) in the gulf of Thessaloníki.

The eight shells will be the remains of animals that were consumed in the houses. The species is still consumed in Mediterranean countries today after being boiled (Davidson, 1981: pp. 192–193). The small number of shells found in the houses precludes the possibility that they are the remains of purple dye production (Uerpmann, 1972); experiments by D. Ruscillo, presented at the 8th ICAZ Conference, 2002).

Hexaplex trunculus, rock murex (*Greek: porphýra*): Nineteen shells/shell fragments of the rock murex, with a total weight of 72 g, were found in houses 1,

3, 4, 5 and 6 (table 4.6). The four measurable shells show that small, *i.e.* young, as well as adult, but medium-sized animals were collected (table 4.21) (*cf.* Poppe & Goto, 1991: p. 136: 40–80 mm in height). The species prefers muddy substrates under 1–100 m of water (Poppe & Goto, 1991: p. 136). Empty shells of rock murex are regularly found on the two aforementioned beaches of the Pagasitikós gulf today (table 4.22). Delamotte & Vardala-Theodorou (1994: p. 227) mention the species as occurring in central Greek waters. Sakellaríou (1957: pp. 203–204) found the species in the gulf of Thessaloníki. The rock murex is another source of purple dye. The shells from the houses of New Halos will however be the remains of rock murexes that were consumed (Davidson, 1981: p. 192).

Buccinulum corneum: A single shell of this species, with a weight of 9 g, was found in room 2 of house 4. The shell derives from a medium-sized animal (table 4.21) (*cf.* Poppe & Goto, 1991: p. 144: 30–70 mm in height). The species lives in rocky habitats in shallow water (Poppe & Goto, 1991: p. 144). Its shells

Table 4.22. Marine molluscs collected as empty shells or valves on the beaches of Amaliápolis and Almirós on the Pagasitikós gulf in July 1993 (two weeks) and June 1994 (three weeks) by the author, C. Kruyshaar and M. Huisman, their presence or absence in Hellenistic New Halos and the numbers for each species counted by Sakellaríou (1957) in the Gulf of Thessaloníki (she also found species not present in the Pagasitikós Gulf). Amaliápolis and Almirós 1993/94 beach finds: +: rare; ++: found in small numbers; +++: common; -: not found. New Halos: numbers of shells and valves in the six houses (cf. table 4.6). Gulf of Thessaloníki: numbers of living animals.

MARINE GASTROPODS	1993/94	New Halos	Gulf of Thessaloníki
Haliotis tuberculata	++	-	-
Diodora italica	++	-	9
Patella caerulea	+++	41	8
Patella rustica	+	-?	-
Patella ulyssiponensis f. *bonardii*	+	-?	-
Clanculus corallinus	+	-	-
Gibbula divaricata	+++	1	-
Gibbula albida	++	2	8
Gibbula fanulum	+	-	-
Monodonta articulata	+++	-	-
Monodonta turbinata	+++	-	-
Vermetus sp.	+	-	-
Cerithium rupestre	+	-	-
Cerithium vulgatum	+++	10	110
Bittium sp.	+	-	-
Aporrhais pespelecani	+	-	8
Luria lurida	+	1	6
Tonna galea	-	1	28
Bolinus brandaris	++	8	68
Hexaplex trunculus	++	19	112
Ocenebra erinaceus	+	-	-
Thais haemastoma	+	-	33
Buccinulum corneum	+	1	-
Columbella rustica	++	-	5
Nassarius incrassatus	+	-	-
Nassarius reticulatus	+	-	22
Fusinus syracusanus	+	-	2
Conus ventricosus	+	2	30
Epitonium clathrus	+	-	5
Bulla striata	+	-	-
Planaxis sulcata (recent intrusion from Red Sea)	+	-	-
MARINE BIVALVES	1993/94	New Halos	Gulf of Thessaloníki
Arca noae	+++	38	250
Barbatia barbata	+	-	58
Striarca lacteus	+	-	-
Glycymeris insubrica	+	4	-
Lithophaga lithophaga	+	-	-
Mytilus edulis f. *galloprovincialis*	++	1	81
Pinna nobilis	++	3	23
Chlamys cf. *glabra*	+	2	185
Chlamys varia	++	-	15
Spondylus gaederopus	++	9	73
Anomia ephippium	++	-	172
Limaria inflata	+	-	10
Ostrea edulis	++	50	470
Crassostrea gigas	+	-	-
Ctena decussata	+	-	-
Loripes lucinalis	+	-	15
Psoedochama gryphina	+++	1	28
Chama gryphoides	+++	-	8
Cardites antiquata	+	-	137
Acanthocardia aculeata	+	-	5
Acanthocardia tuberculata	+++	22	490
Cerastoderma glaucum	-	188	578

Table 4.22. (continued).

Parvicardium sp.	+	-	-
Laevicardium oblongum	+	-	-
Mactra stultorum	-	3	670
Spisula subtruncata	+	-	3500
Donacilla cornea	++	4	730
Donax venustus	++	-	35
Gari depressa	+	-	-
Glossus humanus	+	-	2
Callista chione	+++	1	7
Chamelea gallina	++	-	12
Clausinella brongniartii	+	-	-
Irus irus	+	-	-
Paphia aurea	+	-	670
Tapes decussatus	+++	31	134
Venus verrucosa	+++	50	882
CEPHALOPODA			
Sepia officinalis	+	-	

are rare on the beaches of the Pagasitikós gulf today (table 4.22). Delamotte & Vardala-Theodorou (1994: p. 231) found the species in central and western Greek waters. Sakellaríou (1957) did not find the species in the gulf of Thessaloníki.

Conus ventricosus, cone shell (*Greek: kónos*): A shell of this species was found in room 2 adjacent to the main living room of house 2 and another in room 5 of house 4. The two shells together weigh 9 g. The one from house 4 is small, that from house 2 medium-sized (table 4.21) (*cf.* Poppe & Goto, 1991: p. 176: average specimens 25 mm, maximum 73 mm). The species lives on rocks covered with algae, in shallow, quiet bays or in deep water (Poppe & Goto, 1991: p. 176). Delamotte & Vardala-Theodorou (1994: p. 237) mention the species as occurring in all Greek waters. It was found in small numbers in the gulf of Thessaloníki (Sakellaríou, 1957: pp. 209–210).

4.9. Marine bivalvia

Arca noae, Noah's ark or turkey wing (*Greek: kalógnomi*): As many as 38 valves or valve fragments, with a total weight of 80 g, were found in house 1 (no indication of the room(s)), in the main living rooms of houses 3, 5 and 6, and in room 5 adjacent to the main living room and in the courtyard of house 6. The largest number, 32, was found in house 6, 30 of which in room 5. They are large fragments of five left and four right valves and twenty smaller fragments of the same valves and possibly one or two more valves. At least five animals were represented by the remains in room 5 of house 6. The valve/valve fragments found in houses 1, 3, 5 and 6 are mostly (fragments of) left valves. The ventral margin of the shell found in room 3 adjacent to the main living room of house 6 was

slightly damaged (broken). Many other shells were broken. The edges had not been rounded by abrasion over a long period of time in the sea or on the beach. So the valves presumably broke some time while they were buried in the soil.

The measurable shells are medium-sized with lengths of 52.4–59.4 mm (n = 4) (table 4.21) (*cf.* Poppe & Goto, 1993: p. 42: 40–70 mm in length). Noah's ark lives with its byssus attached to hard substrates, from the low tide line to water with depths of up to 119 m (Poppe & Goto, 1993: p. 42). Empty shells are found in large numbers on the beaches of the Pagasitikós gulf today (table 4.22). Delamotte & Vardala-Theodorou (1994: p. 242) mention the species as occurring in central and southern Greek waters. Sakellaríou (1957: pp. 136–137) found the species in fairly large numbers in the gulf of Thessaloníki. The species is still consumed in Mediterranean countries today, usually raw (Davidson, 1981: p. 196).

Glycymeris insubrica (*Greek: melokídono*): Valves of this species, with a total weight of 71 g, were found in room 1 of house 5, adjacent to the main living room (two valves), and in room 5, also adjacent to the main living room, and the courtyard of house 6 (table 4.6). The measurable valves indicate that medium-sized animals were brought to the houses (table 4.21) (*cf.* Poppe & Goto, 1993: p. 46: 30–70 mm in length). *Glycymeris insubrica* lives on muddy sand in the infralittoral zone (Poppe & Goto, 1993: p. 46). Shells of the species are to be found in very small numbers on the beaches of the Pagasitikós gulf today (table 4.22). Delamotte & Vardala-Theodorou (1994: p. 244) mention the species as occurring in the western and central Greek waters. It has not been found in the gulf of Thessaloníki (Sakellaríou, 1957). Another species of the genus, *G. glycymeris*, is regularly consumed raw in Mediterranean countries (Davidson, 1981: p. 196).

Mytilus (*edulis* f.) *galloprovincialis, mussel* (*Greek: mýdi*): A small shell fragment of this species, weighing less than 1 g, was found in room 1 of house 3. The species lives on hard substrates, from the intertidal to water with depths of 40 m (Poppe & Goto, 1993: p. 52). Valves of the species are to be found in very small numbers on the Amaliápolis and Almirós beaches today (table 4.22). Delamotte & Vardala-Theodorou (1994: p. 244) found the species in all Greek waters. Sakellariou (1957: pp. 141–142) found it in the gulf of Thessaloníki. Mussels are regularly consumed in Mediterranean countries today, sometimes raw, but mostly cooked (Davidson, 1981: p. 198).

Pinna nobilis, noble, or rough pen shell, fan shell (*Greek: pínna*): Three small valve fragments of this species, with a total weight of 9 g, were found in house 1 (room 3) and house 5 (two fragments in the main living room). The species lives half-buried in sand, mud or gravel, from the low tide line to water with depths of 60 m (Poppe & Goto, 1993: p. 55). Valves or valve fragments are found in small numbers on the Pagasitic beaches today (table 4.22). Delamotte & Vardala-Theodorou (1994: p. 246) report the occurrence of the species in central and western Greek waters. The species is also to be found in the gulf of Thessaloníki (Sakellaríou, 1957: p. 143). After the animal has been extracted from the valves it is eaten raw or cooked (Davidson, 1981: p. 199).

Chlamys cf. *glabra, cf. smooth scallop*: This species was represented by a complete valve in room 5, and a valve fragment in room 1 of house 6. Both rooms are adjacent to the main living room. The total weight of the two items is 19 g. The complete valve belonged to a medium-sized animal (table 4.21) (*cf.* Poppe & Goto, 1993: pp. 61–62: 30–70 mm in length). The species lives on sandy, muddy or rocky substrates in water with depths of 6 to 900 m (Poppe & Goto, 1993: p. 61). Shells and shell fragments of the species are rarely found on the Amaliápolis and Almirós beaches today (table 4.22). It is a common species in the Greek waters (Sakellaríou, 1957: pp. 147–147). Delamotte & Vardala-Theodorou (1994: p. 247) mention the species as occurring in central, western and eastern (= along the Anatolian coast) Greek waters. The species is still frequently consumed in Mediterranean countries today. The best way of preparing it is by cooking (Davidson, 1981: p. 201).

Spondylus gaederopus, spiny or thorny oyster (*Greek: spóndylos*): Nine valves/valve fragments, with a total weight of 509 g, were found in houses 1 (rooms 3, 4 and 5), 3 (main living room), 5 (main living room)

and 6 (room 5 and courtyard) (table 4.6). The measurable shells derive from small to medium-sized animals (table 4.21) (*cf.* Poppe & Goto, 1993: p. 72: 60-125 mm in length). The spiny oyster lives cemented to rocky substrates below 7–50 m of water (Poppe & Goto, 1993: p. 68). Shells and shell fragments of the spiny oyster are nowadays found in small numbers on the two studied beaches of the Pagasitikós gulf (table 4.22). Delamotte & Vardala-Theodorou (1994: p. 249) mention the species as occurring in central and northern Greek waters. Sakellaríou (1957: pp. 148-149) found the species in the gulf of Thessaloníki. The colonies declined substantially in the 1980s (Poppe & Goto, 1993: p. 73).

Six valves/valve fragments derive from convex, right valves. One fragment presumably comes from a right valve. Two fragments are parts of the flat, left valves. The convex, right (or lower) valve of the spiny oyster is attached to the rock. The left (or upper) valve is moveable in the hinge. The greater part of the animal is to be found in the right valve. The raw spiny oyster is most easily consumed from the convex right, lower valve. The preponderance of right valves is an argument for assuming that live spiny oysters, containing the animal, were cut from the rock and brought to the houses as food. The spiny oyster was considered a delicacy in ancient Greece (Keller, 1913: pp. 561–562).

Ostrea edulis, oyster (*Greek: strídi*): As many as fifty oyster valves and valve fragments, with a total weight of 325 g, were found in houses 1 (part of courtyard), 3 (room 1, main living room and courtyard), 5 (room 1) and 6 (courtyard) (table 4.6). The majority of the oyster valves, 44, were found in the courtyard of house 3. The measurable left and right oyster valves show that most of the oysters consumed in Hellenistic Halos were small animals (table 4.21) (*cf.* Poppe & Goto, 1993: p. 79: 60–100(150) mm in length). The oyster lives with its left valve attached to all types of substrates from shallow water to water with depths of up to 90 m (Poppe & Goto, 1993: p. 79). Shells and shell fragments of mostly small oysters are found in large numbers on the shores of the Pagasitikós gulf today (table 4.22). Delamotte & Vardala-Theodorou (1994: p. 250) mention the species as occurring in central and western Greek waters. Sakellaríou (1957: pp. 151–152) found it in the gulf of Thessaloníki.

Twenty of the 50 New Halos shells are convex, left (or lower) valves, or parts of left valves. Eight shells/shell fragments are (parts of) flat, right (or upper) valves. One shell fragment is presumably part of a right valve. The remaining 21 fragments were too small to allow us to identify the valve type. A left and a right valve from the courtyard of house 3 (inventory number HA 262μ/18) were presumably paired. They are assumed to derive from an oyster that was brought to the house alive to be consumed. As with

Table 4.23. Hellenistic New Halos. Dimensions in mm of beads or rings made from ivory, mollusc shell or bone.

House	1	1	4	6
Room	7	5	2	3
	main living room	adjoining main living room		main living room
Inv.No.	X 332	X 78	HA21 104/4	-
Material	ivory	shell	bone	bone
Cross-section	lens-shaped	flat	flat	flat
Diameter	19.2	12.7	21.2	19.8
(Maximum) thickness	6.7	5.2	4.2	8.1
Diameter of hole	5.9	6.6	8.4	8.5
Distance hole-edge	7.1	3.5	7.0	6.3

the spiny oyster, the preponderance of the convex, lower, left valves can be seen as an argument for assuming that complete, live animals were dug from the seabed and brought to the houses for consumption. The flat, right valve was removed from the convex, left valve before consumption. The other 42 valves and valve fragments from the courtyard of house 3 were 17 left valves, 6 right valves and 19 valve fragments of unidentified side. This suggests that oysters were consumed in the courtyard. Oysters were popular among the ancient Greeks (Keller, 1913: pp. 562–568) and are still popular in Mediterranean countries today; they are consumed raw (Davidson, 1981: p. 197).

Psoedochama gryphina: A single valve of this species, with a weight of 3 g, was found in room 2 of house 5, adjacent to the main living room. It belonged to a small animal (table 4.21) (*cf.* Poppe & Goto, 1993: p. 84: 24–40 mm in diameter = height). *Psoedochama gryphina* lives with its right valve fixed to hard substrates below 10–60-m-deep water (Poppe & Goto, 1993: p. 84). Shells of this species are very numerous on the Amaliápolis and Almirós beaches today (table 4.22). Delamotte & Vardala-Theodorou (1994: p. 253) mention the species as occurring in central Greek waters. Sakellaríou (1957: pp. 157–158) found it in the gulf of Thessaloníki. The species is not consumed nowadays (Bini, 1965).

Acanthocardia tuberculata, red-nosed, rough or knotted cockle (Greek: methýstra, kydóni): In total, 22 valves or valve fragments of the red-nosed cockle weighing 93 g (table 4.6) were found in four houses. This cockle species lives in sand, mud or gravel substrates from the intertidal zone to 100-m-deep water (Poppe & Goto, 1993: p. 94). Shell fragments of the species are very common on the Amaliápolis and Almirós beaches at present (table 4.22). Delamotte & Vardala-Theodorou (1994: p. 254) mention the species as occurring in central, northern and western

Greek waters. Sakellaríou (1957: pp. 158–159) found the species in large numbers in the gulf of Thessaloníki. It is often eaten in Mediterranean countries, either raw or cooked in its shell until it opens (Davidson, 1981: p. 202). The consumed red-nosed cockles were small to medium-sized (table 4.21) (Poppe & Goto, 1993: p. 94: 30–90 mm in length, average 50 mm).

Cerastoderma glaucum (Poiret, 1789), *lagoon cockle (Greek: methýstra, kydóni)*. Valves or valve fragments of the lagoon cockle were found in each of the six houses (table 4.16). They are the most numerous mollusc species among the remains found at New Halos. In total, 188 valves or valve fragments were found, with a total weight of 371 g. The lagoon cockle lives in shallow water, especially on sandy or muddy substrates (Poppe & Goto, 1993: pp. 95–96). It prefers lagoonal conditions (Ivell, 1979; Nicolidou *et al.*, 1988). Delamotte & Vardala-Theodorou (1994: p. 255) mention the species only in relation to central and western Greek waters. Sakellaríou (1957: pp. 162–163) found the species in large numbers in the gulf of Thessaloníki. Lagoon cockle valves are not to be found on the Almirós and Amaliápolis beaches, but are found in large numbers in Soúrpi Bay (table 4.22).

The lagoon cockles consumed at New Halos were small, medium-sized and large (table 4.22) (*cf.* Poppe & Goto, 1993: p. 95: length of valve 20–35 mm, exceptionally 50 mm). The lagoon cockle shells from New Halos are significantly smaller than those from the Middle Bronze Age site of Magoúla Pavlína dating from *c.* 2000 BC that lies 2.5 km from New Halos (range of valve lengths 18.0–48.0 mm, mean 33.2 mm, n = 92; range of valve widths 17.2–45.5 mm, mean 31.6 mm, n = 144; in both cases p<0.001) (Prummel in press). This difference in size is to be attributed to a more intensive exploitation of lagoon cockles to feed the large population of Hellenistic New Halos in comparison with the less densely populated Magoúla Pavlína site.

The thickness of the valve walls, and hence the robustness of the valves, of the shells from the two sites also differ. The lagoon cockle valves from the Middle Bronze Age site are much more robust than those from Hellenistic New Halos. The wall thickness of the Middle Bronze Age lagoon cockle valves measured at the pallial line is approx. 3–4 mm, that of the Hellenistic ones approx. 2–3 mm. The disappearance, at the end of the Bronze Age, of the lagoon that lay near the Middle Bronze Age site (Van Straaten, 1988) may explain why the lagoon cockle valves from the Hellenistic period are less robust than those from the Middle Bronze Age.

Mactra stultorum (Linnaeus, 1758), *through shell* (*Greek: achivadáki*): Three valve fragments, with a total weight of 10 g, were found in house 3 (table 4.6), two of them in room 3, the third in room 6 adjacent to the main living room. The two measurable valves derive from medium-sized animals (table 4.21) (*cf.* Poppe & Goto, 1993: p. 101: 25–65 cm in length). The species lives in clean sand, from the low tide zone to 60-m-deep water (Poppe & Goto, 1993: p. 101; Sakellaríou, 1957: pp. 175–176). Valves of the species were not found on the Amaliápolis and Almirós beaches in 1993/94 (table 4.22). Delamotte & Vardala-Theodorou (1994: p. 255) mention the species as occurring in central Greek waters. Sakellaríou (1957: pp. 175–176) found the species in large numbers in the gulf of Thessaloníki.

Donacilla cornea (Poli, 1795), *wedge shell* (*Greek: kochíli*): Four complete valves, with a total weight of less than 2 g, were found in room 1 of house 3 and in the main living room of house 5 (three valves) (table 4.6). All the valves come from large animals (table 4.21) (*cf.* Poppe & Goto, 1993: p. 103: 10–15 mm in length, occasionally up to 28 mm). The wedge shell lives buried in coarse sand in the intertidal zone and just below it (Poppe & Goto, 1993: p. 103). Wedge shell valves are to be found in small numbers on the Pagasitic beaches today (table 4.22). Delamotte & Vardala-Theodorou (1994: p. 256) mention the species as occurring in central Greek waters. The species lives in the gulf of Thessaloníki in large numbers (Sakellaríou, 1957: pp. 179–180).

Callista chione (Linnaeus, 1758), *smooth venus* (*Greek: ghialisterí achiváda*): A complete valve of this species, with a weight of 13 g, was found in room 2 of house 5 adjacent to the main living room. It belonged to a medium-sized animal (table 4.21) (Poppe & Goto, 1993: p. 120). The smooth venus lives in fine, clean sand, from the extreme low tide line to 180-m-deep water (Poppe & Goto, 1993: p. 120). Valves of the species were found in large numbers on the two studied beaches of the Pagasitikós gulf (table 4.22). Delamotte & Vardala-Theodorou (1994: p. 262) men-

tion the species as occurring in central, northern and southern Greek waters. The species is found in small numbers in the gulf of Thessaloníki (Sakellaríou, 1957: pp. 170–171). The smooth venus is considered a very good species for consumption and is eaten raw (Davidson, 1981: p. 206). The small number of valves found in the houses of New Halos is striking.

Tapes decussatus, carpet-shell (*Greek: chávaro*): Two almost complete valves and 29 valve fragments, with a total weight of 9 g, were found in room 5 adjacent to the main living room of house 6 (29 fragments) and in the courtyard of house 3 (one fragment). The 30 valves and valve fragments found in room 5 of house 6 are six left valves, 11 right valves and 13 valves of unknown side. This means that at least 11 animals of the species were represented in this room. The four measurable valves came from small to medium-sized animals (table 4.21) (*cf.* Poppe & Goto, 1993: p. 124: 20–76 mm in length, which is 1.6 times the width of the valve). The carpet-shell lives in sand or muddy gravel in quiet waters, from below mid-tide level to water of a few m depth (Poppe & Goto, 1993: p. 124). Its valves are found in large numbers on the beaches of the Pagasitikós gulf today (table 4.22). Delamotte & Vardala-Theodorou (1994: p. 264) mention the species as occurring in central Greek waters. Sakellaríou (1957: pp. 173–174) characterises it as a common species of the Greek coasts and a well-known consumption species in Mediterranean countries. It is usually eaten raw, but is also cooked (Davidson, 1981: p. 204).

Venus verrucosa Linnaeus, 1758, *warty venus* (*Greek: kýdoni*): Valves or valve fragments of the warty venus were found in each of the six excavated houses of New Halos (table 4.6), in total 24 (almost) complete valves and 26 valve fragments, with a weight of 327 g. The largest number, 28, was found in house 6, in the main living room, the adjacent rooms 1 and 5, the courtyard and room 7. Three complete or almost complete valves from house 6 have intact rims. This is a strong indication that warty venuses were collected alive to be consumed. The rims of six (almost) complete valves from houses 1, 2 and 6 are slightly damaged. This damage, and the fracturing of other valves presumably occurred while the valves were buried in the soil. An almost complete valve with slightly worn rims shows several perforations caused by a natural agent, such as a predatory gastropod. This valve was definitely brought into the house empty. All the (almost) complete valves come from medium-sized animals (table 4.21) (Poppe & Goto, 1993: p. 126: 20–72 cm in length). The warty venuses consumed by the inhabitants of the six houses show hardly any differences in size (table 4.21).

The warty venus lives on all types of substrates, but especially on gravel, from the intertidal zone to water with a depth of about 100 m (Poppe & Goto, 1993: p. 126). Valves and valve fragments are found in large numbers on the beaches of the Pagasitikós gulf today (table 4.22). Delamotte & Vardala-Theodorou (1994: p. 264) mention the species as occurring in central Greek waters. The species lives in the gulf of Thessaloníki in large numbers (Sakellariou, 1957: pp. 168–189). It is frequently consumed in Mediterranean countries, usually raw (Davidson, 1981: p. 203).

4.10. Terrestrial gastropoda

Helix figulina: Shells of this gastropod species were found in houses 3 (six shells in the courtyard), 4 (room 6), 5 (main living room) and 6 (two shells in room 5 adjacent to the main living room). Their total weight is 4 g. The animals will have been consumed. *Helix figulina* is one of the edible species of the genus *Helix* that live in Greece. The species still occurs in the surroundings of the site today.

Lindholmia lens: Two shells of this species were found in house 3. Their total weight is less than 1 g. These shells are present-day intrusions among the remains of this house. The species is commonly found in this part of Greece today.

Helicella sp.: Shells of a *Helicella* were found in houses 3 and 5. Their total weight is less than 1 g. The shells are present-day intrusions. Numerous shells of this genus are to be found around the site today.

4.11. Worked bones

Animals were not only used for consumption in New Halos. Their bones were processed into artefacts. The small number of worked bones suggests that bone working was not an important activity in this town.

Knucklebones: A cattle astragalus found in house 2 contained a hole. The hole may have been filled with metal, but this is not certain. Reese (1985: pp. 387–388) gives a survey of astragali with holes dating from various periods that have been found in eastern Mediterranean countries and were filled with metal, usually lead. The aim of the metal filling was obviously to make the astragali heavier for use in games or divination. The holed cattle astragalus, a sheep astragalus from house 4 and two sheep astragali from house 5 show signs of use.

Beads and rings: Small beads made from materials of animal origin were found in house 1 (three), house 4 (one) and house 6 (one). A lenticular bead of what is thought to be ivory was found in the main living room of house 1 (inventory number X 332; not in Reinders, 1988: Catalogue). The bead is lathe-turned.

Marks of the turning are visible at the edge of the bead. A slightly smaller, flat bead was found in room 5 adjacent to the main living room of house 1 (inventory number X 78) (table 4.23). This bead was made from a mollusc shell (Reinders, 1988: Catalogue 35.38). A ring of bone from room 5 of house 1 (inventory number X 118; Reinders, 1988: Catalogue 35.37) could not be studied. It was of the same size as the shell bead found in the same room (Reinders, 1988: p. 261).

The two beads or small rings that were found in the main living room of house 3 (no inventory number) and room 2 of house 4 (inventory number HA 104/04) were made from the long bones of cattle/horse-sized mammals. Both beads are flat. The bead or ring that was found in house 4 is burned. These beads are all of the same size as the ivory bead from room 7 of house 1, but they are of a slightly different shape – they contain larger holes and are flat instead of lenticular.

The beads or rings from house 1 may have belonged to the same string. Four of the five beads or rings were found in a main living room or a room adjacent to a main living room. The strings of beads or rings may have been stored or used mainly in or near the main living rooms.

Game dish: A circular flat dish made from a bone from a cattle/horse-sized mammal was found in the courtyard of house 3 (inventory number HA 240/5). Its diameter is 15.7 mm, its thickness 5.0 mm. A vague dot is visible at the centre of one side, the other side contains two depressions, a large, deep central one and a smaller, shallower depression off-centre. The central dot and depression may have been formed in lathe-turning. The dish may have been used in a (board) game.

Unknown object: A worked fragment of a long bone of a sheep/goat/pig-sized mammal was found in room 7 of house 6 (inventory number H23 24). Its length is 29.4 mm, its height 9.1 mm, its weight less than 1 g. The object is superficially charred and broken.

4.12. Charred and calcined bones

Black, charred mammal and tortoise bones were found in houses 1, 4, 5 and 6 and white, calcined mammal bones in houses 3 and 6. Charred and calcined bones account for 7.2% of the vertebrate remains found in the six houses (table 4.24). Charring occurs in fires with temperatures of between 400 and 700°C. Bone calcines only in fires with temperatures of over 700°C (Lyman, 1994: pp. 384–391).

Among the charred and calcined bones are those of cattle, sheep, goat, red deer, roe deer, tortoise and unidentified mammals (table 4.24). None of the mol-

Table 4.24. Hellenistic New Halos. Numbers and proportions of charred or burned, and calcined animal remains. Nbr: number of charred or burned bones; Nca: number of calcined bones; %br: proportion of charred or burned bones among the bones of the concerned species in the given house (cf. table 4.6); %ca: proportion of calcined bones among the bones of the concerned species in the given house (cf. table 1). The proportions of charred/burned bones in the Total-column are valid for the six Halos houses (cf. table 4.6).

	House 1		House 3		House 4		House 5		House 6		Total		
	Nbr	%br	Nca	%ca	Nbr	%br	Nbr	%br	Nbr	Nca	%br+ %ca	Nbr+ Nca	%br+ %ca
Cattle	1	4.3	-	-	2	8.7	1	20.0	5	-	55.6	9	12.2
Sheep	-	-	-	-	2	28.6	1	50.0	15	-	78.9	18	52.9
Goat	-	-	-	-	1	100.0	-	-	2	-	66.7	3	50.0
Sheep/goat	-	-	-	-	3	3.5	-	-	4	-	16.7	7	4.0
Sheep+goat+sheep/goat	-	-	-	-	6	6.5	1	5.3	21	-	45.7	28	13.0
Red deer	-	-	-	-	3	75.0	-	-	-	-	-	3	30.0
Roe deer	-	-	-	-	-	-	-	-	-	2	100.0	2	100.0
Unidentified mammal remains													
of cattle/horse size	-	-	2	3.2	6	16.2	-	-	1	1	25.0	10	8.3
of sheep/goat/pig size	-	-	1	2.3	6	2.6	-	-	1	-	2.6	8	2.3
of unknown size	-	-	2	3.3	1	33.3	-	-	-	-	-	3	3.8
Tortoise	-	-	-	-	2	15.4	-	-	-	-	-	2	15.4
Total	1		5		26		2		28	3		65	
% of vertebrate remains		1.0		2.2		6.4		4.3			29.0		7.2

lusc shells shows any signs of having been in contact with fire. Shells are however transformed into lime at high temperatures, and thus become unrecognisable. The highest proportions of charred and calcined bones are of roe deer (100%) and red deer (30%). The proportions of charred bones of tortoise (15.4%), cattle (12.2%) and sheep and goat (with sheep/goat: 13%) are well above the proportion of burned bones among all the vertebrate remains (7.2%) (table 4.24).

The charred and calcined mammal bones include various skeletal elements: antler, carpus, femur, astragalus, metatarsus, phalanx 1 and phalanx 2 (table 4.25). More than half of the antler fragments (83.3%) and the astragali (63.3%) is charred or calcined. The proportions of charred or calcined carpus bones (16.7%), femur (8.3%), metatarsus (6.7%) and phalanx 1 (12.5%) and 2 (16.7%) are much lower (table 4.25). All the charred astragali are from cattle, sheep or goat. They were found in houses 4, 5 and 6. The burned antler fragments were found in houses 4 and 6 (table 4.25).[11] All the identified charred or calcined postcranial bones, except the femur, are feet elements.

There are three ways in which bones may turn black or white. The first is in a fire in a house, which will stain any bones in the rooms where the fire is. The second is when a joint of meat or part of it falls into the fireplace, *i.e.* in a cooking accident. Any bones in the joint may then be burned or calcined. The third is when bones are thrown into the fireplace after a meal, or when bones are deliberately burned, for instance to get rid of waste. Black or white staining of bone is never the direct result of the cooking or roasting of meat. Meat is cooked or roasted at a maximum temperature of 280°C. In roasting, the meat protects the bones from the flames.

A fire in the house is the least likely explanation for the burned bones, considering the small proportions of burned bones and the range of charred and calcined species and elements. It is improbable that only bones of certain species, and only certain elements should have been affected by a fire (tables 4.24 and 4.25). The distal-most ends of the legs of cattle, sheep and goat, comprising the phalanges, metapodia (*i.e.* metatarsus), carpus and tarsus (*i.e.* astragalus) bones, may have fallen from a spit into the flames and have thus been stained black or white. But in that case we would expect approximately equal numbers of charred or calcined metatarsi and astragali of the didactylous species cattle, sheep and goat and the number of charred or calcined phalanges 1, 2 and 3 to be about four times (forefeet and hindfeet) that of charred or calcined astragali. This was not the case in the houses of New Halos. Moreover, some of the astragali had been used before being burned. So cooking accidents can be excluded as the cause of the blackening of the charred cattle, sheep and goat astragali (table 4.25). The bones other than the astragali and the antlers probably became charred or calcined because they were thrown into the fireplace as waste. The finds concerned do however include a bead and a worked bone fragment of an unknown function.

Table 4.25. Hellenistic New Halos. Skeletal elements represented by the charred/burned and calcined bones of domestic mammals and wild mammals. Nbr: number of charred or burned bones; Nca: number of calcined bones; %br: proportion of charred or burned bones among the bones of the concerned skeletal element of each species in the given house; %ca: proportion of calcined bones among the bones of the relevant skeletal element of each species in the given house (cf. tables 4.12, 4.14-4.16 and 4.19). The last row shows the absolute and relative amount of charred/burned or calcined bones for each element for the remains of all species (including those without any burned bones and unidentified remains).

	Antler		Carpus		Femur		Astragalus		Metatarsus		Phalanx 1		Phalanx 2		Total for the species
	Nbr+Nca	%br+%ca	Nbr	%br	Nbr	%br	Nbr	%br	Nbr	%br	Nbr	%br	Nbr	%br	Nbr+Nca
House 1															
Cattle	-	-	1	25.0	-	-	-	-	-	-	-	-	-	-	1
House 4															
Cattle			-		-		2	100.0	-		-		-		2
Sheep			-		-		2	100.0	-		-		-		2
Goat			-		-		1	100.0	-		-		-		1
Sheep/goat			-		1	14.3	-		1	20.0	1	33.3	-		3
Red deer	3	100.0	-		-		-		-		-		-		3
House 5															
Cattle			-		1	100.0	-		-		-		-		1
Sheep			-		-		1	100.0	-		-		-		1
House 6															
Cattle			-		-		4	100.0	-		-		1	100.0	5
Sheep			-		-		15	93.8	-		-		-		15
Goat			-		-		2	100.0	-		-		-		2
Sheep/goat			-		-		4	80.0	-		-		-		4
Roe deer	2	100.0	-		-		-		-		-		-		2
Total for each element	5	83.3	1	16.7	2	8.3	31	63.3	1	6.7	1	12.5	1	16.7	

The cattle, sheep and goat astragali must have been exposed to fire deliberately. They were all burned superficially (charred). Cattle, sheep and goat astragali were commonly used as knucklebones in games, divination and other rituals in ancient Greece (Rohlfs, 1963). Rohlfs does not mention any deliberate exposure of knucklebones to fire. In her ethnographic study of the use of sheep and goat astragali in dice games and rituals among Mongolian shepherds, Kabzinska-Stawarz (1985: p. 241) however reports that some groups sacrificed astragali of their sheep by committing them to fire along with the kneecap, the Achilles tendon and fat. The purpose of this New Year sacrifice was 'to secure for the family a multiplication of every good'.

Burned or charred sheep and goat astragali were also found in the Kabirion near Thebai in Viotía (Boessneck, 1973), at the altar of Aphrodite Ourania in Athens (5th century BC) (Reese, 1989: p. 64) and at the Hellenistic sanctuary of Athena in Stymphalos on the Pelopónnisos (D. Ruscillo, personal communication). And burned cattle astragali were found in the Kabirion near Thebai in Viotía (Boessneck, 1973). These indisputably ritual contexts suggest that the charred astragali from houses 4, 5 and 6 of New Halos also served some ritual purpose.

The red deer and roe deer antlers were presumably likewise deliberately burned. They were found in two of the houses that also contained burned astragali (4 and 6). In house 6, the charred and calcined antlers lay in the same room (room 5) as the charred astragali. The burning of the antlers was possibly connected with that of the knucklebones.

4.13. Discussion

Two main types of meat were consumed in the main living rooms and the courtyards of the six houses of New Halos between 302 and 265 BC: meat of mammals and meat of molluscs. Fish and birds were either hardly ever or never eaten, or their bones were not preserved in the houses.

Mammals: Beef, mutton, goat's meat, meat of horse and ass and pork were regularly eaten. The proportions of consumed mammal meat were: beef 48%, mutton and goat's meat 29%, horse and ass meat 20% and pork 2% (table 4.7). The fact that equid bones, including bones of young animals, were found in four of the six houses implies that horse meat was not irregularly consumed. An occasional red deer, hare and badger supplemented the diet with some game.

The ranges of domestic mammals consumed in the six houses vary little in composition (tables 4.6–4.7; figs 4.12–4.13). The higher proportions of sheep and goat bones in houses 4, 5 and 6 will be mainly attributable to overrepresentation of astragali that were used as knucklebones in games and divination.

The substantial underrepresentation of cattle skull elements (table 4.12) means that cattle were not slaughtered in the houses. Postcranial parts of cattle were brought into the houses for consumption. The cattle may have been killed in two kinds of places. The first are ordinary slaughter places inside the town, for instance near or at an *agora*, or outside the town. The skulls, bearing only little meat, were not taken to the houses.

Sacrifices: The second possibility is that the cattle were killed at a sanctuary where animal sacrifices were made. From inscriptions and representations we know that animal sacrifice was the central act in Hellenistic religion, as in archaic and classical Greek religions (Nilsson, 1961: pp. 67–82; Burkert, 1977: pp. 101–105; Van Straten, 1995). Cattle, sheep, goat and pig were the main sacrificial animals. The sacrifices were made on an altar, which stood before or near a sanctuary. The ritual comprised two elements: the slaughtering of the animal and the shedding of its blood followed by the consumption of the meat. The entrails, the tail and parts of the meat and the skeleton were burned to enable them to reach the god or gods for whom the sacrifice was intended. Part of the meat and sometimes the animal's skin were for the priest. The rest of the meat was consumed by the participants of the sacrifice (Nilsson, 1961: pp. 67–82; Burkert, 1977: p. 103; Van Straten, 1995). Inscriptions forbidding people to take meat away from sanctuaries and prescribing that all meat was to be consumed there have come down to us from the classical period (Burkert, 1977: p. 103). From the fact that there was evidently a need for such bans we may presumably infer that meat from sacrifices was sometimes cooked and consumed at home.

Public sacrifices were a major source of mammalian meat for the inhabitants of Athens in the 4th century (Rosivach, 1994: pp. 65–67). Inscriptions from that period reveal that public sacrifices involved such large quantities of meat and were so regularly made that the citizens could consume mammalian meat every 8th or 9th day. Cattle, sheep and goats of various ages were the main sacrificial animals. Rosivach (1994: pp. 92–94) concluded that these species were hardly ever killed just for their meat, but almost always in a sacrificial ceremony. Any excess young animals and animals that were no longer productive were used as sacrificial animals. Rosivach (1994: p. 145) assumes that most of the meat was distributed among the population, especially the poorer part of it.

Inscriptions brought Jameson (1988) to the conclusion that there was a close link between animal husbandry and religion in classical Greece. In his opinion, the Greeks of that period obtained virtually all their mammalian meat from sacrificial animals. The public sacrifices increased in size under Hellenistic rule (Wes *et al.*, 1978: p. 161). This may imply that the proportion of meat from sacrificial animals in the diet even increased from the classical period to Hellenistic times.

Animal sacrifices will definitely have been made in a Hellenistic town like New Halos (Van Straten, 1993). The meat of the sacrificed animals contributed to the meals of the town's inhabitants. Part of it was consumed at or near the place where the sacrifice was made. Other parts may have been taken to the houses and consumed there. The horned cattle head, the *bucranium*, had a symbolic meaning in archaic and classical Greek religion (Burkert, 1977: p. 114). Cattle heads may have been kept or displayed in the sanctuary for ritual purposes. Where the sacrifices were made is not known; perhaps in the upper town of New Halos and at open-air sanctuaries, for instance in the Óthris mountains.

Two foetal sheep/goat bones found in the courtyard of house 3 could imply that a pregnant ewe or mother-goat was slaughtered in that courtyard in winter or early spring. If the foetal bones however come from a ewe or a mother-goat that suffered an abortion in the courtyard, this would imply that pregnant sheep and/or goats were kept in the courtyard in winter or spring. A soil sample from the courtyard of this house contained no mites, which are indicative of the faeces of domestic animals (Schelvis, this volume). From this we may conclude that animals were not regularly kept in this courtyard. So it's more likely that a pregnant ewe or mother-goat was killed in the courtyard of house 3 than that an animal stalled there aborted her lamb or kid. The underrepresentation of lower limb bones of sheep and goats (table 4.16) however suggests that most sheep and goats were killed at a central place, for instance a sanctuary, and were thus sacrificial animals, parts of whose meat were taken to the houses.

Animal products: The advanced age at which most of the sheep, goats and cattle were killed suggests that these animals were milked. Milk and cheese production will have been the main economic aim of the sheep and goat husbandry. Adult cattle, but not necessarily oxen, were used as draught animals for transporting loads and as plough-animals in the fields (contrary to Hodkinson, 1988, and Halstead, 1995). Some cow's milk may have been available for human consumption. The sheep were kept for their wool as well as for milk and meat. The ceramic strainers that were found in most of the houses and a mortar indicate the processing of milk into dairy products (see Ch. 3: 1). Most of the pigs were slaughtered before or shortly after they had reached maturity, as is common practice with a species that is bred for its meat. The horses and donkeys were used as saddle animals or draft animals before waggons, for either military or civilian purposes. Their meat was consumed. Dog meat was probably not consumed.

The fat, hides, skins and horny sheets of the animals were used, though evidence of such use is scarce. The skin and/or the horn of a billy goat were/was certainly used. The only artefacts made of such materials found in the houses are a perforated cattle astragalus (knucklebone), some beads or rings and a game dish. It is unlikely that bone or shell were worked in the studied houses as no semi-finished products or waste of such activities have been found. Bone was apparently not often used to make objects at New Halos. The same holds for Kassope, where the only worked bone object seemed to be a perforated cattle astragalus (Friedl, 1984: p. 36). This contrasts markedly with the evidence from the town of Mytilene on Lésvos, where an extensive bone and horn working industry flourished in Hellenistic times (Ruscillo, 1993: pp. 201–202). Goat horn is also known to have been processed in Hellenistic Eleutherna (Vila, 1994).

Astragali from sheep, goats and cattle were taken from the butchered animals and used as knucklebones in games, either pastime games or games with a ritual or magic meaning, for example in divination. The many charred knucklebones that were found together with charred red deer and roe deer antlers suggest that knucklebones and antlers featured in myths and divination in which fire played a role (Luther, 1987: p. 42). That knucklebones played a role in religion and in fire rituals in Hellenistic times is illustrated by the many sheep and goat astragali, several of which were burned, which were deposited as votive offerings in the Kabirion in Thebes (Boessneck, 1973: pp. 9–10).

About 4% of the domestic animals in the territory of the town of Hellenistic New Halos were equids, 5% were pigs, 23% were cattle and 68% were sheep and goats (compare table 4.6, dogs excluded). Sheep and goats were kept in a ratio of one goat to almost six sheep.

The town's large population (which is estimated to have been about 9,000) necessitated large-scale breeding of domestic animals for the production of milk, meat and other products. Who owned these animals and where were they kept? There were no areas inside the town where large flocks of sheep and goats and large herds of cattle may have been kept overnight (this in contrast to Hodkinson, 1988: p. 47).

In the early days of the town's short existence of about 35 years and possibly later, too, most of the inhabitants were soldiers or belonged to military families. They may have kept equids and dogs that stayed in the town. The moderate size of a dog whose remains were found in house 3 will have made it unsuitable for herding livestock and protecting it from wolves. The absence of mites indicative of dung (see above) means that no cattle, sheep, goats or pigs were kept in the houses. If inhabitants of New Halos possessed cattle, sheep, goats and pigs, their byres, sheepfolds and pens will have been located outside the town.

The sickles that were found in most of the houses and in large numbers in the House of the Geometric *Krater* may have been used either for gathering (cutting) wild vegetable plants for consumption or for harvesting arable products. Wild vegetables are still gathered in the surroundings of New Halos in spring and autumn today. Only two hoes, a spade or shovel and four shear halves were found in the houses (see Ch. 3: 4). This does not necessarily imply that the town's inhabitants practised animal husbandry or crop cultivation, especially as no arable farming implements like ploughs or plough shares were found in the houses.

It is more likely that the animals were owned by people who lived outside the town. A controversy exists between Halstead (1987) and Hodkinson (1988) on the one hand and Georgoudi (1974), Skydsgaard (1988) and Reinders (1994) on the other as to how the animal husbandry of sheep, goats, cattle and pigs in archaic, classical and Hellenistic Greece was organised. Halstead (1987) and Hodkinson (1988) are of the opinion that farmers practised mixed agro-pastoral farming in the archaic, classical and Hellenistic periods, that they worked on very small acreages where they practised predominantly crop cultivation and that they possessed only a few sheep, goats, cattle and/or pigs that were pastured on fallow and stubble and fed with waste of arable farming, grape and olive crops and even with cultivated fodder plants. These farmers are assumed to have had no time to move around with their animals in search of good pasture, for instance at higher elevations where green vegetation and water were available in hot seasons.

Herding: Georgoudi (1974), Skydsgaard (1988) and Reinders (1994) are on the other hand of the opinion that animal husbandry in classical and Hellenistic Greece frequently, but not necessarily always, involved the movement of herds of animals such as sheep, goats and cattle, but also pigs (for acorns; see Georgoudi, 1974: p. 180), between summer and winter pastures, depending on where the best pasture was available in each season. In the parts of the year when these people did not practise crop cultivation they will have had more time to move around with their animals to secure the best pastures. Various combinations of almost all the members of crop-cultivating families, including the children, may have tended the herds in various seasons. Animals of various owners may have been combined and tended together.

Skydsgaard (1988: p. 76) argued that this dispute is largely attributable to the absence of clear definitions. It is also my opinion until the two parties specify what they understand by a 'small' and a 'large' flock and by a 'small' and 'large' distance and difference in altitude, their statements concerning size and distance are meaningless. An argument against the theory revolving around farmers who practised exclusively mixed agro-pastoral farming with very small

herds is that mixed farming communities in Hellenistic times will not have had enough animals to meet the demand for wool, hides and sacrificial animals.

Whether the domestic animals in the territory of New Halos were produced by specialised pastoralists (Reinders, 1994) or by mixed farmers (Halstead, 1987; Hodkinson, 1988) cannot be readily inferred from their bones (Grant, 1991; Reinders & Prummel, 1998). Halstead (1996) sees Neolithic-Bronze Age animal bone assemblages of various domestic species in Greece as evidence for small-scale herding by mixed farmers. This will most probably be true as far as these periods are concerned, but it does not necessarily hold for the complex economies of the Hellenistic period. Alcock *et al.* (1994: p. 148) suggested that intensive mixed farming and extensive specialised pastoralism were practised in classical Greece side by side by different families, or alternately by the same families (see also Hodkinson, 1988: p. 42). This would mean that these methods of arable farming and animal husbandry did not exclude one another, contrary to what has been suggested in recent literature (Halstead, 1987; Hodkinson, 1988).

Pastures: Let's see what pastures were available for the various animal species in the territory of New Halos. The Almirós and Soúrpi plains and the coastal marsh afforded pasture for cattle, sheep and goats in the winter, spring and autumn (fig. 4.9). These plains were covered with an open parkland vegetation of grasses and herbs, arable fields and small woodlets comprising deciduous oaks and other trees (Bottema, 1988: p. 224; 1994: pp. 62–64). Pastureland for livestock in the plains will have included the grassland, the stubble fields and fallow arable fields. The fields were manured by the animals' droppings during the pasturing. The riverbeds in the plains and the flats along the coast nowadays dry up in hot summers. From the corrosive effects observable in pollen, Bottema (1988: p. 228) concluded that the flats dried up in at least some summers during the occupation of New Halos. There are only four perennial springs in the Almirós and Soúrpi plains and the foothills of the Óthris (pers. comm. H.R. Reinders) (fig. 4.9). So the riverbeds and coastal marsh will have provided only small amounts of food, if any, for animals in summer in Hellenistic times.

In dry periods water and pasture for the many cattle, sheep and goats that were needed for the population of the town of New Halos were to be found in the Óthris mountains, where there are perennial springs and streams carrying water throughout the year (fig. 4.9). The average July temperature in the Óthris mountains is below 25°C, that in the plains is above that temperature. Water, being the largest component of milk, is of essential importance in dairy farming. The lowest springs in the Óthris are to be found at 600 m altitude. This height is reached 9 km from the town of New Halos. Many springs and streams are to be found above 1200 m altitude (pers. comm. H.R. Reinders), which is reached about 12 km from the town of New Halos; an altitude of 1400 m is reached about 14 km from the site (after Reinders, 1994: fig. 4). Animals and their herdsmen will have been able to climb from the plains to the altitude of 600 m in about one day; it will have taken them three days to reach that of 1300 m.

The Mavrikopoúla pollen diagram, covering a span from *c.* 3000 to 0 BP (Woldring, this volume) shows that the greater part of the forest of deciduous (*Quercus-robur*-type species, in this case *Q. frainetto* and *Q. pubescens*) and sclerophyllous (*Q. coccifera*) oaks that lay at an altitude of approx. 1200 m in the Óthris mountains had been cut by the beginning of the Iron Age (the beginning of zone 1a in his diagram: spectrum 2). An open vegetation of solitary oaks, patches of bracken (*Pteridium*) and grasses (Gramineae) prevailed at that altitude during Hellenistic times (Woldring, this volume, approx. spectrum 5) (also: Bottema, 1988: p. 223). That vegetation might have been suitable pasture for cattle, sheep, goats and pigs in summer. The snow cover of the Óthris in winter will have precluded year-round grazing in the mountains (Georgoudi, 1974: pp. 169–170). So at the beginning of winter any herds that had spent the summer season in the mountains had to be brought back to the plains in one to three days. Only moderate use will have been made of the pastureland in the Óthris in the Hellenistic period and there will have been no question of any over-grazing. Indications of over-grazing at 1200 m altitude in the Óthris mountains are observable in the Mavrikopoúla pollen diagram only in the past 1000 years (Woldring, this volume, zone 2).

The herbaceous vegetations in the Almirós and Soúrpi plains contained the most suitable pastures for equids. The small woods in the plains and the Óthris with their various species of acorn-producing oaks (Bottema, 1988: pp. 223–224 and fig. 110) will have constituted suitable feeding grounds for pigs in the autumn. Pigs could however also be fed on household waste, so they may have been kept near the town, too.

Hunting: The citizens of New Halos hunted some game on the slopes of the Óthris or in the Almirós and Soúrpi plains. The small and large arrowheads found in the houses (see Ch. 3: 4) were possibly used in hunting. Red deer, roe deer, fox and badger will have found suitable habitats in the plains and the mountains. Brown hares were to be found in the plains. More or less the same small percentages of red deer and roe deer remains were found at 3rd-century Kassope in Ípiros and at 3rd–2nd-century Messene on the Pelepónnisos. The animal remains of Hellenistic Kassope and Messene also include a few bones of wild boar (*Sus scrofa*) (Friedl, 1984: p. 218; Nobis, 1994: p. 299). The antlers of red and roe deer were used for ritual purposes in New Halos.

Husbandry: Let's now see whether the animal husbandry practised in the territory of New Halos may be termed representative of that of Hellenistic Greece. The proportions of remains of pig, sheep and goat recovered from the third-century layers of house 5 of Kassope, the only area at this site where animal remains were collected in a stratigraphic context (Friedl, 1984: Tab. 63; Boessneck, 1986: Abb. 137), differ strikingly from those of New Halos. That of pig, being 25.3% of all the domestic mammal remains, is much higher than the 5.0% found for New Halos, whereas that of sheep and goat, 47.8%, is much lower than that of New Halos (66.8%). The proportions of remains of equids and cattle found at 3rd-century Kassope (3.0% and 20.5%, respectively) are about the same as those of New Halos (4.0% and 23.0%, respectively).

Pig husbandry was evidently far more important in the surroundings of Kassope in the 3rd century BC than in those of New Halos. This may be due to the higher levels of precipitation in Ípiros.[12] The surroundings of Kassope contained suitable habitats for both pigs and wild boars. Sheep and goat husbandry was much more important in 3rd-century-BC Thessaly than along the west coast of Greece.

The animal remains from the slightly younger Hellenistic occupation site in Eleutherna on the northern slope of Mount Psilorítis (Ídi) on Kríti (end of 2nd century–beginning of 1st century BC) include a very high proportion of sheep and goat remains: 73% of the domestic mammal remains excluding those of dog. Pig is well represented, in a proportion of 14% of the remains of the aforementioned species. Cattle and equids are each represented by 6% of the remains of domestic mammals excluding dog (Vila, 1994: Table 1).

The proportion of pigs at 3rd–2nd-century Messene in the western half of the Pelepónnisos, being 30–44% of the remains of cattle, sheep/goat and pig, is even higher than at Epirotical Kassope. Sheep and goat, which are represented by 15–28% of the remains of cattle, sheep/goat and pig (Nobis, 1994: p. 299), were far less numerous at Messene than at New Halos and Kassope. Cattle was very well represented in the 3rd–2nd-century layers of this site, in a proportion of 41–46% of the remains of cattle, sheep/goat and pig (Nobis, 1994: p. 299). The 3rd–2nd-century-BC animal remains of Messene were however recovered from areas around that site's temples of Artemis Ortheia and Asklepios (Nobis, 1994: p. 297). The proportions may consequently have been influenced by the sacrificial activities associated with these temples. They suggest that cattle were the most important sacrificial animals for the gods of these temples. So whether cattle were indeed more numerous than pigs and sheep/goats in the surroundings of Messene in the 3rd–2nd century BC is uncertain.

The proportions of remains found in the Hellenistic layers of the Kabirion, a sanctuary near Thebai in Viotía (13% cattle, 85% sheep/goat (one third of which were astragali) and 1% pig) (Boessneck, 1973:

p. 4), directly reflect the composition of the sacrificial animals and the votive offerings (astragali) and possibly indirectly the composition of the herds in that part of Boiotia in Hellenistic times.

In conclusion, the bones of domestic animals from Hellenistic layers at New Halos, Kassope, Eleutherna, Messene and Thebai represent strongly varying ranges of species. This variation may be caused by both differences in local environmental and climatic conditions and differences in the individual sites' socio-cultural or religious functions (houses *versus* sacrificial sites, *etc.*). More research is needed to identify patterns of animal husbandry and obtain a better understanding of the use of domestic animals in Hellenistic times.

The present composition of the livestock in the surroundings of New Halos differs substantially from that in Hellenistic times. In the 1980s, villagers of Áyos Ióannis, Vrínena, Kokotí, Kofí and Anávra together possessed 39,260 animals, which grazed in the Óthris mountains and the Almirós and Soúrpi plains. No less than 84.0% of these animals were sheep and goats, 13% were cattle and 3% were pigs. Pigs were only kept by inhabitants of Anávra. The cattle roamed free in the mountains and were not milked. Goat and sheep were kept in a ratio of 1:1.5 (Reinders & Prummel, 1998).

The much smaller proportion of cattle in the present stock may be partly associated with the present use of tractors for ploughing instead of the cattle that were used for this purpose in Hellenistic times. Goats are better adapted to mountainous habitats and to pastures containing many trees and shrubs than sheep, which prefer more supple plants. Nowadays, pastures for sheep and goats are in this area to be found almost exclusively in the mountains (Reinders, 1994). The plains have been virtually devoid of trees since about AD 1500 (Bottema, 1994: pp. 62–64). In the past fifteen years they have increasingly often been used for the cultivation of cotton alongside grain. Cotton cultivation precludes pasturing in winter. The much stronger preponderance of sheep over goats in Hellenistic times could mean that livestock keepers had better access to the plains and the coastal marsh than present-day shepherds. A second possible explanation for the much larger number of sheep than goats in Hellenistic times is a high demand for wool for clothing.

Molluscs: Molluscs constitute the second type of animal food that was regularly consumed in the six houses of New Halos. Ten marine gastropods, fifteen marine bivalves and a terrestrial gastropod have been identified. Four marine gastropods (*Patella caerulea, Cerithium vulgatum, Bolinus brandaris* and *Hexaplex trunculus*), twelve bivalves (*Arca noae, Glycymeris insubrica, Mytilus galloprovincialis, Pinna nobilis, Chlamys* cf. *glabra, Spondylus gaederopus, Ostrea edulis, Acanthocardia tuberculata, Cerastoderma glaucum, Callista chione, Tapes decussatus* and *Venus verrucosa*) and the terrestrial gastropod *Helix figulina* are well-known for the quality of their meat in Mediterranean countries nowadays (Bini, 1965; Davidson, 1981; Poppe & Goto, 1991; 1993). The meat of most of the other marine species may have been consumed too. Mollusc shells were used to make beads.

Nine of the 25 marine molluscs (5 gastropods, 4 bivalves) live on hard, rocky substrates, thirteen live on sandy, muddy or gravel substrates (3 gastropods, 10 bivalves) and the other three marine species (2 gastropods and a bivalve) may live on all kinds of substrates. As many as 21 of the identified marine species live or may live in shallow water: from the intertidal to water with depths of a few – 4 or 5 – metres. Thirteen of these species may also live in much deeper waters, with depths of up to 40, 100 or 200 m. Four species live only in deep water. The minimum depth of the waters in which these species live varies from 6 to 20 m (table 4.26). No less than 97.4% of the marine shells or shell fragments that were found in the six houses (NR = 480) derive from the 21 mollusc species that live or may live in shallow water. The remaining 2.6% are from the four species that live in waters with a minimum depth of between 6 and 20 m (tables 4.26, 4.27).

The marine molluscs were gathered in Soúrpi Bay (Órmos Soúrpis), which is part of the Pagasitikós gulf (Pagasítikos Kólpos) (fig. 4.9). The present length of Soúrpi Bay from north to south is 3 km, its present width from west to east is 2.5–3 km. The town of New Halos lay about 1.5 km to the west of the southern shore of Soúrpi Bay. This shore lay about 300 m further inland in Hellenistic times (Reinders, 1988: fig. 20). The water table of Soúrpi Bay was about 1.5 m lower than at present (Reinders, 1988: fig. 19).

The western and southern shores of Soúrpi Bay are gravelly, sandy and muddy. The bay's floor declines very gradually from the western shore in an easterly direction. Water with a depth of 5 m is to be found about 150 m from the present western shore, water with a depth of 10.8 m about 550 m from the shore. The southern part of the bay is even shallower. There, water with a depth of 5 m is to be found only about 325 m from the present southern shore and water with a depth of 10.8 m lies about 1125 m from the shore. The eastern shore of Soúrpi Bay along the Mitzéla peninsula is rocky and steep and has small gravel beaches. Water with a depth of 10.8 m is here to be found about 55 m from the coast. The deepest point of Soúrpi Bay, which is now 27.5 m, is to be found at the mouth of the bay, where it opens into the Pagasitikós gulf, close to its eastern shore. The shortest distance from this point to the present western shore near the site of New Halos is about 3.8 km (fig. 4.9).

The mollusc species collected in shallow water (ten or more preserved shells or shell fragments) are *Patella caerulea* and *Arca noae* from rocky shores,

Table 4.26. Hellenistic New Halos, marine molluscs. Break-down of demonstrated species by the type of habitat and the depth of water where they may occur (data after Poppe & Goto 1991; 1993).

Depth of water	Habitat			total	%
	rocky, hard	mud, sand or gravel	all bottoms		
Shallow, intertidal to a few m	3	5	-	8	32
Intertidal or 1 m, up to 40–100 m	3	5	2	10	40
Intertidal or 1 m, up to 200 m	1	2	-	3	12
From 6, 7 or 10, up to 50, 60 or 900 m	2	1	-	3	12
From 20 up to 80 m	-	-	1	1	4
Total	9	13	3	25	100
%	36	52	12	100	

Table 4.27. Hellenistic New Halos, marine molluscs. Break-down of shells or shell fragments by the type of habitat and depth of water where the species may occur (data after Poppe & Goto 1991; 1993).

Depth of water	Habitat			total	%
	rocky, hard	mud, sand or gravel	all bottoms		
Shallow, intertidal to a few m	43	237	-	280	56.8
Intertidal or 1 m, up to 40–100 m	4	97	52	153	31.0
Intertidal or 1 m, up to 200 m	38	9	-	47	9.6
From 6, 7 or 10, up to 50, 60 or 900 m	10	2	-	12	2.4
From 20 up to 80 m	-	-	1	1	0.2
Total	95	345	53	493	100.0
%	19.3	70.0	10.7	100.0	

Cerithium vulgatum, *Hexaplex trunculus*, *Acanthocardia tuberculata*, *Cerastoderma glaucum*, *Tapes decussatus* and *Venus verrucosa* from muddy, sandy or gravelly substrates, and *Ostrea edulis* from rocky or soft substrates. The shallow parts of the depth ranges of these species overlap: from the intertidal to waters with depths of 1 to about 5 m. This is presumably where most molluscs were collected for consumption. The area concerned measured about 2.5 km².[13] Most of this area lay in the southern part of the bay, so close to the town of New Halos. The molluscs will have been collected in this shallow water by dredging, picking and cutting (the last method will have been employed for *Arca noae* in particular). The fishermen will have waded through the shallow waters close to the shore and have dived or dredged up the molluscs from a boat on waters with depths of more than 2 m. A few empty shells were brought into the houses along with the molluscs intended for consumption.

The most numerous of the four species that live in waters with depths of more than 6 m is *Spondylus gaederopus*, which lives attached to rocky substrates in waters with depths of 7 to 50 m. These molluscs will have been cut from the rocks in waters with depths of between 7 and about 12 m, the maximum depth for diving without modern equipment (Poppe & Goto, 1991: p. 23). *Arca noae* may have been collected along with *Spondylus gaederopus* in these waters. *Tonna galea*, which lives in waters with depths of 20-80 m, shows that some molluscs living in deep waters were caught with nets. This species can be caught at the mouth of Soúrpi Bay. The part of Soúrpi Bay with waters with depths of 6 m or more had an area of about 4 km² in Hellenistic times.

In conclusion, the fishermen did not have to leave Soúrpi Bay to catch the mollusc species identified in the houses of New Halos. The fact that the fishermen went out to waters with depths of at least 20 m to catch some of the marine species implies that molluscs were a relatively important component of the diet. Other types of seafood that may have been consumed, but which rarely leave behind traces, are cephalopods, lobsters and sea-urchins. It is hard to believe that no, or virtually no fish will have been caught in the waters near New Halos. It could be that fish (class Pisces) was an expensive type of food, beyond the reach of the inhabitants of the six studied houses. Other possibilities are that the waste of fish meals was discarded away from the houses on account of its smell or that the remains have not survived. Fish remains were however rare or absent at all the Hellenistic sites studied so far (Boessneck, 1973; Friedl, 1984; Uerpmann *et al.*, 1992; Ruscillo, 1993; Nobis, 1994; Vila, 1994). Trantalidou (1990: pp. 402–403) suggested that fishing became important in Greece only in the Roman imperial period. In earlier days it was perhaps practised only in times of famine.

The majority of the molluscs consumed at New Halos were medium-sized, relative to the size range of the species concerned. Small molluscs were consumed in smaller numbers (but see Ch. 4: 4.9, *Cerastoderma glaucum*). Large individuals are rare. The preponderance of medium-sized and small molluscs reflects fishing in a population with a natural age composition: many young, small individuals, fewer medium-aged individuals and only a few old, large animals. Medium-sized molluscs were obviously preferred over small ones for consumption.

The proportions in which the marine mollusc shells were represented in the houses of New Halos correspond only partly to the frequency distribution of the mollusc shells that are to be found on the southwestern beaches of the Pagasitikós gulf nowadays (table 4.22). The most important difference between the frequencies of beach finds today and in the Hellenistic town of New Halos is the present absence of certain species (at least during the summers of 1993, 1994 and 1999) abundantly represented among the remains from the Hellenistic town (tables 4.6 and 4.22). Shells of the *Monodonta* species and *Callista chione*, which were not represented, or represented only in the form of a single valve at New Halos, are on the contrary quite numerous on these beaches nowadays. Why the last two edible species were not consumed in large(r) numbers in New Halos is uncertain. Perhaps the species were not as common in this part of the Pagasitikós gulf in Hellenistic times. Changes in water conditions between the Hellenistic period and the present may be responsible for these differences. The other species commonly found nowadays that were not identified at New Halos (table 4.22) are on the whole small, and fairly unattractive for consumption. Several of the species found among the remains of New Halos and on the present beaches have been living in the Pagasitikós gulf since at least 7000 BP (Van Straaten, 1988).

Molluscs constituted a slightly smaller part of the diet at Epirotical Kassope. 37% of all the animal remains found at New Halos were mollusc shells as against 21% at Kassope (Friedl, 1984: pp. 197–199; but for the Hellenistic and Roman periods combined). Various marine and two terrestrial gastropods, various marine bivalves and a freshwater bivalve were consumed at Kassope. Various marine gastropods and

bivalves were consumed at Mytilene on Lesbos, which was occupied from the late classical period to Byzantine times (Ruscillo, 1993).

Marine molluscs, excluding cephalopods, are not fished by commercial fishermen in the waters near New Halos nowadays. A few inhabitants of Amaliápolis, a short distance from the site of New Halos, collect *Monodonta* gastropods from the rocks for consumption (pers. comm. Judith Dimitríou-Hoogendijk, Amaliápolis). Collecting marine gastropods and bivalves was not a commercial fishing activity in Greek waters at the end of the 19th century (Apostolidès, 1883). The terrestrial *Helix figulina* is still frequently collected for consumption in the Almirós and Soúrpi plains nowadays.

4.14. Acknowledgements

R. Moolenbeek of the Zoological Museum (Department of Malacology) of the Institute for Systematics & Biology of Populations of the University of Amsterdam assisted in the identification of the molluscs.

4.15. Notes

1. Age data on equids: a distally unfused metatarsus, an erupted M^1 or M^2, a femur and a tibia, both proximally fused, and a distinctly worn M^1 or M^2.
2. Dental age data on pig: an almost erupting M_3 and an erupting M_3.
3. Epiphyseal age data on pig: a distally unfused metacarpus IV, an unfused calcaneum and a proximally unfused femur from pigs that were aged younger than 24 months, 24–30 and 42 months, respectively.
4. One of four cattle astragali from house 6 is a left one.
5. Age data on sheep: an astragalus of a juvenile sheep, a distally fused humerus, a distally fused tibia and the proximally fused calcaneus. Age data on goat: a distally fused humerus and a proximally fused radius (compare table 4.18, in which the age data on sheep, goat and sheep/goat have been combined).
6. The dental and epiphyseal age data differ in the proportions of sheep and goats killed when they were less and more than 2 years old (dental age data: 13% less than 2 years old, 87% more than two years old (table 4.17); epiphyseal age data: 25% less than 2 years old, 75% more than two years old or even more than 3½ years old (table 4.18)). The slightly different age classes employed in dental and epiphyseal age analyses may be the main cause of this difference. The ages of the sheep that have been classified as having been killed when they were more than 2 years old on the basis of dental evidence may have been slightly overestimated as loose molars from broken maxillae and mandibles of adult sheep or goats were counted separately (table 4.17).
7. The sheep astragali from New Halos are a little smaller than those from Kassope. However, most sheep astragali from New Halos are charred and some are worn due to handling. These factors will have led to some reduction in the size of the bones.
8. The inventory numbers of the carapace fragments are HA 115/12, HA 150/15 (two fragments), HA 150/14 (six fragments), HA 150/20, HA 178, HA 178/10. The latter is from an

abdominal carapace. The inventory number of the pelvis fragment is HA 115/3.

9. They are fragments HA 115/15, HA 178 and HA 178/10.

10. The English and Greek mollusc names are after Davidson, 1981; Bini, 1965; Reese, 1982 and Reese, 1992.

11. The burned cattle carpus bone from house 1 was found in room 4 of that house. The calcined unidentified mammal bones from house 3 were found in the main living room, room 1 and the courtyard. Four charred bones from house 4 were found in room 2 (the goat and a cattle astragalus – the goat astragalus is worn – and a burned bead made from an unidentified bone), seven in room 3 (two red deer antler fragments, the sheep/goat femur and metatarsus and two unidentified remains), three in room 4 (a sheep astragalus, the sheep/goat phalanx 1, and an unidentified bone), five in room 5 (unidentified bones), three in room 6 (the tortoise carapace fragments and an unidentified bone), three in open area 7 (unidentified bones); which room yielded a third burned antler fragment is not certain. The burned cattle femur from house 5 was found in room 4 adjacent to the main living room; the used and burned sheep astragalus was found in the main living room itself. Twenty-eight of the charred or calcined bones from house 6 were found in room 5 (the two roe deer antlers, all the burned astragali from this house and the cattle phalanx 2); a burned and a calcined unidentified bone were found in the courtyard, the other burned bone, which was also worked, in room 7 of house 6 (cf. table 4.25).

12. Present-day levels of precipitation in Kassope are twice to four times as high as those in Thessaly.

13. The facts that in Hellenistic times the water table was about 1.5 m lower than today and that the shoreline lay 300 m further to the west were taken into account in calculating this figure.

4.16. References

ALCOCK, S.E., J.F. CHERRY & J.L. DAVIS, 1994. Intensive survey, agricultural practice and the classical landscape in Greece. In: I. Morris (ed.), *Classical Greece: ancient histories and modern archaeologies.* Cambridge University Press, Cambridge, pp. 137–170.

APOSTOLIDÈS, N.C., 1883. *La Pêche en Grèce.* Ministry of Inner Affairs of Greece, Athens.

BINI, G., 1965. *Catalogue of Names of Fishes, Molluscs and Crustaceans of Commercial Importance in the Mediterranean.* Vito Bianco Editore, Milan.

BOESSNECK, J., 1973. *Die Tierknochenfunde aus dem Kabirenheiligtum bei Theben (Böotien).* Munich.

BOESSNECK, J., 1986. Zooarchäologische Ergebnisse an den Tierknochen- und Molluskenfunden. In: W. Hoepfner & E.-L. Schwandner (eds), *Haus und Stadt im klassischen Griechenland* (= Wohnen in der klassischen Polis 1). Deutscher Kunstverlag, Munich, pp. 136–140.

BOTTEMA, S., 1988. A reconstruction of the Halos environment of the basis of palynological information. In: H.R. Reinders (ed.), *New Halos, a Hellenistic town in Thessalía, Greece.* Hes Publishers, Utrecht, pp. 216–226.

BOTTEMA, S., 1994. The prehistoric environment of Greece: A review of the palynological record. In: P.N. Kardulias, *Beyond the site. Regional studies in the Aegean Area.* University Press of America, Lanham, New York and London, *etc.,* pp. 45–68.

BURKERT, W., 1977. *Griechische Religion der archaischen und klassichen Epoche.* Verlag W. Kohlhammer, Stuttgart, *etc.*

BÜTZLER, W., 1986. Cervus elaphus Linnaeus, 1758 – Rothirsch. In: J. Niethammer & F. Krapp (eds), *Handbuch der Säugetiere Europas* 2/II. AULA-Verlag, Wiesbaden, pp. 107–139.

CORBET, G.B. & S. HARRIS, 1991. *The Handbook of British Mammals,* 3rd ed. Blackwell Scientific Publications, Oxford, *etc.*

CRIEL, D., A. ERVYNCK & J. STEVENS, 1997. *De Das in Vlaanderen: een Verhaal in Zwart en Wit.* Van de Wiele, Brugge.

DAVIDSON, A., 1981. *Mediterranean Seafood,* 2nd ed. Penguin Books, London, *etc.*

DELAMOTTE, M. & E. VARDALA-THEODOROU, 1994. *Kochulia apo tis Ellinikes Thalasses (Shells from the Greek Seas).* Mouseío Goulandrí Phusikís Istoriás, Kifisiá.

DIJKSTRA, Y., 1994. De opgravingscampagne 1993 in Hellenistisch Halos. *Paleo-aktueel* 5, pp. 65–68.

ENGELMANN, W.-E., J. FRITZSCHE, R. GÜNTHER & F.J. OBST, 1986. *Lurche und Kriechtiere Europas.* Ferdinand Enke Verlag, Stuttgart.

FRIEDL, H., 1984. *Tierknochenfunde aus Kassope/Griechenland (4.–1. Jh.v.Chr.).* Doctoral thesis Munich.

GEORGOUDI, S., 1974. Quelques problèmes de la transhumance dans la Grèce ancienne. *Revue des Études Grecques* 87, pp. 155–185.

GRANT, A., 1991. Identifying and understanding pastoralism and transhumance: an archaeozoological approach. *Rivista di Studi Liguri* 57, pp. 13–20.

HAAGSMA, M.J., 1991. Halos 1991. A preliminary report. *Newsletter of the Netherlands Institute at Athens* 4, pp. 1–12.

HABERMEHL, K.-H., 1985. *Altersbestimmung bei Wild- und Pelztieren,* 2nd ed. Paul Parey, Hamburg, Berlin.

HALSTEAD, P., 1987. Traditional and ancient rural economy in Mediterranean Europe: plus ça change? *Journal of Hellenic Studies* 107, pp. 77–87.

HALSTEAD, P., 1995. Plough and power: the economic and social significance of cultivation with the ox-drawn ard in the Mediterranean. *Bulletin Sumerian Agriculture* 8, pp. 11–22.

HALSTEAD, P., 1996. Pastoralism or household herding? Problems of scale and specialization in early Greek animal husbandry. *World Archaeology* 28, pp. 20–42.

HODKINSON, S., 1988. Animal husbandry in the Greek polis. In: C.R. Whittaker (ed.), *Pastoral economies in classical antiquity.* The Cambridge Philological Society, Cambridge, pp. 35–75.

IVELL, R., 1979. The biology and ecology of a brackish lagoon bivalve, *Cerastoderma glaucum* Bruguière, in Lago Lungo, Italy. *The Journal of Molluscan Studies* 45, pp. 364–382.

JAMESON, M.H., 1988. Sacrifice and animal husbandry in classical Greece. In: C.R. Whittaker (ed.), *Pastoral economies in classical antiquity.* The Cambridge Philological Society, Cambridge, pp. 87–119.

KABZINSKA-STAWARZ, I., 1985. Mongolian games of dice. Their symbolic and magic meaning. *Enthnologia Polonia* 11, pp. 237–263.

KELLER, O., 1913. *Die Antike Tierwelt II.* Wihelm Engelmann, Leipzig.

KRAPP, F., 1982. *Microtus nivalis* (Martins, 1842) – Schneemaus. In: J. Niethammer & F. Krapp (eds), *Handbuch der Säugetiere Europas* 2/I. Akademische Verlagsgesellschaft, Wiesbaden, pp. 261–283.

LEHMANN, E. VON & H. SÄGESSER, 1986. *Capreolus capreolus* Linnaeus, 1758 – Reh. In: J. Niethammer & F. Krapp (eds), *Handbuch der Säugetiere Europas* 2/II. AULA-Verlag, Wiesbaden, pp. 233–268.

LÜPS, P. & A.I. WANDELER, 1993. *Meles meles* (Linnaeus, 1758) – Dachs. In: J. Niethammer & F. Krapp (eds), *Handbuch der Säugetiere Europas* 5/II. AULA-Verlag, Wiesbaden, pp. 856–906.

LUTHER, M.H., 1987. *Hellenistic Religions: an Introduction.* Oxford University Press, New York, *etc.*

LYMAN, R.L., 1994. *Vertebrate Taphonomy*. Cambridge University Press, Cambridge.

NICOLAIDOU, A., F. BOURGOUTZANI, A. ZENETOS, O. GUELORGET & J.-P. PERTHUISOT, 1988. Distribution of Molluscs and Polychaetes in Coastal Lagoons in Greece. *Estuarine, Coastal and Shelf Science* 26, pp. 337–350.

NIETHAMMER, G., 1942-1949. Über das Schicksal des Peloponnesos-Rehes. *Zeitschrift für Säugetierkunde* 17, pp. 123–126.

NIETHAMMER, J., 1982a. *Microtus guentheri* Danford et Alston, 1880 – Levante-Wühlmaus. In: J. Niethammer & F. Krapp (eds), *Handbuch der Säugetiere Europas* 2/I. Akademische Verlagsgesellschaft, Wiesbaden, pp. 331–339.

NIETHAMMER, J., 1982b. *Microtus thomasi* Barrett-Hamilton, 1903 – Balkan-Kurzohrmaus. In: J. Niethammer & F. Krapp (eds), *Handbuch der Säugetiere Europas* 2/I. Akademische Verlagsgesellschaft, Wiesbaden, pp. 485–490.

NILSSON, M.P., 1961. *Geschichte der Griechischen Religion II, Die Hellenistische und Römische Zeit*[2] (= Handbuch der Altertumswissenschaft 5,2). C.H. Beck'sche Verlagsbuchhandlung, München.

NOBIS, G., 1994. Die Tierreste aus dem antiken Messene – Grabung 1990/1991. In: M. Kokabi & J. Wahl (eds), *Beiträge zur Archäozoologie und prähistorischen Anthropologie* (= Forschungen und Berichte zur Vor- und Frühgeschichte in Baden-Württemberg 53). Konrad Theiss Verlag, Stuttgart, pp. 297–313.

POPPE, G.T. & Y. GOTO, 1991. *European Seashells I (Polyplacophora, Caudofoveata, Solenogastra, Gastropoda)*. Verlag Christa Hemmen, Wiesbaden.

POPPE, G.T. & Y. GOTO, 1993. *European Seashells II (Scaphopoda, Bivalvia, Cephalopoda)*. Verlag Christa Hemmen, Wiesbaden.

PRUMMEL, W., in press. Middle Bronze Age and Hellenistic mollusc meals from Soúrpi Bay (Thessaly, Greece). *Bulletin of the Florida Museum of Natural History*.

REESE, D.S., 1982. The molluscs from Bronze Age to post-geometric Asine in Argolid, Greece. In: S. Dietz (ed.), *Asine II. Results of the excavations east of the Acropolis 1970–74*, Fasc. 1 (= Skrifter utg. av Svenska Institutet i Athen, 4⁰, 24:1). Stockholm, pp. 139–142.

REESE, D.S., 1985. The Kition astragali. Appendix VIII(C) in V. Karageorghis, *Excavations at Kition 5.2*. Nicosia, Department of Antiquities, pp. 382–391.

REESE, D.S., 1989. Faunal remains from the altar of Aphrodite Ourania, Athens. *Hesperia* 58, pp. 63–70.

REESE, D.S., 1992. Recent and fossil invertebrates (with a note on the nature of the MHI fauna). In: W.A. McDonald & N.C. Wilkie (eds), *Excavations at Nichoria in Southwest Greece* II. *The Bronze Age occupation*. The University of Minnesota Press, Minneapolis, pp. 770–778 and pp. 964–965.

REINDERS, H.R., 1988. *New Halos, a Hellenistic Town in Thessalía, Greece*. Hes Publishers, Utrecht.

REINDERS, H.R., 1989. Het onderzoek van een huis in Hellenistisch Halos. *Paleo-aktueel* 1, pp. 48–52.

REINDERS, H.R., 1994. *Schaarse Bronnen*. Inaugural lecture, University of Groningen.

REINDERS, H.R. & W. PRUMMEL, 1998. Transhumance in Hellenistic Thessaly. *Environmental Archaeology* 3, pp. 81–95.

ROHLFS, G., 1963. *Antikes Knöchelspiel im Einstigen Grossgriechenland*. Max Niemeyer Verlag, Tübingen.

ROSIVACH, V.J., 1994. *The System of Public Sacrifice in Fourth-Century Athens* (= American Classical Studies 34). Scholars Press, Atlanta, Georgia.

ROWLEY-CONWY, P., 1994. Dung, dirt and deposits: site formation under conditions of near-perfect preservation at Qasr Ibrim, Egyptian Nubia. In: R. Luff & P. Rowley-Conwy (eds), *Whither environmental archaeology?* (= Oxbow Monograph 38). Oxbow Books, Oxford, pp. 25–32.

RUSCILLO, D., 1993. Faunal remains from the Acropolis site, Mytilene. *Echos du Monde Classique/Classical Views* 37, pp. 201–210.

SAKELLARÍOU, H.G., 1957. Les mollusques vivants du golfe de Thessaloniki et leurs contribution à la stratigraphie. *Annales Géologiques des Pays Helléniques*, 1e série, 8, pp. 135–221.

SCHELVIS, J., this volume.

SKYDSGAARD, J.E., 1988. Transhumance in ancient Greece. In: C.R. Whittaker (ed.), *Pastoral economies in classical antiquity*. The Cambridge Philological Society, Cambridge, pp. 75–86.

STORCH, G., 1982. *Microtus majori* Thomas, 1906. In: J. Niethammer & F. Krapp (eds), *Handbuch der Säugetiere Europas* 2/I. Akademische Verlagsgesellschaft, Wiesbaden, pp. 452–462.

STRAATEN, L.M.J.U. VAN, 1988. Mollusc shell assemblages in core samples from ancient Halos (Greece). In: H.R. Reinders (ed.). *New Halos, a Hellenistic town in Thessalía, Greece*. Hes Publishers, Utrecht, pp. 227–235.

STRATEN, F.T. VAN, 1993. Images of Gods and Men in a changing society: Self-identity in Hellenistic Religion. In: A. Bullock, E.S. Gruen, A.A. Long & S. Stewart (eds), *Images and ideologies, self-definition in the Hellenistic world*. University of California Press, Berkeley, pp. 248–284.

STRATEN, F.T. VAN, 1995. *Hierà Kalá: Images of Animal Sacrifice in Archaic and Classical Greece*. Brill, Leiden.

TRANTALIDOU, C., 1990. Animals and human diet in the prehistoric Aegean. In: D.A. Hardy (ed.), *Thera and the Aegean World III. Vol. 2: Earth Sciences*. The Thera Foundation, London, pp. 392–405.

UERPMANN, H.P., K. KÖHLER & E. STEPHAN, 1992. Tierreste aus den neuen Grabungen in Troia, Erster Bericht. *Studia Troica* 2, pp. 105–121.

UERPMANN, M., 1972. Archäologische Auswertung der Meeresmolluskenreste aus der westphönizischen Faktorei von Toscanos. *Madrider Mitteilungen* 13, pp. 164–171.

VILA, E., 1994. Les vestiges osseux animaux de l'habitat hellénistique d'Eleutherna. In: T. Kalpaksis, A. Furtwängler & A. Schnapp (eds), *Eleutherna, tomeas II, 2. Ena ellinistiko spiti ("Spiti A") sti thesi Nisi* (contributions in Greek, German and French). Ekdosis Panepistimiou Kritis, Rethumno, pp. 193–209.

WANDELER, A.I. & P. LÜPS, 1993. *Vulpes vulpes* (Linnaeus, 1758) – Rotfuchs. In: J. Niethammer & F. Krapp (eds), *Handbuch der Säugetiere Europas* 5/I. AULA-Verlag, Wiesbaden, pp. 137–193.

WES, M.A., H.S. VERSNEL & E.CH.L. VAN DER VLIET, 1978. *De Wereld van de Oudheid*. Wolters-Noordhoff, Groningen.

5. ANALYSIS OF THE ARTHROPOD REMAINS

JAAP SCHELVIS

5.1. Introduction

The study of arthropod remains from archaeological contexts is relatively new. During the 1960s British researchers discovered that remains of quaternary insects in general, and beetles (Coleoptera) in particular, are good indicators of past climates and environments (Coope, 1967). At first, arthropod remains were considered exclusively in geological contexts, but their value in archaeological research was soon acknowledged, too, again primarily by British researchers, in this case Buckland (1974), Kenward (1974), Osborne (1973) and others. Since then many researchers all over the world have adopted this 'new' science and quaternary entomology and entomo-archaeology have become established disciplines (Elias, 1994). At first only beetles were considered, but nowadays many other groups of arthropods are successfully used to reconstruct past ecological conditions. One of these groups is the relatively unknown one of mites (*Acari*) within the Arthropoda phylum (fig. 4.15). The only representatives of this group known to the general public are the small number of species which have an adverse effect on human health and welfare such as ticks, chiggers, scabies mites and household-dust mites. This has led to the false impression that all mites are harmful parasites. Most of the tens of thousands of species concerned do however not live parasitically, and in fact play an important role in essential processes such as biodegradation and soil formation.

Mites are closely related to better known groups within the class of Arachnida such as spiders, harvestmen and scorpions. Mites range in size from 0.1 mm to approx. 3.0 mm, although ticks swollen with the blood of their hosts may attain lengths of over 1 cm. Besides by their small size, mites are characterised by the absence of spinnerets, poison glands and body segmentation. Mites, like all arthropods, have an external skeleton which in their case consists of chitin, a highly resistant organic matter. Mites are moreover wingless, which is of great importance with respect to archaeological reconstructions, as it implies that their remains generally reflect strictly local ecological conditions.

Mites are one of the most successful groups of arthropods in terms of species diversity and distribution. Mites are to be found in practically every available habitat from the arctic to the tropics and from the highest mountain tops to the deep sea. Tens of thousands of species have evolved, which are all more or less critical in the choice of their habitats. This evolution however took place over a long period of time and the rate of morphogenesis in certain groups of mites is indeed very low. The oldest known fossil of a mite was found near Gilboa, New York, in a middle Devonian deposit with an age of more than 375 million years (Shear *et al.*, 1984). The most amazing aspect of this find is not its old age but the fact that the mite could be identified as a representative of the extant family of Ctenacaridae.

The earliest records of subfossil mite remains date from the mid-19th century and refer to finds in amber (Koch & Berendt, 1854). Although mite remains in quaternary deposits were described by Nordenskiöld in 1901 already, their role as ecological in-

dicators has been recognised only in the past few decades. Earlier reports usually consist only of a list of species without further ecological data. It was only when Coope (1967) discussed the use of insect faunas in paleoclimatological reconstructions that the first attempts were made to use mite remains as tools in quaternary paleoecology. Some researchers doubt the value of mites as paleoecological indices (Taylor & Coope, 1985), whereas others, such as Krivolutsky & Druk (1986) and Erickson (1988), are enthusiastic about their potential.

The first serious attempts at using mites in archaeological analyses were made by Denford (1976), who identified the mite remains found in Roman deposits in York (UK). These pioneer studies, as well as the Dutch research they inspired (Schelvis, 1992a; 1992b), were based on the good preservation conditions which are usually found at northwestern European waterlogged archaeological sites. In more arid areas conditions for the preservation of chitinous remains are considerably less favourable. Arillo *et al.* (1992) however discovered (identifiable) mite remains in Spanish rock shelters and cave deposits dating from as early as 34,800 BP.

In order to study the possibilities of more detailed analysis of the invertebrate remains found at Mediterranean archaeological sites, the department of archaeology of Groningen University launched a pilot study in 1992. This study was carried out by Sanz Breton of the *Universidad Autónoma de Madrid* during an Erasmus exchange programme. He studied samples from two sites in Spain: the fortified Bronze Age settlement Peñalosa in the province of Jaén and the Iron Age settlement Cerro de la Cruz near Almedinilla in the province of Córdoba.

The results of this methodological pilot study (Sanz Breton & Schelvis, 1994) indicate that the extraction techniques developed for arthropod remains in northwestern Europe can also be used in Mediterranean contexts. The study furthermore showed that allowance must be made for the possibility of (sub)recent contamination by burrowing representatives of the Dichosomata cohort in interpreting archaeological mite faunas from such arid deposits (fig. 4.16). The overall conclusion, however, was that the study of arthropod remains found at Mediterranean archaeological sites can yield as much valuable information to enhance our understanding of the past as it does in northwestern Europe. Subsequent studies (Schelvis, unpublished results) of remains from Byzantine deposits in Turkey and Egypt have further corroborated this conclusion.

5.2. Aim, material and methods

During the 1993 excavations conducted by the department of archaeology of Groningen University at the Hellenistic town of New Halos several samples were taken to assess the potential of arthropod research at this particular site. A 500-g sample taken from the fill of a pit in the courtyard of the House of the Snakes yielded some remains of arthropods. As at the Spanish sites, a relatively large number of representatives of burrowing species were found. However, since they were readily identified as such and the sample contained small numbers of other arthropod remains, too, it was decided to study a larger sample from the same feature. To minimise the risk of recent contamination, a second sample (No. 266) was taken from the deepest deposits in the same pit from which the first had been taken. The origin and function of this somewhat enigmatic pit were unknown, but it was considered not to be a well (see chapter 2).

Ethnographic, palynological and osteo-archaeological results led Reinders & Prummel (1998) to assume that the inhabitants of New Halos practised transhumance, in other words that specialised pastoralists took their herds into the nearby Óthris mountains during the hot, dry summer. This will most probably indeed have been essential, as the relatively high age of death of the cattle, sheep and goats demonstrated by the bones strongly suggests that these species were kept primarily for milk production. Pigs and horses may however have been kept in the town for part of the year at least. Although not a single bird bone was found in the six excavated houses this does not necessarily imply that the inhabitants of Halos did not keep poultry in their backyards. The aim of the study of arthropods from the pit was to find out whether livestock or poultry had been kept in the backyard of the house concerned. If they had, this would most probably be revealed by the arthropod death assemblage, by the presence of either remains of specific ectoparasites or of species which are considered to be characteristic of droppings of domestic mammals or poultry.

All the arthropod remains were extracted from the sample using an adapted version of the Paraffin Flotation Method described by Kenward *et al.* (1980). Because of their small size, mite remains have to be retrieved using a sieve with very fine meshes of 106 μm. After the flotation the remains were separated from the small amount of botanical material under a low-power stereomicroscope. The arthropod remains were subsequently transferred to a strong (80%) lactic acid solution. This solution not only preserves the remains, but also makes the exoskeleton more or less transparent, which is essential for identification. The specimens were then individually studied under a light microscope at a magnification of at least 400x using Grandjean's half-open-slide technique described by Balogh & Mahunka (1983). The mite remains were identified on the basis of Siepel (in prep.) for the Oribatida and Karg (1971; 1989) for the Gamasida. Since subfossil mite remains are usually incomplete when found, and mouth parts and legs with identifi-

able features are generally missing, the use of a reference collection is essential in the identification process.

The most important group of mites in paleoecology and archaeology is usually that of the order of Oribatida. Oribatids are small soil dwellers which may occur in very high densities of up to 400,000 individuals per square metre in forest soils. This order of mites has been studied in relatively close detail in the past century. More than 7000 species have been described and the ecological preferences of many of them have been extensively studied. This makes Oribatid remains a powerful tool in the reconstruction of local natural environments (Schelvis, 1990).

The interpretation of remains of Gamasida (mesostigmatic predatory mites) in archaeological samples has proved to be more problematic as the ecological information available on this group was obtained predominantly in research carried out in natural habitats. Acarologists studying gamasids had apparently preferred to study them in a nature reserve rather than in a pigsty, cesspit or some other foul place. But as archaeologists are often interested in the whereabouts of livestock and their excrement, I have developed a method for identifying archaeological dung deposits on the basis of remains of gamasid predatory mites (Schelvis, 1992b). In natural deposits or archaeological samples devoid of remains of dung-indicating gamasids the predatory mites must be interpreted on the basis of the ecological data on the individual species available in the literature, for example in Karg (1971; 1989).

5.3. Results

The method developed for extracting arthropod remains from samples from archaeological sites in northwestern Europe, the Paraffin Flotation Method, indeed proved to be efficient in extracting the chitinous remains from the sample from this arid Mediterranean site. However, the density of the arthropod remains was significantly lower than what is usually found at archaeological sites in northwestern Europe. The question whether this is a general phenomenon in Mediterranean contexts or a characteristic of this particular pit cannot be answered until many more samples from Mediterranean sites have been studied. Preservation conditions in Greece are of course different, and presumably less favourable for chitinous remains, than in the countries bordering the North Sea.

Like the samples in the Spanish pilot study, the sample from Halos yielded a mite death assemblage with a high abundance of representatives of the oribatid cohort Dichosomata. Only a few dichosomatic mite species are found in northwestern Europe. In the Mediterranean and especially the (sub)tropics this cohort comprises hundreds of species. These species all

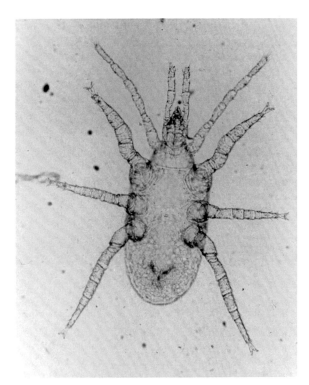

Fig. 4.15. A recent individual of a ♀ ...of *Macrocheles glaber* (Müller, 1860), a gamasid predatory mite illustrating the typical acarine habitus (length 745 μm).

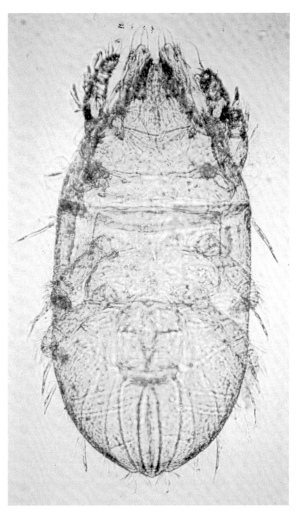

Fig. 4.16. This Spanish 'subfossil' specimen of *Papillacarus chamartinensis* Pérez-Iñigo, 1967 (length 495 μm) may have entered the Iron Age deposit in which it was found from the overlying younger deposits as the elongated shape of its body enabled it to penetrate the soil to a considerable depth.

live relatively deep down in the soil, a way of living known as euedaphic. This is relevant with respect to the interpretation of Mediterranean mite death assemblages as it means that the remains of these euedaphic species are not necessarily contemporary with the deposits in which they are found. Species that live on top of the soil and in litter layers are known as epedaphic and hemiedaphic, respectively. Since these species cannot contaminate underlying deposits they may safely be regarded as contemporary with the deposits in which they are found. The first step in interpreting any arthropod death assemblage should therefore be to exclude the euedaphic species. They can only be used in analyses when there are absolutely no doubts concerning their origins, for instance in the case of completely sealed features.

From table 4.28 it is clear that sample 266 from a pit in the House of the Snakes is dominated by the remains of oribatid mites. This is usually the case in archaeological samples with the possible exception of samples taken from extremely dirty places, in which the Gamasida sometimes outnumber the Oribatida. Table 4.28 also shows that almost a third of all the oribatids found (41 of the 132 individuals) are euedaphic, and should therefore be regarded as possibly non-contemporary contaminants. In this case this did not cause serious problems since the aim of this study was to search for indications of the presence of live-

stock in the backyard. Oribatid remains usually reflect the past ecological conditions of the natural environment and provide only limited information on anthropogenic environments. There are in fact only very few oribatids which are considered to be synanthropes. One of these species is *Ramusella clavipectinata*, of which a single individual was found among the few epedaphic or hemiedaphic oribatids identified to species level. This may indicate at least a certain degree of foulness of the sampled site.

Gamasida are generally found in places with a high degree of biological turnover such as cesspits, middens, dung heaps, *etc.* Their remains can therefore be regarded as evidence for 'dirty' environments. Some gamasids are so specific in their choice of habitat that they are considered characteristic of specific types of excrement (Schelvis, 1992b). Considering the aim of this study it was therefore obvious that the

Table 4.28. Number of remains (N) of each of the identified taxa found in sample 266 from a pit in the House of the Snakes. Whether each taxon is euedaphic (Eu) or hemiedaphic/epedaphic (He) is also indicated.

Acari, Oribatida	N	Eu/He
Lohmannia hungarorum/loebli	6	Eu
Papillacarus ondriasi Mahunka, 1974	3	Eu
Cryptacarus promecus Grandjean, 1950	3	Eu
Epilohmannia gigantea Berlese, 1916	1	Eu
Passalozetes cf. africanus Grandjean, 1932	1	He
Mongaillardia cf. callitoca Grandjean, 1961	1	He
Ramusella clavipectinata (Michael, 1885)	1	He
Scheloribates sp.		
Galumna sp.		Eu 28
Sphaerochtonius sp.	116	
Scutovertex? sp.		He 88
Indet. Oribatida		
Oribatida total	132	

Acari, Gamasida		N	Eu/He
Cosmolaelaps miles (Berlese, 1892)		1♀	He
Hypoaspis oblonga (Halbert, 1915)		1♀	He
Uropoda minima Kramer, 1882		1♂	He
Urodiaspis sp.	1♀	9♂♂	He
Pachylaelaps sp. approx. 800 μm		2♂♂	He
Pachylaelaps sp. approx. 400 μm		1♂	He
Pachylaelaps sp. approx. 520 μm		4♂♂	He
Indet. Uropodina		2	He
Gamasida total		22	

Insecta	N	Eu/He
Hymenoptera; Formicidae	4	Eu
Coleoptera; Carabidae	3	He
Hemiptera; Cicadidae (larvae)	2	Eu
Insecta total	9	

gamasids would be the most useful group to study in this particular mite death assemblage, especially since there are only very few euedaphic Gamasida.

Table 4.28 shows that, indeed, none of the gamasids found are considered to be euedaphic. The table also shows that only three gamasid species could be identified to species level:
- *Cosmolaelaps miles*, a rare species found in decomposing wood, rotting chaff and nests of rodents and ants;
- *Hypoaspis oblonga*, found under the bark of dead trees and in mosses;
- *Uropoda minima*, a common species found in litter, decomposing wood, mosses and in the nests of ants and moles.

None of these species are among the General Dung Indicating (GDI) or Possible Dung Indicating (PDI) species I defined in a previous work (Schelvis, 1992b) on the basis of a series of samples taken from recent dung deposits in combination with a survey of the literature on dung-inhabiting mites. The genus *Pachylaelaps* is moreover well represented in this sample whereas the genus *Macrocheles* of the same Macrochelinae subfamily is completely absent. This is significant since representatives of the *Pachylaelaps* genus are seldom, if ever, found on animal excrement. The *Macrocheles* genus is on the other hand usually to be found in such high densities on animal excrement that their remains are considered to be reliable indicators of archaeological dung deposits.

Several taxonomic groups of insects may also be regarded as indicative of the presence of excrement of domestic animals (Kenward & Hall, 1977). The most obvious are of course the well-known dung beetles (Scarabaeidae) and dung flies (Scatophagidae). No dipterous remains were however found in the sample and the few beetle remains that were found are all from carabid beetles. Although carabids are usually regarded as the most useful family of beetles to be used in archaeological analyses (Ervynck *et al.*, 1994) their strength lies mainly in the information they provide on past natural environments rather than in what they can tell us about anthropogeneous eco-systems (much the same as oribatids in acaro-archae-ology).

The other insect remains are considered to repre-sent intrusive species, for instance the larval cicads (Cicadidae) and the ants (Formicidae). The latter group may have influenced the species distribution of the mites since two of the identified gamasids are also known to occur in ant nests.

An ecological rather than a taxonomical group of insects which I have successfully used (Schelvis & Koot, 1995) to establish the presence (and identify the producer) of animal excrement in archaeological de-posits is that of ectoparasites. Remains of ectopara-sites such as fleas and lice are strong indications of the presence of specific animals at a site. The sample from New Halos, however, did not contain any re-mains of these groups.

5.4. Conclusion

The analysis of the arthropod remains found in the pit in the courtyard of the House of the Snakes in the Hel-lenistic town of New Halos yielded no evidence im-plying that livestock or poultry were kept in this courtyard. The arthropod remains found are not char-acteristic of animal excrement, nor were any remains of specific livestock ectoparasites found. The mite remains seem to indicate the presence of some decom-posing organic matter in the pit. The exact nature of the pit fill, and hence the function of the pit, remains a mystery.

5.5. Acknowledgements

I would like to thank Dr Jørn Zeiler (ArchaeoBone), Leeuwarden, for his comments on a draft version of this paper. Dr Anton Ervynck of the Institute for the Archaeological Heritage (IAP) of Flanders kindly as-sisted in the preparation of the SEM recording at the Laboratory of Palaeontology of the University of Ghent. The research was financially supported by the Thessalika Erga foundation in Groningen. The iden-tifications mentioned in this text were checked through direct comparison with the reference collec-tion of Scara*b* (Subfosil, Contemporary & Archaeo-logical Research of Arthropods Bureau) in Wirdum. The facilities of the department of archaeology of Groningen University were essential for the success of this study.

5.6. References

ARILLO, A., J. GIL-MARTÍN & L.S. SUBÍAS, 1992. Acaros Oribatidos subfosiles de Galicia. *Boletim da Sociedade Por-* *tuguesa de Entomologia* 2, Suplemento 3 (= Actas do V congreso Ibérico de entomologia), pp. 491–498.

BALOGH, J. & S. MAHUNKA, 1983. *The Soil Mites of the World.* 1. *Primitive Oribatids of the Palearctic Region.* Akadémiai Kiadó, Budapest.

BUCKLAND, P.C., 1974. Archaeology and environment in York. *Journal of Archaeological Science* 1, pp. 303–316.

COOPE, G.R., 1967. The value of Quaternary insect faunas in the interpretation of ancient ecology and climate. In: E.J. Cushing & H.E. Wright (eds), *Quartenary palaeoecology.* New Haven, pp. 359–380.

DENFORD, S.M., 1976. *The Archaeology of York.* 14. *The Past Environment of York. Environmental Evidence from Roman Deposits in Skeldergate.* Council for British Archaeology, Lon-don, pp. 139–141.

ELIAS, S.A., 1994. *Quaternary Insects and their Environments.* Washington.

ERICKSON, J.M., 1988. Fossil oribatid mites as tools for quater-nary paleoecologists: preservation quality, quantities and taphonomy. In: R.S. Laub, N.G. Miller & D.W. Steadman (eds), Late Pleistocene and Early Holocene paleoecology and archeology of the eastern Great Lakes Region. *Bulletin of the Buffalo Society of Natural Sciences* 33, pp. 207–226.

ERVYNCK, A., K. DESENDER, M. PIETERS & J. BUNGE-NEERS, 1994. Carabid beetles as palaeo-ecological indicators in archaeology. In: K. Desender *et al.* (eds), *Carabid beetles: ecology and evolution.* Dordrecht/Boston, Kluwer Academic Publishers, pp. 261–266.

KARG, W., 1971. *Die Freilebenden Gamasina, Raubmilben* (= Die Tierwelt Deutschlands, 59. Teil. Acari, Milben. Unterordnung Anactinochaeta (Parasitiformes)). Gustav Fischer, Jena.

KARG, W., 1989. *Uropodina Kramer, Schildkrötenmilben* (= Die Tierwelt Deutschlands, 67. Acari (Acarina), Milben. Unterord-nung Parasitiformes (Anactinochaeta)). Gustav Fischer, Jena.

KENWARD, H.K., 1974. Methods for palaeo-entomology on site and in the laboratory. *Science and Archaeology* 13, pp. 16–24.

KENWARD, H.K. & A.R. HALL, 1997. Enhancing bioarchaeo-logical interpretation using indicator groups: stable manure as a paradigm. *Journal of Archaeological Science* 24, pp. 663–673.

KENWARD, H.K., A.R. HALL & A.K.G. JONES, 1980. A tested set of techniques for the extraction of plant and animal macro-fossils from waterlogged archaeological deposits. *Science and Archaeology* 22, pp. 3–15.

KOCH, C.L. & G.C. BERENDT, 1854. Die im Bernstein befind-lichen Crustaceen, Myriapoden, Arachniden und Aptoren der Vorwelt. *Organische Reste im Bernstein* 1 (2), pp. 1–124.

KRIVOLUTSKY, D.A. & A.Y.A. DRUK, 1986. Fossil oribatid mites. *Annual Review of Entomology* 31, pp. 533–545.

NORDENSKIÖLD, E., 1901. Zur Kenntnis der Oribatidenfauna Finnlands. *Acta Societatis pro Fauna et Flora Fennica* 21(2), pp. 1–35.

OSBORNE, P.J., 1973. Insects in Archaeological deposits. *Science and Archaeology* 10, pp. 4–6.

REINDERS, H.R. & W. PRUMMEL, 1998. Transhumance in Hel-lenistic Thessaly. *Environmental Archaeology* 3, pp. 81–95.

SANZ BRETÓN, J.L. & J. SCHELVIS, 1994. De mijten (Acari) uit Peñalosa en Cerro de la Cruz (Spanje) en Halos (Grieken-land). De mogelijkheden van acaro-archeologisch onderzoek in het Mediterrane gebied. *Paleo-aktueel* 5, pp. 47–49.

SCHELVIS, J., 1990. The reconstruction of local environments on the basis of remains of oribatid mites (Acari; Oribatida). *Jour-nal of Archaeological Science* 17, pp. 559–571.

SCHELVIS, J., 1992a. *Mites and Archaeozoology. General Meth-*

ods; *Applications to Dutch Sites.* Ph.D. thesis, Groningen.

SCHELVIS, J., 1992b. The identification of archaeological dung deposits on the basis of remains of predatory mites (Acari; Gamasida). *Journal of Archaeological Science* 19, pp. 677–682.

SCHELVIS, J. & C. KOOT, 1995. Sheep or goat? *Damalinia* deals with the dilemma. *Experimental and Applied Entomology. Proceedings of the Netherlands Entomological Society* 6, pp. 161–162.

SHEAR, W.A., P.M. BONAMO, J.D. GRIERSON, W.D. IAN

ROLFE, E.L. SMITH & R.A. NORTON, 1984. Early land animals in North America: Evidence from Devonian Age Arthropods from Gilboa, New York. *Science* 224, pp. 492–494.

SIEPEL, H., in prep. *De Westeuropese Mosmijten (Acari: Oribatida).*

TAYLOR, B.J. & G.R. COOPE, 1985. Arthropods in the quaternary of East Anglia – Their role as indices of local palaeo-environments and regional palaeoclimates. *Modern Geology* 9, pp. 159–185.

H. REINDER REINDERS

1. DATING EVIDENCE CONCERNING THE CITY'S FOUNDATION

1.1. Introduction

The beginning and end of the occupation of the city of New Halos are discussed in this chapter. The hypothesis that the city was occupied for only a short period of time, from 302 until 265 BC, was postulated in the first publication (Reinders, 1988).

The hypothesis that New Halos was founded shortly after 302 BC is based on the assumption that the city did not exist in 302 BC, when the armies of Kassandros and Demetrios Poliorketes opposed one another in the Krokion plain. The account of this engagement by Diodoros Sikoulos mentions the cities of Larisa Kremaste, Antron and Pteleon, but not New Halos, along the route followed by Demetrios Poliorketes before he encamped opposite Kassandros. Presumably the city of New Halos was founded after Demetrios Poliorketes' unexpected departure to Asia Minor to assist his father Antigonos in his struggle against Lysimachos in 302 BC. Marzolff (pers. comm.), however, is of the opinion that Kassandros was the founder of New Halos.

The excavation of the six houses yielded no information whatsoever on the exact date of the city's foundation. No coins were found in the trenches that held the limestone blocks of the walls' foundations. The number of coins of Kassandros (306–297 BC) is relatively large, certainly when compared with the number of coins of Demetrios I Poliorketes (294–287 BC), but these quantities of coins do not tell us when or by whom the city was founded. Coins of Kassandros have been found at many archaeological sites, generally in a good state of preservation. Most of the datable pottery found at the site of New Halos dates from the last quarter of the 4th century BC; only a few items date from the first quarter of the 3rd century BC (*cf.* Ch. 3: 1.4).

1.2. A coin hoard

The only evidence relating to the city's foundation is a small hoard of silver coins that was found inside the city's southern wall, in a trench a few metres from the Southeast Gate during that gate's excavation in 1995 (Reinders *et al.*, 1996: pp. 131–132). Soil from the fill of the wall between the two rows of facing blocks must have spilled out behind the wall. The coins must have been hidden or lost some time during the wall's construction.

In other Thessalian coin hoards more than 50% of the coins are usually coins of Hellenistic kings. The hoards generally contain only few coins struck by Thessalian or other cities. As an example of a coin hoard from *c.* 310–300 BC, Oeconomides (1994: p. 337) quotes the hoard of Áyi Theódori (IGCH 93, found in 1901/2). About 50% of this hoard consists of tetradrachms of Philippos II and Alexandros III, 25% are tetradrachms from Athens and the remaining 25% are coins from Lokris, Thebai, Histiaia and Sikyon. The hoard includes no coins of Thessalian cities. A small hoard found at Guerli near Larisa in 1884 (IGCH 80) consists of two tetradrachms of Philippos II and ten of Alexandros III. According to Oeconomides, in Thessaly, tetradrachms of Philippos II are only found together with tetradrachms of Alexandros III, which would mean that coins of Philippos II were not in circulation in Thessaly during his lifetime.

The coin hoard from the Southeast Gate of Halos is of a distinctly different nature: it includes only two drachms of Alexandros III as against thirteen coins struck by cities: Ephesos (1 tetradrachm), Tenos (1 tetradrachm, 1 didrachm), Karystos (2 didrachms), Chalkis (1 drachm), Histiaia (6 tetrobols), Boeotia, federal mint (1 hemidrachm). The coins of Histiaia are excellently preserved. They will have to be studied in detail to determine the period in which they were minted. Assuming that these tetrobols belong to the first group of Histiaian tetrobols, they must date from 338–304 BC (Picard, 1979: Tableau). The low amount of wear of the coins suggests that the tetrobols of Histiaia are the youngest coins of this hoard, and for the time being we suggest that they were hidden or lost some time between 310 and 300 BC.

Does the composition of the hoard reflect the approximate route followed by the individual who buried or lost these coins some time during the construction of the wall near the Southeast Gate? As already mentioned above, opinions differ as to who founded New Halos: Demetrios Poliorketes (Reinders, 1988) or Kassandros, as suggested by Marzolff (pers. comm.). What we do know for sure is that the hoard includes no coins from the cities of Thessaly. Coins from Thessalian cities encountered in hoards are usually coins from the city of Larisa. There is no evidence to suggest that the coins or their owner came from the north, considering the small number of Alexandrian coins. This traveller's hoard may well support the hypothesis that the city was founded by Demetrios Poliorketes, because we know that he sailed from Rhodos along the Cyclades to Chalkis in 304 BC and on to Larisa Kremaste and the Krokion plain in 302 BC. A great number of datable pottery types (with dates in the last quarter of the 4th century BC) agree with this date (see Ch. 3: 1).

2. DATING EVIDENCE CONCERNING THE END OF THE CITY'S OCCUPATION

2.1. Coin finds from the houses

Dating evidence for the end of the city's occupation is provided by both coins and pottery. We will resume the dating evidence of the coins, already discussed in Ch. 3: 5, and add some preliminary results of the excavation of the site of the Southeast Gate, which was inhabited for some time after the city was abandoned.

In addition to coins of the Thessalian and other Greek cities (tables 3.18 and 5.1), 37 coins of Hellenistic kings were found in the six houses. The survey in table 5.2 lists the names of the kings and their regnal years along with the type and the number of coins in question found in the six excavated houses. The coins struck by Ptolemaios II and Antigonos II Gonatas are of particular interest with respect to the date of the city's abandonment.

In Ch. 3: 5 two possible explanations were given for the letter between the eagle's legs on the Ptolemaian coins: regnal year or issue numbering. If we assume that the letters represent the king's regnal years, the coin with the letter Φ, which is indeed less worn than the other coins, would be the youngest coin and would provide a *terminus post quem* of 265/4 BC for the abandonment of New Halos. However, a letter Φ indicating the 21st regnal year seems to be quite uncommon. If the letter on the contrary refers to the number of the issue, the series of large Ptolemaian coins could have been produced over a period of between five and ten years. The different amounts of wear of the coins bearing the letters Ο, Λ, Τ and Φ at least suggest a chronological sequence. From Ptolemaian coins from sites in Attica we know that the large Ptolemaian bronzes were in circulation during the Chremonidean war (268/7–262/1). Later bronze emissions of Ptolemaios II which came into circulation around 261/0 BC were not found in the houses of New Halos (Reinders, 1997).

If the letters represent a kind of issue numbering, the entire series would according to Mørkholm have been produced within a period of five to ten years, perhaps 285–275 BC. As already mentioned above, these coins were still in circulation during the Chremonidian war. Antigonos returned to Macedonia in 276, but Macedonia was invaded by Pyrrhos in 274 and 273; it was not until after 272 that Antigonos'

position was firmly established. Silver tetradrachms showing the head of Pan on the reverse were in circulation after 270 BC, "remembering the help of the god when he caused a panic terror among the Gauls during Antigonos' battle with them near Lysimachia in 277" (Mørkholm, 1991: p. 134). Bronze coins showing Pan erecting a trophy are therefore generally attributed to Antigonos II Gonatas. Emissions of these coins circulated in the parts of Greece which were under Macedonian control. Although they were very copious issues, not a single coin showing Pan erecting a trophy on the reverse was found in the houses of New Halos.

However, two coins of Antigonos II Gonatas of a different type, bearing a Macedonian shield and the monogram ANTI on the obverse and a Macedonian helmet on the reverse, were found. In the past, there was some disagreement as to whether these coins were struck by Antigonos II Gonatas or Antigonos III Doson (229–221 BC), but the excavations at New Halos have shown beyond doubt that the coins in question were struck by Antigonos II Gonatas. According to Furtwängler (1990: p. 131), there is a connection between the monograms of Antigonos' tetradrachms and the monograms on the shield/helmet bronzes. He places the twelve groups of the latter series at the beginning of the reign of Antigonos, in the 270s and early 260s BC. The two coins from the houses of New Halos belong to Furtwängler's group 9, bearing three monograms.

The datable, *i.e.* imported, pottery from the six houses represents only a small proportion of the total amount of pottery recovered, but it includes no types that were introduced in Athens around 260 BC (*cf.* Ch.3: 1.4). This would agree with the assumption that the town's occupation ended before 260 BC.

2.2. Coin finds from the Southeast Gate

Around 250 BC the large bronze emissions of Antigonos II Gonatas – Athena wearing a Corinthian helmet/Pan erecting a trophy – were in circulation. Coins of this type were not found in the six houses of New Halos, but they were found in the occupation layers within, on and around the Southeast Gate complex. The coin finds of the 1995 and 1997 excavation of the Southeast Gate are of a quite different nature than the coin finds from the houses, as can be seen in table 5.1: so far, they comprise 48 coins of Hellenistic kings and only thirteen coins of Greek cities. The survey of coins of Hellenistic kings in table 5.3 lists the name of the king, his regnal years and the type and number of coins concerned. Among the coins struck by Greek cities is a coin from the city of Halos, which was found in the lowest layer of one of the trenches and can be associated with the period in which the gate was in use.

Layers of occupation remains were found in trench O in the guardroom between the gate's two spur-walls. Among the finds are loomweights, pottery, nails and

Table 5.1. Survey of the Hellenistic coins found in the six houses and at the site of the Southeast Gate (after the excavations of 1995 and 1997, excluding the coin hoard; AE = bronze, AR = silver).

House	Gate		Houses	Gate
MAKEDONIA				
Alexandros III (336–332 BC)		AR Heracles/Zeus enthroned	-	2
		AE Heracles/Bow in case and club	-	1
Kassandros (316–197 BC)		AE Heracles/Horseman	10	6
		AE Heracles/Seated lion	1	-
Demetrios I (294–288 BC)		AE Shield with monogram/Helmet	6	-
		AE Demetrios/Prow	2	-
Antigonos II (277–239 BC)		AE Shield with monogram/Helmet	2	-
		AE Athena/Pan erecting trophy	-	25
Unknown (monogram illegible)		AE Shield/Helmet	-	3
AIGYPTOS				
Ptolemaios II (285–246 BC)		AE Zeus/Eagle on thunderbolt (small)	1	1
		AE Zeus/Eagle on thunderbolt (large)	13	10
EPEIROS				
Pyrrhos (295–272 BC)		AE Shield with monogram/Thunderbolt	2	-
THRAKE	Orthagoreia	AE Apollo/Helmet	1	-
KORKYRA	Korkyra	AE Amphora/Bunch of grapes	1	-
PERRHAIBIA	Orthe	AE Athena/Forepart of horse	1	-
PELASGIOTIS	Gyrton	AE Zeus/Horse	1	-
	Larisa	AE Nymph/Horseman	3	-
		AE Nymph/Horse feeding	-	1
	Skotoussa	AE Athena/Horse prancing	4	-
HESTIAIOTIS	Trikka	AE Nymph/Asklepios feeding serpent	1	-
PHTHIOTIS	Thebai	AE Demeter/Protesilaos	1	-
	Peuma	AE Male head/XA	7	-
	Halos	AE Zeus/Phryxos, series 6	2	-
		AE Zeus/Phryxos, series 7	21	1
		AE Zeus/Phryxos, series 8–20	5	3
	Ekkara	AE Zeus/Artemis	3	-
	Larisa	AE Male head/Thetis on hippocamp	3	2
		AE Nymph/Harpa	8	2
MALIS	Lamia	AE Nymph/Philoktetes	4	-
LOKRIS	Lokris	AR Demeter/Ajax	3	1
		AE Apollon, Athena/Bunch of grapes	7	1
EUBOIA	Chalkis	AE Female head/Eagle and serpent	15	-
	Histiaia	AE Maenad/Bull walking	4	-
		AE Maenad/Forepart of bull	5	-
		AE Heracles/Bull's head	-	1
		AE Maenad/Bull standing in front of a vine	-	1
	League	AE Bull standing/Bunch of grapes	1	-
PHOKIS	Phokis	AE Athena/Ö	1	-
BOEOTIA	Thebai	AR Shield/Kantharos	2	-
	Thespiai	AR Shield/Aphrodite	1	-
	Federacy	AE Shield/Trident	1	-
Uncertain mint			1	-
KORINTHIA	Korinthos	AR Pegasus/Nymph	1	-
SYKIONIA	Sykion	AE Dove/Olive wreath	1	-
Illegible			4	14

Table 5.2. Survey of the coins of Hellenistic kings found in the six houses.

Name of king	Regnal years	Type and number
Kassandros	316–297 BC	Naked youth on horse (10), seated lion (1)
Demetrios Poliorketes	294–288 BC	Crested helmet (6), prow (2)
Pyrrhos	295–272 BC	Thunderbolt (2)
Ptolemaios II	285–246 BC	Eagle on thunderbolt (14)
Antigonos II Gonatas	277–239 BC	Crested helmet (2)

Table 5.3. Survey of the coins of Hellenistic kings found in the Southeast Gate, after the excavations of 1995 and 1997.

Name of king	Regnal years	Type and number
Alexandros III	336–332 BC	Zeus enthroned (AR, 2), Bow in case and club (AE, 1)
Kassandros	316–197 BC	Horseman (6)
Antigonos II Gonatas	277–239 BC	Pan erecting trophy (25)
Ptolemaios II	285–246 BC	Eagle on thunderbolt (11)

coins. The coins indicate when this area was occupied. Layer 3 yielded three coins of Ptolemaios II, one of them bearing the letter Λ. The other coins are too worn to allow determination of the letters between the eagle's legs. Another three coins of Ptolemaios II were found together with five coins of Antigonos II Gonatas in layer 2, above layer 3. The Ptolemaian coins of layer 2 were all found at a lower level than the coins of Antigonos II Gonatas.

The coins of Antigonos II Gonatas belong to the issue showing Athena wearing a Corinthian helmet/Pan erecting a trophy. It is not clear when these bronze coins came into circulation. Furtwängler relates the bronze issues of the Pan erecting a trophy type to the Pan tetradrachms of Antigonos Gonatas, considering the similarity of the monograms on both issues. Mørkholm (1991) is of the opinion that the Pan tetradrachms circulated in the 260s: the bulk of the coinage dates from the 260s, when the Chremonidian war made heavy demands on the finances of Macedon, whereas the last two decades of Gonatas' reign were relatively quiet and peaceful. If the relation between the silver and bronze Pan issues of Antigonos II Gonatas is correct, we may assume that the Pan erecting a trophy bronzes were already in circulation in the late 260s (tables 5.4 and 5.5).

The occupation layers in, on and around the Southeast Gate indicate that the gate was destroyed and the passageway was blocked by walls of buildings that were occupied in several phases shortly after the gate had been destroyed. The gate's passageway was blocked with *poros* blocks and no longer provided access to the city. The same was observed at one of the minor gates along the southern wall (Reinders, 1988: pp. 92–95). The coins of Ptolemaios II and Antigonos II Gonatas suggest that the buildings erected at the site of the Southeast Gate were occupied already

in the 260s. This occupation phase did not last very long; burned artefacts show that it ended with a fire.

The destruction of the Southeast Gate supports the hypothesis that the city of New Halos was abandoned after a sudden disaster. The combined chronological evidence of the coins from the houses and the Southeast Gate leads to a chronological order for the dates of issue (table 5.4). For the time being we may conclude that the occupation of the six houses in the town of New Halos came to an end in the mid-260s and that the site of the Southeast Gate was reoccupied in the late 260s. After the city had been abandoned, some of its former inhabitants lived in houses and buildings along the city's southern wall and buried their deceased in a small cemetery south of the gate (Malakasioti *et al.*, 1993).

3. EARTHQUAKES IN THE ALMIRÓS PLAIN AND THE ABANDONMENT OF NEW HALOS

3.1. A disaster

For some time we racked our brains over the question as to what kind of accident may have caused the destruction of the houses of New Halos and the blockage of the Southeast Gate until a disaster occurred in the Almirós area. On the 9th of July 1980, at two o'clock in the morning, an earthquake shook the Almirós plain in Thessaly. We had left Almirós after a successful excavation campaign only three days before. The artefacts found in previous campaigns had been stored in the local museum after being studied and restored. The shocks of the earthquake, which had a magnitude of 6.5 on the Richter scale, caused severe damage to the museum building and the restored artefacts from New Halos were once again destroyed.

Table 5.4. Survey of the dates of issue of the coins of Ptolemaios II and Antigonos II Gonatas.

Name of king	Type	Date of issue
Ptolemaios II	Eagle on thunderbolt (letter indicating issue numbering)	285–275 BC
Ptolemaios II	Eagle on thunderbolt (letter indicating regnal year)	275–265/4 BC
Antigonos II Gonatas	Macedonian shield monogram/crested helmet	late 270s, early 260s
Antigonos II Gonatas	Athena wearing a helmet/Pan erecting trophy	late 260s

During a visit to the Almirós area in September 1980 we found that certain houses of the town of Almirós and surrounding villages had been damaged by the earthquake, mostly houses with walls made of mudbricks, cobbles or irregularly worked stones. The nature of the damage caused to the houses in the Almirós plain and the frequency of earthquakes in this area led to the hypothesis that the Hellenistic city of New Halos was abandoned because it, too, was destroyed by an earthquake (Reinders, 1988; Reinders 1992).

The excavation of five more houses after the publication in 1988 has yielded further evidence on the site-formation process and the end of the town's occupation. Moreover, a number of articles have been published on seismicity in Greece in general, and on the earthquake of 1980 in particular. The aims of this section are to present the information available on seismicity in the Almirós plain and to discuss the abandonment of the city of New Halos and the possibility that the city was destroyed by an earthquake.

3.2. Volcanic and tectonic activity in the Almirós plain

Two phenomena in the Almirós plain are of interest to us here: the volcanic centres at Mikrothíve and Néa Anchíalos and the two Zerélia lakes. It is thought that the Zerélia lakes formed in craters resulting from phreatic eruptions. Stiros and Papayeoryiou (1994: p. 31) are of the opinion that such phreatic eruptions took

Table 5.5. New Halos. Bronze coins struck by the Antigonids (A = found in the houses, B = found at the site of the Southeast Gate, C = not found; after the excavations of 1995 and 1997).

ANTIGONOS I MONOPHTHALMOS (306–301 BC)
- no bronze coins known

DEMETRIOS I POLIORKETES (306–283 BC)
- Demetrios wearing Corinthian helmet/Prow, BA **[A]**
- Macedonian shield, monogram ΔHMHTPH/Macedonian helmet, BA ΣI; struck by Demetrios II according to Mørkholm **[A]**
- Poseidon/Athena Promachos, ΒΑΣΙΛΕΩΣ ΔHMHTPIOY **[C]**
- Poseidon/Prow, BA **[C]**

ANTIGONOS II GONATAS (277–239 BC)
- Macedonian shield, monogram ANTI/Macedonian helmet; struck by Antigonos III according to some authors **[A]**
- Athena wearing Corinthian helmet/Pan erecting trophy, monogram ANTI **[B]**
- Heracles/Horseman, monogram ANTI; struck by Antigonos III according to some authors **[C]**

DEMETRIOS II (239–229 BC)
- Macedonian shield, ΔHMHTPH/Macedonian helmet; struck by Demetrios I according to some authors [**A, see Demetrios I**]

ANTIGONOS III DOSON (229–221 BC)
- Heracles/Horseman, monogram ANTI; struck by Antigonos II according to some authors [**A, see Antigonos II**]
- Macedonian shield, monogram ANTI/Macedonian helmet; struck by Antigonos II Gonatas according to some authors [**A, see Antigonos II**]

PHILIPPOS V (221–179 BC)
- Heracles/Horseman, ΦI BA **[C]**
- Sixteen other emissions of bronze coins **[C]**

PERSEUS (179–168 BC)
- Perseus wearing a helmet/Eagle on plough or thunderbolt, BA **[C]**

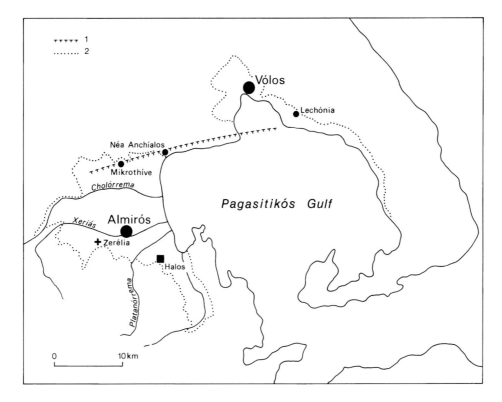

Fig. 5.1. The neotectonic fault in the northern part of the Pagasitikós gulf (1 = neotectonic fault; 2 = limits of the plain).

place in the 'past few thousand years'. The craters' almost perfectly conical shapes suggest that they are relatively young; such a shape is not preserved for a long period of time. On the other hand it is hard to imagine that people lived on the huge Magoúla Zerélia from the 4th millennium BC until well into the 2nd millennium BC without a major water source in the immediate surroundings (Wace & Thompson, 1912). A 10-m-long core taken from one of the lakes yielded information on the process of sedimentation and the vegetation over the past 1000 years only (Bottema, 1994; cf. Ch. 4: 3). Unfortunately, we were at the time unable to perform deeper coring for technical reasons.

Thessaly is a tectonically active area. Four earthquakes with magnitudes of between 6.5 and 7.0 on the Richter scale have occurred here since 1950, three of them between 1954 and 1957, when many buildings in the city of Vólos were damaged. A large number of smaller earthquakes, with magnitudes up to 6.5, were recorded over a longer period, i.e. between 1911 and 1980 (Papazachos et al., 1983). Although indications of seismic faulting were observed in the neighbourhood of Néa Anchíalos after the 1980 earthquake, the neogene and quaternary sediments of the Almirós plain contain very little evidence for seismic faulting at the surface. The Almirós plain was however found to have subsided 25 cm along a 20-km-long levelling

line after 1980 (Stiros & Papageorgiou, 1994: pp. 30–31).

A major fault line has been found in the northern part of the Almirós plain and the Pagasitikós gulf, extending from the Lechónia plain, southeast of the city of Vólos, to the Mikrothíve region in the Almirós plain (fig. 5.1). This tectonic fault is about 15 km long, which is a common length for a fault segment. Similar lengths have been recorded for 'normal' faults all over the world. Normal-fault earthquakes usually generate shocks with magnitudes of between 6.0 and 6.5 (Ambraseys & Jackson, 1990: p. 9). That would mean that the seismic risk in the Almirós plain is not that great.

There is another fault in the southern part of the Almirós plain (Ambraseys & Jackson, 1990: fig. 1). It is not indicated on the Anávra sheet of the geological map and no data are available as to whether or not this fault is active. The formation of the two Zerélia lakes is evidently connected with the presence of this fault (Institute of Geological and Mining Research, pers. comm.).

3.3. The 1980 earthquake

The earthquake that shook the Almirós plain in 1980 occurred on July 9, at 02.11.57 a.m. The epicentre lay

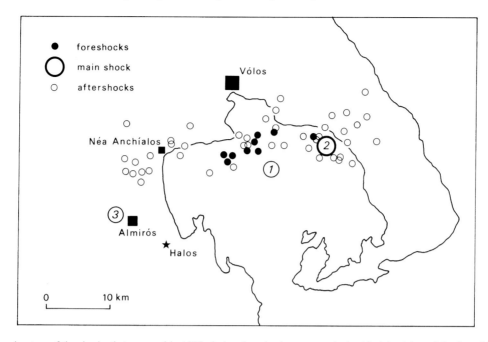

Fig. 5.2. The epicentres of the shocks that occurred in 1980, during the seismic sequence in the Almirós plain and the Pagasitic gulf (after Papazachos *et al.*, 1983). The three successive shocks on July 9 are numbered 1, 2 and 3.

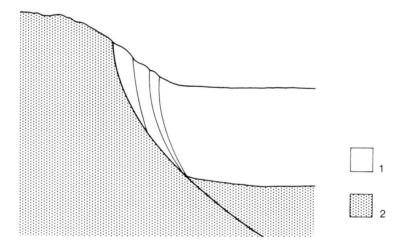

Fig. 5.3. Cross-section of the neotectonic fault showing ruptures in the quaternary deposits in the village of Néa Anchíalos (1. Alluvial deposits; 2. Bedrock; after Papazachos *et al.*, 1983).

at 39.28 N and 23.11 E, in the Pagasitikós gulf, southeast of the city of Vólos, near the shore. The shock had a magnitude of 6.5. There was a foreshock of 5.6 1.6 minutes before the main shock, which was followed by an aftershock with a magnitude of 6.0 after 24 minutes. The epicentre of the aftershock was near the town of Almirós, at a depth of 12 km. The average focal depth of all the shocks was 9 km (Papazachos *et al.*, 1983).

The seismic activity in the Almirós area in 1980 was however not restricted to these three shocks.

Papazachos *et al.* (1983) give a list of no fewer than 45 shocks with magnitudes of >4.2. Seismic activity was already perceptible in January of 1980. The activity culminated in the month of July. There was another shock, of 4.2, on the 20th of September. The epicentres show a ENE–WSW trend approximately following the neotectonic fault extending from Mikrothíve to Lechónia along the northern shore of the Pagasitikós gulf (fig. 5.2).

After the main shocks a number of fractures, ruptures and cracks were observed in the alluvial sedi-

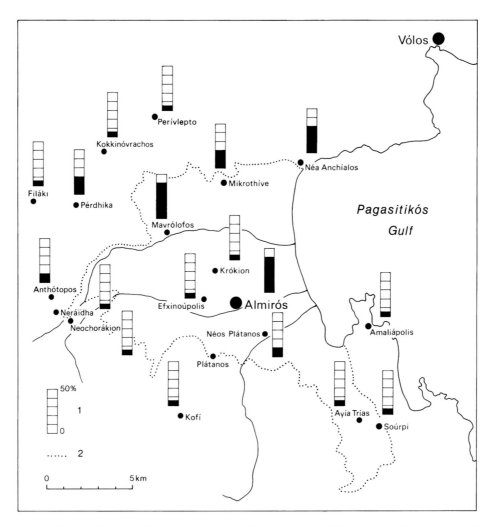

Fig. 5.4. Percentages of destroyed houses in the villages on the Almirós plain after the 1980 earthquake (1 = percentage of houses damaged beyond repair; 2 = limits of the Almirós plain).

ments in the village of Néa Anchíalos. The largest of these fractures had a length of about 2 km. Some of the buildings along this fracture had been destroyed or damaged. On the other hand, the free-standing columns of a basilica in Néa Anchíalos had remained upright and had suffered no damage whatsoever (Ambraseys & Jackson, 1990: p. 42). In and near Néa Anchíalos ruptures were observed especially at the transition from the alluvial sediments to the bedrock (Papazachos et al., 1983: pp. 160–161; Ambraseys & Jackson, 1990: p. 42). These ruptures and cracks were filled up with sediments later (fig. 5.3).

The earthquake caused severe damage, especially in the northern part of the Almirós plain, because the epicentre of the main aftershock was near the town of Almirós (fig. 5.4). According to Papazachos & Papazachos (1989), 5,222 houses collapsed or were damaged beyond repair, while 14,726 houses suffered severe damage and 10,688 were slightly damaged. Ac-

cording to the Thessaloníki newspaper of the 10th of July 1980, the number of destroyed houses was smaller: about 600. Perhaps the newspaper did not have accurate information on the actual damage at the time. Another explanation for this discrepancy could be that a large number of houses that were initially damaged by the main shocks were destroyed in the following days. The area continued to suffer the consequences of seismic activity throughout the remaining part of the month, with many shocks with magnitudes exceeding 4.5.

None of the inhabitants of the collapsed houses were killed; only 24 were injured. The relatively low number of injured inhabitants is probably attributable to the fact that the cluster of three main shocks was preceded by a series of foreshocks. The foreshock with a magnitude of 5.6 and the main shock with a magnitude of 6.5 moreover had offshore epicentres (Ambraseys & Jackson, 1990: p. 9). The collapsed buildings were mainly old structures, built from mudbricks, cobbles from nearby riverbeds or roughly worked blocks of limestone. Many buildings had already been affected by previous earthquakes. The Neo-Classical local museum in Almirós, the only remaining 19th-century building, was damaged. It was restored and reopened in 1998, eighteen years after the 1980 earthquake.

3.4. The abandonment of New Halos

Only six of the original 1,440 houses of the city of New Halos have been excavated. These six houses lay in three different housing blocks. Six houses of course represent only a very small proportion of the city. Nevertheless, we assume that the entire city was abandoned after an event which occurred around 265 BC. For a start, no indications of later occupation whatsoever were found in the inhabited area within the city walls, with the exception of a single medieval coin. Evidence for later habitation was found only at the site of the Southeast Gate. The excavation of the site of the former gate revealed that the Southeast Gate already lay in ruins, no longer providing access to the city, by the time of the reign of Antigonos Gonatas (277–239 BC).

The fact that the occupation layer within the houses was relatively thin (5–40 cm) caused surprise and it was first thought that part of the topsoil, in which any remains from a later occupation phase would have been embedded, had been washed away. But there is no evidence suggesting that this is the case. The Pleistocene soil of the Almirós and Soúrpi plains is rather stable. During a survey we found that remains of settlements in the flat parts at the centre of the plain and on the ridges extending into the plain from the foothills of the Óthris mountains were relatively undisturbed and lay concentrated within restricted areas; virtually no remains were found on the slopes of the ridges.

Large stretches of the area within the enceinte of New Halos have been disturbed by agricultural activity. At the beginning of the twentieth century arable land in the eastern part of the city was used for the cultivation of cotton. Blocks of limestone and other stones were removed from this land and virtually no building remains or artefacts have consequently been found in these areas. In the western part of the city the remains of housing blocks and houses survived relatively intact until the 1980s at least. Only small parts of the land were under cultivation, the rest lay waste. Although many remains have in recent years been destroyed and foundation blocks have been removed, there is no evidence whatsoever to suggest that a layer containing remains from a later occupation phase has disappeared.

At present, we have no evidence that can help us answer the question why the inhabitants of New Halos abandoned their city. If a blaze of the whole city

had been the reason for their departure, the artefacts would have shown traces of burning and charcoal and layers of ash would have been found in the houses. Tiny pieces of charcoal and small patches of ash have been encountered here and there, but they have been interpreted as the remains of cooking or small fires. Charred antler fragments and astragali were probably burned during rituals (*cf.* Ch. 4: 4.12). The effects of a fire were observed during the excavation of the Southeast Gate in 1995. Black and red discolourations in the soil, black, crumbly artefacts and burned mudbricks were clear indications that the occupation of the site of the Southeast Gate had come to an end after a fire. If the houses in the city had likewise been destroyed by a fire, the same burned artefacts and mudbricks would have been found there too. It is equally unlikely that the citizens moved elsewhere because part of their city had been destroyed following a siege, for then, too, we would most probably have found traces of burning because victorious besiegers of cities usually razed the besieged cities to the ground by setting fire to the houses.

I also considered the possibility that the inhabitants abandoned their houses in order to settle elsewhere. We studied this possibility by investigating the abandoned village of Plátanos, a former *dhímos*, some 5 km from New Halos. Plátanos' 1000 inhabitants abandoned their village in the late 1950s in order to settle at Néos Plátanos, some 3.5 km further northeast.

Some of them had already built houses at the site that was to become their new village along a new road close to their fields in the Voulokalíva area. A good reason for moving were the narrow, winding streets in the old village, through which it was virtually impossible to transport modern agricultural machinery and equipment. Another reason seems to have been that the village was given money to repair the houses that had sustained damage during the earthquakes of 1955 and 1957. The villagers then decided to build a new village rather than repair the houses in the old village. As Platanos lay close to the epicentres of the 1955 and 1957 earthquakes, many of the houses must have been damaged beyond repair.

After the 1980 earthquake the houses of Platanos collapsed. The houses in the abandoned village are now in different stages of decay. In 1994 we started to map the village and its remains and measure the houses. Only few of the inhabitants had refused to leave their houses. Hardly any artefacts had been left behind in the houses, because most villagers had departed with all their belongings. The storage bins, tools and furniture in the few houses that had remained occupied after the 1960s had been removed by relatives – or, in a few cases, looters – after the occupants died.

If the inhabitants of the city of New Halos had likewise moved to a new site nearby, fewer artefacts would have been found during the excavation of the six houses.

3.6. An earthquake in 265 BC?

The amount of broken pottery lying *in situ* in the six excavated houses of New Halos suggests that the citizens abandoned their houses after some disaster. It is highly unlikely that such large numbers of objects would have been left behind in the case of a gradual drift to a different city. The earthquake of 1980 inspired the hypothesis that a similar earthquake was responsible for the end of the occupation of New Halos. However, Papazachos & Papazachos (1989) not mention an earthquake in Achaia Phthiotis in the 260s BC in their list of earthquakes in Greece, based mainly on written records relating to the period before the 18th century. The only recorded earthquake in the neighbourhood of Almirós occurred in 1773, when a series of shocks on the 4th, 5th and 14th of March destroyed houses, towers and churches. Like that on so many similar disasters in Greece, the information on this earthquake is inaccurate and based on vague reports; indeed, only for the earthquakes that have occurred in Greece since 1845 do we have more detailed reports.

What evidence does the earthquake that shook the Almirós plain in 1980 provide to support the hypothesis that the city of New Halos was abandoned after a similar disaster? In the preceding paragraphs I mentioned an active neotectonic 15-km-long fault extending from Lechónia to Mikrothíve. During the earthquake, fractures, ruptures and cracks were formed along this fault at the transition from the bedrock to the alluvial sediments; they were filled up with sediments within a relatively short period of time. It is unlikely that such signs of disturbance are to be found elsewhere in the plain. During the excavations at New Halos no gaps were observed in the foundations of the houses.

It is not known whether the tectonic faults in the Almirós plain near Néa Anchíalos and Zerélia were active in Hellenistic times, but it is unlikely that earthquakes with a magnitude exceeding 6.5 ever occurred in this area. Of importance with respect to the question regarding the abandonment of New Halos are the three earthquakes consisting of series of foreshocks, main shocks and aftershocks that have occurred in this area since 1950 (table 5.6), the offshore positions of some of the epicentres and the fact that many houses with mudbrick walls were damaged beyond repair.

During our excavations we observed a sequence in the 5–40-cm-thick occupation layer, which was more or less the same everywhere: hardpan, crushed pottery and other artefacts, roof tiles and topsoil. The coins found in the excavated houses support the hypothesis that the occupation came to an abrupt end. There is no evidence for any major fires.

What happened? Let's imagine a day in the mid-260s BC. Seismic activity has been perceptible for some time already; some families decide to leave the city. All of a sudden the inhabitants of New Halos are alarmed by a powerful shock. They run out of their houses into the crowded streets. A few minutes later another shock, with a magnitude of 6.5, follows. The upper parts of the mudbrick walls of the houses and the tiled roofs begin to fall down. The inhabitants try to leave the city, but they have difficulty passing through the gates. A third shock brings down more walls and roofs, but only a few of the 9,000 inhabitants are injured; no skeleton remains were found in the houses. After some days it is clear that almost all the houses have been damaged beyond repair and the people decide not to rebuild them. The mudbrick walls of the former houses are gradually dissolved by the rain, the mud forming a layer on top of the stone foundations.

So the 9,000 inhabitants abandoned their city and settled elsewhere, somewhere within the city's territory. In 1990 we started a survey in the Almirós plain and the foothills of the Óthris mountains, but so far we have found only a small number of sites that were occupied in Hellenistic and Roman times. We expect to find more evidence for the consequences of an earthquake at the site of the Southeast Gate, where we started an excavation in 1995.

4. FINAL REMARKS

The excavation of the six houses, the city's west wall and the Southeast Gate confirmed the results of the investigation conducted in the late 1970s (Reinders, 1989), but also yielded new evidence enhancing our knowledge of Hellenistic New Halos. In this section a few results and hypotheses and new research questions will be briefly presented.

Founder. After the excavation of the six houses we are still in doubt as to who founded the city of New Halos: Kassandros or Demetrios Poliorketes (Reinders, 1988). A small coin hoard found during the excavation of the city's Southeast Gate suggests that Demetrios was the founder (Ch. 5: 1.2). I assume that this hoard was hidden or lost during the construction of the enceinte of New Halos and that the coins, which come from Ephesos, Tenos, Karystos, Chalkis and Hiastiaia, reflect the approximate route followed by the individual who buried or lost them. The hoard may well support the hypothesis that the city was founded by Demetrios Poliorketes, because we know that he sailed from Rhodos along the Cyclades to Chalkis in 304 BC, and on to Larisa Kremaste and the Krokion Plain in 302 BC (Reinders *et al.*, 1996).

Overseas contacts. In the first volume on the city of New Halos I emphasised the city's military character and its strategic location on the coastal route from northern to central Greece. Coin finds from the six houses however indicate that New Halos also main-

Table 5.6. List of the main shocks from the earthquakes of Lechónia (1955), Velestíno (1957) and Almirós (1980).

		Magnitude	Time	Date
Lechónia, 1955	foreshock	4.9	19.46	21 February 1955
	main shock	6.2	16.47	19 April 1955
	aftershock	5.8	07.18	21 April 1955
Velestíno, 1957	foreshock	6.5	12.14	8 March 1957
	main shock	6.8	12.21	8 March 1957
	aftershock	6.0	23.35	8 March 1957
Almirós, 1980	foreshock	5.4	02.10	9 July 1980
	main shock	6.5	02.11	9 July 1980
	aftershock	6.1	02.35	9 July 1980

tained overseas contacts with cities on Euboia and in Lokris. A survey of the data available on the sea route from Peiraieus to Thessaly revealed the importance of the ports along the northern and western coast of the Pagasitikós gulf from Classical times until the beginning of the Ottoman domination (Ch. 1: 2). Preponderance shifted from one port to another. In my opinion New Halos may well have functioned as a port of call in Hellenistic times despite the fact that it lay 1.5 km inland from the Pagasitikós gulf.

The powerful springs that originate between Kefálosis and Áyos Márkos nowadays empty into the Pagasitikós gulf via a canal which was dug in the 1930s, when the backswamp east of New Halos was drained. The river Amphrysos, whose course is still visible in the backswamp, was presumably navigable up to the city –for small craft at least – in Hellenistic times.

City wall. When, between 1998 and 2002, the national road was reconstructed, the greater part of the western wall of the lower town of New Halos was excavated by archaeologists of the 13th Ephorate of Prehistoric and Classical Antiquities at Vólos. In most places three to four courses of the wall were found to be intact and along some parts of the wall another five to six courses of limestone blocks lay collapsed on the eastern side.

The question remains whether the wall was built entirely of limestone blocks or consisted of a limestone foundation surmounted by a mudbrick upper structure (Reinders, 1988: p. 72). Two or three courses will have sufficed as a foundation and two or three courses of limestone blocks have indeed survived along most sections of the walls of New Halos. The eight to ten courses of the western wall point to a city wall built entirely of limestone blocks. On the other hand, large quantities of dissolved mudbricks were found during the excavation of the Southeast Gate, which point to a mudbrick upper structure. This problem deserves further attention in the near future.

Gates. The investigation of the Northwest Gate in 1982

and the ongoing excavation of the Southeast Gate of New Halos resulted in the plans of two major gates of the courtyard type. The discovery of another gate of the courtyard type during the excavation of the western wall by archaeologists of the 13th Ephorate at Vólos between 1998 and 2002 caused some surprise. The Greek archaeologists excavated the remains of a square tower at the southwest corner of the city wall, part of a courtyard gate, the South Gate, which provided access to the upper town, but not to the lower town of New Halos. Only the foundations of this gate have survived; the remains were not visible at the surface when we surveyed the city in the 1970s.

The passageway between the two towers of the South Gate provided access to a courtyard. Behind this courtyard the passageway narrowed between two pairs of spur walls. Also excavated was a short stretch of the southern wall of the upper town which was connected to the west tower of the South Gate. The question whether the walls of the upper and lower towns date from the same period is now solved (Reinders, 1988: p. 58). The walls of the upper and lower towns and the South Gate were constructed *'aus einem Guss'*: the two eastern spur walls of the gate were bonded in the west wall of the lower town. The existence of a South Gate at the southwest corner raises new questions about the layout of the enceinte and access from the surrounding area to the different parts of the city. A new reconstruction of the layout of New Halos is presented in figure 5.5.

Access and communication lines. People travelling from northern to central Greece will have crossed the Almirós plain to the Kefálosis spring and entered the city of New Halos through the Northwest Gate. They will have followed their route via Avenue A along the west wall and will have reached the Southeast Gate via the Main Avenue and Avenue D, from where they will have left the city and continued on their journey southwards, as I assumed in previous publications. Smaller gates lay in line with Avenues B, C and D. They were presumably sally gates, though they may

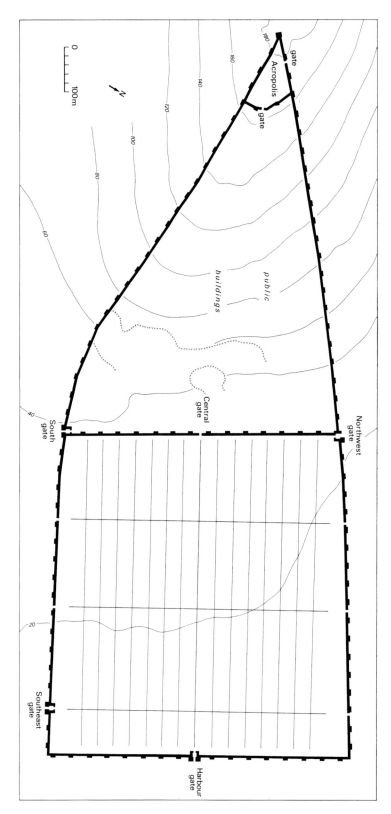

Fig. 5.5. Plan of New Halos: reconstruction.

also have provided access to the lower town in times of peace.

Access to the upper town was provided by the recently excavated South Gate. I presume that the northern wall of the upper town was connected to the northwest corner of the lower town. The question arises whether a courtyard gate at this point may have provided access to the upper town in the same way as the South Gate does on the southern side of the upper town. The asymmetrical layout of the Northwest Gate however suggests a simple connection between the wall and the tower rather than a double gate *(dipylon)*.

If this hypothesis is correct, there must certainly have been another gate: a Central Gate in the west wall of the lower town for communication between the lower town with its residential area and the upper town with its public buildings. This Central Gate must have been located at the west end of the Main Avenue. Unfortunately this part of the west wall lies buried beneath the national road and the Almirós/Soúrpi junction. It is likely that this gate was very simple, a mere opening in the wall near tower 62 (Reinders, 1988: map 2), because the west wall functioned as a second line of defence and the remains of a large courtyard gate would have been indicated in Stählin's map (1928). Stählin indeed observed some minor gates and it seems strange that a gate between the two parts of the city should have left no traces at the surface. If there was indeed a Central Gate at the west end of the Main Avenue, a short communication line between the Almirós plain in the north and the Soúrpi plain in the south will have been provided via the Northwest, Central and South Gate, avoiding the route through the lower town via the Main Avenue and Avenue D to the Southeast Gate.

At the site of the acropolis we found a small gate in the northern wall. It consisted of a simple opening in the wall flanked by tower 95 (Reinders, 1988: map 1). This gate, which provided access from the Almirós plain via a steep slope of the foothills of Óthris mountain to the acropolis, is the only gate in the entire acropolis wall discovered so far. If the upper town was accessible from the Soúrpi plain via the South Gate and from the lower town via the Central Gate, there must also have been a connection between the upper town and the acropolis. A detour to the acropolis from the upper town through the lower town and then uphill to the Acropolis Gate seems highly unlikely. No signs of any opening were found during the survey of the acropolis area in 1976 and unfortunately the east wall of the acropolis is not depicted in the map drawn by Stählin (1928). There may well have been a simple gate between towers 98 and 99.

All the remains of the city's east wall at the east end of the Main Avenue were removed in the 1830s. In my reconstruction of the east wall I placed a tower at the end of the Main Avenue (Reinders, 1988: map 2), but if we consider the possibilities that New Halos

functioned as a port of call and that the river Amphrysos was navigable for small craft, we would expect a 'Harbour' Gate at this position. No limestone blocks were observed at the end of the Main Avenue, but the foundations of a (courtyard?) gate may lie buried beneath the surface at this point.

In the reconstruction of the city of New Halos in figure 5.5 the enceinte is even more impressive than presented in previous publications. If the hypothesis of the existence of a Harbour Gate and a Central Gate is correct, the axiality of the city's layout will have been striking, in particular from the viewpoint of a sailor entering Soúrpi Bay. The hypothetical gates will also have provided better access to the city and more sophisticated communication lines between the lower town, the upper town and the acropolis, and also between the Almirós and Soúrpi plains.

Housing blocks and houses. The space within the enceinte of the lower town of Halos was filled with housing blocks with the same width and a varying length, which is a rather uncommon feature in Greek town planning in Classical and Hellenistic times. The housing blocks were longitudinally divided into two rows of house plots. In the first volume on New Halos it was observed that the housing blocks were filled with house plots of more or less the same depth and a varying width. The widths of the excavated houses presented in this volume are approx. 12.50, 13.75 and 15 m.

Housing block 6.4 is the only block in which four houses have been excavated and in which the arrangement of four houses with a width of 12.50 m between plots with a width of 15 m could be determined (fig. 5.6). The arrangement of the plots in the other housing blocks deserves further analysis. The width of the house of the Geometric *Krater*, for instance, has been reconstructed as approx. 12.50 m, but a width of 15 m is also possible in the housing block's layout (chapter 2). Unfortunately we were unable to uncover the remains of the three adjacent houses in housing block 6.4 in a single campaign, so they were separately measured and documented. The plans of the House of the Snakes, the House of the Ptolemaic Coins and the House of Agathon have been joined in figures 5.7 and 5.8. From these plans it is clear that the foundations of the houses were constructed *'aus einem Guss'* because the stones of the walls of adjacent houses were bonded and adjacent houses shared a common partition wall. Re-excavation and re-measurement will be an objective of future research.

From the excavation we know that the houses had a simple layout: a large room with two smaller rooms on either side and a courtyard situated to the south of the living quarters (chapter 2). The walls consisted of a limestone foundation surmounted by a mudbrick upper structure; the living quarters were covered with a roof consisting of large tiles resting on a wooden

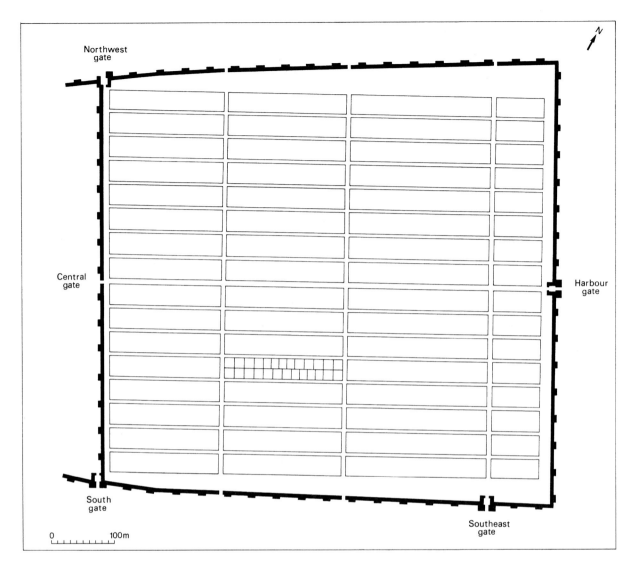

Fig. 5.6. Plan of the lower town of New Halos: reconstruction.

structure. The reconstruction of these houses is how-ever still a matter of discussion and uncertainties re-main concerning the height of the foundation, the tim-ber roof structure, the shape of the roof, how water was discharged, whether there was an *andron* and a *pas-tas*, the houses' water supply, *etc.* What we do know for certain is that the houses of New Halos had a sim-ple layout and were simply furnished.

Inhabitants. What do we know about the inhabitants of New Halos? I assumed that non-combatants of Demetrios Poliorketes' army and the inhabitants of Classical Halos who lived in the Krokion plain after the destruction of their city in 346 BC were among the first settlers of New Halos (Reinders, 1988: pp. 182-183). Although the city's enceinte is impressive, it is unlikely that a large proportion of the inhabitants was involved in warfare or defence at the time when the

city was abandoned; the number of weapons found in the six houses is too small to represent a population of warriors.

The simple structure of the houses and the nature of the artefacts presented in chapter 3 suggest an urban society without much luxury, in contrast to other cities like Olynthos, Priene, Delos, Kassope and Eretria.

The excavations of the houses in New Halos yield-ed a good survey of the pottery types that were in use around 265 BC. Each house contained a number of large storage vessels, kitchenware and tableware. The greater part of the pottery was of local manufacture and only few good-quality imports were found (Ch. 3: 1).

Figurines and moulds yielded an indication of the profession of the inhabitant of one of the excavated houses: a coroplast. Agricultural implements, such as sickles and shears, suggest that the inhabitants of some

N

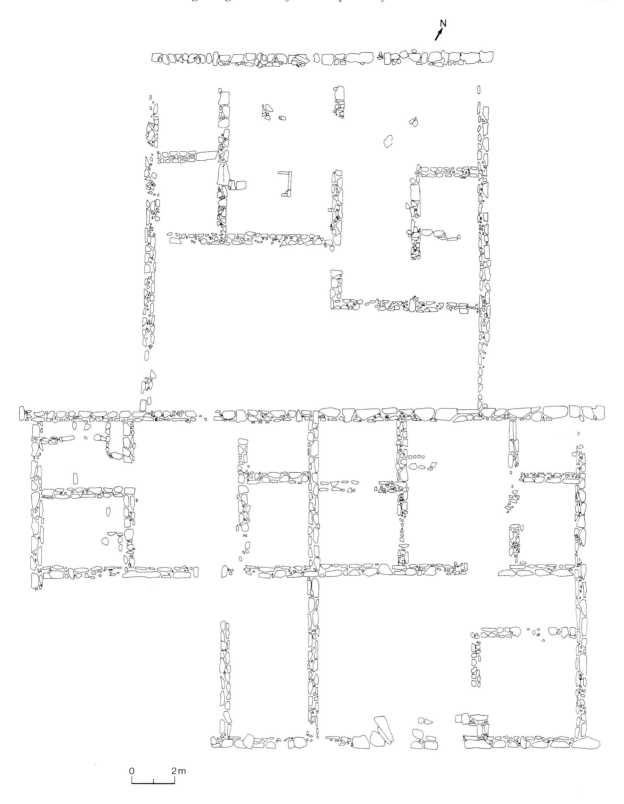

0 2m

Fig. 5.7. Foundations of the House of the Snakes (top), the House of the Ptolemaic Coins (bottom, left) and the House of Agathon (bottom, right).

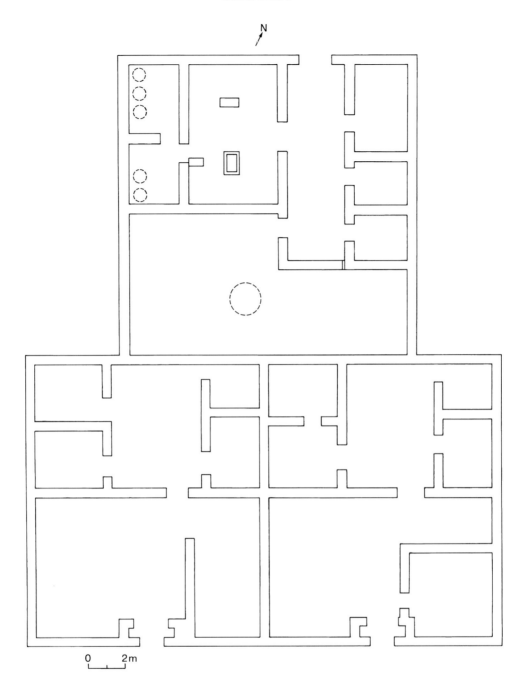

Fig. 5.8. Plan of the House of the Snakes, the House of the Ptolemaic Coins and the House of Agathon.

of the houses at least were farmers. It could well be that a large proportion of the 9,000 inhabitants was in one way or another involved in farming. However, six houses is too small a proportion of the original number of 1500 houses in the lower town to allow hard statements about the nature of the community of New Halos.

Coin circulation. The citizens of New Halos struck their own coins with denominations of approx. 2 and

6 g. They showed their independence by depicting their legendary ancestor Phryxos and the ram with the golden fleece on the reverse of the coins. Although it is generally accepted that the circulation of a city's bronze coinage was restricted to the *polis* itself in Hellenistic times, the numerous finds of bronze coins of other cities in the houses of New Halos prove that such coins also circulated outside the *polis* where they had been struck (Ch. 3: 5). At least three denominations of bronze coinage of approx. 2, 6 and 15 g were

used in New Halos. It is tempting to assume that these denominations represent values related to silver coinage, but this hypothesis cannot be proved due to the absence of value marks on the coins.

Transhumance. Chapter 4: 4 discusses animal remains, animal husbandry and in particular transhumance. There was no room for any milking pens or byres inside the enceinte of the lower town. It is also unlikely that such structures were to be found between or to the west of the public buildings in the upper town, in spite of the fact that the South Gate provided access to the upper town; some room may have been reserved here for animals in times of danger. In all probability cattle, sheep and goat were kept in the plains in winter, spring and autumn.

There is an ongoing discussion about the question as to whether the pastoralists of the cities and villages in the plains in Greece pastured their sheep and goats in the pastures of the mountain areas in the summer in Hellenistic times. We gathered data on the present-day pastoralists of the Almirós area and studied indications of transhumance in written sources. We also studied the present vegetation and analysed a core from a small swamp on Óthris mountain in order to reconstruct the vegetational history of the mountain area and the impact of human activity on the landscape.

The vegetation on Óthris mountain started to change as a result of anthropogenic influence around 1000 BC (Ch. 4: 3). In all probability the summer pastures of Óthris mountain were used by the pastoralists of the towns around this mountain in Hellenistic times. The short distance of approx. 10–16 km between the plain and the mountain, the small difference in altitude of less than 1700 m, the favourable winter conditions in the Almirós plain along the shore of the Pagasitikós gulf and the presence of urban markets support the hypothesis of short-distance transhumance and husbandry by specialised pastoralists.

End of the occupation. The excavation of the six houses in the Hellenistic city of New Halos revealed that the citizens abandoned their houses around 265/4 BC, after a disaster during which many objects in the houses were crushed. Occupation remains dating from shortly after 265 BC found at the site of the Southeast Gate indicate that the gate no longer provided access to the city by this date. No traces of a large fire or enemy action have been found, nor any evidence suggesting a different reason for the city's sudden abandonment. The Almirós plain is a tectonically active area. The pattern of the seismic phenomena that occurred in the Almirós plain in 1980 – a series of foreshocks, two main shocks with offshore epicentres and aftershocks – makes it plausible that the 9,000 inhabitants of New Halos abandoned their city after an earthquake (Ch. 5: 3).

5. REFERENCES

AMBRASEYS, N.N. & J.A. JACKSON, 1990. Seismicity and associated strain of Central Greece between 1890 and 1988. *Geophysical Journal International* 101, pp. 1–46.

BOTTEMA, S., 1994. The prehistoric environment of Greece: A review of the palynological record. In: P.N. Kardulias (ed.), *Beyond the site. Regional studies in the Aegean area.* University Press of America, Lanham, pp. 45–68.

HAAGSMA, M. & R. REINDERS, 1991. Verwoesting door een aardbeving. *Tijdschrift voor Mediterrane Archeologie* 8, pp. 16–25.

MALAKASIOTI, Z., V. RONDIRI & E. PAPPI, 1993. Prosfates anaskafikes erevnes sti archaia Alo: mia parousiasi ton evrimaton. In: V. Kondonatsios (ed.), *Achaiaphthiotika* 1. Ores, Almiros, pp. 93–101.

FURTWÄNGLER, A.E., 1990. *Demetrias, eine makedonische Gründung im Netz hellenistischer Handels- und Geldpolitik.* Habilitationsschrift. Heidelberg.

ICGH = Thompson, M. *et al.*, 1973.

MØRKHOLM, O., 1991. *Early Hellenistic Coinage from the Accession of Alexander to the Peace of Apamea (336–188 BC).* University Press, Cambridge.

OECONOMIDES, M., 1994. Les trésors Thessaliens du Musée Numismatique d' Athènes. In: R. Misdrahi-Kapon (ed.): *La Thessalie. Quinze anées de recherches archéologiques, 1975–1990. Bilans et perspectives,* Vol. B. Kapon, Athènes, pp. 335–338.

PAPAZACHOS, B.C., D.G. PANAGIOTOPOULOS, T.M. TSAPANOS, D.M. MOUNTRAKIS & G.CH. DIMOPOULOS, 1983. A study of the 1980 summer seismic sequence in the Magnesia region of Central Greece. *Geophys. J.R. astr. Soc.* 75, pp. 155–168.

PAPAZACHOS, B. & K. PAPAZACHOS, 1989. *I sismi tis Elladas.* Ziti, Thessaloniki.

PICARD, O., 1979. *Chalcis et la Confédération Eubéenne. Étude de Numismatique et d'Histoire (IVve–Ier siècle).* De Boccart, Paris.

REINDERS, H.R., 1988. *New Halos, a Hellenistic Town in Thessalía, Greece.* Hes, Utrecht.

REINDERS, H.R., 1992. The end of the occupation of Hellenistic Halos. An earthquake about 260 BC? Proceedings of *Diethnes sinedrio yia tin Archaia Thessalia sti mnimi tou Dimitri R. Theochari (Volos 1987).* Athens, pp. 430–436.

REINDERS, H.R., 1997. Ta nomismata tis Alou. In: V. Kondonatsios (ed.), *Achaiophthiotika* 2. Almiros, pp. 103–120.

REINDERS, H.R., Y. Dijkstra, V. Rondiri, S.J. Tuinstra & Z. Malakasioti, 1996. The Southeast Gate of New Halos. *Pharos* 4, pp. 121–138.

STIROS, S. & S. PAPAGEORGIOU, 1994. Post Mesolithic evolution of the Thessalian Landscape. In: R. Misdraki-Kapon (ed.): *La Thessalie. Quinze années de recherches archéologiques, 1975–1990. Bilans et perspectives,* Vol. A. Kapon, Athènes, pp. 29–36.

THOMPSON, M., O. MØRKHOLM & C. M. KRAAY, 1973. *An Inventory of Greek Coin Hoards* (=ICGH). The American Numismatic Society, New York.

WACE, A.J.B., & M.S. THOMPSON, 1912. *Prehistoric Thessaly.* Cambridge.

APPENDIX 1. CATALOGUE OF POTTERY

COLETTE BEESTMAN-KRUYSHAAR

Abbreviations
H.   = Height
Pres. H.   = Preserved height
Rec. H.   = Reconstructed height
Th.   = Thickness
Dim.   = Dimensions
D.   = Diameter
All measurements are in millimeters.

COOKING VESSELS

Chytra P1–73

Type 1. Convex base, globular body, everted rim, convex lip. Vertical strap handle.
1a. Medium size, medium coarse orange ware.

P1   House 1, Inv. No. X 644, Fig. 6.1. Complete profile (= Reinders, 1988: Cat. No. 33.09). Dim.: H. 123; Th. wall 4; D. rim 89; D. body 130.

P2   House 1, Inv. No. X 601, Fig. 6.1. Complete profile. Slight angle halfway body (= Reinders, 1988: Cat. No. 33.07). Dim.: H. 108; Th. wall 4; D. rim 90; D. body 123.

P3   House 1, Inv. No. X 643, Fig. 6.1. Complete profile. Slight angle halfway body. Convex lip with ridge on inside (= Reinders, 1988: Cat. No. 33.08). Dim.: H. 91; Th. Wall 4-5; D. rim 74; D. body 100.

P4   House 4, Inv. No. HA 127/42. Complete pot. Concentric grooves from the base to the widest diameter of the body. Dim.: H. 78; Th. wall 2–4; D. rim 120; D. body 230.

P5   House 4, Inv. Nos. HA 136/11 + HA 149/3. Almost complete profile, base missing. Flattened lip, thickened on the outside, transition marked with groove. Dim.: Pres. H. 126; Th. wall 4–6; D. rim 95; D. body 155.

P6   House 6, Inv. No. H23 5/104. Complete profile. Flattened lip, thickened on the outside. Dim.: H. 80; Th. wall 3–4; D. rim 60; D. body 90.

P7   House 6, Inv. Nos. H23 5/125+127a. Complete profile. Dim.: H. 101; Th. wall 3–4; D. rim 120; D. body 178.

P8   House 3, Inv. Nos. HA 197/18+24 + HA 202/2b+27+28 + HA 250. Rim fragment. Everted rim with internal angle, lip bevelled on the outside. Dim.: Pres. H. 63; Th. wall 4; D. rim 150.

P9   House 1, Inv. No. X 309. Rim fragment (= Reinders, 1988: Cat. No. 37.04). Dim.: Pres. H. 98; Th. Wall 5; D. rim 140.

P10   House 4, Inv. No. HA 106/48. Rim fragment. Dim.: Pres. H. 20; Th. wall 5.

P11   House 4, Inv. No. HA 127/16a. Rim fragment. Dim.: Pres. H. 101; Th. wall 3–4; D. rim 130.

P12   House 2, Inv. No. H21 3/19. Rim and base fragments. Dim.: Pres. H. 46; Th. wall 3–4; D. rim 88.

P13   House 6, Inv. No. H23 1/104-1. Rim fragment. Dim.: Pres. H. 42; Th. wall 5; D. rim 95.

P14   House 6, Inv. No. H23 1/108. Rim fragment. Dim.: Pres. H. 36; Th. wall 3; D. rim 130.

P15   House 6, Inv. No. H23 1/114a-1. Rim fragment. Dim.: Pres. H. 22; Th. wall 3; D. rim 130.

P16   House 6, Inv. No. H23 1/114a-2. Rim fragment. Dim.: Pres. H. 27; Th. wall 3; D. rim 105.

P17   House 6, Inv. No. H23 1/128. Rim fragment Dim.: Pres. H. 33; Th. wall 7.

P18   House 6, Inv. No. H23 1/5b. Rim fragment. Dim.: Pres. H. 22; Th. wall 6; D. rim 100.

P19   House 6, Inv. No. H23 3/78. Rim fragment. Dim.: Pres. H. 19; Th. wall 4; D. rim 130.

P20   House 6, Inv. No. H23 3/85. Rim and handle fragments. Dim.: Pres. H. 25; Th. wall 7.

P21   House 6, Inv. No. H23 5/103. Rim and handle fragments. Dim.: Pres. H. 265; Th. wall 4; D. rim 120; D. body 260.

P22   House 6, Inv. No. H23 5/106a. Rim fragment. Dim.: Pres. H. 33; Th. wall 4; D. rim 130.

P23   House 6, Inv. No. HA 4/12. Rim fragment. Dim.: Pres. H. 18; Th. wall 5; D. rim 100.

P24   House 6, Inv. No. HA 42/5. Rim fragment. Dim.: Pres. H. 25; Th. wall 4; D. rim 90.

P25   House 6, Inv. No. HA 68/7. Rim fragment. Dim.: Pres. H. 54; Th. wall 5; D. rim 125.

P26   House 5, Inv. Nos. HA 14/1+4. Rim fragment. Dim.: Pres. H. 65; Th. wall 5; D. rim 120.

P27   House 5, Inv. No. HA 14/20. Rim fragment. Dim.: Pres. H. 29; Th. wall 4; D. rim 80.

P28   House 5, Inv. No. HA 26/3. Rim fragment. Dim.: Pres. H. 22; Th. wall 4; D. rim 110.

P29   House 5, Inv. No. HA 39/12. Rim fragment. Dim.: Pres. H. 21; Th. wall 4; D. rim 130.

P30   House 5, Inv. No. HA 41/3. Rim fragment. Dim.: Pres. H. 32; Th. wall 3; D. rim 110.

P31   House 5, Inv. Nos. HA 55/21+22+24+39. Rim and handle fragments. Dim.: Pres. H. 91; Th. wall 3; D. rim 74; D. body 95.

P32   House 5, Inv. No. HA 85/5. Rim fragment. Dim.: Pres. H. 51; Th. wall 5; D. rim 70.

P33   House 5, Inv. No. HA 51/12. Rim fragment. Dim.: Pres. H. 26; Th. wall 3; D. rim 80.

P34   House 5, Inv. No. HA 53/14. Rim fragment. Dim.: Pres. H. 18; Th. wall 3; D. rim 80.

P35   House 1, Inv. No. X 659. Base fragment (= Reinders, 1988: Cat. No. 33.16). Dim.: Pres. H. 54; Th. wall 3–4; D. body 120.

P36   House 4, Inv. No. HA 155/6. Base fragment. Dim.: Pres. H. 63; Th. wall 3–4; D. body 140.

P37   House 1, Inv. No. X 109. Wall fragment (= Reinders, 1988: Cat. No. 35.09). Dim.: Pres. H. 89; Th. wall 6; D. body 160.

P38   House 4, Inv. No. HA 103/6. Wall fragment. Dim.: Th. wall 4.

1b. Medium size, medium coarse red ware.

P39   House 3, Inv. Nos. HA 193/14+22, Fig. 6.1. Complete profile. Two incisions on lower body. Dim.: H. 120; Th. wall 2–9; D. rim 100; D. body 120.

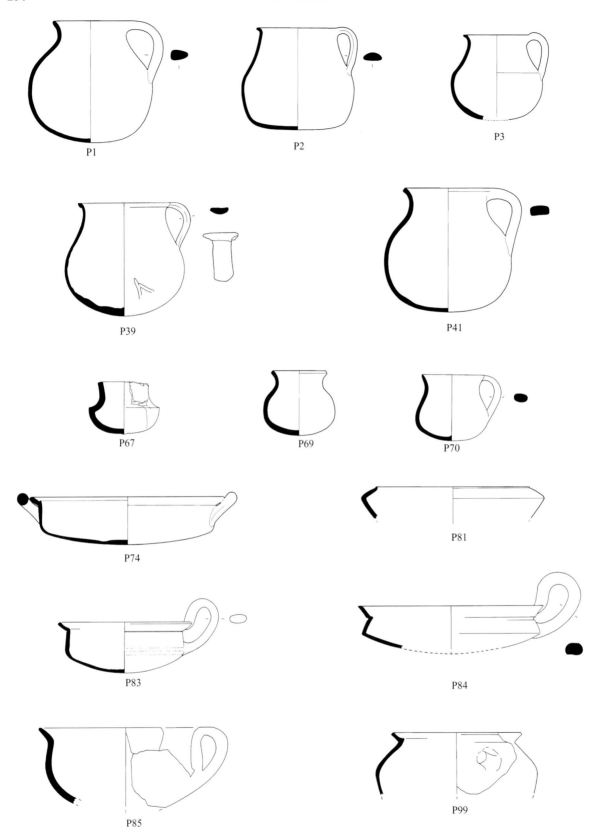

Fig. 6.1. Cooking vessels (scale 1:4).

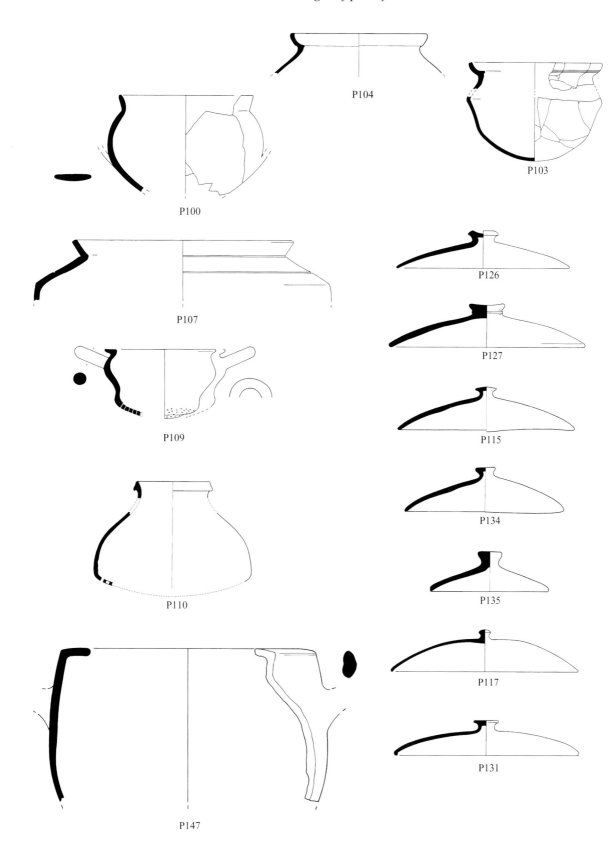

Fig. 6.1. Cooking vessels (continued, scale 1:4).

P40 House 3, Inv. Nos. HA 197/3+10a + HA 202/2b+4+5+ 24+29+36. Complete profile. Dim.: H. 123; Th. wall 2–5; D. rim 80; D. body 120.

P41 House 1, Inv. No. X 103, Fig. 6.1. Almost complete profile, base missing (= Reinders, 1988: Cat. No. 35.08). Dim.: Pres. H. 148; Th. wall 5; D. rim 90; D. body 140.

P42 House 1, Inv. No. X 306. Rim fragment (= Reinders, 1988: Cat. No. 37.09). Dim.: Pres. H. 94; Th. wall 4; D. rim 110.

P43 House 4, Inv. No. HA 106. Rim and handle fragments. Dim.: Pres. H. 40; Th. wall 4; D. rim 140.

P44 House 4, Inv. Nos. HA 113/4 + HA 129/4. Rim and handle fragments. Dim.: Pres. H. 34; Th. wall 4; D. rim 110.

P45 House 6, Inv. No. H23 3/41a. Rim fragment. Dim.: Pres. H. 32; Th. wall 4; D. rim 90.

P46 House 6, Inv. No. H23 3/72. Rim fragment. Dim.: Pres. H. 60; Th. wall 5; D. rim 110.

P47 House 6, Inv. No. H23 6/109a. Rim fragment. Dim.: Pres. H. 26; Th. wall 4; D. rim 80.

P48 House 6, Inv. No. H23 5/109b. Rim fragment. Dim.: Pres. H. 45; Th. wall 3; D. rim 100.

P49 House 6, Inv. No. H23 6/133. Rim fragment. Dim.: Pres. H. 36; Th. wall 5–8; D. rim 110.

P50 House 6, Inv. No. H23 6/175. Rim fragment. Dim.: Pres. H. 29; Th. wall 3; D. rim 90.

P51 House 6, Inv. No. H23 6/75. Rim fragment. Dim.: Pres. H. 20; Th. wal 6; D. rim 200.

P52 House 6, Inv. No. H23 6/85. Rim fragment. Dim.: Pres. H. 35; Th. wall 4; D. rim 110.

P53 House 6, Inv. Nos. HA 4/38 + HA 43/15. Rim fragment. Dim.: Pres. H. 59; Th. wall 6; D. rim 150.

P54 House 5, Inv. No. HA 26/1. Rim fragment. Dim.: Pres. H. 24; Th. wall 4; D. rim 70.

P55 House 5, Inv. No. HA 41/5. Rim fragment. Dim.: Pres. H. 27; Th. wall 5; D. rim 90–130.

P56 House 5, Inv. No. HA 60/17. Rim fragment. Dim.: Pres. H. 33; Th. wall 3; D. rim 80.

P57 House 5, Inv. Nos. HA 66/13+16+17+20+22+40. Rim fragment. Dim.: Pres. H. 19; Th. wall; D. rim 110.

P58 House 2, Inv. Nos. H21 3/38+43. Base fragment. Dim.: Pres. H. 119; Th. wall 5–7; D. body 200.

P59 House 5, Inv. No. HA 66/43. Base and handle fragments. Dim.: Pres. H. 156; Th wall 4; D. body 160.

1c. Medium size, New Halos ware.

P60 House 3, Inv. No. HA 202/25. Base and rim fragments. Angular shoulder. Dim.: Pres. H. base: 14; Pres. H. rim: 68; Th. wall 2–3; D. rim 112; D. body 150.

P61 House 6, H23 5/154. Complete profile. Dim.: H. 89; Th. wall 3; D. rim 77; D. body 120.

P62 House 3, Inv. Nos. HA 197/5+10a + HA 202/4+24+29+36. Rim fragment. Everted rim with internal angle. Dim.: Pres. H. 63; Th. wall 3; D. rim 81; D. body 110.

P63 House 1, Inv. No. X 96A. Rim fragment (= Reinders, 1988: Cat. No. 33.21). Dim.: Pres. H. 69; Th. wall 4; D. rim 120; D. body 103.

P64 House 4, Inv. Nos. HA 128/17+18. Rim fragment. Dim.: Pres. H. 41; Th. wall 3; D. rim 120.

P65 House 3, Inv. No. HA 197/5. Base and body fragments. Dim.: Pres. H. 78; Th. wall 3; D. body 117.

P66 House 3, Inv. Nos. HA 202/24+25+38. Rim, body and base fragments. Angular shoulder. Dim.: Rec. H. 49; H. base 14; Th. wall 2–4; D. rim 112; D. body 150.

Type 2. Small size. Convex base, angular body, everted rim, convex lip. Vertical strap handle.
2a. Medium coarse red ware.

P67 House 6, Inv. No. H23 2/51, Fig. 6.1. Complete profile. Dim.: H. 56; Th.wall 5; D. rim 53; D. body 78.

2b. New Halos ware.

P68 House 4, HA 168/2. Base fragment. Groove on transition of shoulder to rim. Dim.: Pres. H. 24; Th. wall 4–5; D. body 70.

Type 3. Small size. Convex base, globular body, everted rim, convex lip. Vertical strap handle. Medium coarse orange ware.

P69 House 1, Inv. No. X 336, Fig. 6.1. Complete profile (= Reinders, 1988: Cat. No. 37.06). Dim.: H. 71; Th. wall 4–6; D. rim 60; D. body 80.

P70 House 2, Inv. No. H21 2/50, Fig. 6.1. Complete pot. Dim.: H. 70; Th. wall 5; D. rim 64; D. body 80.

P71 House 6, Inv. No. H23 5/146. Base fragment. Dim.: Pres. H. 35; Th. wall 4; D. body 80.

P72 House 5, Inv. No. HA 49/3. Rim fragment. Dim.: Pres. H. 64; Th. wall 4; D. rim 60; D. body 80.

Type 4. Large size. Convex base, globular body, everted rim, convex lip. Vertical strap handle. New Halos ware.

P73 House 5, Inv. Nos. HA 36/3 + HA 59/9. Almost complete profile, part of base missing. Dim.: Pres. H. 284; Th. wall 2–8; D. rim 290; D. body 360.

Lopas P74–98

Type 1. Convex base, straight wall, flaring rim with flange for lid, convex lip. Two horizontal ring handles, circular in section.
1a. Medium coarse orange ware, smoothened surface.

P74 House 1, Inv. X 89, Fig. 6.1. Complete pot (= Reinders, 1988: Cat. No. 35.07). Dim.: H. 51; Th. wall 3–6; D. rim 213; D. body 190.

1b. Medium coarse red ware.

P75 House 3, Inv. Nos. HA 192/7 + HA 193/16 + HA 195/6. Complete profile. Convex lip, thickened on inside and outside. Dim.: H. 58; Th. wall 4; D. rim 180; D. body 160.

P76 House 5, Inv. Nos. HA 55/2+12. Rim fragment. Dim.: Pres. H. 19; Th. wall 5; D. rim 150.

Type 2. Splayed wall, flaring rim with flange for lid, convex lip. Two horizontal ring handles with circular section. Medium coarse orange ware.

P77 House 3, Inv. Nos. HA 207/1+7+9 + HA 256/1. Complete profile. Flat base. Dim.: H. 44; Th. wall: 3–4; D. rim 200–210; D. body 184.

P78 House 4, Inv. No. HA 130/32. Rim fragment. Dim.: Pres. H. 25; Th. wall 5; D. rim 130.

Type 3. Angular wall, slightly everted rim. Two horizontal ring handles.
3a. Medium coarse red ware.

P79 House 2, Inv. No. H21 2/10. Rim fragment. Convex lip, thickened on the outside. Dim.: Pres. H. 22; Th. wall 4; D. rim 170.

P80 House 5, Inv. No. HA 85/1, Fig. 6.1. Rim fragment. Lip bevelled on the inside. Dim.: Pres. H. 58; Th. wall 5; D. rim 150.

3b. Medium coarse orange ware.

P81   House 1, Inv. No. X 727, Fig. 6.1. Rim fragment (= Reinders, 1988: Cat. No. 34.07). Dim.: Pres. H. 16; Th.wall 4; D. rim 168; D. body 178.

P82   House 4, Inv. No. HA 163/11a. Rim fragment. Dim.: Pres. H. 28; Th. wall 4; D. rim 180; D. body 200.

Type 4. Convex base, straight wall, flaring rim with flange for lid, convex lip, slightly thickened on the inside. Vertical strap handle, rising from base, arching above and down to lip. Medium coarse orange ware.

P83   House 3, Inv. Nos. HA 197/8+10a + HA 202/32, Fig. 6.1. Complete pot. Dim.: H. 54; Th. wall 3–4; D. rim 146; D. body 130.

Type 5. Convex base, angular wall, flaring rim with flange for lid, flattened lip, slightly thickened on the outside. Groove on transition from body to rim. Vertical strap handle, rising from base, arching above and down to lip. New Halos ware.

P84   House 4, Inv. No. HA 127/62, Figs 3.1 and 6.1. Complete profile. Dim.: H. 45; T. wall 3; D. rim 200; D. body 192.

Type 6. Splayed wall, flaring rim, convex lip. Vertical strap handle, rising from base to lip. Medium coarse orange ware, probably slibbed. Irregularly shaped. This vessel may not be Hellenistic.

P85   House 2, Inv. No. H21 3/34, Fig. 6.1. Rim fragment. Dim.: Pres. H. 62; Th. wall 3–5; D. rim 140; D. body 130.

7. Rim fragments. Medium coarse orange ware.

P86   House 3, Inv. No. HA 253/43a. Dim.: Pres. H. 30; Th. wall 5; D. rim 200.

P87   House 4, Inv. No. HA 146/9. Dim.: Pres. H. 24; Th. wall 4; D. rim 190–230; D. body 158–198.

P88   House 4, Inv. No. HA 127/40c. Dim.: Pres. H. 18; Th. wall 4; D. rim 170.

P89   House 4, Inv. No. HA 130/90. Dim.: Pres. H. 21; Th. wall 5; D. rim 250–285.

P90   House 6, Inv. No. HA 62/7. Dim.: Pres. H. 40; Th. wall 5; D. rim 180.

P91   House 6, Inv. No. H23 3/33a. Dim.: Pres. H. 27; Th. wall 4; D. rim 200.

P92   House 5, Inv. No. HA 55/15. Dim.: Pres. H. 18; Th. wall 4; D. rim 160.

8. Rim fragments. Medium coarse red ware.

P93   House 4, Inv. No. HA 153/27. Dim.: Pres. H. 20; Th. wall 4; D. rim 170–200.

P94   House 4, Inv. No. HA 129/10. Dim.: Pres. H. 38; Th. wall 4; D. rim 200; D. body 180.

P95   House 2, Inv. No. H21 2/70. Dim.: Pres. H. 35; Th. wall 4; D. rim 150.

P96   House 2, Inv. Nos. H21 2/3+11. Dim.: Pres. H 40; Th. wall 5; D. rim 170–190.

P97   House 5, Inv. Nos. HA 54/1+2. Almost complete profile, handle missing. Dim.: Pres. H. 51; Th. wall 6; D. rim 240; D. body 207.

9. Rim fragment. New Halos ware.

P98   House 2, Inv. No. H21 2/56, Fig. 6.1. Almost complete profile, handle missing. Dim.: Pres. H. 48; Th. wall 4–5; D. rim 160; D. body 145.

Stewpot P99–108

Type 1. Small size.
1a. Medium coarse orange ware.

P99   House 6, Inv. No. H23 3/41, Fig. 6.1. Almost complete profile, base and rim not fitting. Convex base, globular body with slight angle on widest part, flaring rim with flange for lid, bevelled lip. Horizontal ring handle, circular in section, on shoulder. Dim.: Pres. H. rim 66; Th. wall 44; D. rim 140; D. body 171.

P100   House 2, Inv. Nos. H21 5/14+21, Fig. 6.1. Rim and body fragments. Handle missing. Globular body, flaring rim with internal angle, convex lip. Attachment of vertical strap handle, flattened in section, on widest part of body. Very fine temper, reddish yellow to light red colour, traces of slip or glaze. This vessel may not be Hellenistic. Dim.: Pres. H. 76; Th. wall 4; D. rim 110; D. body 130.

P101   House 6, Inv. No. H23 6/101. Rim fragment. Dim.: Pres. H. 17; Th. wall 5; D. rim 100.

P102   House 6, Inv. No. H23 6/109. Wall fragment with vertical strap handle attachment, flattened in section. Dim.: Pres. H. 15; Th. wall 4.

1b. New Halos ware.

P103   House 1, X 636, Fig. 6.1. Complete profile. Convex base, slightly angular transition to angular body, flaring rim with flange for lid, bevelled lip. Two parallel grooves on the outside of the rim. No handles preserved (= Reinders, 1988: Cat. No. 33.31). Dim.: H. 102; Th. wall 3–4; D. rim 110; D. body 140.

1c. Medium coarse red ware.

P104   House 1, Inv. No. X 632, Fig. 6.1. Rim fragment. Straight shoulder, flaring rim with flange for lid, flattened lip (= Reinders, 1988: Cat. No. 33.06). Dim.: Pres. H. 43; Th. wall 3–6; D. rim 154.

Type 2. Large size
2a. New Halos ware.

P105   House 1, Inv. No. X ?. Rim fragment. Dim.: Pres. H. 30; Th. wall 5–6; D. rim 190.

P106   House 6, Inv. No. H23 5/122a. Rim fragment. Dim.: Pres. H. 32; Th. wall 5–6; D. rim 228.

2b. Medium coarse red ware.

P107   House 3, Inv. No. HA 269/2, Fig. 6.1. Rim fragment. Dim.: Pres. H. 64; Th. wall 6; D. rim 240; D. body 320.

P108   House 1, Inv. Nos. X 606 + X 634. Rim and handle fragments. Vertical ring handle, oval in section. (= Reinders, 1988: Cat. No. 33.05). Dim.: Pres. H. 78; Th. wall 4; D. rim 230; D. body 270.

Strainer P109–114

Type 1. Convex base with pierced holes, with angular transition to body, splayed body with slightly angular shoulder, everted rim with flattened top, convex lip. Two horizontal ring handles attached to shoulder, circular in section. Medium coarse orange ware.

P109   House 4, Inv. No. HA 106/1, Figs 3.2 and 6.1. Complete profile. Dim.: H. 74; Th. wall 4; D. rim 130; D. body 132; D. base 109; D. holes 1.

Type 2. Convex base with pierced holes, angular transition to body, broad tapering body, everted rim, overhanging convex lip. Medium coarse red ware.

P110   House 1, Inv. No. X 721, fig. 6.1. Almost complete profile, part of base missing (= Reinders, 1988: Cat. No. 34.04). Dim.: Pres. H. 124; Th. wall 4; D. rim 80; D. base 170; D. holes 3–4.

Type 3. Base fragments with pierced holes.
3a. Medium coarse red ware.

P111   House 6, Inv. Nos. HA 17/14+15. Dim.: Th. wall 5; D. holes 6.

P112   House 5, Inv. Nos. HA 59/30-32. Dim.: Th. wall 5; D. holes 4.

P113   House 5/6, Inv. Nos. HA 64/15+16. Dim.: Th. wall 4; D. holes 2.

3b. Medium coarse orange ware.

P114   House 5, Inv. No. HA 60/16. Dim.: Th. wall 4; D. holes 4.

Lid P115–146

Type 1. Domed or conical lid, small conical knob.
1a. New Halos ware.

P115   House 1, Inv. No. X 603, Fig. 6.1. Complete profile. Slightly domed (= Reinders, 1988: Cat. No. 33.10). Dim.: H. 46; Th. wall 4; D. rim 194; D. knob 23.

P116   House 6, Inv. No. H23 1/117. Complete profile. Conical. Dim.: H. 41; Th. wall 5; D. rim 150; D. knob 17.

P117   House 6, Inv. No. H23 5/156, Fig. 6.1. Complete profile. Domed. Dim.: H. 45; Th. wall 3; D. rim 205; D. knob 14.

P118   House 6, Inv. No. H23 2/83. Knob. Dim.: Pres. H. 17; D. knob 32.

P119   House 6, Inv. No. H23 2/58a. Knob. Dim.: Pres. H. 11; D. knob 18.

P120   House 6, Inv. No. H23 3/96. Knob. Dim.: Pres. H. 15; D. knob 30.

1b. Medium coarse red ware.

P121   House 6, Inv. No. H23 5/161. Complete profile. Slightly domed. Dim.: H. 32; Th. wall 4; D. rim 134; D. knob 20.

P122   House 2, Inv. No. H21 3/22. Knob. Dim.: Pres. H. 10; D. knob 14.

P123   House 5, Inv. No. HA 54/5. Knob. Dim.: Pres. H. 19; Th. wall 3; D. knob 30.

1c. Fine pink ware.

P124   House 3, Inv. No. HA 247/35. Complete profile. Conical. Dim.: H. 38; Th. wall 4; D. rim 180; D. knob 20.

P125   House 4, Inv. No. HA 171/4. Knob. Dim.: Pres. H. 43; D. knob 24.

Type 2. Slightly domed lid, knob with concave top.
2a. Medium coarse red ware.

P126   House 3, Inv. No. HA 237/3b, Fig. 6.1. Complete profile. Dim.: H. 40; Th. wall 4; D. rim 190; D. knob 31.

P127   House 5, Inv. No. HA 51/133, Fig. 6.1. Complete profile. Dim.: H. 51; Th. wall 6; D. rim 210; D. knob 36.

P128   House 2, Inv. No. H21 5/19. Knob. Dim.: Pres. H. 18; D. knob 37.

2b. Medium coarse orange ware.

P129   House 4, Inv. Nos. HA 106/1+9. Rim fragment and knob. Dim.: H. rim 15; H. knob 20; Th. wall 6; D. rim 160; D. knob 35.

P130   House 6, Inv. No. HA 78B/13. Knob. Dim.: Pres. H. 17; Th. wall 5; D. knob 25.

Type 3. Slightly domed lid, knob with flattened top.
3a. New Halos ware.

P131   House 6, Inv. No. H23 3/30, Fig. 6.1. Complete profile. Dim.: H. 37; Th. wall 5; D. rim 200; D. knob 28.

P132   House 4, Inv. No. HA 148/4. Knob. Dim.: Pres. H. 11; D. knob 24.

P133   House 2, Inv. No. H21 1/21. Knob. Dim.: Pres. H. 11; D. knob 23.

3b. Medium coarse orange ware.

P134   House 1, Inv. No. X 660, Fig. 6.1. Complete profile. Slightly domed (= Reinders, 1988: Cat. No. 33.11). Dim.: H. 47; Th. wall 5; D. rim 175; D. knob 22.

P135   House 1, Inv. No. X 728, Fig. 6.1. Complete profile. Conical (= Reinders, 1988: Cat. No. 34.05). Dim.: H. 42; Th. wall 5; D. rim 129; D. knob 27.

3c. Medium coarse red ware.

P136   House 6, Inv. No. HA 2/8. Knob. Dim.: Pres. H. 29; D. knob 28.

Type 4. Domed lid, rim fragments.
4a. Medium coarse red ware.

P137   House 4, Inv. No. HA 153/18. Dim.: Pres. H. 14; Th. wall 2–3; D. rim 190.

P138   House 6, 2 H23 1/100b. Dim.: Pres. H. 34; Th. wall 4–5; D. rim 220.

P139   House 6, Inv. No. H23 1/149. Dim.: Pres. H. 14; Th. wall 6; D. rim 190.

P140   House 5, Inv. No. HA 71. Dim.: Pres. H. 32; Th. wall 6; D. rim 120.

P141   House 6, Inv. No. HA 70/2. Dim.: Pres. H. 19; Th. wall 5; D. rim 190.

4b. New Halos ware.

P142   House 3, Inv. Nos. HA 193/37+41. Conical. Dim.: Pres. H. 13; Th. wall 4–5; D. rim 240.

P143   House 4, Inv. No. HA 127/16. Conical. Dim.: Pres. H. 8; Th. wall 4; D. rim 130.

P144   House 4, Inv. Nos. HA 130/24a + HA 148. Dim.: Pres. H. 32; Th. wall 3; D. rim 200–270.

4c. Medium coarse orange ware.

P145   House 4, Inv. Nos. HA 181 + HA 182/3. Dim.: Pres. H. 9; Th. wall 3–4; D. rim 230.

P146   House 6, Inv. No. HA 67/1. Dim.: Pres. H. 17; Th. wall 4; D. rim 180.

Brazier

P147   House 5, Inv. Nos. HA 38/1a + HA 71/12 + HA 81/7, Fig. 6.1. Rim fragment. Splayed lower body, almost straight upper body, inturned rim with external angle, flattened top, convex lip. Two parallel grooves on the outside of the body along the rim. On upper body, attachment for horizontal strap handle, oval in section. Medium coarse orange ware, bright orange colour. Dim.: Pres. H. 230; Th. wall 7-11; D. rim 220; D. body 280.

CLOSED VESSELS

Jug P148–164

Type 1. Raised flat base, wide splayed lower body, slightly convex shoulder, short concave neck, everted rim. Vertical strap handle, double in section, rising from shoulder, arching above and down to lip.
1a. Overhanging convex lip, thickened on the outside. New Halos ware.
P148   House 1, Inv. Nos. X 325+318, Figs 3.3 and 6.2. Reconstructed profile (= Reinders, 1988: Cat. No. 37.02). Dim.: Rec. H. 182; Th. wall 5–6; D. rim 86; D. body 184; D. base 114.

1b. Lip thickened on the outside, undercut, groove on edge. Medium coarse orange ware.
P149   House 4, Inv. Nos. HA 120/3 + HA 147/10+23 + HA 150 + HA 178/23+24, Fig. 6.2. Complete profile. Dim.: H. 214; Th. wall 4; D. rim 80; D. body 196; D. base 90.

Type 2. Convex base, globular body, narrow, slightly concave neck, everted rim, convex lip. Vertical strap handle with central rib, attached to shoulder and lip. Medium coarse orange ware.
P150   House 6, Inv. Nos. H23 5/24+152+164, Fig. 6.2. Complete profile. Dim.: H. 315; Th. wall 4–7; D. rim 91; D. body 280; D. base 190.

Type 3. Convex base, globular body, wide neck, everted rim, convex lip. Vertical strap handle with central rib, attached to shoulder and lip. Medium coarse orange ware.
P151   House 6, Inv. Nos. H23 5/103+125. Complete profile. Dim.: H. 305; Th. wall 4–6; D. rim 115; D. body 250; D. base 148.

Type 4. Raised flat base, globular body, wide straight neck, everted rim, convex lip. Vertical strap handle, oval in section, attached to shoulder and lip. Medium coarse orange ware.
P152   House 6, Inv. Nos. H23 5/127+140. Complete profile. Dim.: H. 322; Th. wall 5; D. rim 103; D. body 260; D. base 130.
P153   House 6, Inv. No. H23 5/144. Rim fragment. Dim.: Pres. H. 52; Th. wall 4–6; D. rim 130.

Type 5. Convex shoulder, narrow, slightly concave neck, everted rim, convex lip. Vertical strap handle with central rib, attached to shoulder and lip.
5a. Medium coarse red ware.
P154   House 6, Inv. No. H23 3/5. Rim and handle fragments. Dim.: Pres. H. 120; Th. wall 5; D. rim 100.
P155   House 6, Inv. No. H23 3/79. Rim and handle fragments. Thickened lip. Dim.: Pres. H. 130; Th. wall 6; D. rim 108.

5b. New Halos ware.
P156   House 6, Inv. Nos. HA 74/7+10. Rim and handle fragments. Thickened lip. Dim.: Pres. H. 70; Th. wall 4; D. rim 97.

Type 6. Straight neck and rim, convex lip, thickened on the outside, undercut.
6a. Medium coarse red ware.
P157   House 4, Inv. No. HA 129/18. Rim fragment. Dim.: Pres. H. 67; Th. wall 5–7; D. rim 150.

6b. Medium coarse orange ware.
P158   House 5/6, Inv. No. HA 64/17. Rim fragment. Dim.: Pres. H. 57; Th. wall 5; D. rim 164.

Type 7. Slightly convex shoulder, narrow, slightly spreading neck, everted rim, top flattened, convex lip, undercut. Vertical strap handle, oval in section, attached to shoulder and lip. Medium coarse orange ware.
P159   House 5, Inv. Nos. HA 75/3+69+76. Rim and handle fragments. Dim.: Pres. H. 130; Th. wall 4–5; D. rim 98.

Type 8. Slightly concave neck, everted rim, convex lip with groove on the edge. Fine temper, brown colour.
P160   House 2, Inv. No. H21 5/22. Rim fragment. Dim.: Pres. H. 38; Th. wall 3; D. rim 88.

Type 9. Straight neck, everted rim, convex, thickened lip, undercut. Medium coarse temper, pink colour.
P161   House 4, Inv. No. HA 130/27. Rim fragment. Dim.: Pres. H. 43; Th. wall 4–6; D. rim 90.

Type 10. Slightly spreading neck, everted rim, convex lip. Medium coarse temper, burnt.
P162   House 5, Inv. No. HA 86/11. Rim fragment. Dim.: Pres. H. 34; Th. wall 4; D. rim 106

Type 11. Slightly spreading neck, everted rim, thickened lip, bevelled on the outside. Medium coarse orange ware.
P163   House 5, Inv. No. HA 39/10. Rim fragment. Dim.: Pres. H. 74; Th. wall 7; D. rim 114.

Type 12. Narrow neck, everted rim, convex lip. Medium coarse red ware.
P164   House 5, Inv. No. HA 55/20. Rim fragment. Dim.: Pres. H. 32; Th. wall 7; D. rim 74.

Lekythos P165–174

Type 1. Raised flat base, ovoid body, narrow concave neck, carinated rim, thickened lip. Vertical ring handle, circular in section.
1a. New Halos ware.
P165   House 5, Inv. No. HA 51/123a, Fig. 6.2. Complete profile. Dim.: H. 220; Th. wall 4; D. rim 65; D. body 183; D. base 90.

1b. Medium coarse red ware.
P166   House 3, Inv. Nos. HA 247/26 + HA 232. Rim and body fragments. Dim.: Pres. H. 150; Th. wall 3–5; D. rim 61; D. body 163.
P167   House 5, Inv. No. HA 39/1. Rim fragment. Dim.: Pres. H. 58; Th. wall 6; D. rim 70.

1c. Medium coarse orange ware.
P168   House 4, Inv. No. HA 116/4. Rim fragment. Dim.: Pres. H. 73; Th. wall 5; D. rim 67.

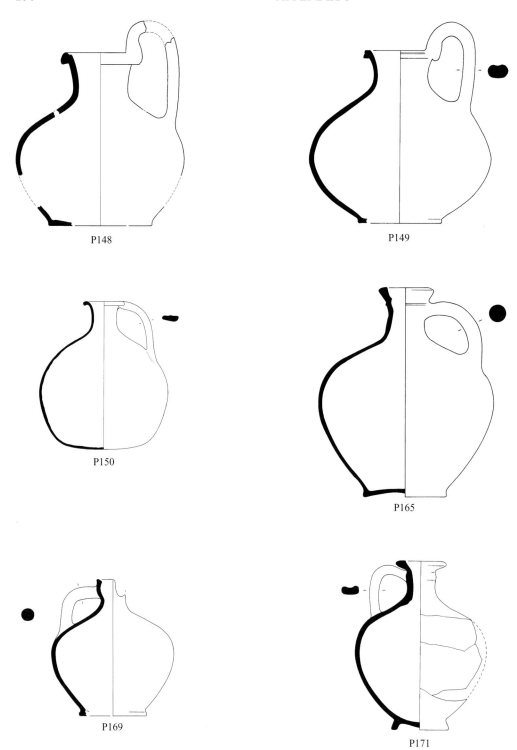

Fig. 6.2. Closed vessels (scale 1:4).

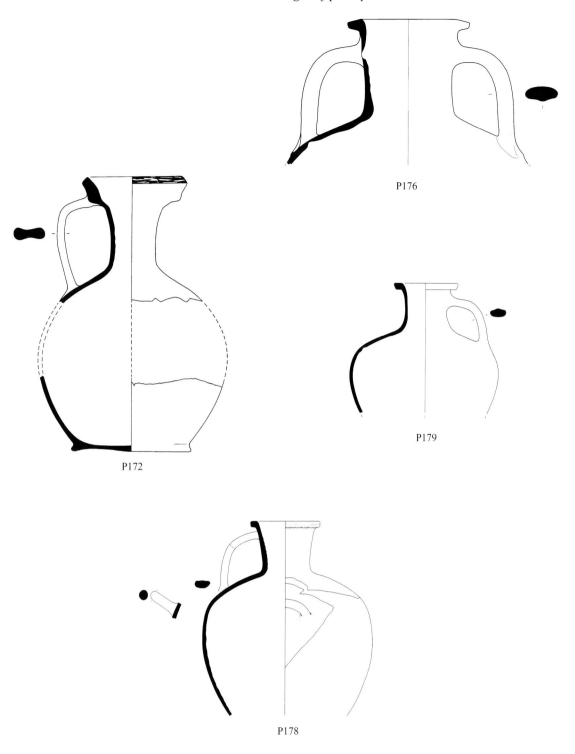

P176

P172

P179

P178

Fig. 6.2. Closed vessels (continued, scale 1:4).

Type 2. Raised flat base, ovoid body, convex neck, everted rim, lip bevelled on the inside. Vertical ring handle, circular in section. Medium coarse orange ware.

P169   House 6, Inv. No. H23 1/148, Fig. 6.2. Complete profile. Dim.: H. 145; Th. wall 4; D. rim 38; D. body 132; D. base 74.

Type 3. Convex shoulder, narrow concave neck, everted rim, convex lip, thickened on the outside. Vertical ring handle, oval in section. Medium coarse orange ware.

P170   House 6, Inv. No. H23 1/104. Rim and body fragments. Dim.: Pres. H. 146; Th. wall 5; D. rim 70; D. body 200.

Type 4. Splayed base ring with straight edge, ovoid body, carinated neck, everted carinated rim, convex lip. Vertical double strap handle. Fine temper, black glazed ware.

P171   House 5, Inv. No. HA 36/2, Fig. 6.2. Complete profile. Dim.: H. 179; Th. wall 5; D. rim 53; D. body 140; D. base 61.

Type 5. Raised flat base, splayed wall, narrow concave neck, wide everted rim. Vertical double strap handle. Painted with a reddish brown to very dark grey slip, on the inside and outside of the rim. Medium fine temper, reddish yellow colour.

P172   House 4, Inv. Nos. HA 92/11 + HA 126/1 + HA 166/3+15+20, Fig. 6.2. Reconstructed profile. Dim.: H. rim 134; H. base 82; Th. wall 3–4; D. rim 103.

Type 6. Squat lekythos. Wide convex base, slightly tapered body, convex shoulder, narrow concave neck, everted rim, thickened lip, bevelled on the outside. Vertical strap handle, flattened oval in section, attached to lip and shoulder. Medium coarse orange ware, dark grey glaze.

P173   House 1, Inv. No. X 657, Fig. 6.2. Complete profile (= Reinders, 1988: Cat. No. 33.02). Dim.: H. 90; Th. wall 4; D. rim 36; D. base 116.

Type 7. Squat lekythos. Wide convex base, tapered body, narrow neck. Attachment of vertical strap handle on body, oval in section. Medium coarse orange ware, dark grey glaze.

P174   House 5, Inv. Nos. HA 14/11-15 + HA 25/5+5a. Base and body fragments. Dim.: Pres. H. 62; Th. wall 5–6; D. base 138.

Hydria P175-182

Type 1. Ovoid body, straight neck. Vertical strap handle, two horizontal ring handles. Medium coarse orange ware.
1a. Splayed base ring, rounded edge, splayed wall.

P175   House 3, Inv. Nos. HA 220/9a + HA 242/3. Base fragment. Dim.: Pres. H. 37; Th. wall 4–5; D. rim 152.

1b. Convex shoulder, straight neck, everted rim, flattened top, slightly bevelled lip. Vertical strap handle with central rib, attached to shoulder and below rim.

P176   House 1, Inv. No. X 302, Fig. 6.2. Rim fragment (= Reinders, 1988: Cat. No. 37.03). Dim.: Pres. H. 156; Th. wall 4–6; D. rim 134.

Type 2. Ovoid body, slightly convex shoulder, straight neck, slightly everted rim, convex lip, thickened on the outside, undercut. Vertical strap handle, oval in section, attached to shoulder and below rim. Horizontal ring handle, circular in section, on widest part of the body. New Halos ware.

P177   House 5, Inv. No. HA 75/77. Rim and body fragments. Dim.: Pres. H. 320; Th. wall 5; D. rim 140; D. body 330.

Type 3. Ovoid body, slightly convex shoulder, slightly splayed neck, everted rim with flattened top, convex lip with groove. Vertical strap handle with central rib, attached to shoulder and below rim, horizontal ring handle, circular in section, on the shoulder. Medium coarse red ware.

P178   House 2, Inv. Nos. H21 3/10+12+21+29, Fig. 6.2. Rim and body fragments. Dim.: Pres. H. 410; Th. wall 6–10; D. rim 152; D. body 360.

Type 4. Wide ovoid body, slightly convex shoulder, straight neck, everted rim with flattened top, bevelled lip. Vertical strap handle with central rib, attached to shoulder and below rim. Medium coarse red ware.

P179   House 3, Inv. No. HA 226/15, Fig. 6.2. Rim and body fragments. Dim.: Pres. H. 278; Th. wall 4; D. rim 140; D. body 326.

Type 5. Splayed neck, everted rim, bevelled lip, undercut. Vertical strap handle, oval in section, attached below rim. Medium coarse red ware.

P180   House 2, Inv. No. H21 2/42. Rim and handle fragments. Dim.: Pres. H. 57; Th. wall 5; D. rim 135.

Type 6. Raised flat base, straight edge, splayed wall. Convex shoulder, straight neck; everted carinated rim, lip flattened and thickened on the outside. Vertical strap handle, oval in section, attached to shoulder and below rim. Medium coarse orange ware.

P181   House 6, Inv. Nos. H23 5/24+152+164. Rim and base fragments. Dim.: H. base 40; H. rim 150; Th. wall 3–6; D. rim 110; D. base 115.

Type 7. Everted rim, convex top, thickened lip with groove. Top of rim painted with ovules in white slip. Very fine temper, reddish yellow colour.

P182   House 1, Inv. No. X 648. Rim fragment. Dim.: D. rim 170.

Water pitcher P183–185

Type 1. Convex shoulder, wide straight neck with sharp angle on transition to shoulder, everted rim with internal angle, top concave, convex lip. Vertical triple strap handle attached to lip and shoulder. Fine to medium fine temper, reddish yellow colour.

P183   House 3, Inv. No. HA 262a/11. Rim fragment. Dim.: Pres. H. 123; Th. wall 3; D. rim 120.

P184   House 4, Inv. Nos. HA 130/91 + HA 132/8. Rim fragment. Dim.: Th. lip 7; D. rim 130.

P185   House 6, Inv. No. H23 2/25. Rim fragment. Dim.: Th. lip 7; D. rim 130.

Lagynos P186–187

Type 1. Splayed base ring, splayed lower body, angular convex shoulder, narrow, slightly concave neck. Diverging strap handle, oval on section. Medium coarse light ware, traces of very pale brown slip. Provenance: Chios?

P186   House 4, Inv. Nos. HA 127/1+7a+11+27+43+51 + HA 117/2, Fig. 6.2. Almost complete profile, rim missing. Ref.: Grace, 1961: No. 50; Edwards, 1963: No. 50 (PR 42). Date: 3rd cent. BC. Dim.: Pres. H. 265; Th. wall 3–8; D. body 304; D. base 155.

Type 2. Flat base, splayed wall, angular transition to convex shoulder. Medium coarse orange ware, traces of black glaze.

P187   House 1, Inv. No. X 113. Base and body fragments. Dim.: Pres. H. 107; Th. wall 3–6; D. body 170; D. base 86.

Olpe P188–190

Type 1. Flat base, slightly splayed, almost straight wall. Irregular shape, probably turned on a slow wheel.
1a. Medium coarse orange ware.

P188   House 4, Inv. No. HA 127/18. Base fragment. Dim.: Pres. H. 44; Th. wall 7; D. body 59; D. base 46.

1b. Medium fine temper, reddish yellow colour, dark grey glaze on interior.

P189   House 4, Inv. Nos. HA 127/20+65+102+116. Base fragment. Dim.: Pres. H. 56; Th. wall 5; D. body 62; D. base 40.

Type 2. Raised flat base, high and narrow, slightly splayed wall, convex shoulder, narrow neck. Medium coarse temper, reddish yellow to brownish yellow colour.

P190   House 3, Inv. Nos. HA 232/33 + HA 247, Fig. 6.2. Base and body fragments. Dim.: Pres. H. 112 (base); Th. wall 3–10; D. body 62; D. base 30.

Pelike P191–194

Type 1. Raised flat base, rounded edge, globular body, concave shoulder, wide straight neck, carinated rim, lip slightly bevelled on the outside. Two vertical double strap handles attached to rim and shoulder. Medium temper, pale brown colour.

P191   House 6, Inv. No. H23 5/143, Fig. 6.2. Complete profile. Dim.: H. 175; Th. wall 4–8; D. rim 79; D. body 137; D. base 78.

P192   House 6, Inv. No. H23 1/136.Dim.: Pres. H. 31; Th. wall 4; Th. lip 6; D. rim 80.

P193   House 6, Inv. No. H23 3/46. Dim.: Pres. H. 34; Th. wall 3; Th. lip 8; D. rim 90.

P194   House 6, Inv. No. H23 6/137. Dim.: Pres. H. 30; Th. wall 3; Th. lip 3.

Amphora

P195   House 1, Inv. No. X 610, Fig. 6.2. Complete profile (= Reinders, 1988: Cat. No. 33.04). Dim.: H. 101; D. rim 65; D. body 92; D. base 60. Raised flat base, splayed wall, slightly concave shoulder, wide concave neck, moulded rim. Vertical strap handles, slightly oval in section, attached to shoulder and below rim. New Halos ware.

OPEN VESSELS

Lekanis P196–236

Type 1. Raised flat base with rounded edge, splayed wall, everted rim with internal angle, flattened top, bevelled lip.
1a. Medium coarse orange ware.

P196   House 1, Inv. No. X 316, Fig. 6.3. Almost complete profile, no handles preserved (= Reinders, 1988: Cat. No. 37.05). Dim.: H. 68; Th. wall 4–5; D. rim 170; D. base 50.

P197   House 3, Inv. Nos. HA 220 + HA 259. Rim fragment. Bevelled lip, thickened on the outside. Horizontal ring handle, circular in section. Dim.: Pres. H. 69; Th. wall 7; D. rim 350.

P198   House 3, Inv. No. HA 262/16. Rim fragment. Dim.: Pres. H. 77; Th. wall 7; D. rim 32.

P199   House 4, Inv. No. HA 135/4a. Rim fragment. Bevelled lip, thickened on the outside. Dim.: Pres. H. 20; Th. wall 6–8; D. rim 310.

P200   House 4, Inv. No. HA 130/82. Rim fragment. Convex lip, thickened on the outside. Dim.: Pres. H. 35.

P201   House 4, Inv. No. HA 153/25. Rim fragment. Two parallel rills on top of rim. Dim.: Pres. H. 28; Th. wall 4.

P202   House 4, Inv. No. HA 153/42. Rim fragment. Two parallel rills on top of rim. Dim.: Pres. H. 13; D. rim 360–400.

P203   House 4, Inv. No. HA 153/45. Rim fragment. Dim.: Pres. H. 25; Th. wall 8; D. rim 180.

P204   House 6, Inv. No. H23 6/21. Rim fragment. Rim pierced with two holes. Dim.: Pres. H. 24; Th. wall 9; D. rim 230.

1b. Medium coarse red ware.

P205   House 4, Inv. Nos. HA 97/1 + HA 115/47+88 + HA 150 + HA 178, Fig. 6.3. Reconstructed profile, handles missing. Dim.: Rec. H. 124; Th. wall 4–7; D. rim 320; D. base 110.

P206   House 6, Inv. No. HA 4/15, Fig. 6.3. Complete. Rim pierced with two holes. Dim.: H. 100; Th. wall 4; D. rim 280; D. base 96.

P207   House 1, Inv. No. X 49 A, Fig. 6.3. Rim fragment (= Reinders, 1988: Cat. No. 39.03). Dim.: Pres. H. 80; Th. wall 5; D. rim 370.

P208   House 1, Inv. Nos. X 131 A+B. Rim fragment. Convex lip with incision. Dim.: Dim.: Pres. H. 66; Th. wall 4–6; D. rim 340.

P209   House 4, Inv. No. HA 117/19. Rim fragment. Rill on top of rim, running parallel to lip. Dim.: Pres. H. 38; Th. wall 4; D. rim 270–310.

P210   House 4, Inv. No. HA 139/5.Rim fragment. Dim.: Pres. H. 31; Th. wall 3; D. rim 210.

P211   House 2, Inv. No. H21 3/9. Rim fragment. Rill on lip. Dim.: Pres. H. 38; T-W: 5; D. rim 230.

P212   House 6, Inv. No. HA 2/13. Rim fragment. Dim.: Th. wall 6; D. rim 230.

P213   House 5/6, Inv. Nos. HA 4/35 + HA 63/16, Fig. 6.3. Rim fragment. Rim pierced with two holes. Dim.: Pres. H. 98; Th. wall 4; D. rim 300.

1c. New Halos ware.

P214   House 1, Inv. No. X 325. Almost complete profile, handles missing. Dim.: H. 63; Th. wall 5; D. rim 190; D. base 80.

P215   House 4, Inv. No. HA 135/50. Rim fragment. Dim.: Pres. H. 34; Th. wall 9; D. rim >350.

P216   House 4, Inv. No. HA 129/9. Rim fragment. Lip with groove. Dim.: Pres. H. 73; Th. wall 7; D. rim 340.

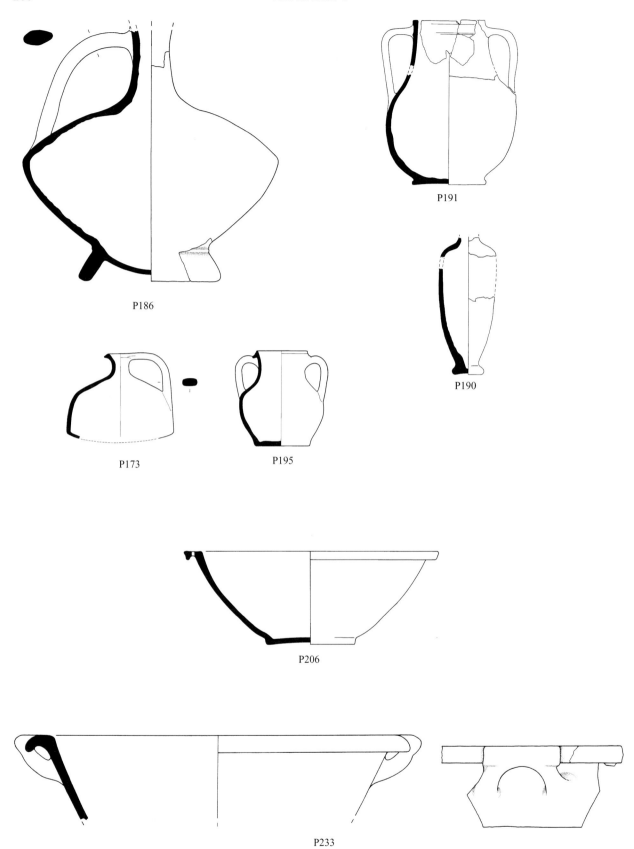

Fig. 6.3. Open vessels (scale 1:4).

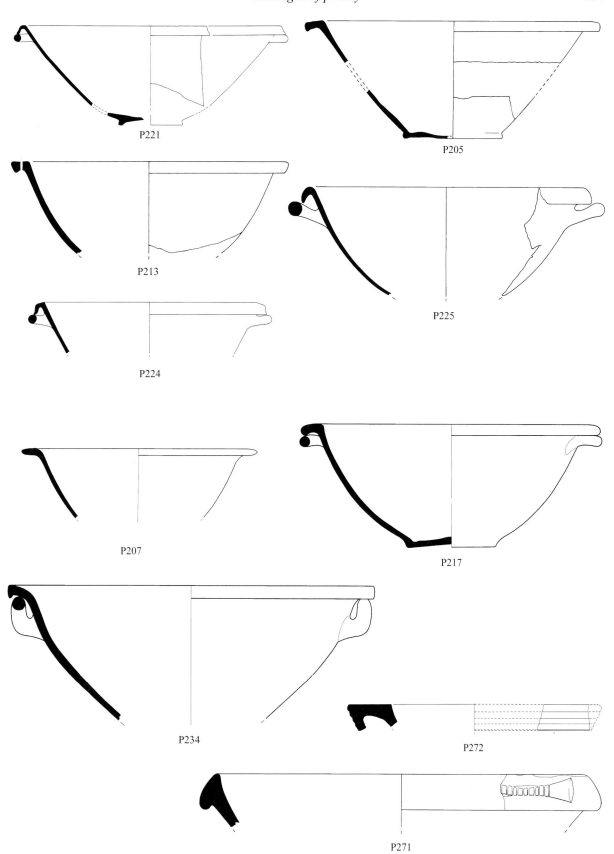

Fig. 6.3. Open vessels (continued, scale 1:4 except P272: scale 1:8).

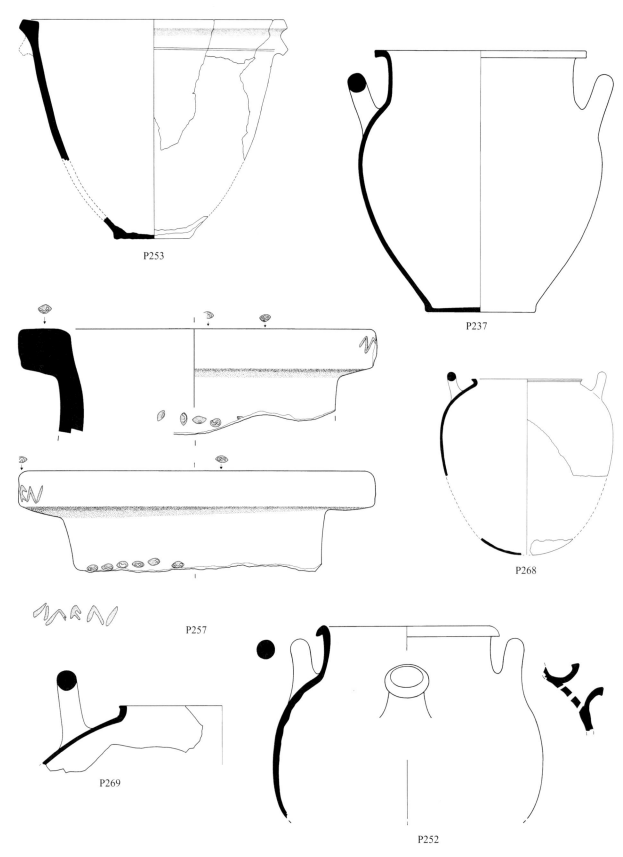

Fig. 6.3. Open vessels (continued, scale 1:4).

Type 2. Splayed wall, everted, overhanging rim with internal angle, convex top. Two horizontal ring handles, circular in section, just below rim.

2a. Medium coarse red ware.

P217 House 1, Inv. No. X 88, Fig. 6.3. Complete profile. Raised base (= Reinders, 1988: Cat. No. 35.05). Dim.: H. 130; Th. wall 5; D. rim 330; D. base 100.

P218 House 1, Inv. No. X 94. Rim fragment (= Reinders, 1988: Cat. No. 31.14). Dim.: Pres. H. 28; Th. wall 7; D. rim 240.

P219 House 6, Inv. Nos. H23 4/10+11. Rim fragment. Dim.: Pres. H. 42; Th. wall 8; D. rim 570.

P220 House 5/6, Inv. No. HA 64/27. Rim fragment. Dim.: Pres. H. 31; Th. wall 8; D. rim 390.

2b. Medium coarse orange ware.

P221 House 5, Inv. No. HA 65/4, Fig. 6.3. Reconstructed profile. Dim.: Th. wall 7–9; D. rim 600; D. base 140.

P222 House 5, Inv. Nos. HA 50/13+13a. Rim and base fragments. Splayed base ring, flattened edge, splayed wall, collar rim, everted lip, thickened on the outside. Dim.: Th. wall 4; D. rim 600; D. base 140.

P223 House 3, Inv. No. HA 246/2b. Rim fragment. Dim.: Pres. H. 52; Th. wall 6; D. rim 250.

P224 House 5, Inv. Nos. HA 51/1 + HA 71/5, Fig. 6.3. Rim fragment. Dim.: Pres. H. 110; Th. wall 7; D. rim 530.

P225 House 5, Inv. Nos. HA 75/52+75, Fig. 6.3. Rim fragment. Dim.: Pres. H. 28; Th. wall 4; D. rim 305.

Type 3. Splayed wall, everted rim, overhanging lip. Two horizontal ring handles just below rim, bent towards the rim.

3a. Medium coarse red ware.

P226 House 3, Inv. No. HA 237/1. Rim and handle fragments. Dim.: Pres. H. 97; Th. wall 6–8; D. rim 460.

P227 House 3, Inv. No. HA 230/4. Rim fragment. Dim.: Pres. H. 70; Th. wall 7.

P228 House 2, Inv. No. H21 2/77. Rim fragment. Dim.: Pres. H. 84; Th. wall 6; D. rim 345.

P229 House 6, Inv. Nos. H23 6/12+13+80. Rim fragment. Dim.: Pres. H. 90; Th. wall 8; D. rim 360.

P230 House 5/6, Inv. Nos. HA 64/6+12. Rim fragment. Dim.: Pres. H. 16; Th. wall 6–7; D. rim 260.

3b. Medium coarse orange ware

P231 House 6, Inv. No. H23 2/3. Rim fragment. Dim.: Pres. H. 23; Th. wall 5; D. rim 320.

P232 House 6, Inv. No. H23 2/54. Rim fragment. Dim.: Pres. H. 16; Th. wall 7; D. rim 320.

Type 4. Splayed wall, everted, overhanging rim with internal angle, convex top. Horizontal ring handle, attached to the rim with a rectangular strip of clay. Medium coarse red ware/New Halos ware.

P233 House 4, Inv. Nos. HA 115/45 + HA 120/7+17, Fig. 6.3. Rim fragment. Dim.: Pres. H. 89; Th. wall 9; D. rim 420.

Type 5. Splayed wall, everted rim, flattened lip, thickened on the outside. Horizontal ring handle, circular in section, bent towards the wall, just below the rim. Medium coarse temper, pink colour, medium coarse light ware.

P234 House 1, Inv. No. X 711, Fig. 6.3. Rim fragment (= Reinders, 1988: Cat. No. 34.03). Dim.: Pres. H. 92; Th. wall 9; D. rim 410.

Type 6. Raised flat base, rounded edge, splayed wall, slightly everted rim, thickened lip, bevelled on the outside. New Halos ware.

P235 House 2, Inv. Nos. H21 1/7+16. Almost complete profile, part of body missing. Dim.: Pres. H. rim 46; Pres. H. base 40; Th. wall 5–8; D. rim 272; D. base 148.

Type 7. Splayed wall, straight rim, thickened, slightly overhanging bevelled lip. Medium coarse red ware.

P236 House 3, Inv. No. HA 247/2b. Rim fragment. Dim.: Pres. H. 35; Th. wall 7; D. rim 435.

Krater P237-252

Type 1. Raised flat base, ovoid body, convex shoulder, straight neck, everted rim, flattened top, lip thickened on the outside. Two vertical ring handles, circular in section, on shoulder.

1a. Bright orange colour. Medium coarse orange ware.

P237 House 1, Inv. No. X 87, Fig. 6.3. Complete profile (= Reinders, 1988: Cat. No. 35.02). Dim.: H. 270; Th. wall 5; D. rim 230; D. body 260; D. base 120.

P238 House 6, Inv. Nos. H23 1/96+114a. Complete profile. Spreading neck. Dim.: Pres. H. rim 92; Th. wall 4–6; D. rim 235; D. body 236; D. base 100.

P239 House 6, Inv. Nos. H23 1/6+124+155+156. Complete profile. Dim.: Pres. H. rim 24; Th. wall 6; D. rim 310.

P240 House 6, Inv. Nos. H23 1/14+158+159. Rim fragment. Dim.: Pres. H. 63; Th. wall 5; D. rim 285.

P241 House 6, Inv. No. H23 1/112. Rim fragment. Dim.: Pres. H. 9; Th. wall 4; D. rim 210.

P242 House 6, Inv. No. H23 1/114. Rim fragment. Dim.: Pres. H. 55; Th. wall 5; D. rim 230.

P243 House 6, Inv. No. HA 4/18. Rim fragment. Dim.: Pres. H. 26; Th. wall 5; D. rim 220.

P244 House 6, Inv. No. HA 4/23. Rim fragment. Dim.: Pres. H. 31; Th. wall 3.

P245 House 6, Inv. No. HA 57/3. Rim fragment. Dim.: Pres. H. 11; Th. wall 8; D. rim 190.

1b. New Halos ware

P246 House 1, Inv. Nos. X 82+85+86+87+104+105+202. Reconstructed profile. Raised flat base, ovoid body, convex shoulder, straight neck, everted rim, flattened top, lip thickened on the outside. Two vertical ring handles, circular in section, on shoulder. Two parallel grooves on shoulder (= Reinders, 1988: Cat. No. 35.01; 35.03; 35.04; 35.20). Dim.: Rec. H. 260; Th. wall 4–5; D. rim 210; D. body 280; D. base 111.

P247 House 3, Inv. No. HA 228/11. Rim fragment. Dim.: Pres. H. 10; Th. lip 10; D. rim 280.

P248 House 4, Inv. No. HA 157/7. Rim fragment. Dim.: Pres. H. 20; Th. wall 5; D. rim 150.

P249 House 6, Inv. No. HA 62/6. Rim fragment. Dim.: Pres. H. 50; Th. wall 6; D. rim 240.

1c. Medium coarse red ware

P250 House 5, Inv. No. HA 50/12. Rim fragment. Dim.: Pres. H. 10; Th. wall 5; D. rim 220

Type 2. Raised flat base, splayed wall, convex shoulder, wide slightly concave neck, everted rim. Spout on shoulder. Two vertical ring handles, circular in section, attached to shoulder.
2a. Medium coarse red ware.
P251   House 6, Inv. Nos. H23 1/14+38+56+57+134+135+ 137+138+156+159. Rim and base fragments, with spout. Rim with slightly convex top, overhanging lip, flattened and thickened. Dim.: Pres. H. rim 98; Pres. H. base 60; D. rim 180; D. base 130; D. spout 62.

2b. Medium coarse orange ware.
P252   House 4, Inv. Nos. HA 136/1 + HA 149/14, Figs 3.4 and 6.3. Rim fragment. Rim with convex, overhanging lip. Dim.: Pres. H. 203; D. rim 200; D. body 295; D. stem 47.

Tub P253–267

Type 1. Raised flat base, splayed wall, everted rim with internal angle, flattened top, bevelled lip. Rectangular lug below rim. Medium coarse red ware.
P253   House 6, Inv. No. H23 5/23, Fig. 6.3. Complete profile. Dim.: H. 466; Th. wall 140; D. rim 590; D. base 175.

2a. Medium coarse red ware.
P254   House 1, Inv. No. X 95. Rim fragment. Everted rim, flattened top, bevelled lip (= Reinders, 1988: Cat. No. 31.17). Dim.: Pres. H. 48; Th. wall 140; D. rim 440.

2b. Coarse orange ware.
P255   House 5, Inv. Nos. HA 46/14–17. Rim fragment. Splayed wall, everted rim with internal angle, flattened top, bevelled lip. Dim.: Pres. H. 89; Th. wall 130; D. rim 420.

Type 3. Straight wall, everted rim, bevelled on the inside, flattened top, flattened lip. Medium coarse orange ware.
P256   House 6, Inv. No. H23 3/85a. Rim fragment. Dim. Pres. H. 308; Th. wall 24; D. rim >610; D. body >540.

Type 4. Straight wall, everted rim, flattened top, flattened lip. Stamped on the wall and top of rim, illegible incisions on lip. Coarse red ware.
P257   House 4, Inv. No. HA 99/3, Fig. 6.3. Rim fragment. Dim.: Pres. H. 150; Th. wall 34; D. rim 390.

Type 5. Straight wall, everted rim, top flattened, bevelled lip. From rectangular vessel. Medium coarse orange ware.
P258   House 5, Inv. No. HA 46/3. Rim fragment. Dim.: Pres. H. 95; Th. wall 34.

Type 6. Splayed wall, everted rim, top flattened, convex lip.
6a. Coarse red ware.
P259   House 5, Inv. Nos. HA 36/1b + HA 59/34. Rim fragment. Dim.: Pres. H. 235; Th. wall 15; D. rim 759.

6b. Coarse orange ware.
P260   House 4, Inv. No. HA 130/3. Rim fragment. Dim.: Pres. H. 80; Th. wall 23; D. rim >560.

Type 7. Everted rim, flattened lip. Coarse orange ware.
P261   House 6, Inv. No. HA 2/1. Rim fragment. Dim.: Pres. H. 69; Th. wall 37; D. rim 380.

Type 8. Splayed base ring, rounded edge. Coarse orange ware.
P262   House 4, Inv. No. HA 116/2. Base fragment. Dim.: Pres. H. 30; Th. wall 9–12; D. base 123.

Type 9. Raised flat base, rounded edge. Coarse orange ware.
P263   House 4, Inv. No. HA 119/2. Base fragment. Dim.: Pres. H. 57; Th. wall 9; D. base 195.

Type 10. Raised flat base, splayed wall. Coarse red ware.
P264   House 1, Inv. No. X 650. Base fragment (= Reinders, 1988: Cat. No. 33.15). Dim.: Pres. H. 84; Th. wall 10; D. base 188.
P265   House 6, Inv. No. HA 78E/13. Base fragment. Dim.: Pres. H. 40; Th. wall 28; Th. base 15; D. base 200.

Type 11. Splayed wall, rounded base, wall fragment with triangular spout. Coarse red ware.
P266   House 5, Inv. Nos. HA 65/5 + HA 85/3+4. Base and wall fragments. Dim.: Pres. H. 244; Th. wall 20–30.

Type 12. Straight wall, with repair of lead strips. Coarse red ware.
P267   House 3, Inv. No. HA 253/35. Wall fragments. Dim.: Th. wall 13–19.

Lebes P268–270

Type 1. Rounded base, globular body, convex shoulder, everted rim, flattened lip, thickened on the outside. Two horizontal ring handles, circular in section, attached to shoulder and arching above rim. New Halos ware.
P268   House 3, Inv. No. HA 238/5, Fig. 6.3. Reconstructed profile. Dim.: Pres. H. rim 210; Pres. H. base 40; Th. wall 5; D. rim 220; D. body 360.

Type 2. Convex shoulder, upright rim, flattened lip, thickening on the outside. Two horizontal ring handles, circular in section, attached to shoulder and arching above rim. Medium coarse red ware.
P269   House 1, Inv. No. X 84a, Fig. 6.3. Rim fragment (= Reinders, 1988: Cat. No. 35.06). Dim.: Pres. H. 70; Th. wall 4; D. rim 226.

Type 3. Slightly pointed base, oval body, convex shoulder, upright rim, thickened lip, bevelled on the inside. Two horizontal ring handles, circular in section, attached to shoulder and arching above rim. Two parallel grooves on shoulder.
P270   House 1. Almost complete profile, base missing. Dim.: Pres. H. 328; Th. wall 5–6; D. rim 210; D. body 400.

Mortar

P271   House 6, Inv. No. HA 2/2. Rim fragment. Splayed wall, slightly everted rim, thickened convex, overhanging lip. Bolster hand grip: thin strip of clay with regular incisions representing beads. Medium coarse temper, pale yellow colour. Provenance: Corinth. Ref.: Anderson-Stojanović, 1993: p. 276, No. 21 (IP 883). Date: before 275 BC. Dim.: Pres. H. 56; Th. wall 11; D. rim 440.

Louterion

P272   House 6, Inv. No. H23 6/97. Rim and wall fragments. Splayed wall, everted rim with internal angle, top flattened, overhanging lip with four broad grooves. Coarse temper, pink colour. Provenance: Corinth.

Ref.: Iozzo, 1987: No. 126 (IPG 1970-69); No. 131 (IPG 1970–74 a,b). Date: mid 5th cent. BC / 5th or 4th cent. BC. Sparkes & Talcott, 1970: No. 1861 (P 24252). Date: 5th–4th cent. BC. Dim.: Pres. H. 59; Th. wall 22; D. rim 560.

STORAGE VESSELS

Amphora P273–301

Type 1. Large pointed amphora with a short button knob, slight nipple and rounded edge, wide ovoid body, concave shoulder, long straight neck, convex lip, thickened on the outside, undercut. Two slightly diverging strap handles, oval in section, attached just below lip and to shoulder. Medium coarse temper, reddish yellow colour.

P273   House 4, Inv. Nos. HA 117/13 + HA 127/2+3+7-9, Figs 3.5 and 6.4. Complete profile. Dim.: H. 825; Th. wall 5–19; D. rim 128; D. body 498; D. base 55; Volume 59.37 l (neck included); 57.84 l (neck excluded).

P274   House 4, Inv. Nos. HA 117/13 + HA 127/8+27 + HA 137/3. Reconstructed profile. Dim.: Th. wall 6–16; D. rim 100; D. base 58.

P275   House 5, Inv. No. HA 37/4, Fig. 6.3. Reconstructed profile. Dim.: Rec. H. 745; Th. wall 7–16; D. rim 108; D. body 400; D. base 55; Volume 29.12 l (neck excluded).

P276   House 5, Inv. Nos. HA 51/122–124, Figs 3.5 and 6.4. Complete. Dim.: H. 770; Th. wall 10; D. rim 104; D. body 420; D. base 55; Volume 37.83 l (neck included); 36.79 l (neck excluded).

P277   House 4, Inv. Nos. HA 117/2+3+10+13 + HA 127/5+7+10. Rim and base fragments. Dim.: Pres. H. rim 348; Pres. H. base 160; Th. wall 6–25; D. rim 110; D. body 460; D. base 60.

P278   House 3, Inv. Nos. HA 276/9+10. Base fragment. Dim.: Pres. H. 171; Th. wall 13–15; D. knob 60.

P279   House 2, Inv. No. H21 4/13. Base fragment. Dim.: Pres. H. 61; Th. wall ; D. base 52.

P280   House 6, Inv. No. HA 78B/18. Base fragment. Dim.: Pres. H. 34; D. base 62.

P281   House 5, Inv. No. HA 66/38. Base fragment. Dim.: Pres. H. 73; Th. wall 19; D. base 56.

P282   House 3, Inv. No. HA 191/1. Rim fragment. Dim.: Pres. H. 219; Th. wall 4–9; D. rim 114.

P283   House 1, Inv. Nos. X 710+704+714+724+737, Fig. 6.4. Rim fragment (= Reinders, 1988: Cat. No. 34.02). Dim.: Pres. H. 270; Th. wall 7–12; D. rim 114.

Type 2. Long solid foot, almost completely circular dome on outside knob, lower part of edge is concave, a broad ridge runs around the upper part runs a broad ridge, narrow body with a spreading, slightly everted wall, sharp angle on transition to concave shoulder, short neck, convex lip, thickened on the outside. Two strap handles attached just below lip, sloped slightly in towards the shoulder. Medium coarse temper, pale brown colour on the exterior and reddish yellow on interior and core. Provenance: Chios (?)

P284   House 6, Inv. Nos. H23 5/153+165, Fig. 6.4. Complete.

Dim.: H: 470; D. rim 110; D. body 270; D. base 50–55; Volume 8.04 l (neck included); 7.01 l (neck excluded).

P285   House 4, Inv. No. HA 130/9. Base fragment. Dim.: Pres. H. 33; D. base 54.

P286   House 6, Inv. No. H23 6/78. Base fragment. Dim.: Pres. H. 37; D. base 50.

Type 3. Small solid foot with straight edge, ovoid body, convex shoulder, straight neck, high slender lip. Two vertical strap handles attached below lip and to shoulder. Medium coarse temper, pink colour. Provenance: Rhodos (?).

P287   House 3, Inv. Nos. HA 210/3 + HA 224/3, Fig. 6.4. Complete. Dim.: H. 468; Th. wall 5–14; D. rim 110; D. body 287; D. base 38; Volume 10.44 l (neck included); 9.51 l (neck excluded).

P288   House 3, Inv. No. HA 242/4. Rim fragment. Dim.: Pres. H. 61; Th. wall 7; D. rim 110.

Type 4. Straight neck, convex lip thickened on the outside, undercut. Vertical, double barred handle attached below lip. Medium coarse temper, yellowish white colour. Provenance: Kos.

P289   House 1, Inv. No. X 318. Rim fragment. Dim.: Pres. H. 70; Th. wall 9–11; D. rim 130.

Type 5. Solid foot, outside knob concave, rounded edge, spreading wall, slightly concave shoulder, straight neck, convex lip, thickened on the outside, undercut. Slightly diverging vertical strap handles, oval in section, attached just below lip and to shoulder. Medium coarse temper, reddish yellow colour.

P290   House 6, Inv. No. H23 5/135, Fig. 6.4. Almost complete profile, part of neck missing. Dim.: Pres. H. base 584; Pres. H. rim 88; Th. wall 6–16; D. rim 120; D. body 412; D. base 60; Volume 29.24 l (neck and shoulder excluded)

Type 6. Ovoid body, straight narrow neck. Medium coarse temper, light reddish brown to reddish yellow colour.

P291   House 4, Inv. Nos. HA 117/2 + HA 127/10, Fig. 6.4. Rim and body fragments. Dim.: Pres. H. 565; Th. wall 8–12; D. rim 105; D. body 360.

P292   House 6, Inv. Nos. H23 1/42+77+96a+100+113. Base and body fragments. Solid foot, knob concave on the outside, edge bevelled. Dim.: Pres. H. 720; Th. wall 5–16; D. body 392; D. base 49.

Type 7. Slender ovoid body, straight narrow neck, convex lip, thickened on the outside. Medium temper, reddish yellow colour.

P293   House 6, Inv. Nos. H23 5/153+157, Fig. 6.4. Almost complete profile, base knob missing. Dim.: Pres. H. 686; Th. wall 8–34; D. rim 114; D. body 338.

Type 8. Solid foot, outside concave, edge bevelled, slender spreading wall. Medium fine to medium temper, reddish yellow colour.

P294   House 1, Inv. No. X 619. Base fragment (= Reinders, 1988: Cat. No. 33.27). Dim.: Pres. H. 122; Th. wall 20; D. base 41.

P295   House 6, Inv. No. H23 6/96. Base fragment. Dim.: Pres. H. 182; Th. wall 11; D. base 60.

Type 9. Angular, concave shoulder, long narrow neck, convex lip, thickened on the outside. Two strap handles, oval in section, attached below rim, sloped in towards shoulder. Medium coarse temper, reddish yellow colour.

P296   House 6, Inv. Nos. H23 5/106+126. Rim fragment. Dim.: Pres. H. 344; D. rim 110.

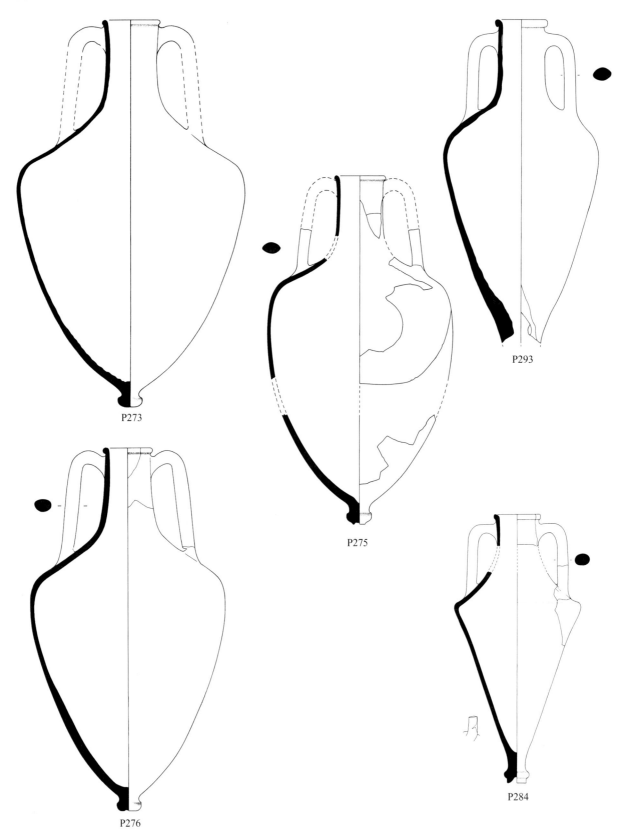

Fig. 6.4. Storage vessels (scale 1:8).

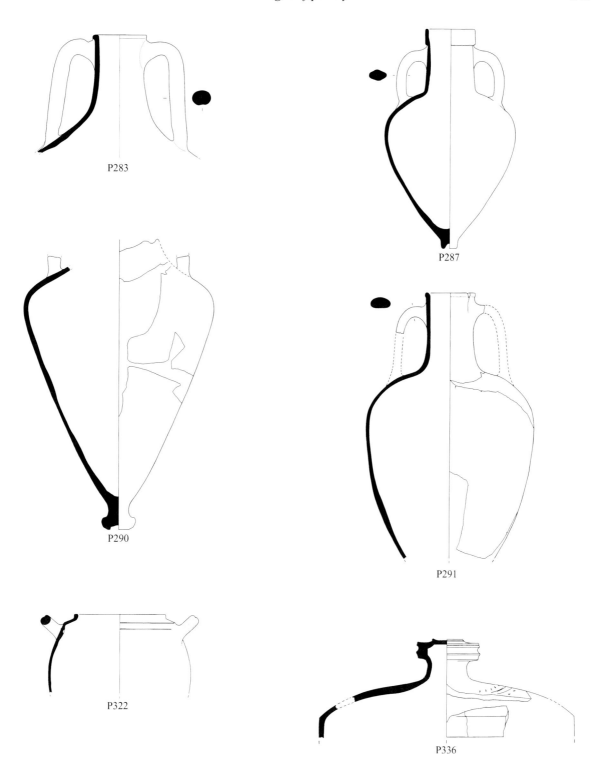

Fig. 6.4. Storage vessels (continued, scale 1:8).

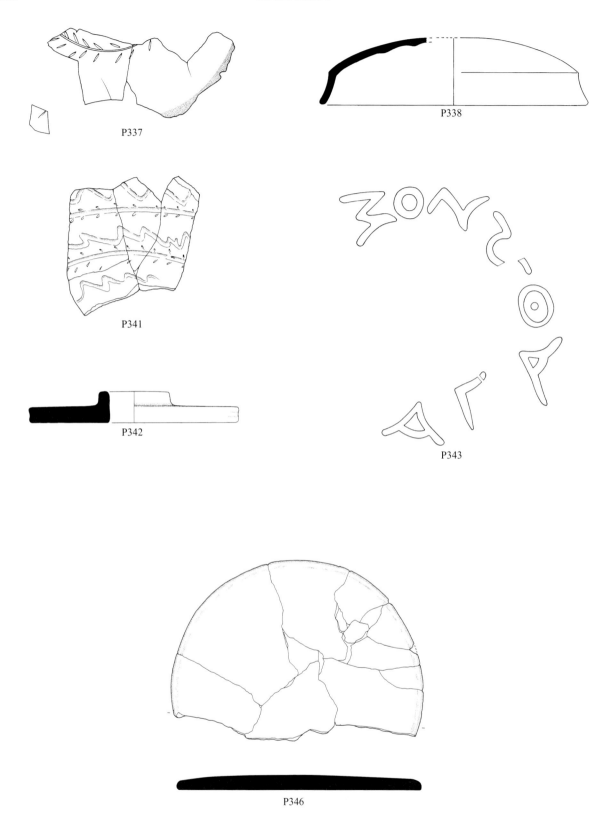

Fig. 6.4. Storage vessels (continued, scale 1:8).

Type 10. Cilindrical neck, convex lip, thickened on the outside, undercut. Attachment of vertical strap handle, oval in section. Base knob with flattened underside, rounded edge. Medium coarse temper, reddish yellow colour.

P297   House 2, Inv. Nos. H21 5/2+8. Rim fragment and base knob. Dim.: Pres. H. rim 210; Pres. H. base 29; Th. wall 10; D. rim 106; D. knob 48.

Type 11. Cilindrical neck, slightly everted rim, lip thickened and bevelled on the outside. Vertical strap handle, oval in section. Medium coarse temper, reddish yellow colour.

P298   House 2, Inv. Nos. H21 5/3+7. Rim fragment. Dim.: Pres. H. 108; Th. wall 14; D. rim 120.

Type 12. Concave underside, rounded edge, slender, spreading wall. Medium coarse temper, reddish yellow colour.

P299   House 5/6, Inv. No. HA 64/25. Base knob. Dim.: Pres. H. 54; D. knob 50.

Type 13. Concave underside, rounded edge, wide spreading wall. Medium coarse temper, pink colour.

P300   House 6, Inv. Nos. H23 5/150+151. Base knob. Dim.: Pres. H. 49; D. knob 70.

Type 14. Raised base, concave underside, straight edge, wide spreading wall.

P301   House 6, Inv. No. H23 1/5. Base fragment. Dim.: Pres. H. 27; Th. wall 9; D. base 69.

Jar P302–326

Type 1. Splayed base ring, splayed wall, convex shoulder, short everted rim, lip thickened on the inside. Horizontal ring handle on shoulder. Medium size.
1a. Medium coarse orange ware.

P302   House 3, Inv. No. HA 253/28-1. Rim fragment. Decorated with incised floral pattern on shoulder. Dim.: Pres. H. 39; Th. wall 5; D. rim 140.

P303   House 4, Inv. No. HA 153/40. Rim fragment. Dim.: Pres. H. 15; Th. wall 5.

P304   House 6, Inv. No. H23 1a/3. Rim fragment. Dim.: Pres. H. 18; Th. wall 5; D. rim 135.

P305   House 6, Inv. No. H23 1/98. Rim fragment. Dim.: Pres. H. 32; Th. wall 6; D. rim 135.

P306   House 6, Inv. No. H23 1/94. Rim fragment. Dim.: Pres. H. 29; Th. wall 4; D. rim 150.

P307   House 6, Inv. Nos. H23 4/12a+14a. Rim fragment. Dim.: Pres. H. 32; Th. wall 6; D. rim 160.

1b. Medium coarse red ware.

P308   House 5, Inv. No. HA 20/3. Rim fragment. Dim.: Pres. H. 17; Th. wall 5; D. rim 160.

Type 2. Convex shoulder, short everted rim. Large size.
2a. Medium coarse red ware.

P309   House 6, Inv. No. H23 5/8. Rim, handle and base fragments. Splayed base ring, pointed edge, splayed wall, convex lip. Horizontal ring handle, circular in section. Groove on transition to rim and two grooves running along shoulder. Dim.: Pres. H. rim 57; Pres. H. base 57; Th. wall 4–10; D. rim 220; D. base 160.

P310   House 2, Inv. No. H21 1/18. Rim fragment. Thickened flattened lip. Dim.: Pres. H. 38; Th. wall 5; D. rim 190.

P311   House 2, Inv. No. H21 4/19b-11. Rim fragment. Thickened flattened lip. Dim.: Pres. H. 17; Th. wall 6; D. rim 180.

P312   House 2, Inv. No. H21 4/4+10. Rim fragment. Thickened flattened lip. Dim.: Pres. H. 25; Th. wall 7; D. rim 314.

P313   House 5, Inv. No. HA 63/8. Rim fragment. Dim.: Pres. H. 48; Th. wall 7; D. rim 375.

P314   House 5, Inv. No. HA 20/1. Rim fragment. Dim.: Pres. H. 28; Th. wall 6; D. rim 250.

P315   House 5, Inv. No. HA 30/1. Rim fragment. Dim.: Pres. H. 35; Th. wall 7; D. rim 210.

2b. Reddish yellow colour.

P316   House 6, Inv. No. H23 3/77. Rim and base fragments. Splayed base ring, rounded edge, splayed wall. Horizontal ring handle, circular in section. Dim.: Pres. H. rim 45; Pres. H. base 56; Th. wall 7; D. rim 200; D. base 120.

P317   House 6, Inv. No. H23 1a/2. Rim fragment. Dim.: Pres. H. 28; Th. wall 7; D. rim 300.

Type 3. Everted rim, flattened top. Medium coarse orange ware.

P318   House 1, Inv. No. X 714. Rim fragment (= Reinders, 1988: Cat. No. 34.06). Dim.: Pres. H. 35; Th. wall 4; D. rim 200.

P319   House 6, Inv. No. HA 78E/5. Rim fragment. Dim.: Pres. H. 22; Th. wall 5; D. rim 260.

Type 4. Small size. Convex shoulder, upright rim. Medium coarse orange ware.

P320   House 2, Inv. No. H21 4/19. Rim and body fragments. Dim.: Pres. H. 60; Th. wall 7; D. rim 86; D. body 130.

P321   House 5, Inv. No. HA 53/1. Rim fragment. Dim.: Pres. H. 7; Th. wall 4; D. rim 60.

Type 5. Splayed wall, angular shoulder, upright rim, thickened convex lip. Medium coarse red ware.

P322   House 3, Inv. Nos. HA 232 + HA 247/2+2b+10b + HA 248/4, Fig. 6.4. Rim and body fragments. Two horizontal double ring handles, attached just below shoulder. Small ridge just below shoulder. Dim.: Pres. H. 168; Th. wall 5–8; D. rim 200; D. body 308.

P323   House 3, Inv. No. HA 278/3. Rim fragment. Dim.: Pres. H. 133; Th. wall 3–7; D. rim 320.

Type 6. Everted rim, convex lip. Coarse orange ware.

P324   House 5, Inv. No. HA 36/4. Rim fragment. Dim.: Pres. H. 60; Th. wall 7; D. rim 250.

Type 7. Large globular body, upright rim, convex lip. Attachment of horizontal ring handle on shoulder. Thick sherd, medium coarse orange ware.

P325   House 1, Inv. No. X 707. Rim and body fragments. Dim.: Pres. H. 215; Th. wall 12–19; D. rim 220; D. body 500.

Type 8. Upright rim, flattened lip, slightly thickened and bevelled on the outside. Thick sherd, coarse red ware.

P326   House 4, Inv. No. HA 117/10. Rim fragment. Dim.: Pres. H. 65; Th. wall 20; D. rim 230.

Pithos P327–335

Type 1. Convex shoulder, everted thickened rim, flattened top.
1a. Coarse orange ware.

P327   House 4, Inv. No. HA 122/9. Rim fragment. Dim.: Pres. H. 46; D. rim 600.

P328   House 4, Inv. No. HA 96/1a. Rim fragment. Dim.: Pres. H. 67; Th. wall 30–34.

P329   House 4, Inv. Nos. HA 128/32-35. Rim fragment. Dim.: Pres. H. 72; Th. wall 20–25; D. rim>560.

P330   House 2, Inv. No. H21 5/10. Rim fragment. Dim.: Pres. H. 168; Th. wall 28; D. rim 540.

P331   House 6, Inv. No. HA 57/1. Rim fragment. Dim.: Pres. H. 82; Th. wall 36; Th. lip 110; D. rim 640–700.

P332   House 5, Inv. Nos. HA 36/1 + HA 59/17. Rim fragment. Dim.: Pres. H. 178; Th. wall 28; Th. lip 120; D. rim 650.

P333   House 5, Inv. No. HA 45/1. Rim fragment. Dim.: Pres. H. 57; Th. wall 37; Th. lip 69; D. rim 420.

1b. Coarse red ware.

P334   House 6, Inv. No. H23 3/75. Rim fragment. Dim.: Pres. H. 100; Th. wall 27; Th. lip 100; D. rim 560.

1c. Coarse ware, pinkish white colour.

P335   House 3, Inv. No. HA 222/8. Rim fragment. Dim.: Pres. H. 78; Th. wall 39; Th. lip 97; D. rim 770.

Large (storage jar) lid P336–341

Type 1. Domed lid, straight rim with external angle. Decorated with incised floral pattern. New Halos ware.

P336   House 6, Inv. No. H23 5/151, Figs 3.6 and 6.4. Complete profile. Dim.: H: 104; D. rim 260–280; D. knob 66.

P337   House 3, Inv. Nos. HA 221 + HA 230 + HA 231/2 + HA 246/10+10a + HA 262, Fig. 6.4. Rim fragment. Dim.: Pres. H. 24; Th. wall 6; D. rim >24.

Type 2. Domed lid, flaring rim with external angle. Medium coarse red ware.

P338   House 6, Inv. No. H23 5/142, Fig. 6.4. Rim and wall fragments, knob missing. Dim.: Pres. H. 71; Th. wall 7; D. rim 290.

P339   House 5/6, Inv. No. HA 64/22. Knob. Dim.: Pres. H. 21; D. knob 63.

Type 3. Small size. Domed lid, straight rim with external angle, expanding knob with slightly concave top. New Halos ware.

P340   House 3, Inv. Nos. HA 197/7+9+27 + HA 202/5. Complete profile. Dim.: H. 40; Th. wall 4–5; D. rim 180.

Type 4. Flat lid, thick sherd. Decorated with incised floral pattern. New Halos ware.

P341   House 4, Inv. Nos. HA 98/3 + HA 157/., Fig. 6.4. Wall fragment. Dim.: Th. wall 11–12.

Pithos lid P342-347

Type 1. Flat disk, flattened lip, hole in the centre, upright rim with flattened lip around hole.
1a. Coarse orange ware.

P342   House 4, Inv. No. HA 109/2, Figs 3.7 and 6.4. Complete profile. Inscribed clockwise with ΑΓΑΘΩΝΟΣ around hole. Dim.: H: 73; Th. wall 4–4.5; D. rim 450; D. hole 150.

P343   House 6, Inv. No. H23 1/. Complete. Inscribed anticlockwise with ΑΓ [...] around hole. Dim.: H. ; Th. wall 2–25; D. rim 370–380; D. hole 140–145.

P344   House 4, Inv. No. HA 117/9. Rim fragment. Dim.: Pres. H. 45; Th. wall 25; D. rim 275; D. hole 100.

1b. Coarse red ware.

P345   House 4, Inv. No. HA 110/6. Rim fragment. Three incised lines on the underside. Dim.: Pres. H. 33; Th. wall 25; D. rim 380.

Type 2. Flat disk. Coarse red ware.

P346   House 5, Inv. Nos. HA 59/15+16, Fig. 6.4. Rim fragment. Dim.: Th. wall 22–31; D. rim 530.

P347   House 5, Inv. No. HA 75/51. Rim fragment. Dim.: Th. wall 30; D. rim 575.

TABLE WARE

Echinos bowl P348–369

Type 1. Splayed base ring, rounded edge, splayed wall, incurved rim, convex lip.
1a. Thick wall.

P348   House 3, Inv. No. HA 262a/12, Fig. 6.5. Complete profile. Medium fine temper, reddish yellow colour, very dark grey glaze. Dim.: H. 46; Th. wall 5; D. rim 116; D. body 126; D. base 54.

P349   House 3, Inv. Nos. HA 197/7 + HA 202/5, Fig. 6.5. Complete profile. Slightly angular wall. Medium coarse temper, reddish yellow colour, reddish yellow to black glaze. Dim.: H. 45; Th. wall 4; D. rim 102; D. body 110; D. base 47.

P350   House 2, Inv. No. H21 1/4, Fig. 6.5. Complete profile. Fine temper, light red colour, very dark grey glaze. Dim.: H. 48; Th. wall 4; D. rim 115; D. body 120; D. base 54.

P351   House 6, Inv. No. H23 5/149, Fig. 6.5. Complete. Very fine temper, pink colour, black glaze. Rouletting. Dim.: H. 51; Th. wall 4; D. rim 101; D. body 113; D. base 68.

P352   House 6, Inv. Nos. HA 4/4-8+10+11, Fig. 6.5. Complete profile. Fine temper, very pale brown colour, very dark grey glaze. Dim.: H. 44; Th. wall 4; D. rim 110; D. body 120; D. base 61.

P353   House 6, Inv. No. HA 62/1, Fig. 6.5. Complete. Very fine temper, pink colour; black glaze. Dim.: H. 51; Th. wall 5; D. rim 110; D. body 120; D. base 57.

P354   House 6, Inv. No. H23 5/159, Fig. 6.5. Complete profile. Very fine temper, pink colour, very dark grey glaze. Dim.: H. 49; Th. wall 6; D. rim 110; D. body 121; D. base 53.

P355   House 6, Inv. Nos. H23 5/85+134. Complete profile. Medium fine temper, very pale brown colour, very dark grey glaze. Dim.: H. 47; Th. wall 4; D. rim 105; D. body 120; D. base 54.

1b. Thin wall.

P356   House 3, Inv. No. HA 202/3, Fig. 6.5. Complete profile. Medium fine temper, reddish yellow colour and glaze. Dim.: Rec. H. 49; Th. wall 3; D. rim 120; D. body 130; D. base 49.

Type 2. Splayed base ring, rounded edge, high splayed wall, incurved rim, convex lip. Medium fine temper, reddish yellow colour, red glaze.

P357   House 3, Inv. No. HA 234/3, Fig. 6.5. Reconstructed profile. Dim.: Rec. H. 55; Th. wall 3–4; D. rim 104; D. body 108; D. base 38.

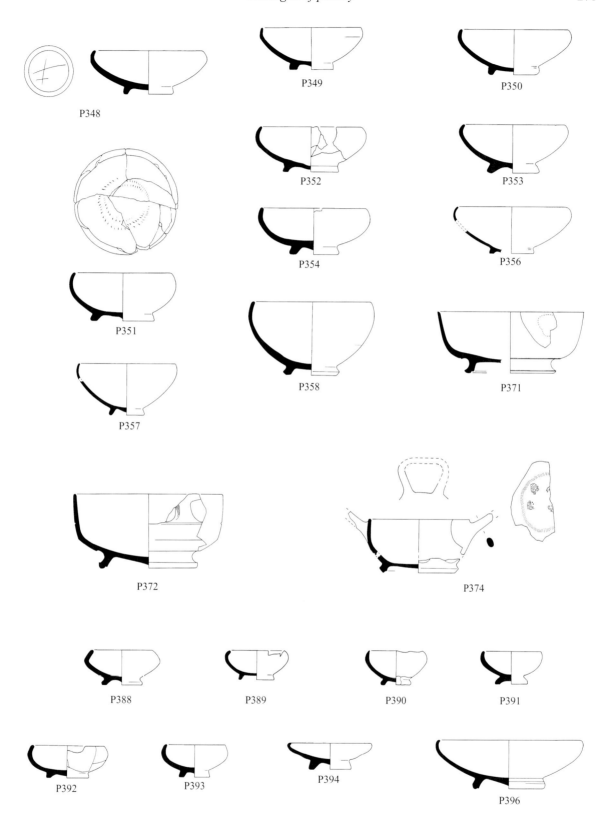

Fig. 6.5. Table ware (scale 1:4).

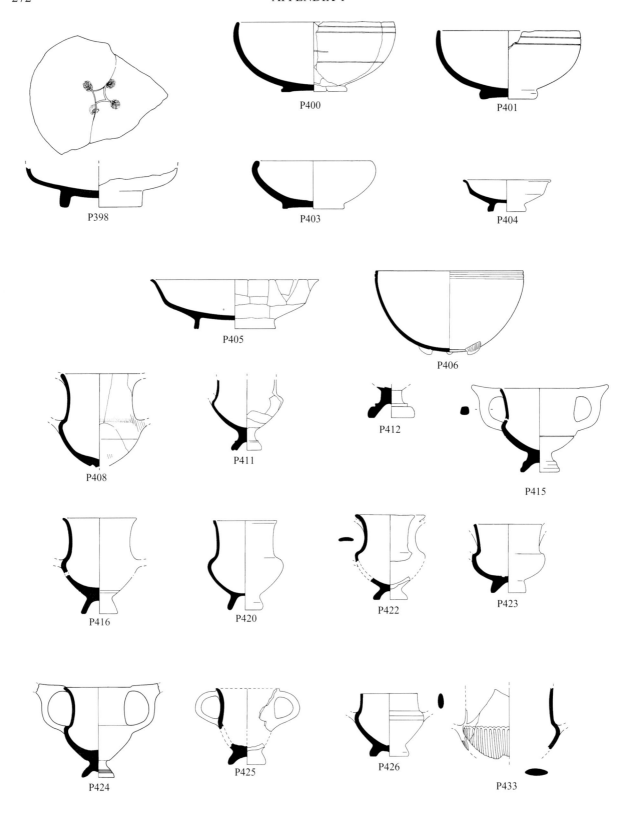

Fig. 6.5. Table ware (continued, scale 1:4).

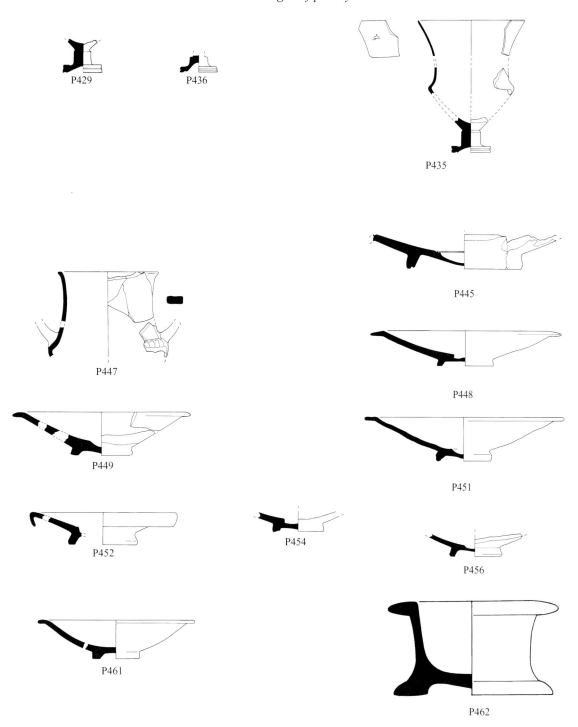

Fig. 6.5. Table ware (continued, scale 1:4).

Type 3. Splayed base ring, bevelled edge, high splayed wall, incurved rim, convex lip. Medium fine temper, reddish yellow colour, very dark grey glaze. Dipped.

P358   House 2, Inv. Nos. H21 2/54+58+63, Fig. 6.5. Complete profile. Dim.: H. 76; Th. wall 6; D. rim 120; D. body 136; D. base 56.

Type 4. Base fragments, decorated. Splayed base ring, rounded edge.

P359   House 6, Inv. No. H23 6/112. No base ring preserved. Fine temper, reddish yellow colour, red to very dark grey glaze. Four seven-petalled palmettes. Dim.: Th. base 7.

P360   House 6, Inv. Nos. H23 6/93+106. Groove on inside resting surface. Fine temper, dark reddish brown colour, very dark grey glaze. Five horse shoe stamps. Dim.: Th. base 3; D. base 60.

P361   House 6, Inv. No. HA 42/1. Very fine temper, reddish yellow to grey colour, dark grey glaze. Four seven-petalled palmettes, around impressed circle in centre. Dim.: Pres. H. 15; Th. base 5; D. base 62.

P362   House 4, Inv. No. HA 173/30. Very fine temper, reddish yellow colour, very dark grey glaze. Rouletting. Dim.: Pres. H. 19; Th. wall 6; Th. base 7; D. base 130.

P363   House 4, Inv. No. HA 176/6. No base ring preserved. Fine temper, reddish yellow colour, very dark grey glaze. Rouletting. Dim.: Pres. H. 12; Th. wall 6; Th. base 7–8.

P364   House 6, Inv. No. H23 6/117a. Fine temper, pink colour, very dark grey glaze. Rouletting and two seven-petalled palmettes. Dim.: Pres. H. 14; D. base 60.

P365   House 5, Inv. No. HA 45/8. Very fine temper, pinkish grey to white colour, dark grey glaze. Rouletting around incised circle. Dim.: Pres. H. 17; Th. wall 7; Th. base 6; D. base 50.

P366   House 5/6, Inv. No. HA 64/1c. No base ring preserved. Very fine temper, pink to pinkish grey colour, dark grey glaze. Rouletting. Dim.: Pres. H. 28; Th. base 4–5.

P367   House 4, Inv. No. HA 116/24. Groove on resting surface. Very fine temper, burnt, black glaze. Rouletting and four seven-petalled palmettes. Dim.: Pres. H. 18; Th. wall 4; Th. base 2; D. base 60.

P368   House 4, Inv. Nos. HA 127/107 + HA 146/11. No base ring preserved. Fine temper, very pale brown colour, brown glaze. Rouletting and one seven-petalled palmette. Dim.: Th. base 6.

P369   House 4, Inv. No. HA 156/27. No base ring preserved. Fine temper, reddish yellow colour, very dark grey glaze. Seven-petalled palmettes. Dim.: Th. base 9.

P370   House 5, Inv. Nos. HA 36/18+19. No base ring preserved. Very fine temper, reddish yellow to light brown colour, black glaze. Rouletting between concentric double circles, and fragmentary palmette stamp. Dim.: Th. base 5.

Bolsal P371–387

Type 1. Large bolsal, base ring with a wide concave edge, groove on inside resting surface, splayed wall, straight rim, convex lip. Two horizontal strap handles below lip.

P371   House 6, Inv. Nos. HA 42/2+3, Fig. 6.5. Complete profile. Rouletting, painted with ivy and grape garland below rim in buff clay. Reserved groove on transition from base to wall. Handles missing. Very fine temper, reddish yellow colour, very dark grey glaze. Powdery fabric. Ref.: Rotroff, 1997: No. 169 (P 443). Date: 350–300 BC. Dim.:

H. 65; Th. wall 4; Th. base 3–4; D. rim 160; D. base 100.

P372   House 5, Inv. Nos. HA 51/3+139 + HA 71/32, Figs 3.8 and 6.5. Complete profile. Angular transition from lower to upper body. Handles missing. Fine temper, reddish yellow colour, black glaze. Dim.: H. 78; Th. wall 4–5; Th. base 4; D. rim 161; D. base 106.

P373   House 6, Inv. No. HA 78B/16. Base fragment. Fine temper, reddish yellow colour, very dark grey glaze. Dim.: Pres. H. 19; Th. wall 6; Th. base 6; D. base 110.

Type 2. Base ring, groove on resting surface, bevelled lower and upper edge, straight rim, convex lip. Two horizontal strap handles below lip, arching above rim.

P374   House 5/6, along wall, Inv. No. HA 64/1, Fig. 6.5. Base and rim fragments, not fitting. Rouletting and four nine-petalled palmette stamps. Very fine temper, reddish yellow colour, dark grey glaze. Dim.: Pres. H. 38 (rim), 14 (base); Th. wall 5; Th. base 4, D. rim 110; D. base 75.

P375   House 6, Inv. No. HA 172/7. Base fragment. Very fine temper, very pale brown colour, black glaze. Rouletting. Dim.: Pres. H. 17; Th. wall 3; Th. base 3; D. base 90.

P376   House 6, Inv. No. H23 6/110. Base fragment. Fine temper, burnt, very dark grey glaze. Two circles of rouletting and four eleven-petalled palmettes in centre. Same workshop as P384. Dim.: Pres. H. 17; Th. wall 5; Th. base 6; D. base 65.

P377   House 6, Inv. No. HA 78B/14. Base fragment. Fine temper, pink colour, very dark grey glaze. Rouletting. Same workshop as P383. Dim.: Pres. H. 19; Th. wall 5; Th. base 5; D. base 110.

P378   House 6, Inv. Nos. H23 6/101a +113a. Base fragment. Very fine temper, reddish yellow colour, very dark grey glaze. Dim.: Pres. H. 10; D. base 80.

P379   House 6, Inv. No. HA 2/5. Base fragment. Fine temper, light red colour, very dark grey glaze. Dim.: Pres. H. 14; Th. wall 3; Th. base 8; D. base 70.

Type 3. Ring base, rounded edge.

P380   House 4, Inv. No. HA 127/19b. Base fragment. Fine temper, very pale brown colour, dark grey glaze. Rouletting. Ref.: Rotroff, 1997: No. 168 (P 29119). Date: 325–300 BC. Dim.: Pres. H. 23; Th. wall 6; Th. base 6; D. base 100.

P381   House 6, Inv. Nos. HA 17/17+19+24+29. Base fragment. Fine temper, reddish yellow colour. Dim.: Pres. H. 36; Th. wall 6; D. base 110.

P382   House 6, Inv. No. H23 6/86. Base fragment. Groove on resting surface. Splayed wall. Fine temper, pink colour, black glaze. Ref.: Rotroff, 1997: No. 168 (P 29119). Date: 325–300 BC. Dim.: Pres. H. 25; Th. wall 5; Th. base 6; D. base 100.

P383   House 5 and 6, along wall, Inv. No. HA 64/2. Base fragment. Very fine temper, pink colour, dark grey glaze. Dim.: Pres. H. 19; Th. wall 5; D. base 100.

Type 4. Ring base, rounded lower edge, concave upper edge. Fine temper, reddish yellow colour. Good quality glaze, but very worn fragments.

P384   House 3, Inv. No. HA 228/12. Base fragment. Very fine temper, pink colour, black glaze. Dim.: Pres. H. 29; Th. wall 6; Th. base 5; D. base 80.

P385   House 4, Inv. No. HA 130/65. Base fragment. Fine temper. Dim.: Pres. H. 20; Th. wall 7.

P386   House 4, Inv. No. HA 115/52. Base fragment. Fine temper, pink colour, very dark grey glaze. Dim.: Pres. H. 16; Th. wall 4; D. base 90–100.

P387   House 6, Inv. No. HA 43/3. Base fragment. Fine temper, reddish yellow colour, black glaze. Rouletting. Dim.: Pres. H. 18; Th. wall 5; Th. base 4; D. base 80.

Small bowl P388–395

Type 1. Base ring, splayed wall, incurved rim.
P388   House 3, Inv. No. HA 216/21, Fig. 6.5. Complete profile. Angular wall. Fine temper, reddish yellow colour, red glaze. Dim.: H. 37; Th. wall 4–5; Th. base 4–6; D. rim 66; D. body 82; D. base 45.
P389   House 4, Inv. No. HA 126/11, Fig. 6.5. Complete profile. Fine temper, reddish yellow colour, grey glaze. Dim.: H. 31; Th. wall 3; Th. base 2; D. rim 60; D. body 70; D. base 40.
P390   House 4, Inv. No. HA 126/30, Fig. 6.5. Complete profile. Fine temper. Dim.: H. 36; Th. wall 3-5; Th. base 4; D. rim 62; D. body 68; D. base 32.
P391   House 2, Inv. No. H21 2/24, Fig. 6.5. Complete. Fine temper, pink colour, black glaze. Dim.: H. 35; Th. wall 4; Th. base 6; D. rim 60; D. body 70; D. base 35.
P392   House 6, Inv. No. H23 5/95, Fig. 6.5. Complete profile. Fine temper, pink colour, grey glaze. Dim.: H. 34; Th. wall 4; Th. base 5; D. rim 82; D. body 88; D. base 47.
P393   House 6, Inv. No. H23 5/158, Fig. 6.5. Complete. Very fine temper, pink colour, traces of glaze. Dim.: H. 35; Th. wall 4; Th. base 7; D. rim 60; D. body 72; D. base 33.

Type 2. Splayed base ring, low splayed wall, incurved rim with slight angle. Fine temper, very pale brown colour, dark grey glaze.
P394   House 6, Inv. Nos. H23 5/149+149a, Fig. 6.5. Complete profile. Dim.: H. 28; Th. wall D. rim 80; D. body 92; D. base 45. Ref.: Vatin *et al.*, 1976: No. 115*bis*4. Date: 300–250.

Type 3. Ring base, splayed wall, incurved rim. Fine temper, yellow colour, not glazed.
P395   House 5, Inv. Nos. HA 55/8+29. Reconstructed profile. Dim.: D. rim 60; D. body 68.

Large glazed bowl P396–399

Type 1. Splayed base ring, rounded edge, groove on resting surface, splayed wall, incurved rim, convex lip. Fine temper, light red colour, black glaze.
P396   House 3, Inv. No. HA 197/3a, Figs 3.9 and 6.5 Complete. Dim.: H. 53; Th. wall 4; D. rim 152; D. body 162; D. base 75.

Type 2. Splayed wall, incurved rim, convex lip. Decorated with palmette stamps and rouletting. Repaired with strips of lead, drilled holes. Very fine temper, reddish yellow colour, dark grey glaze. Classical Attic.
P397   House 1, Inv. Nos. X 635A + 688. Base and rim fragments, not fitting. Base ring broken off (= Reinders, 1988: Cat. No. 33.30). Dim.: D. rim 230; D. attachment of base 120.

Type 3. Base ring with straight edge, splayed wall.
P398   House 6, Inv. Nos. HA 78E/1 + HA 84/1, Fig. 6.5. Base fragment. Fine temper, reddish yellow colour, grey glaze.

Four seven-petalled palmettes, linked with interlocking arcs. Dim.: Pres. H. 54; Th. wall 6; Th. base 8; D. base 90.
P399   House 6, Inv. No. HA 42/2a. Base fragment. Groove on resting surface and on transition from base ring to wall. Very fine temper, reddish yellow colour, very dark grey glaze. Rouletting and two fir-tree-like stamps. Dim.: Pres. H. 20; Th. wall 7; Th. base 5; D. base 90.

Large unglazed bowl P400–403

Type 1. Raised flat base, rounded edge, splayed wall, incurved rim, convex lip. New Halos ware.
P400   House 6, Inv. No. H23 5/25, Figs 3.10 and 6.5. Complete profile. Dim.: H. 75; Th. wall 5; D. rim 162; D. body 174; D. base 70.
P401   House 6, Inv. No. H23 5/145, Fig. 6.5. Complete profile. Dim.: H. 73; Th. wall 5; D. rim 147; D. body 156; D. base 66.
P402   House 5, Inv. No. HA 51/125. Base fragment. Raised concave base, rounded edge. Dim.: Pres. H. 21; Th. wall 9; D. base 70.

Type 2. Raised, slightly concave base, straight edge, splayed wall, incurved rim, slightly thickened, convex lip. Medium coarse temper, reddish yellow colour.
P403   House 6, Inv. No. H23 5/100, Fig. 6.5. Complete profile. Dim.: H. 50; Th. wall 7; D. rim 127; D. body 138; D. base 66.

Bowl with everted rim P404–405

Type 1. Splayed base ring, rounded edge, splayed wall with slight angle, everted rim, convex lip. Medium fine temper, reddish yellow colour, dark grey glaze.
P404   House 3, Inv. No. HA 236/4, Fig. 6.5. Complete profile. Dim.: H. 34; Th. wall 3–6; D. rim 95; D. base 42.

Type 2. Slightly splayed base ring, groove on resting surface, straight edge, splayed wall with slight angle, everted rim, convex lip. Stacking circle. Rouletting and four seven-petalled palmette stamps. Fine temper, pink colour, dark grey glaze. Attic.
P405   House 6, Inv. Nos. H23 1/118+123, Fig. 6.5. Complete profile. Dim.: H. 52; Th. wall 3; D. rim 190; D. base 90.

Hemispherical bowl

Type 1. Rounded base with three shell feet, semi-circular body, convex lip. Decorated with two incised lines below lip. Very fine temper, very pale brown colour, black glaze.
P406   House 1, Inv. No. X 612H, Fig. 6.5. Complete profile (= Reinders, 1988: Cat. No. 33.01). Ref.: Smettana-Scherrer, 1982: No. 523; Rotroff, 1997: No. 311 (P 27986). Date: 285-275 BC. Dim.: H. 88; Th. wall 2–3; D. rim 152.

Kantharos P407–434

The *kantharos* has a plain body, a slightly everted rim, a convex lip and spur handles, unless otherwise stated.

Type 1. Small circular stem. Thin wall, hard fabric. Very fine to fine temper, pink to very pale brown colour.

P407   House 3, Inv. No. HA 220/8. Rim and wall fragments. Sharply angular shoulder. Dim.: Pres. H. 50; Th. wall 3–5; D. body 60; D. stem 19.

P408   House 4, Inv. Nos. HA 110/1 + HA 136/2+3 + HA 149/13 + HA 181, Fig. 6.5. Body and rim fragments, handles missing. Decorated with incised line on the lower body. Dim.: Pres. H. 98; Th. wall 4–10; D. rim 85; D. body 90; D. stem 20.

P409   House 2, Inv. No. H21 2/11. Rim fragment. Dim.: Pres. H. 75; Th. wall 4; D. rim 74; D. body 82

P410   House 6, Inv. No. HA 4/24. Rim fragment. Dim.: Pres. H. 26; Th. wall 3–5; D. rim 140

P411   House 5, Inv. No. HA 86/23, Fig. 6.5. Base and body fragments. Dim.: H. 76; Th. wall 3–7; D. body 76; D. base 34; D. stem 18.

P412   House 3, Inv. No. HA 281/1, Fig. 6.5. Base fragment. Ref. for base: Rotroff, 1997: Bolster cup No. 166 (P 7760). Date: 300–280 BC. Dim.: Pres. H. 36; D. base 50; D. stem 22.

P413   House 6, Inv. No. H23 3/98. Base fragment. Dim.: Pres. H. 24; Th. wall 8; D. base 43; D. stem 23.

P414   House 5, Inv. Nos. HA 14/5+7+9+10. Base fragment. Dim.: Pres. H. 45; Th. wall 4; D. stem 17.

Type 2. Thick wall, soft powdery fabric. Fine temper, reddish yellow colour.

P415   House 6, Inv. Nos. H23 1/107-1+114+129a+161, Figs 3.11 and 6.5. Complete profile. High splayed base ring with broad ridge. Incised line on lower body. Dim.: H. 88; Th. wall 4–6; D. rim 84; D. body 85; D. base 42.

P416   House 2, Inv. Nos. HA 1/3+5+9, Fig. 6.5. Complete profile. High splayed base ring, rounded edge. Dim.: H. 102; Th. wall 4; D. rim 80; D. body 85; D. base 39.

P417   House 3, Inv. No. HA 246/5. Rim, wall, handle and stem fragments. Dim.: Th. wall 4; D. rim 80; D. stem 19.

P418   House 5, Inv. Nos. HA 18/3 + HA 26/2. Rim fragments. Dim.: Pres. H. 54; Th. wall 5–7; D. rim 70; D. body 80.

P419   House 4, Inv. Nos. HA 166/16+22+23+26 + HA 104/30 + HA 146/15. Base, wall and rim fragments. Dim.: Th. wall 3; D. base 44.

Type 3. Splayed base ring. Fine temper, reddish yellow colour, very good black glaze.

P420   House 4, Inv. Nos. HA 136/8+12 + HA 149/8, Fig. 6.5. Almost complete profile, handles missing. Dim.: H. 94; D. rim 70; D. body 85; D. base 40.

P421   House 6, Inv. No. HA 62/3. Base fragment. Dim.: Pres. H. 27; D. base 36.

Type 4. High, splayed base ring, slight groove on resting surface. Strap handles attached below lip and shoulder. Fine temper, very pale brown colour, very dark grey to weak red glaze.

P422   House 4, Inv. No. HA 99/5 + HA 106/2, Fig. 6.5. Reconstructed profile. Dim.: H. 91; D. rim 75; D. body 75; D. base 40.

Type 5. Handle attached below lip and on shoulder. Medium fine temper (with large quartz inclusion on inside base), bright reddish yellow colour, glaze fired black and red.

P423   House 2, Inv. Nos. H21 2/21+40+71+78, Fig. 6.5. Almost complete profile, handles missing. Dim.: H. 78; D. rim 76; D. body 78; D. base 38.

Type 6. Low stemmed ring base. Medium fine temper with mica, pinkish white colour, dark grey glaze.

P424   House 1, Inv. No. X 344, Fig. 6.5 Complete profile (= Reinders, 1988: Cat. No. 37.01). Dim.: H. 96; Th. wall 3–8; D. rim 76; D. body 73; D. base 35.

Type 7. High, splayed base ring, rounded edge. Vertical strap handle, flattened in section (without spur). Medium temper with lime and quartz, traces of reddish brown glaze.

P425   House 2, Inv. No. H21 2/68, Fig. 6.5. Rim and base fragments. Dim.: Pres. H. rim 41; Pres. H. base 20; Th. wall 3-5; D. rim 72; D. body 72; D. base 41.

Type 8. Splayed base ring, slightly bevelled edge. Medium fine temper with lime, reddish yellow colour, dark grey glaze, shifting to red. Very powdery fabric.

P426   House 3, Inv. Nos. HA 277/17 + HA 278/21, Fig. 6.5. Base, body and rim fragments. Dim.: Pres. H. 67; Th. wall 3-6; D. rim ND; D-N: 64; D. body 74; D. base 41.

Type 9. High, splayed base ring, rounded edge. Medium fine temper, reddish yellow colour, dark grey glaze, shifting to red.

P427   House 1, Inv. No. X 672. Base fragment (= Reinders, 1988: Cat. No. 33.20). Dim.: Pres. H. 20; Th. wall 6; D. base 42.

Type 10. High, splayed base ring, rounded edge. Medium fine temper, reddish yellow colour, dark grey glaze.

P428   House 1, Inv. No. X 335. Base fragment (= Reinders, 1988: Cat. No. 37.23). Dim.: Pres. H. 31; Th. wall 5; D. base 40.

Type 11. Stemmed base ring, moulded edge. Groove on resting surface with red miltos. Fine temper, very pale brown colour, dark grey glaze, shifting to red.

P429   House 1, Inv. No. X 604, Fig. 6.5. Base fragment (Reinders, 1988: Cat. No. 33.12). Ref. for base: Rotroff, 1997: Bolster cup No. 109 (P 7765) and 110 (P 6948). Date: 290–275 BC. Dim.: Pres. H. 30; Th. wall 4; D. base 43.

Type 12. Splayed base ring, moulded edge. Medium coarse temper with lime, reddish yellow colour, red glaze.

P430   House 5, Inv. No. HA 80/1. Base fragment. Dim.: Pres. H. 21; Th. wall 6; D. base 49.

Type 13. High, splayed base ring, rounded edge. Very fine temper, light red colour, dark grey glaze.

P431   House 5, Inv. No. HA 73/4. Base fragment. Dim.: Pres. H. 21; Th. wall 7; D. base 39.

Type 14. Ribbed lower body, transition to neck marked with groove. Fine temper, reddish yellow colour, dark grey glaze.

P432   House 4, Inv. No. HA 115/78. Wall fragment. Pres. H. 18; Th. wall 3–4; D. body 80

Type 15. Ribbed lower body. Handle oval in section. Very fine temper, light red colour, black glaze.

P433   House 5, Inv. Nos. HA 44/9+15 + HA 53/9, Fig. 6.5. Wall fragment with handle attachment. Dim.: Pres. H. 70; Th. wall 4; D-N: 80; D. body 100.

Type 16. Ribbed lower body. Very fine temper, very pale brown colour, dark grey glaze.

P434   House 5, Inv. No. HA 86/23a. Wall fragment. Dim.: Pres. H. 32; Th. wall 4; D. body 100.

Cup-kantharos P435–437

Type 1. Stemmed base with moulded disk foot, groove on resting surface. Wide everted rim. Miltos in grooves on resting surface and lower body. Painted ivy leaves in white slip below the rim. Very fine temper, pink colour, dark grey glaze.

P435 House 1, Inv. Nos. X 612 A, B, C, E, Figs 3.12 and 6.5. Reconstructed profile, handles missing (= Reinders, 1988: Cat. No. 33.14; 33.22; 33.23; 33.25). Ref.: Braun 1970: No. 10 (KER 9722); Miller 1974: No. 15 (P 27970); Reinders, 1988: No. 54.14; Rotroff, 1997: No. 109 (P 7765). Date: 290-275 BC. Dim.: Rec. H. 132; Th. wall 3–6; D. rim 113; D. body 92; D. base 43; D. stem 19.

Type 2. Stemmed base with moulded disk foot. Fine temper, reddish yellow colour, dark grey glaze.

P436 House 4, Inv. No. HA 127/70, Fig. 6.5. Base fragment. Ref.: Braun 1970: No. 10 (KER 9722); Miller 1974: No. 15 (P 27970); Reinders, 1988: No. 54.14; Rotroff, 1997: No. 109 (P 7765). Date: 290–275 BC. Dim.: Pres. H. 19; D. base 41.

Type 3. Large kantharos, ribbed body. Very fine temper.

P437 House 5, Inv. No. HA 49/1, Fig. 6.5. Rim and shoulder fragments. Dim.: Pres. H. 88; D. rim 110; D. body 128; S-1: 8; S-2: 34.

Bowl-kantharos P438–439

Type 1. Straight rim, convex lip. Vertical strap handle, oval in section, with ivy leaf thumb rest. Ref.: Reinders, 1988: No. 54.14. Date: c. 300 BC; Rotroff, 1997: No. 141 (P 27970). Date: 290–275 BC.

P438 House 6, Inv. No. H23 6/80a. Handle fragment. Fine temper, very dark grey glaze. Dim.: Pres. H. 33; S-1: 15; S-2: 6.

P439 House 6, Inv. Nos. H23 6/136+170. Handle fragment. Fine temper. Dim.: Pres. H. 16; S-1: 13; S-2: 6.

Skyphos P440–441

P440 House 5, Inv. No. HA 51/5. Base fragment. Splayed base ring, flattened edge, straight wall. Very fine temper, dark grey glaze, outside base unglazed. Dim.: Pres. H. 22; Th. wall 6; D. base 80.

P441 House 2, Inv. Nos. H21 4/4+14b. Rim fragment with handle attachment. Splayed wall, everted rim, convex lip. Horizontal ring handle, circular in section. Very fine temper, pale brown colour, dark grey glaze. Dim.: Pres. H. 24; Th. wall 4; S-1: 10.

Fish plate P442–460

Type 1. Splayed base ring, rounded edge, deep depression, spreading wall, overhanging, down-turned rim. Groove along depression and edge rim. Fine temper, reddish yellow colour, black glaze of good quality. Provenance: Attic.

P442 House 1, Inv. No. X 738A, Figs 3.13 and 6.5. Complete profile. Groove on inside resting surface and rim with traces of red miltos (= Reinders, 1988: Cat. No. 34.01). Ref.: Sparkes & Talcott, 1970: No. 1073. Date: c. 325 BC.

Rotroff, 1997: No. 713 (P 23538). Date: 310–290 BC. Dim.:H. 35; Th. wall 7; D. rim 190; D. base 103; D. depression 48.

P443 House 4, Inv. Nos. HA 97/5+8 + HA 115/2. Complete profile. Ref.: Rotroff, 1997: No. 715 (P 20318). Date: 310–290 BC. Dim.: H. 38; Th. wall 6; D. rim 170; D. base 84; D. depression 44.

P444 House 6, Inv. No. H23 6/125+150+157. Almost complete profile. Dim.: H. 40; Th. wall 9–11; D. rim 250; D. base 140.

P445 House 4, Inv. Nos. HA 126/28 + HA 166/10, Fig. 6.5. Base + wall fragment, reconstructed profile. Two grooves along depression. Dim.: H. 36; Th. wall 7–10; D. rim 210; D. base 126; D. depression 52.

P446 House 2, Inv. No. H21 2/25. Rim fragment. Dim.: Pres. H. 16; Th. wall 7–9; D. rim 137.

Type 2. Splayed base ring, edge of base with shallow ridge towards resting surface, wide, slightly offset rim. Medium temper with white parts, light brown colour.

P447 House 6, Inv. Nos. H23 1/101 + 114b2, Figs 3.14 and 6.5. Complete profile. Dim.: H. 33; Th. wall 5; D. rim 180; D. base 52; D. depression 30.

Type 3. Splayed base ring, flattened edge, shallow depression with incurved edge, splayed wall, offset rim, tapering lip. Medium fine temper with lime, dark grey glaze, soft powdery fabric, burnt?

P448 House 6, Inv. Nos. H23 1/115+116, Fig. 6.5. Complete profile. Dim.: H. 39; Th. wall 7; D. rim 210; D. base 65; D. depression 24.

Type 4. Splayed base ring, spreading wall, slightly offset rim. Shallow depression. The fragments vary in thickness. Fine temper, light reddish yellow colour.

P449 House 6, Inv. No. HA 87/1, Fig. 6.5. Complete profile. Ref.: Reinders, 1988: No. 52.04. Date: c. 300 BC. Dim.: H. 45; D. rim 192; D. base 67; D. depression 24.

Type 5. Splayed base ring, shallow depression. Fine temper; light reddish yellow colour, red glaze.

P450 House 2, Inv. No. H21 2/20. Base fragment. Dim.: Pres. H. 24; Th. wall 5; D. base 59; D. depression 18.

Type 6. Straight base ring, spreading wall, slightly offset rim. Depression marked with ridge. Medium temper with quartz, reddish yellow colour, grey glaze.

P451 House 3, Inv. Nos. HA 247/7c+10 + HA 248/7c, Fig. 6.5. Complete profile. Dim.: H. 47; D. rim 220; D. base 60; D. depression 26.

Type 7. Splayed base ring, deep depression, spreading wall, overhanging, down-turned rim. Fine temper, reddish yellow colour, light red glaze.

House 3, Inv. Nos. HA 207/3+8, Fig. 6.5. Complete profile. Ref.: Hellström, 1971: No. 56. Date: second half of 4th century BC. Dim.: H. 34; Th. wall 5–8; D. rim 160; D. base 74; D. depression 52.

Type 8. Splayed base ring, splayed wall with angle towards rim, offset rim, tapering lip. Very fine temper, very pale brown colour.

P453 House 5, Inv. Nos. HA 14/22+23a + HA 50/3+7. Reconstructed profile, depression on inside base missing. Ref.: Heimberg, 1982: No. 336. Date: mid 3rd century BC. Dim.: D. rim 200; D. base 120.

Type 9. Straight base ring, resting surface bevelled on the inside, depression with straight edge, spreading wall. Fine temper, brown to greyish brown colour, red glaze.

P454   House 3, Inv. No. HA 276/5, Fig. 6.5. Base fragment. Dim.: Pres. H. 21; Th. wall 6–7; D. base 55; D. depression 29.

Type 10. Splayed base ring, rounded edge, shallow depression. Fine temper, reddish yellow colour, light grey to light red glaze.

P455   House 3, Inv. No. HA 277/7. Base fragment. Dim.: Pres. H. 20; D. base 71; D. depression 42.

Type 11. Splayed base ring , bevelled edge, depression with central nipple, spreading wall. Medium fine temper with quartz, pink to very pale brown colour, grey to dark reddish brown glaze. Partly burnt.

P456   House 3, Inv. No. HA 278/6, Fig. 6.5. Base fragment. Dim.: Pres. H. 23; Th. wall 4–5; D. base 58; D. depression 27.

Type 12. Splayed base ring, shallow depression. Medium fine temper, reddish yellow colour, black glaze.

P457   House 4, Inv. No. HA 130/11. Base fragment. Dim.: Pres. H. 15; D. base 59; D. depression 30.

Type 13. Splayed base ring, shallow depression. Medium fine temper, reddish yellow colour, very dark grey glaze, soft powdery fabric.

P458   House 4, Inv. Nos. HA 137/16 + HA 141/38. Base fragment. Dim.: Pres. H. 28; D. base 56; D. depression 29.

Type 14. Splayed base ring, shallow depression. Very fine temper, reddish yellow colour, light reddish brown glaze.

P459   House 2, Inv. No. H21 2/55. Base fragment. Dim.: Pres. H. 15; D. base 61; D. depression 28.

Type 15. Splayed base ring, resting surface bevelled on the inside, shallow depression. Fine temper, very pale brown colour, grey glaze.

P460   House 6, Inv. No. H23 2/26. Base fragment. Dim.: Pres. H. 25; Th. wall 6; D. base 60; D. depression 22.

Saucer

P461   House 3, Inv. Nos. HA 197/5+13 + HA 202/17, Fig. 6.5. Reconstructed profile. Straight base ring, splayed wall, offset rim. Medium temper, very pale brown colour. Dim.: Rec. H. 41; D. rim 170; D. base 50.

Spool saltcellar

P462   House 5, Inv. No. HA 86/17, Figs 3.15 and 6.5. Complete pot. Splayed base ring with rounded edge and broad resting surface. Slightly concave body. Wide everted rim, convex on top. Very fine temper, reddish yellow colour, dark grey glaze. Ref.: Rotroff, 1997: No. 1070 (P 20908). Date: c. 325–295 BC. Dim.: H. 51; Th. wall 4–7; D. rim 92; D. body 62; D. base 86.

MISCELLANEOUS

Oil lamp P463–479

Type 1a. Attic, type Howland 30B. Splayed wall, turned-in rim with external angle, convex lip. Fine temper, completely glazed. Ref.: Howland, 1958: No. 419–421 (L 2890, L 3530, L 1899). Date: mid 4th – first quarter 3rd cent. BC; Rotroff, 1997: p. 500. Date: 325–275.

P463   House 4, Inv. No. HA 106/17. Almost complete profile. Concentric groove on outside base. Dim.: H. 29; Th. wall 3–4; D. rim 38; D. body 60; D. base 50.

P464   House 2, Inv. No. H21 2/46. Rim fragment. Dim.: Pres. H. 24; Th. wall 7; D. rim 50.

P465   House 6, Inv. No. H23 6/177. Base fragment. Concave base. Dim.: Pres. H. 15; Th. base 5; D. base 45.

P466   House 3, Inv. No. HA 229/1. Base and spout fragments. Concave base, rectangular spout, widening a little at end. Dim.: Pres. H. 27; Th. base 4.

P467   House 5, Inv. No. HA 25/6. Rim fragment. Dim.: Pres. H. 23; Th. wall 7; D. rim 40; D. body 51.

P468   House 5, Inv. No. HA 45/33. Rim and spout fragments. Dim.: Pres. H. 31; Th. base 8; D. rim 36; D. body 60; D. base 38.

Type 1b. Raised concave base, splayed wall, turned-in rim with external angle, convex lip. Variation on type Howland 30B.

P469   House 3, Inv. No. HA 234/2, Figs 3.16 and 6.6. Complete lamp. Dim.: H. 30; Th. wall 5; D. rim 26; D. body 60; D. base 42.

P470   House 3, Inv. No. HA 279/2, Fig. 6.6. Complete profile. Dim.: H. 27; Th. wall 3–4; D. rim 29; D. body 59; D. base 34.

P471   House 2, Inv. No. H21 2/80, Fig. 6.6. Complete lamp. Slightly convex base, splayed wall, incurved, overhanging rim, convex lip. Dim.: H. 21; Th. wall ; D. rim 42; D. body 53; D. base 39.

P472   House 4, Inv. No. HA 146/10. Rim and body fragments. Dim.: Pres. H. 25; Th. wall 5; D. rim 30; D. body 50.

P473   House 4, Inv. No. HA 127/72. Base and spout fragments. Flat base. Dim.: Pres. H. 24; Th. wall 3–5; D. body 55; D. base 40.

P474   House 2, Inv. No. H21 2/2. Rim and base fragments. Dim.: Pres. H. 22; Th. wall 6–8; D. rim 30; D. body 60; D. base 23.

1c. Medium coarse orange ware, bright orange colour, traces of black glaze.

P475   House 5, Inv. No. HA 83/1, Fig. 6.6. Complete profile. Concave base, splayed wall, turned-in rim, convex lip. Variation on type Howland 30 B. Dim.: H. 28; Th. wall 5; D. rim 25; D. body 65.

Type 2. Attic, type Howland 30A. Lower wall splayed, with two incised parallel lines, upper wall convex, turned-in rim with slight external angle, convex lip. Ref.: Howland, 1958: No. 418 (L 1873). Date: last quarter 5th cent. BC. Scheibler, 1976: date: 410–370 (HT-date).

P476   House 4, Inv. No. HA 127/12, Fig. 6.6. Rim and body fragments. Dim.: Pres. H. 33; Th. wall 4–5; D. rim 38; D. body 70; D. base 40.

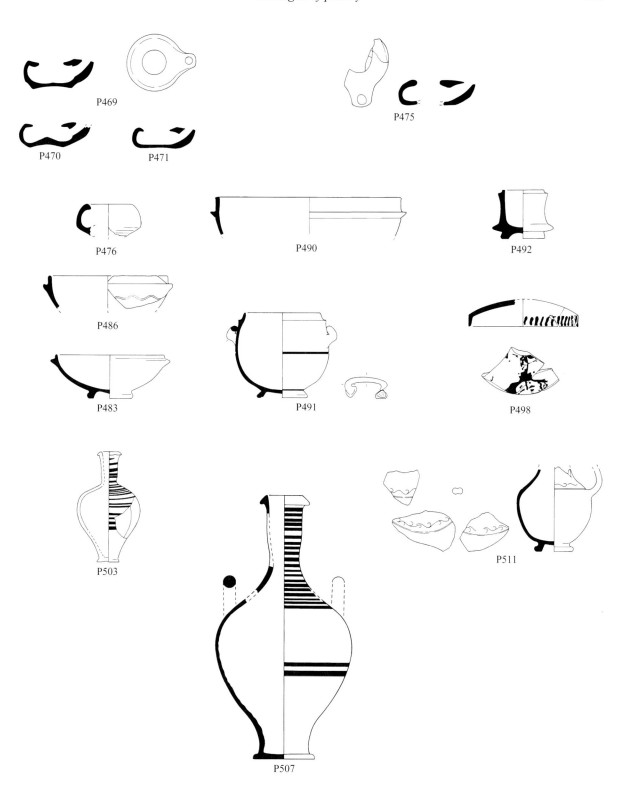

Fig. 6.6. Miscellaneous ceramics (scale 1:4).

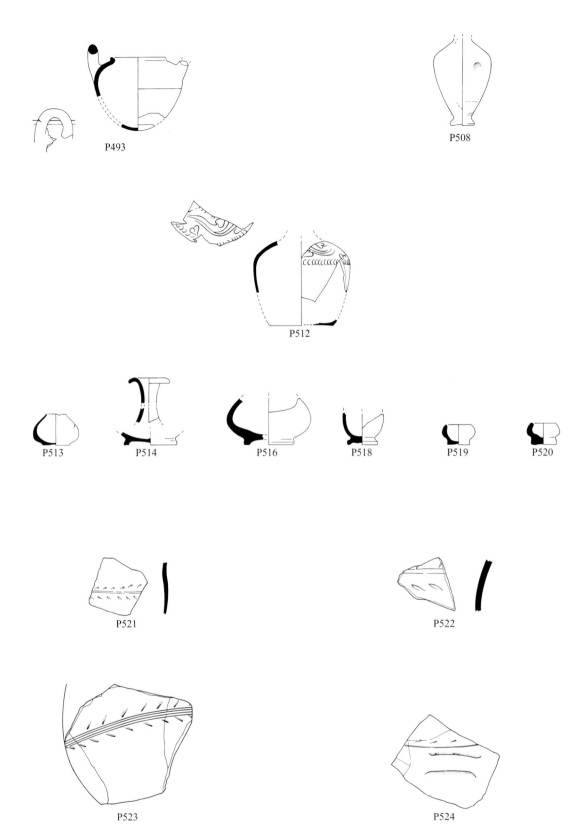

Fig. 6.6. Miscellaneous ceramics (continued, scale 1:4).

Type 3. Splayed wall, turned-in rim with external angle, upturned convex lip. Medium fine temper, reddish yellow colour, dark grey glaze.

P477   House 3, Inv. No. HA 193/4. Rim fragment. Dim.: Pres. H. 17; Th. wall 4; D. rim 30; D. body 55.

Type 4. Convex body, turned-in rim, convex lip. Very fine temper, reddish yellow colour, dark grey glaze.

P478   House 4, Inv. No. HA 73/3. Rim and body fragments. Dim.: H. 15; Th. wall 4.

Type 5. Splayed wall, incurved rim, convex lip. Very fine temper, reddish yellow colour, dark grey glaze.

P479   House 5, Inv. No. HA 51/.. Rim fragment. Dim.: Pres. H. 10; Th. wall 4; D. rim 36.

Lekanis P480–490

Type 1a. Splayed wall, slightly incurved rim with flange for a lid, convex lip. Fine temper, reddish yellow colour, very dark grey to black glaze.

P480   House 3, Inv. No. HA 201. Rim fragment. Dim.: Pres. H. 31; Th. wall 4–6; D. rim 115.

P481   House 4, Inv. No. HA 130/44. Rim fragment. Two parallel grooves below lip. Dim.: Pres. H. 23; Th. wall 4–9; D. rim 170–210.

P482   House 4, Inv. No. HA 130/92. Rim fragment. Two parallel grooves below lip. Dim.: Pres. H. 17; Th. wall 3–10; D. rim 110.

P483   House 4, Inv. No. HA 137/21. Rim fragment. Attachment of horizontal strap handle. Dim.: Pres. H. 29; Th. wall 6–8.

P484   House 6, Inv. No. H23 6/161. Rim fragment. Dim.: Pres. H. 21; Th. wall 4; D. rim 180.

P485   House 5, Inv. No. HA 14/17. Rim fragment. Groove on lower body. Dim.: Pres. H. 32; Th. wall 7; D. rim 180.

P486   House 5, Inv. No. HA 65/1, Fig. 6.6. Rim fragment. Painted on body with a wavy line in light brown slip. Dim.: Pres. H. 35; Th. wall 5; D. rim 140.

P487   House 4, Inv. No. HA 130/60. Wall fragment with knob on flange.

Type 1b. Splayed base ring, rounded edge, splayed wall with flange for lid, slightly incurved rim, convex lip. Medium fine temper, brownish grey colour, dark grey glaze.

P488   House 3, Inv. Nos. HA 197/22 + HA 202b/16, Fig. 6.6. Complete. Dim.: H. 46; Th. wall 3–4; D. rim 120; D. body 130; D. base 50.

Type 2. Splayed wall, straight rim with flange for lid, convex lip. Fine temper, reddish yellow colour, glaze worn off. Ref.: Sparkes & Talcott, 1970, No. 1275 (P 25688). Date: 325–310 BC. Vatin et al., 1976: No. 63.7. Date: 300–250 BC.

P489   House 6, Inv. No. HA 48/3. Rim fragment. Dim.: Pres. H. 28; Th. wall 3–6; D. rim 90.

Type 3. Splayed wall, straight rim with pointed flange, convex lip. Very fine temper, reddish yellow colour, very dark grey and light olive grey glaze.

P490   House 1, Inv. No. X 35, Fig. 6.6. Rim fragment (= Reinders, 1988: Cat. No. 39.01). Dim.: Pres. H. 38; Th. wall 3–5; D. rim 202.

Pyxis P491-494

Type 1. Baggy type: splayed base ring, rounded edge, groove on resting surface; globular body, with groove on the widest part, incurved rim with pointed flange, convex lip. Two horizontal ring handles, circular in section, on the upper body, bent towards the wall. Groove on body painted red.

P491   House 3, Inv. Nos. HA 197/2+5+10b+12 + HA 202/5+23+34, Figs 3.17 and 6.6. Complete profile. Dim.: Rec. H. 92; Th. wall 2–3; D. rim 82; D. body 102; D. base 53.

Type 2. Classical Attic shape. Splayed base ring, rounded edge, transition from base to body marked with wide pointed ridge, concave body, straight rim with flange for lid, flattened lip. Medium fine temper, pink colour, very dark grey glaze. Ref.: Sparkes & Talcott, 1970: No. 1312 (P 7369). Date: 4th cent. BC.

P492   House 4, Inv. No. HA 113/10, Figs 3.18 and 6.6. Complete profile. Dim.: H. 50; Th. wall 75; D. rim 45; D. body 66; D. base 48.

Type 3. Stamnoid pyxis. Rounded base, globular body, convex shoulder, upright rim, lip bevelled on the inside. Two horizontal ring handles, circular in section, attached to shoulder and arching above rim. Groove below handles. Medium fine temper, red colour. Ref.: Almirós Archaeological Museum BE 9244 (Halos, Hellenistic cemetery).

P493   House 6, Inv. No. H23 5/101, Fig. 6.6. Reconstructed profile. Dim.: H. 78; Th. wall 3–4; D. rim 61; D. body 95.

Type 4. Rotroff type D. Flat base, straight wall. Fine temper, very pale brown colour, dark grey glaze. Ref.: Rotroff, 1997: No. 1249 (P 6282). Date: 250–240 BC?

P494   House 5, Inv. No. HA 81/10. Base fragment. Dim.: Pres. H. 38; Th. wall 7; D. base 48; D. body 52.

Glazed pyxis lid P495–502

Type 1. Domed lid with straight edge. Very fine to fine temper.
1a. Greyish brown to light brown colour, dark grey glaze.

P495   House 3, Inv. No. HA 267/18. Rim fragment. Dim.: H. 23; Th. wall 5–6; D. rim 85.

P496   House 1, Inv. No. X 638 B. Rim fragment. Groove on top of the lid, along the edge. Dim.: Pres. H. 22; Th. wall 5; D. rim 155.

P497   House 4, Inv. No. HA 113/12. Rim fragment. Groove on the inside of the lid, along the edge. Dim.: Pres. H. 29; Th. wall 4; D. rim 230.

1b. Convex lip. Painted in red-figure with two facing heads and floral pattern in between. Reddish yellow colour, grey glaze. Attic. Ref.: Blondé, 1985: No. 34. Date: 350 BC.

P498   House 1, Inv. No. X 638 A, Fig. 6.6. Rim fragment (= Reinders, 1988: Cat. No. 33.19). Dim.: Pres. H. 29; Th. wall 6; D. rim 120.

Type 2. Domed lid, incurved rim, lip thickened on the inside, with groove on edge. Fine temper, reddish yellow to light red colour, traces of glaze.

P499   House 6, Inv. Nos. HA 17/7 + HA 17/20. Rim fragment. Dim.: Pres. H. 27; Th. wall 6–7; D. rim 220.

P500   House 6, Inv. No. HA 78E/9. Rim fragment. Dim.: Pres. H. 24; Th. wall 7; D. rim 220.

Type 3. Small domed lid, thickened lip, with groove along the rim. Medium temper, reddish yellow colour, very dark grey glaze.

P501   House 6, Inv. No. H23 1/106. Rim fragment. Dim.: Pres. H. 18; Th. wall 4–5; D. rim 80.

Type 4. Domed lid, everted rim with ridge on edge.

P502   House 1, Inv. No. X 727 B. Rim fragment. Dim.: Pres. H. 33; Th. wall 5; D. rim 140.

Unguentarium P503–510

Type 1. Raised flat base, splayed wall, convex shoulder, narrow slightly spreading neck, straight rim, convex thickened lip, overhanging. Very fine to fine grey ware. Decorated on body, shoulder and neck with dark grey painted bands. Ref.: Schlörb-Vierneisel, 1966: No. 166.1, 170.2 and 174.1. Date: 325-300 BC. Vatin et al., 1976: 75.9-12 and 115.4. Date: c. 300 BC.

P503   House 6, Inv. No. H23 5/166, Fig. 6.6. Complete pot. The inner core has a reddish yellow colour. Dim.: H. 117; Th. wall 3; D. rim 28; D. body 68; D. base 25.

P504   House 3, Inv. No.2 HA 251/4. Base fragment. Dim.: Pres. H. 70; Th. wall 2; D. body 76; D. base 32.

P505   House 3, Inv. No. HA 257/3 (1). Base fragment. Dim.: Pres. H. 36; Th. wall 3–4; D. base 41.

P506   House 3, Inv. No. HA 257/3(2). Wall fragment. May belong to P504. Dim.: Pres. H. 84; Th. wall 2–3; D. body 100–110.

Type 2. Unguentarium/Amphoriskos. Flat base, rounded edge, concave foot, convex body, convex shoulder, long, slightly convex neck, straight rim, lip thickened and bevelled on the outside, undercut. Two vertical ring handles, circular in section, attached to the shoulder. Fine temper, dark greenish grey colour with red core, painted bands in weak red. Ref.: Almirós Archaeological Museum K 2951 VII, Cemetery Phthiotic Thebes. Date: beginning 3rd cent. BC. Hausmann, 1996: No. 58 (K1110). Date: shortly after c. 250 BC.

P507   House 3, Inv. Nos. HA 278/18 + HA 280/12, Figs 3.19 and 6.6. Complete profile. Dim.: H. 280; Th. wall 2–4; D. rim 54; D. body 145; D. base 65.

Type 3. Raised flat, solid base, splayed wall, slightly concave shoulder, narrow neck. Fine temper, reddish yellow colour, black glaze. Ref.: Edwards, 1975: No. 584 (C-48-119). Date: c. 325 BC.

P508   House 5, Inv. No. HA 55/43, Fig. 6.6. Base fragment. Dim.: Pres. H. 91; D. body 62; D. base 27.

Type 4. Very fine temper, reddish yellow colour, no glaze. Very fragmentary.

P509   House 3, Inv. No. HA 253a. Base fragment. Raised flat base, rounded edge, slightly splayed wall. Dim.: Pres. H. 14; Th. wall 2; D. base 20.

P510   House 5, Inv. No. HA 50/6. Rim fragment. Slightly spreading neck, everted rim, bevelled lip, thickening on the outside. Dim.: Pres. H. 26; Th. wall 2; D. rim 25.

Aryballos P511–513

Type 1. Splayed base ring, rounded edge, globular body, slightly concave shoulder, double barred vertical strap handle. Very fine temper, reddish yellow colour, black glaze. Painted on shoulder with band of ivy-leaves in West slope style, partly incised.

P511   House 6, Inv. No. H23 2/62, Fig. 6.6. Base and body fragments. Dim.: Pres. H. 90; Th. wall 3–4; D. body 81; D. base 42.

Type 2. Flat base, splayed wall, convex shoulder, on shoulder attachment for vertical strap handle, rectangular in section. Medium temper, reddish yellow colour, hard egg shell fabric. Decorated on the shoulder with incised ovules and ivy-leaves. Corinthian, Blister ware. Ref.: Edwards, 1975: No. 777 (C-34-1645). Date: 325–300 BC.

P512   House 5, Inv. Nos. HA 41/98 + HA 46/2a + HA 51/134 + HA 71, Fig. 6.6. Base and body fragments. Dim.: Pres. H. body 84; Th. wall 2–4; D. body 100; D. base 80.

Type 3. Raised base, outside slightly concave, splayed wall, concave shoulder.

P513   House 5, Inv. Nos. HA 51/139 + HA 71/4+12, Fig. 6.6. Base and body fragments. Dim.: Pres. H. 37; Th. wall 4; D. body 49; D. base 40.

Guttus P514–516

Type 1. Splayed base ring, rounded edge, groove on resting surface. Narrow concave neck, everted rim, overhanging convex lip. Circular vertical strap handle, rectangular in section.

1a. Medium fine temper, black glaze.

P514   House 3, Inv. Nos. HA 197/4 + HA 202/2b, Fig. 6.6. Base and rim fragments. Ref.: Rotroff, 1997: No. 1142 (P 28028). Date: c. 285 BC. Dim.: Pres. H. base 14; Pres. H. rim 49; Th. wall 4–4.5; D. rim 45; D. base 58.

1b. Very fine temper, black glaze.

P515   House 2, Inv. No. H21 2/52. Rim fragment. Dim.: Pres. H. 35; Th. wall 4; D. rim 39.

Type 2. Guttus/askos. Splayed base ring, rounded edge, splayed, slightly angular body, slightly convex shoulder. Medium fine temper with lime, reddish yellow colour, traces of reddish slip. Powdery fabric.

P516   House 3, Inv. No. HA 249/3, Fig. 6.6. Base and body fragments. Dim.: Pres. H. 51; Th. wall 4–9; D. body 90; D. base 54.

Amphoriskos

P517   House 4, Inv. No. HA 126/29. Base fragment. Dim.: Pres. H. 38; Th. wall 5–12. Pointed base, spreading wall. New Halos ware.

Squat lekythos

P518   House 5, Inv. No. HA 26/4, Fig. 6.6. Base fragment. Dim.: Pres. H. 34; Th. wall 4; D. body 45; D. base 34. Splayed base ring, bevelled edge, globular body. Fine temper, reddish yellow to pale brown colour, dark grey glaze. Ref.: Sparkes & Talcott, 1970: No. 1137 (P 876), 1138 (P 765), 1139 (P 7606), 1140 (P 14811) and 1141 (P 10436). Date: 410–30 BC.

Miniature P519–520

Type 1. Bowl-like shape. Raised flat base, splayed wall, incurved rim. Fine temper, reddish yellow colour, black glaze.

P519   House 4, Inv. No. HA 104/19, Fig. 6.6. Complete profile.

Dim.: H. 22; Th. wall 4; D. rim 26; D. body 32; D. base 24.

P520   House 4, Inv. No. HA 130/50, Fig. 6.6. Complete profile. Dim.: H. 23; Th. wall 5; D. rim ; D. body 36; D. base 25.

Wall fragments of pottery of undeterminable types, with floral pattern decoration.

Type 1. New Halos ware/medium coarse red ware.

P521   House 6, Inv. No. H23 6/62, Fig. 6.6. Dim.: Th. wall 5.

Type 2. Medium coarse red ware.

P522   House 5, Inv. No. HA 71/25, Fig. 6.6. Dim.: Th. wall 7.

P523   House 5, Inv. No. HA 51/120, Fig. 6.6. Shoulder fragment of jug or jar. Dim.: Th. wall 7.

P524   House 3, Inv. No. HA 262/6, Fig. 6.6. Jar? Wall fragment with horizontal ring handle, oval in section. Dim.: Th. wall 5.

Type 3. Medium coarse orange ware, bright orange colour.

P525   House 1, Inv. No. X 665 D. Jug? Dim.: Th. wall 5.

APPENDIX 2. SURVEY OF THE DISTRIBUTION OF TYPES OF VESSELS IN THE SIX HOUSES

COLETTE BEESTMAN-KRUYSHAAR

	house 3	house 1	house 4	house 2	house 6	house 5	5/6
COOKING VESSELS							
Chytra							
Type 1a							
complete	-	P1–3	P4–5	-	P6–7	-	
rim	P8	P9	P10–11	P12	P13–25	P26–34	
base	-	P35	P36	-	-	-	
wall	-	P37	P38	-	-	-	
Type 1b							
complete	P39–40	P41	-	-	-	-	
rim	-	P42	P43–44	-	P45–53	P54–57	
base	-	-	-	P58	-	P59	
Type 1c							
complete	P60	-	-	-	P61	-	
rim	P62	P63	P64	-	-	-	
base	P65–66	-	-	-	-	-	
Type 2a							
complete	-	-	-	-	P67	-	
Type 2b							
base	-	-	P68	-	-	-	
Type 3							
complete	-	P69	-	P70	P71	-	
rim	-	-	-	-	-	P72	
Type 4							
rim	-	-	-	-	-	P73	
Lopas							
Type 1a							
complete	-	P74	-	-	-	-	
Type 1b							
complete	P75	-	-	-	-	-	
rim	-	-	-	-	P76	-	
Type 2							
complete	P77	-	-	-	-	-	
rim	-	-	P78	-	-	-	
Type 3a							
rim	-	-	-	P79	P80	-	
Type 3b							
rim	-	P81	P82	-	-	-	
Type 4							
complete	P83	-	-	-	-	-	
Type 5							
complete	-	-	P84	-	-	-	
Type 6							
rim	-	-	-	P85	-	-	
Type 7							
rim	P86	-	P87–89	-	P90–91	P92	
Type 8							
rim	-	-	P93–94	P95–96	-	P97	
Type 9							
rim	-	-	-	P98	-	-	
Stewpot							
Type 1a							
rim + base	-	-	-	-	P99	-	

	house 3	house 1	house 4	house 2	house 6	house 5	5/6
rim	-	-	-	P100	P101	-	
wall	-	-	-	-	P102	-	
Type 1b							
rim + base	-	P103	-	-	-	-	
Type 1c							
rim	-	P104	-	-	-	-	
Type 2a							
rim	-	P105	-	-	P106	-	
Type 2b							
rim	P107	P108	-	-	-	-	
Strainer							
Type 1							
complete	-	-	P109	-	-	-	
Type 2							
complete	-	P110	-	-	-	-	
Type 3a	-	-	-	-	P111	P112	P113
Type 3b	-	-	-	-	-	P114	
Lid							
Type 1a							
complete	-	P115	-	-	P116–117	-	
knob	-	-	-	-	P118–120	-	
Type 1b							
complete	-	-	-	-	P121	-	
knob	-	-	-	P122	-	P123	
Type 1c							
complete	P124	-	-	-	-	-	
knob	-	-	P125	-	-	-	
Type 2a							
complete	P126	-	-	-	-	P127	
knob	-	-	-	P128	-	-	
Type 2b							
complete	-	-	P129	-	-	-	
knob	-	-	-	-	P130	-	
Type 3a							
complete	-	-	-	-	P131	-	
knob	-	-	P132	P133	-	-	
Type 3b							
complete	-	P134–135	-	-	-	-	
Type 3c							
knob	-	-	-	-	P136	-	
Rim fragments							
MCR	-	-	P137	-	P138–139	P140–141	
New Halos ware	P142	-	P143–144	-	-	-	
MCO	-	-	P145	-	P146	-	
Brazier							
rim	-	-	-	-	-	P147	
CLOSED VESSELS							
Jug							
Type 1a							
complete	-	P148	-	-	-	-	
Type 1b							
complete	-	-	P149	-	-	-	
Type 2							
complete	-	-	-	-	P150	-	

	house 3	house 1	house 4	house 2	house 6	house 5	5/6
Type 3							
complete	-	-	-	-	P151	-	
Type 4							
complete	-	-	-	-	P152	-	
rim	-	-	-	-	P153	-	
Type 5a							
rim	-	-	-	-	P154–155	-	
Type 5b							
rim	-	-	-	-	P156	-	
Type 6a							
rim	-	-	P157	-	-	-	
Type 6b							
rim	-	-	-	-	-	-	P158
Type 7							
rim	-	-	-	-	-	P159	
Type 8							
rim	-	-	-	P160	-	-	
Type 9							
rim	-	-	P161	-	-	-	
Type 10							
rim	-	-	-	-	-	P162	
Type 11							
rim	-	-	-	-	P163	-	
Type 12							
rim	-	-	-	-	P164	-	
Lekythos							
Type 1a							
complete	-	-	-	-	-	P165	
Type 1b							
rim	P166	-	-	-	-	P167	
Type 2							
rim	-	-	P168	-	-	-	
Type 3							
complete	-	-	-	-	P169	-	
Type 4							
rim	-	-	-	-	P170	-	
Type 5							
complete	-	-	-	-	-	P171	
Type 6							
base + rim	-	-	P172	-	-	-	
Type 7							
complete	-	P173	-	-	-	-	
Type 8							
base	-	-	-	-	-	P174	
Hydria							
Type 1a							
rim	P175	-	-	-	-	-	
Type 1b							
rim	-	P176	-	-	-	-	
Type 2							
rim	-	-	-	-	-	P177	
Type 3							
rim	-	-	-	P178	-	-	
Type 4							
rim	P179	-	-	-	-	-	
Type 5							
rim	-	-	-	P180	-	-	

	house 3	house 1	house 4	house 2	house 6	house 5	5/6
Type 6							
rim + base	-	-	-	-	P181	-	
Type 7	-	P182	-	-	-	-	
(Water)pitcher							
Type 1							
rim	P183	-	P184	-	P185	-	
Lagynos							
Type 1							
base + body	-	-	P186	-	-	-	
Type 2							
base + body	-	P187	-	-	-	-	
Olpe							
Type 1a							
base	-	-	P188	-	-	-	
Type 1b							
base	-	-	P189	-	-	-	
Type 2							
base	P190	-	-	-	-	-	
Pelike							
Type 1							
complete	-	-	-	-	P191	-	
rim	-	-	-	-	P192–194	-	
Amphora							
complete	-	P195	-	-	-	-	
OPEN SHAPES							
Lekane							
Type 1a							
complete	-	P196	-	-	-	-	
rim	P197–198	-	P199–203	-	P204	-	
Type 1b							
complete	-	-	P205	-	P206	-	
rim	-	P207–208	P209–210	P211	P212-213	-	
Type 1c							
complete	-	P214	-	-	-	-	
rim	-	-	P215–216	-	-	-	
Type 2a							
complete	-	P217	-	-	-	-	
rim	-	P218	-	-	P219	-	P220
Type 2b							
complete	-	-	-	-	-	P221	
rim + base	-	-	-	-	-	P222	
rim	P223	-	-	-	-	P224–225	
Type 3a							
rim	P226–227	-	-	P228	-	P229	P230
Type 3b							
rim	-	-	-	-	-	P231–232	
Type 4							
rim	-	-	P233	-	-	-	
Type 5							
rim	-	P234	-	-	-	-	
Type 6							
rim + base	-	-	-	P235	-	-	

	house 3	house 1	house 4	house 2	house 6	house 5	5/6
Type 7							
rim	P236	-	-	-	-	-	
Krater							
Type 1a							
complete	-	P237	-	-	P238–239	-	
rim	-	-	-	-	P240–245	-	
Type 1b							
reconstructed	-	P246	-	-	-	-	
rim	P247	-	P248	-	P249	-	
Type 1c							
rim	-	-	-	-	-	P250	
Type 2a							
rim + base	-	-	-	-	P251	-	
Type 2b							
rim	-	-	P252	-	-	-	
Tub							
Type 1							
complete	-	-	-	-	P253	-	
Type 2a							
rim	-	P254	-	-	-	-	
Type 2b							
rim	-	-	-	-	-	P255	
Type 3							
rim	-	-	-	-	P256	-	
Type 4							
rim	-	-	P257	-	-	-	
Type 5							
rim	-	-	-	-	-	P258	
Type 6a							
rim	-	-	-	-	-	P259	
Type 6b							
rim	-	-	P260	-	-	-	
Type 7							
rim	-	-	-	-	P261	-	
Type 8							
base	-	-	P262	-	-	-	
Type 9							
base	-	-	P263	-	-	-	
Type 10							
base	-	P264	-	-	P265	-	
Type 11							
base + rim	-	-	-	-	-	P266	
Type 12							
wall	P267	-	-	-	-	-	
Lebes							
Type 1a							
rim + base	P268	-	-	-	-	-	
Type 1b							
rim	-	P269	-	-	-	-	
Type 2							
rim + body	-	P270	-	-	-	-	
Mortar							
rim	-	-	-	-	-	P271	
Louterion							
rim	-	-	-	-	P272	-	

	house 3	house 1	house 4	house 2	house 6	house 5	5/6
STORAGE VESSELS							
Amphora							
Type 1							
complete	-	-	P273–274	-	-	P275–276	
rim + base	-	-	P277	-	-	-	
base	P278	-	-	P279	P280	P281	
rim	P282	P283	-	-	-	-	
Type 2							
complete	-	-	-	-	P284	-	
base	-	-	P285	-	P286	-	
Type 3							
complete	P287	-	-	-	-	-	
rim	P288	-	-	-	-	-	
Type 4							
rim	-	P289	-	-	-	-	
Type 5							
complete	-	-	-	-	P290	-	
Type 6							
body + rim	-	-	P291	-	-	-	
body + base	-	-	-	-	P292	-	
Type 7							
complete	-	-	-	-	P293	-	
(base knob missing)							
Type 8							
base	-	P294	-	-	P295	-	
Type 9							
rim	-	-	-	-	P296	-	
Type 10							
base knob	-	-	-	P297	-	-	
Type 11							
rim + wall	-	-	-	P298	-	-	
Type 12							
base knob	-	-	-	-	-	P299	
Type 13							
base knob	-	-	-	-	P300	-	
Type 14							
base knob	-	-	-	-	P301	-	
Jar							
Type 1a							
rim	P302	-	P303	-	P304–307	-	
Type 1b							
rim	-	-	-	-	-	P308	
Type 2a							
base + rim	-	-	-	-	P309	-	
rim	-	-	-	P310–312	-	P313–315	
Type 2b							
base + rim	-	-	-	-	P316	-	
rim	-	-	-	-	P317	-	
Type 3							
rim	-	P318	-	-	-	P319	
Type 4							
rim	-	-	-	P320	-	P321	
Type 5							
rim	P322	-	-	-	-	-	P323
Type 6							
rim	-	-	-	-	-	P324	
Type 7							
rim	-	P325	-	-	-	-	

	house 3	house 1	house 4	house 2	house 6	house 5	5/6
Type 8							
rim	-	-	P326	-	-	-	
Pithos							
Type 1a							
rim	-	-	P327–329	P330	P331	P332–333	
Type 1b							
rim	-	-	-	-	P334	-	
Type 1c							
rim	P335	-	-	-	-	-	
Large (storage jar) lid							
Type 1							
complete	-	-	-	-	P336	-	
rim	P337	-	-	-	-	-	
Type 2							
rim	-	-	-	-	P338	-	
knob	-	-	-	-	-	-	P339
Type 3							
wall	-	-	P340	-	-	-	
Type 4							
complete	P341	-	-	-	-	-	
Pithos lid							
Type 1a							
complete	-	-	P342	-	P343	-	
rim	-	-	P344	-	-	-	
Type 1b							
rim	-	-	P345	-	-	-	
Type 2							
complete	-	-	-	-	-	P346–347	
TABLE WARE							
Echinus bowl							
Type 1a							
complete	P348–349	-	-	P350	P351–355	-	
Type 1b							
complete	P356	-	-	-	-	-	
Type 2							
complete	P357	-	-	-	-	-	
Type 3							
complete	-	-	-	P358	-	-	
Bolsal							
Type 1							
complete	-	-	-	-	P371	P372	
base	-	-	-	-	P373	-	
Type 2							
rim + base	-	-	-	-	-	-	P374
base	-	-	P375	-	P376–379	-	
Type 3							
base	-	-	P380	-	P381–382	-	P383
Type 4							
base	P384	-	P385–386	-	P387	-	
Small bowl							
Type 1							
complete	P388–389	-	P390	P391	P392–393	-	
Type 2							
complete	-	-	-	-	P394	-	

	house 3	house 1	house 4	house 2	house 6	house 5	5/6
Type 3							
complete	-	-	-	-	-	P395	
Large glazed bowl							
Type 1							
complete	P396	-	-	-	-	-	
Type 2							
complete	-	P397	-	-	-	-	
Type 3							
base	-	-	-	-	P398–399	-	
Large unglazed bowl							
Type 1							
complete	-	-	-	-	P400–401	-	
base	-	-	-	-	-	P402	
Type 2							
complete	-	-	-	-	P403	-	
Everted rim							
Type 1							
complete	P404	-	-	-	-	-	
Type 2							
complete	-	-	-	-	P405	-	
Hemispherical bowl							
complete	-	P406	-	-	-	-	
Kantharos							
Type 1							
rim	P407	-	P408	P409	P410	P411	
base	P412	-	-	-	P413	P414	
Type 2							
complete	-	-	-	-	P415	-	
base + rim	-	-	-	P416			
rim	P417	-	-	-	-	P418	
base + wall	-	-	P419	-	-	-	
Type 3							
complete	-	-	P420	-	-	-	
Type 4							
base	-	-	-	-	P421	-	
base	-	-	P422	-	-	-	
Type 5							
complete	-	-	-	P423	-	-	
Type 6							
complete	-	P424	-	-	-	-	
Type 7							
base + rim	-	-	-	P425	-	-	
Type 8							
base + body	P426	-	-	-	-	-	
Type 9							
base	-	P427	-	-	-	-	
Type 10							
base	-	P428	-	-	-	-	
Type 11							
base	-	P429	-	-	-	-	
Type 12							
base	-	-	-	-	-	P430	
Type 13							
base	-	-	-	-	-	P431	
Type 14							
wall	-	-	P432	-	-	-	

	house 3	house 1	house 4	house 2	house 6	house 5	5/6
Type 15							
wall	-	-	-	-	-	P433	
Type 16							
wall	-	-	-	-	-	P434	
Cup kantharos							
Type 1							
complete	-	P435	-	-	-	-	
Type 2							
base	-	-	P436	-	-	-	
Type 3							
rim	-	-	-	-	-	P437	
Bowl kantharos							
Type 1							
handle	-	-	-	-	P438	-	P439
Skyphos							
Type 1							
base	-	-	-	-	-	P440	
Type 2							
rim	-	-	-	P441	-	-	
Fish plate							
Type 1							
complete	-	P442	P443	-	P444	-	
base	-	-	P445	-	-	-	
rim	-	-	-	P446	-	-	
Type 2							
complete	-	-	-	P447	-	-	
Type 3							
complete	-	-	-	-	P448	-	
Type 4							
complete	-	-	-	-	P449	-	
Type 5							
complete	-	-	-	P450	-	-	
Type 6							
complete	P451	-	-	-	-	-	
Type 7							
base + rim	P452	-	-	-	-	-	
Type 8							
base + rim	-	-	-	-	-	P453	
Type 9							
base	P454	-	-	-	-	-	
Type 10							
base	P455	-	-	-	-	-	
Type 11							
base	P456	-	-	-	-	-	
Type 12							
base	-	-	P457	-	-	-	
Type 13							
base	-	-	P458	-	-	-	
Type 14							
base	-	-	-	P459	-	-	
Type 15							
base	-	-	-	-	P460	-	
Saucer							
complete	P461	-	-	-	-	-	

	house 3	house 1	house 4	house 2	house 6	house 5	5/6
Spool saltcellar							
complete	-	-	-	-	-	P462	
MISCELLANEOUS CERAMICS							
Oil lamp							
Type 1a							
rim	-	-	P463	P464	-	-	
base	P466	-	-	-	P465	-	
rim + spout	-	-	-	-	-	P467–468	
Type 1b							
complete	P469–470	-	-	P471	-	-	
rim	-	-	P472	-	-	-	
base	-	-	P473	P474	-	-	
Type 1c							
complete	-	-	-	-	-	P475	
Type 2							
rim	-	-	P476	-	-	-	
Type 3							
rim	P477	-	-	-	-	-	
Type 4							
rim	-	-	-	-	-	P478	
Type 5							
rim	-	-	-	-	-	P479	
Lekanis							
Type 1a							
rim	P480	-	P481–483	-	P484	P485–486	
handle	-	-	P487	-	-	-	
Type 1b							
complete	P488	-	-	-	-	-	
Type 2							
rim	-	-	-	P489	-	-	
Type 3							
rim	-	P490	-	-	-	-	
Pyxis							
Type 1							
complete	P491	-	-	-	-	-	
Type 2							
complete	-	-	P492	-	-	-	
Type 3							
body + rim	-	-	-	-	P493	-	
Type 4							
base	-	-	-	-	-	P494	
Glazed pyxis lid							
Type 1a							
rim	P495	P496	P497	-	-	-	
Type 1b							
rim	-	P498	-	-	-	-	
Type 2							
rim	-	-	-	-	P499–500	-	
Type 3							
rim	-	-	-	-	P501	-	
Type 4							
rim	-	P502	-	-	-	-	

	house 3	house 1	house 4	house 2	house 6	house 5	5/6
Unguentarium							
Type 1							
complete	-	-	-	-	P503	-	
base	P504–505	-	-	-	-	-	
wall	P506	-	-	-	-	-	
Type 2							
complete	P507	-	-	-	-	-	
Type 3							
base + body	-	-	-	-	-	P508	
Type 4							
base	P509	-	-	-	-	-	
rim	-	-	-	-	-	P510	
Aryballos							
Type 1							
base	-	-	-	-	P511	-	
Type 2							
base	-	-	-	-	-	P512	
Type 3							
base	-	-	-	-	-	P513	
Guttus							
Type 1a							
rim + base	P514	-	-	-	-	-	
Type 1b							
rim	-	-	-	P515	-	-	
Type 2							
base	P516	-	-	-	-	-	
Amphoriskos							
base	-	-	P517	-	-	-	
Squat lekythos							
base	-	-	-	-	-	P518	
Miniature							
complete	-	-	P519–520	-	-	-	

APPENDIX 3. CATALOGUE OF FIGURINES

GEORGETTE M.E.C. VAN BOEKEL AND BIENEKE MULDER

The terracottas are arranged according to their subjects. The Munsell Soil Color Charts have been used for the colours. All figurines are hollow and mould-made. The section 'Bibliography' (Bibl.) comprises previous publications of terracottas and the section 'References' (Ref.) refers to comparable terracotttas (see Ch. 3: 2 for a list of references).

F1. WOMAN STANDING, NEARLY COMPLETELY PRESERVED BUT DAMAGED

Find pl.: House of Agathon, room 5 (fig. 3.24).

State: Damage: a large hole in the *himation*'s lower half at the front. The statuette is broken into many fragments that have been reconstructed. Missing: small parts of the lower rims and along the break and most probably the plinth.

H.: 24.0 cm.

Techn.: Colour: pale- to reddish-yellow, Munsell 7.5YR 8/2–7.5YR 8/6. Many traces of a white to pale-grey slip are preserved. The clay is of medium hardness and has fine tempering. It is very likely that the head is double moulded, separately from the body, and was attached to the body prior to firing. The front is moulded as is most probably, the largely undetailed back. Retouched: hair, ear-drops, l. hand, folds of the chiton and of the *himation*'s neckline. A large rectangular vent-hole was made in the back before firing. Imprints of thumbs can be seen on the interior wal.

Bibl.: Unpublished.

Ref.: Higgins, 1986: p. 132 fig. 158, figurine from Tanagra, dated most probably between 300 and 275; Uhlenbrock & Thompson, 1990: pp. 50–51, and no. 4.

Descr.: Woman standing, wearing a long *himation* over a long *chiton* that falls to the booted feet leaving their tips free. The weight of the body rests on the l. leg while the knee of the flexed r. leg protrudes. The r. foot is placed slightly back and its tip points obliquely outwards. The body, especially the upper part, inclines strongly backwards. The head is small compared to the large and broad body and is turned a little to the left. Wavy hair frames the oval face, combed back from the high forehead and is parted in the middle. It leaves the ears, provided with ear-drops, free. The hair is gathered up in a chignon at the centre of the back of the head and is similar to that known from the Knidian Aphrodite by the sculptor Praxiteles. A double ribbon encircles the head. The woman has deep-set eyes.

The flexed r. arm, wrapped in the *himation*, protrudes to the right and the concealed hand rests on the r. hip (cf. F2 and the l. arm in F3). The broad l. arm hangs loosely at the side whereby the hand may have held an attribute, attached to a small opening between the fingers (a fan?).

The principal shapes of the body are visible through the garments that have long, smooth sections. The wide *himation* has many parallel folds over the r. arm, in the broad neckline and on the stomach. It hangs over the shins with a curved hem and falls in bundles of folds near the legs.

The smooth back has an angular contour and is undetailed, apart from the head and a few folds of the *himation*'s neckline.

The figurine rests on the rims along its open underside but originally must have had a separately hand-modelled and solid, plaque-shaped base.

The statuette is part of the '*muffled Tanagra*' type that was much favoured among early Hellenistic terracottas.

F2. WOMAN STANDING, FRAGMENTARY UPPER BODY

Find pl.: House of the Coroplast, room 5 (fig. 3.25).

State: Preserved (front): parts of chest, stomach, l. arm (except hand), r. arm, except hand and strip of upper arm). Damage: broken into many parts that have been reconstructed; strongly abraded.

H.: 6.5 cm.

Techn.: Colour: reddish-yellow, Munsell 5YR 6/8. The clay is of medium hardness and has medium to fine tempering. It is unknown whether the back, now missing, was moulded or separately hand-modelled.

Bibl.: Reinders, 1988: no. 35.20.

Ref.: Uhlenbrock & Thompson, 1990: pp. 50–51 and no. 4.

Descr.: Woman standing, wrapped in a *himation* that is draped tightly over the upper body. The fragment is similar to F1, with the same posture for the r. arm and with the partly covered l. arm hanging along the body. However, the fragment's himation has more (parallel) folds and they run mainly in different directions. The now missing r. hand rested on the hip. The *en creux* mark (?) (capital) in the interior of the figurine is perhaps a potter's mark.

The statuette is part of the category of *Tanagra* figurines.

F3. WOMAN STANDING; HEAD, RIGHT ARM AND PARTS OF BACK MISSING

Find pl.: House of the Coroplast, room 1 (fig. 3.26).

State: Missing: head, upper body at front and back (except centre of breast and most of l. forearm), r. arm, r. half of the draperies' hems on the front, a large part of the lower half at the back and the vent-hole's upper edge. The statuette is broken into several fragments that have been reconstructed; some small parts along the break are missing.

H.: 15.7 cm.

Techn.: Colour: pale-red, Munsell 2.5Y 6/8. The coarsely tempered clay is of medium hardness. The *chiton*'s lower folds have been retouched with a sharp tool. A large rectangular vent-hole was made in the back before firing. The front is moulded as is most probably the plain back too.

Bibl.: Reinders, 1988: no. 31.18.

Ref.: Uhlenbrock & Thompson, 1990: p. 50–51.

Descr.: Woman standing, wearing a *himation* over a long *chiton* that covers the feet. The cloak is draped tightly over the upper body and l. arm and crosses the body from l. hip down to r. knee. A bundle of zig-zag folds hangs from the l. hip along the l. side of the l. leg. The concealed l. hand rests on the l. hip whereby the flexed arm protrudes to the left (cf. r. arm in F1). The statuette's back is undetailed. The volume of the folds is enhanced by retouching.

The figure rests on the rims along its open under side but most probably originally provided with a plaque base (cf. F1).

The statuette is part of the Hellenistic category of *Tanagra* types.

F4. WOMAN STANDING ON A ROUND BASE; FRONTAL MOULD'S LOWER HALF

Find pl.: House of the Coroplast, room 3 (fig. 3.27).

State: Preserved: frontal mould's lower half with lower legs, knees, small part of thighs and plinth. The mould was broken into several parts that have been reconstructed and it has traces of chalk on the surface.

H.: 8.9 cm.

Techn.: Colour: reddish-yellow to pink, Munsell 5YR 7/8. The clay is of medium hardness and has fine tempering. The mould's lower contour is nipped in at shin level.

Bibl.: Reinders, 1988: pp. no. 33.46.

Descr.: Woman wearing a long *chiton* and standing on a low, spool-shaped plinth. The weight of the body rests on the r. leg and the knee of the flexed l. leg protrudes strongly. The r. foot is placed obliquely to the right. The principal shapes of the legs are clearly visible through the chiton that falls to the booted feet leaving the tips free.

The spool-base differs from F1, 3, which may be reconstructed with a plaque base.

The figurine may be classified among the category of *Tanagra* statuettes.

F5. HELLE, FRAGMENTS OF UPPER BODY AND OF RAM WITH HELLE'S RIGHT LEG

Find pl.: House of Agathon, room 5, the fragments were found together at the same spot (fig. 3.28).

State: Preserved: the statuette is broken into many fragments that could be partly reconstructed. The three restored parts are the girl's upper body (front and back), the ram and its lower legs. Preserved parts of the girl: head, neck, upper body to the waist (except part of the upper section of the breast), arms (except hands, r. shoulder and part of l. shoulder), her r. leg on the ram's side (except tips of toes) and most of hem of her garment. The lower fragments comprise the ram's upper body and head (except part of its nose), the beginnings of its tail, parts of its lower legs, its back and of the statuette's rim.

H.: 5.8 cm, upper body. H. 9.5 cm, ram.

Techn.: Colour: upper body, reddish-yellow, Munsell 5YR 6/8; ram: reddish-yellow to pale-red, Munsell 2.5YR 6/8–7.5YR 7/6. Traces of white slip remain. The clay of the lower fragments is smudgy and there may be traces of secundary burning on the ram's head. The clay is of medium hardness and has medium-sized tempering. Retouched: the lower fragments and possibly Helle's face. The statuette has no separate plinth but rests on a narrow rim along its open underside. The rim was included in the mould. Sharp mould.

Ref.: Cf. Jacobsthal, 1931: fig. 5, 31, 34, 69–70, 101–102, Helle on Melian terracotta reliefs. In all cited specimens, Helle is sitting side-saddle.

Descr.: Young girl, wearing a long-sleeved *chiton*, sitting astride a walking ram.

The girl's r. arm is flexed in front of the body and her l. arm is raised. Originally, the l. hand probably held on to the ram's neck. Her head is turned slightly to the right. Round curls frame the face and cover the ears. The hair is parted in the middle and is surmounted by a very large and high, convex diadem or *polos* with a band at the base. She has a child-like face with full, round cheeks and a friendly, smiling expression because the corners of her mouth are turned up. Her back and the ram's l. side are undetailed.

The *chiton*'s hem, that curves over the r. thigh and over the ram's back, leaves the flexed, bare leg largely free. The leg rests far forwards against the animal side.

The ram, that has little depth, has a thick, curly pelt and holds its head straight. It has a thick, crescent-shaped horn and almond-shaped eyes. Two of its legs are naturalistically rendered in relief on the otherwise smooth background. Their cloven hooves are clearly indicated and are probably part of the l. hind leg that is placed forwards and the r. foreleg that is set back. The other legs are schematized into a thick rim that also forms the figurine's contours and its lower rim.

F6. BANQUETER, MOST OF THE FIGURINE'S FRONT AND PART OF ITS BACK

Find pl.: House of the Coroplast, room 3 (fig. 3.29).

State: Preserved: front, except feet, r. edge of patera, the wall at either side of the small table and the foot of the *kline*. Preserved of the back: upper body, arms, rhyton and most of *kline*'s l. leg. Damage: the hair on the l. of the head, the *kline*'s cushion and its l. leg; a hole on the r. of the belly half; further extensive damage especially along the breaks. The figurine has been broken into many fragments that have been reconstructed.

H.: 14.2 cm.

Techn.: Colour: pale-red to red, Munsell 2.5YR 6/8–2.5YR 5/6. The clay is of medium hardness and has fine tempering. The statuette has no separate plinth but rests on its rims along the open underside. The figurine was made in the mould F7.

Bibl.: Reinders, 1988: no. 33.41.

Ref.: Dentzer, 1982: pl. 25 fig. 142; pl. 26 fig. 150.

Descr.: Man, semi-nude, bearded and with long moustache, depicted in the conventional half reclining attitude, legs outstretched, trunk erect and head turned; supporting a *rhyton* with his l. arm and holding a *patera* with his r. hand; a wreath crowns the head and a small and low table stands in front of the *kline*.

Head and muscular trunk are represented in three-quarter view. The anatomy is naturalistically and softly modelled.

He has small eyes with thick lids, a broad nose and sharply rendered lips. The long flaxy beard has long corkscrew curls and falls over the breast. The hair framing the face is parted in the middle, is combed back and covers the ears.

The flexed l. arm rests on the thick and protruding cushion on the head of the couch. The hand rests on the long *rhyton* that has a curved lower end and a broad, convex top. His r. arm is stretched out to the r. and he holds the *patera* with *umbo* that stands obliquely on his r. thigh.

A bundle of the *himation*'s folds is draped over the l. shoulder and falls from there along his side and over the l. hip to the *kline*. The cloak covers the lower body and its upper part is thickly rolled up over the abdomen. The garment is pulled tightly over the legs revealing the shape of the flexed l. leg. Some thin folds stretch between the legs.

The long fringe of a cloth that is draped over the *kline* hangs over the edge of the couch. Undeterminable objects stand on the small table of which two rectangular legs are visible.

F7. BANQUETER, FRAGMENTARY MOULD FOR THE FRONT

Find pl.: House of the Coroplast, room 3.

State: Preserved: a large part of the mould for the front, comprising the man's lower body, part of the trunk and of the fringe of the cloth that is draped over the *kline*. The damaged mould was broken into many parts that have been reconstructed.

H.: 8.9 cm.

Techn.: Colour: pale-red to red, Munsell 2.5YR 6/8–2.5YR 5/6. The clay is of medium hardness and has fine tempering. The mould was probably not long in use as evidenced by its sharpness.

Bibl.: Reinders, 1988: no. 33.42.

Descr.: Fragmentarily preserved mould depicting the front of a banqueter on *kline* from which F6 is taken.

F8. HORSE, TWO LARGELY PRESERVED HALF-MOULDS

Find pl.: House of the Coroplast, room 3 (figs 3.30 & 3.31).

State: Preserved: half-mould for the horse's l. side, except small part of the head's crown. Damage: the mould's rims with a very small part of the horse's back. Preserved of the r. half-mould: most of the horse, except head, outer edge of the hindquarters with lower

hind leg, the outer edge of the mould and of the damaged plinth. The moulds were broken into several fragments that have been reconstructed (three fragments of the l. half-mould).
H.: L. half-mould: 8.3 cm. R. half-mould: 7.7 cm.
Techn.: Colour of l. half-mould: reddish-yellow, Munsell 7.5YR 8/6. Colour of r. half-mould: Munsell 5YR 7/6. The soft clay has fine tempering. The moulds include tail and plinth and have registration marks in the outer rims, consisting of small semi-globular protrusions showing the correct way to assemble the two pieces of the mould during the process of making figurines.
Bibl.: Reinders, 1988: no. 33.43–44.
Descr.: Stallion, standing quietly with raised head; on a low, rectangular plinth. The two moulds fit together and were used for moulding the sides of a freestanding horse. In figurines pressed from the mould, each pair of straight legs either forms one entity or the legs were separated from each other before firing. The male genitals are cursorily indicated but, apparently, not at corresponding height. The animal has a long tail and the hoof of its r. foreleg is modelled.

F9. RELIEF OF APHRODITE ON BILLY GOAT, FRAGMENTARY LEFT HALF OF A ROUND MOULD

Find pl.: House of the Coroplast, room 3.
State: Preserved: mould's l. half with Aphrodite's trunk, neck (except l. rim), l. thigh and most of r. arm with part of the attribute in her r. hand, the billy goat's hind quarters, a goat under the feet of the goddess, comprising part of its back and the hindquarters. The mould was broken into several fragments that have been reconstructed.

H.: 14.3 cm.
Techn.: Colour: pale-red, Munsell 2.5YR 6/6 with traces of slip. The clay is of medium hardness and has medium-sized tempering. The mould is sharp and, therefore, does not appear to have been used over long period.
Bibl.: Reinders, 1988: no. 33.45.
Ref.: Knigge, 1982: pl. 31; Hackens & Lévy 1965: figs 20–21; *Lexicon Iconographicum Mythologiae Classicae* II, Zürich-München, 1984., Aphrodite, fig. 947-976; Ninou 1978: pl. 12 fig. 55; *Venus te lijf*, cf. p. 110–111, no. 86 fig. 56, terracotta statuette of Aphrodite on billy goat, provenance Boeotia.

Descr.: Disc-shaped mould for a relief of a semi-nude woman, sitting side-saddle on the r. side of a rearing billy goat that moves to the l., holding an attribute in her raised r. hand and accompanied by a small goat.

Aphrodite *Epitragia* is depicted with a cloak draped over the lower body that falls over the ankles. The upper portion is rolled up over the abdomen in a thick, twisted bundle. The garment is draped firmly over the legs and is therefore largely smooth apart from some folds.

The flexed r. arm is held away from the body and she seems to hold a smooth slightly convex attribute (a mirror?).

The anatomies of the goddess and of the animal are modelled. The billy goat's very short tail, genitals and cloven hooves are rendered. A small cursorily indicated goat jumps towards the l. directly under Aphrodite's feet.

APPENDIX 4. CATALOGUE OF LOOMWEIGHTS

H. REINDER REINDERS

Abbreviations
H.   = Height in mm
D.   = Diameter in mm
Base = Dimensions base in mm
W.   = Weight in grams
Th.   = Thickness

PYRAMIDAL LOOMWEIGHT
Square or rectangular base, pyramidal body, one suspension hole

L1	House 1, Inv. No. X 6		(fragment missing)
L2	House 1, Inv. No. X 343	H. 71; Base 44×44; W. 111	(slightly damaged, worn)
L3	House 1, Inv. No. X 637	H. 61; Base 38×38; W. 54	
L4	House 1, Inv. No. X 662		(upper part missing)
L5	House 1, Inv. No. X 717B	H. 67; Base 40×57; W. 115	(slightly damaged)
L6	House 1, Inv. No. X 717C		(upper part missing)
L7	House 1, Inv. No. X 720 A		(fragment missing, worn)
L8	House 1, Inv. No. X 720 B	H. 45; Base 38×39; W. 59	
L9	House 1, Inv. No. X 720 C		(large fragment missing)
L10	House 1, Inv. No. X 720 D		(large fragment missing)
L11	House 2, Inv. No. H21 1/15	H. 61; Base 40×44; W. 95	
L12	House 2, Inv. No. H21 1/25		(fragment missing, worn)
L13	House 2, Inv. No. H21 1/26	H. 66; Base 40×42; W. 101	(slightly worn)
L14	House 2, Inv. No. H21 2/39	H. 63; Base 43×44; W. ?	
L15	House 2, Inv. No. H21 2/48	H. 69; Base 45×47; W. 110	
L16	House 2, Inv. No. H21 2/49	H. 63; Base 43×50; W. 105	(fig. 3.32)
L17	House 2, Inv. No. H21 2/57	H. 69; Base 44×48; W. 117	
L18	House 2, Inv. No. H21 3/26		(fragment missing, worn)
L19	House 2, Inv. No. H21 3/30	H. 66; Base 41×44; W. 108	
L20	House 2, Inv. No. H21 4/9	H. 71; Base 47×48; W. 108	(fig. 3.32)
L21	House 3, Inv. No. HA 194/2		(damaged)
L22	House 3, Inv. No. HA 194/3		(fragment missing, worn)
L23	House 3, Inv. No. HA 208/3	H. 52; Base 33×36; W. 46	(worn)
L24	House 3, Inv. No. HA 208/8		(damaged)
L25	House 3, Inv. No. HA 216/15		(fragment missing)
L26	House 3, Inv. No. HA 223		(fragment missing)
L27	House 3, Inv. No. HA 233/1		(damaged)
L28	House 3, Inv. No. HA 238/3		(damaged)
L29	House 3, Inv. No. HA 247/1		(damaged)
L30	House 4, Inv. No. HA 90/1	H. 70; Base 38×41; W. 95	
L31	House 4, Inv. No. HA 90/2	H. 71; Base 29×48; W. 85	(worn)
L32	House 4, Inv. No. HA 106/22	H. 59; Base 40×43; W. 82	
L33	House 4, Inv. No. HA 108/1		(fragment missing, worn)
L34	House 4, Inv. No. HA 113/8		(fragment missing, worn)
L35	House 4, Inv. No. HA 113/14		(fragment missing, worn)
L36	House 4, Inv. No. HA 117/1	H. 61; Base 40×41; W. 108	
L37	House 4, Inv. No. HA 117/5	H. 58; Base 43×44; W. 93	
L38	House 4, Inv. No. HA 122/1	H. 64; Base 37×43; W. 80	(worn)
L39	House 4, Inv. No. HA 124/4		(large fragment missing)
L40	House 4, Inv. No. HA 127/3	H. 71	(fragment missing)
L41	House 4, Inv. No. HA 128/3	H. 57; Base 35×36; W. 63	
L42	House 4, Inv. No. HA 129/11	H. 67; Base 41×41; W. 99	
L43	House 4, Inv. No. HA 133/1	H. 59; Base 39×44; W. 91	
L44	House 4, Inv. No. HA 145/1	H. 56; Base 37×39; W. 57	(slightly damaged)
L45	House 4, Inv. No. HA 150/4		(fragment missing, worn)

L46	House 4, Inv. No. HA 170/1	H. 53; Base 29×31; W. 35	(worn)
L47	House 4, Inv. No. HA 173/1		(damaged)
L48	House 5, Inv. No. HA 7/4		(large fragment missing)
L49	House 5, Inv. No. HA 10/9	H. 68; Base 38×40; W. 94	(slightly damaged)
L50	House 5, Inv. No. HA 10/10		(fragment missing, worn)
L51	House 5, Inv. No. HA 14/29	H. 59; Base 34×35; W. 52	
L52	House 5, Inv. No. HA 14/30	H. 74; Base 39×39; W. 83	(slightly worn)
L53	House 5, Inv. No. HA 26/5	H. 53; Base 40×40; W. 75	
L54	House 5, Inv. No. HA 46/13	H. 55; Base 40×41; W. 83	
L55	House 5, Inv. No. HA 53/16	H. 70; Base 39×40; W. 112	
L56	House 5, Inv. No. HA 55/5	H. 69; Base 39×39; W. 97	(damaged)
L57	House 5, Inv. No. HA 59/18	H. 59; Base 42×40; W. 76	
L58	House 5, Inv. No. HA 59/19		(fragment missing, stamped)
L59	House 5, Inv. No. HA 59/21		(large fragment missing)
L60	House 5, Inv. No. HA 59/22		(fragment missing)
L61	House 5, Inv. No. HA 59/23		(fragment missing)
L62	House 5, Inv. No. HA 59/24		(fragment missing)
L63	House 5, Inv. No. HA 59/25	H. 60; Base 46×51; W. 109	
L64	House 5, Inv. No. HA 59/26		(fragment missing)
L65	House 5, Inv. No. HA 59/27		(fragment missing)
L66	House 5, Inv. No. HA 59/28		(fragment missing)
L67	House 5, Inv. No. HA 59/29		(fragment missing)
L68	House 5, Inv. No. HA 66/39	H. 76	(fragment missing)
L69	House 5, Inv. No. HA 68/8	H. 60; Base 41×41; W. 81	(worn)
L70	House 5, Inv. No. HA 71/13	H. 75; Base 48×50; W. 131	(slightly damaged)
L71	House 5, Inv. No. HA 80/3	H. 62; Base 40×42; W. 90	
L72	House 5, Inv. No. HA 80/4	H. 65; Base 42×43; W. 87	(worn)
L73	House 5, Inv. No. HA 85/5	H. 67; Base 44×44; W. 86	(slightly worn)
L74	House 5, Inv. No. HA 86/4	H. 65; Base 43×43; W. 87	
L75	House 5, Inv. No. HA 86/6		(fragment missing)
L76	House 6, Inv. No. H23 1/13	H. 55; Base 36×40; W. 76	(slightly worn)
L77	House 6, Inv. No. H23 1/39		(fragment missing)
L78	House 6, Inv. No. H23 1/97	H. 52; Base 30×33; W. 55	(slightly worn)
L79	House 6, Inv. No. H23 1/105	H. 63; Base 38×38; W. 80	
L80	House 6, Inv. No. H23 1/119	H. 65; Base 43×46; W. 103	
L81	House 6, Inv. No. H23 1/143		(worn, damaged)
L82	House 6, Inv. No. H23 2/6	H. 71; Base 45×50; W. 118	
L83	House 6, Inv. No. H23 2/57	H. 78; Base 52×52; W. 157	
L84	House 6, Inv. No. H23 2/64	H. 66; Base 39×40; W. 88	(slightly damaged)
L85	House 6, Inv. No. H23 2/69	H. 60; Base 42×44; W. 88	
L86	House 6, Inv. No. H23 2/70		(damaged)
L87	House 6, Inv. No. H23 2/71	H. 77; Base 55×55; W. 152	(two holes)
L88	House 6, Inv. No. H23 2/72	H. 70; Base 41×45; W. 85	
L89	House 6, Inv. No. H23 2/73	H. 66; Base 40×41; W. ?	(slightly damaged)
L90	House 6, Inv. No. H23 2/82	H. 56; Base 35×38; W. 56	(slightly damaged)
L91	House 6, Inv. No. H23 2/86	H. 66; Base 39×41; W. 102	
L92	House 6, Inv. No. H23 3/23	H. 62; Base 38×38; W. 76	(slightly worn)
L93	House 6, Inv. No. H23 3/31	H. 58; Base 50×54; W. 112	(incised cross)
L94	House 6, Inv. No. H23 3/40	H. 63; Base 36×38; W. 65	
L95	House 6, Inv. No. H23 3/69	H. 71; Base 40×41; W. 105	(slightly damaged)
L96	House 6, Inv. No. H23 3/73	H. 66; Base 37×37; W. 69	
L97	House 6, Inv. No. H23 3/76	H. 65; Base 41×42; W. 89	(stamped)
L98	House 6, Inv. No. H23 3/79A	H. 74; Base 39×39; W. 87	
L99	House 6, Inv. No. H23 3/83	H. 62; Base 40×40; W. 82	(slightly damaged)
L100	House 6, Inv. No. H23 5/10	H. 64	(damaged)
L101	House 6, Inv. No. H23 5/27	H. 72; Base 39×39; W. 102	(worn)
L102	House 6, Inv. No. H23 5/28	H. 62; Base 45×47; W. 93	(slightly damaged)
L103	House 6, Inv. No. H23 5/37	H. 83; Base 45×46; W. 136	(slightly damaged)

L104	House 6, Inv. No. H23 5/41	H. 70; Base 46×46; W. 125	
L105	House 6, Inv. No. H23 5/42	H. 66; Base 41×45; W. 100	
L106	House 6, Inv. No. H23 5/45	H. 81; Base 49×50; W. 169	
L107	House 6, Inv. No. H23 5/46	H. 84; Base 48×49; W. 161	
L108	House 6, Inv. No. H23 5/47	H. 65; Base 38×39	(damaged)
L109	House 6, Inv. No. H23 5/48	H. 70	(damaged)
L110	House 6, Inv. No. H23 5/49	H. 55; Base 30×34; W. 64	
L111	House 6, Inv. No. H23 5/50	H. 67; Base 36×39; W. 88	
L112	House 6, Inv. No. H23 5/51	H. 62; Base 43×46; W. 99	(incision T)
L113	House 6, Inv. No. H23 5/52	H. 85; Base 48×48; W. 158	
L114	House 6, Inv. No. H23 5/56	H. 68; Base 36×39; W. 89	
L115	House 6, Inv. No. H23 5/57	H. 78; Base 39×40; W. 125	
L116	House 6, Inv. No. H23 5/58	H. 73; Base 43×45; W. 128	
L117	House 6, Inv. No. H23 5/59	H. 70; Base 42×43; W. 114	
L118	House 6, Inv. No. H23 5/60		(damaged)
L119	House 6, Inv. No. H23 5/61	H. 68; Base 39×39; W. 78	
L120	House 6, Inv. No. H23 5/62	H. 65; Base 40×40; W. 92	
L121	House 6, Inv. No. H23 5/63	H. 71; Base 35×37; W. 81	
L122	House 6, Inv. No. H23 5/64	H. 77; Base 40×42; W. 116	(damaged)
L123	House 6, Inv. No. H23 5/65	Base 48×49	(damaged)
L124	House 6, Inv. No. H23 5/66	H. 62; Base 35×35; W. 71	(slightly damaged)
L125	House 6, Inv. No. H23 5/67	H. 72; Base 42×42; W. 109	
L126	House 6, Inv. No. H23 5/68	H. 64; Base 37×42; W. 89	
L127	House 6, Inv. No. H23 5/69	H. 72; Base 44×46; W. 128	
L128	House 6, Inv. No. H23 5/70	H. 52; Base 37×40; W. 62	
L129	House 6, Inv. No. H23 5/71	H. 66; Base 34×38; W. 70	(incision, O)
L130	House 6, Inv. No. H23 5/72	H. 64; Base 40×41; W. 90	
L131	House 6, Inv. No. H23 5/73	H. 65; Base 38×40; W. 86	
L132	House 6, Inv. No. H23 5/74		(top missing)
L133	House 6, Inv. No. H23 5/75	H. 75; Base 40×45; W. 129	
L134	House 6, Inv. No. H23 5/76	H. 66; Base 37×41; W. 90	(slightly damaged)
L135	House 6, Inv. No. H23 5/76A	H. 57; Base 37×37; W. 61	
L136	House 6, Inv. No. H23 5/76B	H. 65; Base 42×42; W. 98	
L137	House 6, Inv. No. H23 5/77	H. 69; Base 38×39; W. 96	(slightly worn)
L138	House 6, Inv. No. H23 5/79	H. 63; Base 39×42; W. 89	(slightly worn)
L139	House 6, Inv. No. H23 5/80		(damaged)
L140	House 6, Inv. No. H23 5/82	H. 62; Base 38×38; W. 69	(slightly damaged)
L141	House 6, Inv. No. H23 5/83	H. 71; Base 42×44; W. 105	
L142	House 6, Inv. No. H23 5/84	H. 76; Base 42×42; W. 120	
L143	House 6, Inv. No. H23 5/85	H. 76; Base 40×40; W. 102	
L144	House 6, Inv. No. H23 5/86	H. 71; Base 42×46; W. 128	
L145	House 6, Inv. No. H23 5/87	H. 68; Base 39×39; W. 90	
L146	House 6, Inv. No. H23 5/88	H. 68; Base 38×40; W. 86	
L147	House 6, Inv. No. H23 5/90	H. 74; Base 41×43; W. 119	
L148	House 6, Inv. No. H23 5/91	H. 64; Base 39×41; W. 89	
L149	House 6, Inv. No. H23 5/92	H. 73; Base 42×44; W. 131	
L150	House 6, Inv. No. H23 5/93	H. 64; Base 38×41; W. 89	(slightly damaged)
L151	House 6, Inv. No. H23 5/102	H. 57; Base 32×34; W. 50	
L152	House 6, Inv. No. H23 5/107	H. 74; Base 42×42; W. 120	
L153	House 6, Inv. No. H23 5/108	H. 72; Base 38×40; W. 83	(slightly worn)
L154	House 6, Inv. No. H23 5/109		(damaged)
L155	House 6, Inv. No. H23 5/110	H. 64; Base 41×41; W. 98	
L156	House 6, Inv. No. H23 5/126	H. 73; Base 51×51	(damaged)
L157	House 6, Inv. No. H23 5/127	H. 82; Base 48×48; W. 157	
L158	House 6, Inv. No. H23 5/127A	H. 67; Base 36×41; W. 90	(slightly damaged)
L159	House 6, Inv. No. H23 5/128	H. 66; Base 40×38; W. 85	(slightly damaged)
L160	House 6, Inv. No. H23 6/10		(fragment)

L161   House 6, Inv. No. H23 6/60   (fragment)
L162   House 6, Inv. No. HA 7/4   (damaged)
L163   House 6; Inv. No. HA 58/1   (fragment missing)

DISCOID LOOMWEIGHT

Circular loomweight, flat or slightly convex sides unless otherrwise stated, two suspension holes (D. = H.)

L164   House 1, Inv. No. X 717 A   D. 80; W. 140   (slightly damaged, both sides convex)
L165   House 1, Inv. No. X 132   D. 87; W. -   (one side flat, one side slightly
                                                                     convex)

L166   House 2, Inv. No. H21 3/15   D. 80; Th. 21; W. 113   (convex/flat)
L167   House 3, Inv. No. HA 195/1   D. 84; Th. 24; W. 153   (damaged, convex/flat)
L168   House 3, Inv. No. HA 247/14   (damaged, both sides flat)
L169   House 3, Inv. No. HA 247/21   D. 71; Th. 17; W. 100   (slightly damaged, both sides flat)
L170   House 3, Inv. No. HA 247/23   D. 73; Th. 18; W. 126   (both sides flat; fig. 3.32)

L171   House 4, Inv. No. HA 127/17   D. 87; W. 138   (both sides flat)
L172   House 4, Inv. No. HA 166/7   D. 80; Th. 25; W. 178   (both sides convex, flattened 'base';
                                                                     fig. 3.32)

L173   House 5, Inv. No. HA 44/1   D. 91; W. 203   (both sides convex, flattened 'base')
L174   House 5, Inv. No. HA 50/32   D. 69; Th. 19; W. 100   (slightly damaged, both sides flat)
L175   House 5, Inv. No. HA 66/42   D. 80; Th. 29; W. 158   (both sides convex)
L176   House 6, Inv. No. H23 2/4   D. 87; Th. 19; W. 137   (both sides flat)
L177   House 6, Inv. No. H23 2/91   D. 81; Th. 28; W. 157   (both sides convex)
L178   House 6, Inv. No. H23 3/97A   D. 72; Th. 20; W. 131   (both sides flat)
L179   House 6, Inv. No. H23 5/27A   D. 82; Th. 20; W. 134   (slightly damaged)
L180   House 6, Inv. No. H23 5/29   D. 78; Th. 18; W. 109

L181   House 6, Inv. No. H23 5/30   D. 64; W. 86   (slightly damaged, both sides flat)
L182   House 6, Inv. No. H23 5/31   D. 72; Th. 21; W. 112
L183   House 6, Inv. No. H23 5/32   D. 71; W. 95   (slightly damaged, both sides convex)
L184   House 6, Inv. No. H23 5/33   D. 76; W. 115   (slightly damaged, both sides convex)
L185   House 6, Inv. No. H23 5/34   D. 73; Th. 19; W. 109
L186   House 6, Inv. No. H23 5/35   D. 80; Th. 19; W. 109   (slightly damaged)
L187   House 6, Inv. No. H23 5/36   D. 79; Th. 24; W. 117   (slightly damaged)
L188   House 6, Inv. No. H23 5/39   D. 76; W. 107   (damaged, both sides convex)
L189   House 6, Inv. No. H23 5/40   D. 79; W. 110   (both sides convex)
L190   House 6, Inv. No. H23 5/43   D. 63; W. 76   (both sides flat)

L191   House 6, Inv. No. H23 5/44   D. 73; Th. 20; W. 115
L192   House 6, Inv. No. H23 5/53   D. 72; Th. 19   (damaged)
L193   House 6, Inv. No. H23 5/54   D. 74; Th. 18; W. 102   (slightly damaged)
L194   House 6, Inv. No. H23 5/55   D. 74; Th. 19; W. 101
L195   House 6, Inv. No. H23 5/81   Th. 18   (fragment missing)
L196   House 6, Inv. No. H23 5/89   D. 73; Th. 18; W. 97
L197   House 6, Inv. No. H23 5/94 (?)   D. 69; Th. 16; W. 91   (both sides flat)
L198   House 6, Inv. No. H23 5/112   (damaged, both sides convex)
L199   House 6, Inv. No. H23 5/113   D. 58; Th. 14; W. 40   (both sides convex)
L200   House 6, Inv. No. H23 5/114   D. 82; Th. 17; W. 109   (both sides slightly convex)

L201   House 6, Inv. No. H23 5/115   D. 75; Th. 20; W. 111
L202   House 6, Inv. No. H23 5/116   D. 66; Th. 15; W. 80
L203   House 6, Inv. No. H23 5/117   D. 76; Th. 22; W. 112
L204   House 6, Inv. No. H23 5/118   D. 77; Th. 18; W. 108   (irregular shape)
L205   House 6, Inv. No. H23 5/119   D. 80; Th. 17; W. 122
L206   House 6, Inv. No. H23 5/123   D. 74; Th. 18; W. 110   (both sides convex)
L207   House 6, Inv. No. H23 5/128A   D. 76; Th. 20; W. 112
L208   House 6, Inv. No. H23 5/129   D. 78; Th. 19; W. 107
L209   House 6, Inv. No. H23 5/130   D. 79; Th. 20; W. 113
L210   House 6, Inv. No. H23 5/132   D. 76; Th. 17; W. 108

L211   House 6, Inv. No. H23 5/133   D. 76; Th. 17; W. 109
L212   House 6, Inv. No. H23 5/134   D. 77; Th. 19; W. 109

L213   House 6, Inv. No. H23 5/172   D. 90; Th. 24; W. 176   (convex/flat)
L214   House 6, Inv. No. H23 6/17B   (damaged, both sides flat)
L215   House 6, Inv. No. H23 6/32   (fragment, convex/flat)
L216   House 6, Inv. No. H23 6/157   (fragment, both sides convex)

TRAPEZOID LOOMWEIGHT
Rectangular base, trapezoid body, two suspension holes

L217   House 2, Inv. No. H21 4b/6   H. 68; Base 34×47; W. 101   (slightly worn)
L218   House 2, Inv. No. H21 4c/11   H. 74; Base 53×43; W. 160   (slightly damaged)
L219   House 3, Inv. No. HA 196/1   H. 66; Base 26×29; W. 94
L220   House 3, Inv. No. HA 214/4   H. 69; W. 96

L221   House 3, Inv. No. HA 215/5   H. 81; Base 33×51; W. 118
L222   House 3, Inv. No. HA 247/27   H. 78; Base 38×35; W. 148
L223   House 3, Inv. No. HA 267/..   H. 70; W. 258   (one suspension hole)
L224   House 4, Inv. No. HA 147/27   H. 79; Base 39×54; W. 142   (slightly worn)
L225   House 4, Inv. No. HA 153/4   H. 78; Base 40×47; W. 126   (slightly damaged)
L226   House 5, Inv. No. HA 14/31   H. 93; Base 43×67; W. 261   (fig. 3.32)
L227   House 5, Inv. No. HA 59/20   (large fragment missing)
L228   House 5, Inv. No. HA 63/9   H. 80; Base 60×37; W. 172
L229   House 6, Inv. No. H23 2/57   H. 81; Base 44×51; W. 157   (stamped, >)
L230   House 6, Inv. No. H23 2/68   H. 74; Base 44×61; W. 163

L231   House 6, Inv. No. H23 2/71   H. 77; Base 42×55; W. 152   (stamped, >)
L232   House 6, Inv. No. H23 3/4   H. 70; Base 34×49; W. 100   (slightly worn)
L233   House 6, Inv. No. H23 5/41   H. 70; Base 49×37; W. 125   (pyramidal ??)
L234   House 6, Inv. No. H23 5/78   (fragment missing)
L235   House 6, Inv. No. H23 5/111   (fragment missing)

CONICAL LOOMWEIGHT
Circular base, conical body, one suspension hole

L236   House 1, Inv. No. X 637   H. 61; D. 47; W.
L237   House 4, Inv. No. HA 100/4   (fragment missing, worn)
L238   House 4, Inv. No. HA 115/35   H. 55; D. 33; W. 35   (fig. 3.32)
L239   House 6, Inv. No. H23 5/38   H. 75; D. 53; W. 126

APPENDIX 5. CATALOGUE OF METAL ARTEFACTS

STEVEN HIJMANS

Abbreviations

Inv. No.   = Inventory number
Dim.   = Dimensions in cm
For a list of references see Ch. 3: 4

AGRICULTURAL IMPLEMENTS

Hoes

M1   House 6, Inv. No. H23 5/m21, Fig. 3.33. Flat, rectangular hoe, in section somewhat concave; the centre of the rear long side is bent upward in a sharp curve and narrows to form the hollow, inward-curving shaft attachment, the end of which curves outward and upward. Dim.: 20.1 (width) ×16.8×0.4–0.5; shaft attachment 11.9×2.9–2.3 (diameter)× 0.2–0.3 (wall). References: Robinson, 1941: pp. 343–344, no. 1635, pl. CVII (more elaborate handle attachment, welded to the blade of the hoe).

M2   House 6, Inv. No. H23 5/m22, Fig. 3.33. *Dikella*, in profile slightly convex, with a large hole as handle attachment in the centre; one half, in section flat rectangular, consists of a single prong, tapering to a wide point; at 2.6 from the hole the other half divides into two, slightly flaring prongs, in section square/rectangular. Dim.: 35.0 (total length)×3.5 (width middle)×2.4 (thickness middle); 2.0 (diameter hole); 14.0×2.0 (width end)×0.6 (thickness end) single prong; 15.4×1.3×0.6 (prong 1 of double end); 14.0×1.3×0.4 (prong 2 of double end).

Spade or shovel

M3   House 6, Inv. No. H23 5/m26, Fig. 3.33. Flat, rectangular (?) shovel, rear corners slightly rounded and bent upwards; the rear part tapers to a round, hollow handle attachment. Dim.: 17.4 (total length)×12×0.7; handle attachment 6.1×2.7/ 2.1 (diameter)×0.6 (wall)×4.9 (depth).

Sickles and pruning knives

Measurements: Length (tip to tip; with sharply curved blades an approximate total length is given between brackets) ×width×thickness. If the transition from blade to handle is clearly defined, length×width×thickness of handle follows seperately; rivets: distance from tip of the handle; length×section. All sickles found in Halos are of iron. References: Isager & Skydsgaard, 1992: pp. 52– 53; White, 1967: pp. 71–103; Boardman, 1971: pp. 136–137; Jope, 1956: pp. 94–97.

M4   House 2, Inv. No. H21 1/14A. Short, sharply curving flat blade, tapering to a point (broken off) at one end and tapering to the straight handle attachment (most broken off) at the other end. Blade broken into two pieces. Dim.: 10.5 (12.5)×2.6-1.4/0.7 (point/handle)×0.4–0.1. References: White, 1967: pp. 96–97, no. 13, fig. 74.

M5   House 2, Inv. No. H21 1/14B. Long narrow blade, almost straight, flat, tapering and curving very slightly towards the rounded point. Near the handle end, a thicker part may be the heavily corroded remnant of one rivet. Dim.: 19.6×3.5– 0.8×0.5–0.2; 'rivet' 0.9×? References: Carapanos, 1878: p. 108 no. 7, pl LIII.

M6   House 6, Inv. No. H23 2/m1. Short, curving flat blade, tapering to a blunt point at one end, and cut back sharply at the other end to the straight handle, in section flat rectangular. Dim.: 11.9 (tot. length)×1.5-2.2×0.3 (blade); 5.3×0.5× 0.3–0.4 (handle).

M7   House 6, Inv. No. H23 5/m27, Fig. 3.33. Curving flat blade, tapering to a blunt point at one end, cut back slightly to the straight handle at the other end; two rivets in the handle. Dim.: 21.5×3.3–1.5 (blade)×0.7; handle 2.7×0.7; rivet 1: 0.5 from end, 2.2×0.7×0.7 (diameter head 1), 1.1 (diameter head 2); rivet 2: 5.5 from end, 2.4×0.5×0.6 (diameter head 1), 0.6 (diameter head 2).

M8   House 6, Inv. No. H23 5/m28, Fig. 3.33. Sharply curving flat blade, tapering to a blunt point at one end, towards a straight handle (not differentiated) at the other end. Dim.: approx. 18.5 (total length)×2.7–0.8 (point)/ 1.2 (handle)×0.6; sharp bend at approx. 19.0.

M9   House 6, Inv. No. H23 5/m29, Fig. 3.33. Large, curving flat blade, tapering to a blunt point at one end, towards the more sharply pointed handle (not differentiated from blade) at the other; rivet and rivet hole in handle. Dim.: approx. 28 (total length)×6.3–1.0 (point)/0.5 (handle); rivet 3.2 from end of handle, 1.5×0.5×0.6 (diameter head); hole 13.8 from end of handle, 0.4 (diameter); sharp bend at approx. 17.0.

M10 House 6, Inv. No. HA 78E/m1, Fig. 3.33. Straight, flat handle, tapering slightly to the end; curving blade, tapering towards a point (broken off). The blade (broken off near the handle, restored) also bends away from the handle, upwards when held in the right hand with the cutting edge to the left. Dim.: 15.4×2.4–1.2 (blade), 1.9–1.1 (handle)×0.3.

M11 House 3, Inv. No. HA 208/5. Very gently curving blade with round point (or point broken off?), other end tapering back sharply to the straight handle (most missing), ending in a hook (fragment preserved). Dim.: 11.0×1.1/0.3 (handle), 1.8–2.0/0.3–0.5 (blade); 2.2 (length preserved part of handle); 2.2×0.7 (seperate fragment of handle tip).

M12 House 3, Inv. No. HA 208/7, Fig. 3.33. Pruning hook; sharply curving blade ending in a rounded point, other end tapering back and widening to form a hollow handle attachment, the end of which is partially broken. Dim.: 19.1 (total length); handle: 5.4×1.4–1.5 (diameter)×0.2, 4.9 (depth); blade 7.1 (handle to curve), 6.6 (curve to point)×2.0–2.4– 1.0×0.3–0.4. References: Raubitschek, 1998: p. 127, no. 448.

M13 House 3, Inv. No. HA 214/5. Long straight handle, thickening towards the end to form a hook, flaring at the other end towards the blade, curving, most broken off. Dim.: 18.7×1.2– 2.2×0.3–0.7.

Shears

Two triangular blades, sometimes with sharp, sometimes with rounded points. Each blade has a shank running from one corner of the short side to a flattened metal arch at a right angle with the blades. This arch connects the two halves and functions as a spring. The shanks are rectangular or irregular in section, sometimes flattened and widened along the outside. The transition from blade to shank can be sharp or gradual. References: Wiegand & Schrader, 1904: p. 390, fig. 514; Manning, 1972: p. 176, no. 44. Measurements: 1. Total length; 2. Blade: length×width×thickness; 3. Shank: length×section; 4. Spring: length (distance from shank to shank)×width×thickness. Note: the shears found in Halos are incomplete, consisting of one blade only and broken at or near the spring.

M14   House 1, Inv. No. X 730. Dim.: 1) 10.1; 2) Blade: 6.5×0.9–1.9×0.2–0.4; 3) Shank: 3.6×0.3–0.5; 4) Spring: missing = Reinders, 1988: 34.22.

M15   House 6, Inv. No. H23 1/m1, Fig. 3.33. Dim.: 1) 15.2; 2) Blade: 8.1×2.9–0.8×0.2–0.4; 3) Shank: 7.2×0.3–1.1; 4) Spring: 3.7×1.9×0.3 (part missing).

M16   House 6, Inv. No. H23 5/m43, Fig. 3.33. Dim.: 1) 16.5; 2) Blade: 11.2×3.3–1.1×0.3; 3) Shank: 5.3×0.6×0.5; 4) Spring: 3.0×1.2×0.3 (part missing). The spring has been bent back against the shank, forming a loop, so that the blade could be used as a knife.

M17   House 5, Inv. No. HA 63/m1, Fig. 3.33. Dim.: 1) 12.3; 2) Blade: 7.0×2.2–0.8×0.3; 3) Shank: 4.1×0.6–0.3×0.3–1.1; 4) Spring: 1.2×1.1–1.3×0.3 (part missing).

TOOLS

Chisels

Type 1. Straight shank, in section rectangular, flattening at one end to a sharp wedge. References: Robinson, 1941: p. 345, nos. 1641–1642, pl. CVIII (both are smaller and more pointed). Measurements: Length×section shank; width×thickness flattened end.

M18   House 6, Inv. No. H23 4/m2, Fig. 3.34. Dim.: 11.6×0.7–0.8; 0.7×0.3.

Type 2, iron. Straight shank, in section rectangular, both ends tapering to a point. References: Robinson, 1941: p. 345, nos. 1645–1646, pl. CVIII. Measurements: Length×section shank.

M19   House 5, Inv. No. HA 45/m1. Dim.: 6.2×0.7–0.2.

M20   House 4, Inv. No. HA 104/7. Dim.: 9.8×1.0/0.6 - 0.3/0.2 (section: middle - ends).

M21   House 4, Inv. No. HA 117/7. Dim.: 5.0×0.7–0.2.

M22   House 3, Inv. No. HA 224/10. Dim.: 5.7×0.7–0.5; identification dubious.

M23   House 3, Inv. No. HA 247/5. Dim.: 5.1×0.8–0.4; broken at one end, identification dubious.

Saws

M24   House 5, Inv. No. HA 39/m1 + HA 41/m1. Rectangular, flat strip of iron, short shank to one side of each short end. Broken, two fragments, not certain whether there was an intermediate piece now missing (corrosion). Dim.: A. 9.2×2.2–1.6×0.3; B. 5.2×2.4–1.6×0.3. Identification uncertain. References: Marić, 1978: p. 33, nos. 144–146, pl. XXXVI (saw teeth clearly preserved).

M25   House 4, Inv. No. HA 104/9A–D. Rectangular, flat strip of iron, short shank to one side of one short end, other short end missing. Broken, four fragments, not certain how many intermediate pieces now missing (corrosion). Dim.: A. 8.0×2.4×0.4; B. 2.1×2.5×0.6; C. 3.4×2.4×0.2–0.3; D. 2.9×2.6×0.2–0.4. Identification uncertain.

WEAPONS

Arrowheads

Type 1, iron. Lozenge-shaped head, in section square, one end tapering to a (blunt) point, the other end tapering to a long, straight shank, in section rectangular or irregular. The transitions from head to shank range from gradual to very marked. References: Robinson, 1941: pp. 392–397, type E; Dörner & Goell, 1963: Taf. 73.1; Sahm, 1994: p. 118, M21, pl. 41 (with references). Measurements: Length×section head×section shank.

M26   House 1, Inv. No. X 616. Dim.: 11.8×1.1–1.2×0.5–0.6; bent, very corroded; Reinders, 1988; not listed.

M27   House 6, Inv. No. H23 1/m7, Fig. 3.35. Dim.: 9.8×1.2–0.4.

M28   House 5, Inv. No. HA 71/m2. Dim.: 6.4×1.1–0.4×0.4.

M29   House 3, Inv. No. HA 193/13. Dim.: 6.1×1.3–0.4×0.5; most of shank broken off.

M30   House 3, Inv. No. HA 238/4. Dim.: 5.7×0.9–0.3×–; shank completely broken off.

M31   House 3, Inv. No. HA 240/8. Dim.: 5.4×1.1/07–0.4×–; shank completely broken off.

M32   House 3, Inv. No. HA 247/29, Fig. 3.35. Dim.: 8.9×1.2–0.4×0.6–0.3; length of shank: 4.7.

M33   House 3, Inv. No. HA 247/59. Dim.: 5.4×0.9–0.2×0.4; most of shank broken off.

M34   House 3, Inv. No. HA 253/23. Dim.: 6.0×0.9/0.8–0.4×0.4; most of shank broken off.

Type 2, bronze. Three-sided, pyramidal arrowhead, base hollow for shaft. References: Robinson, 1941: type GIII, p. 406ff., e.g. 2122. Measurements: height×width.

M35   House 4, Inv. No. HA 83/m1, Fig. 3.35. Dim.: 1.6×0.8.

Spearheads

Type 1. Flat leaf-shaped pointed head, long hollow end, in section round. References: Marić, 1995: pl. 5,5; Robinson, 1941: pp. 412–414, type AII, nos. 2144–2153, pl. CXXVII. Measurements: length×largest width head×thickness head; diameter hollow end×thickness wall×depth.

M36   House 5, Inv. No. HA 33/m1. Dim.: 17.2×2.9×0.6; 2.2–2.3×0.2–0.4×5.7. Broken into various pieces; cleaned and consolidated, but very fragmented; identification uncertain.

Type 2. Flat leaf-shaped pointed head, tapering to a long straight shank, in section rectangular or irregular. Measurements: length×largest width head×thickness head; length shank (if sharply defined in respect to the head)×section shank.

M37   House 6, Inv. No. H23 5/m13+5/m14, Fig. 3.35. Dim.: 8.4×2.0×0.5; 1.8×0.6. It is possible but not certain that the shank 5/m14 (9.4×0.7–0.9) belongs to this spearhead.

Ferrules

Type 1. Solid head, in section square or rectangular, one end tapering to a point, the other end hollow. The smith shaped the hollow end by first hammering out this end of the solid point to a flat plate, then bending the sides round until they met to form a v-shaped joint, not fully closed. Measurements: 1. Length; 2. Section 1 (widest diameter hollow end, measured from the outside)×max. depth hollow end; in all cases the thickness of the wall at the hollow end ranges from 0.1–0.3; 3. Section 2 (solid point).

M38   House 1, Inv. No. X 733, Fig. 3.35. Dim.: 1) 7.0; 2) 1.5×2.9; 3) 0.8–0.4 = Reinders, 1988: 34.23 ('arrowhead').

M39   House 6, Inv. No. H23 2/m22. Dim.: 1) 6.6; 2) 1.1×1.4; 3) 0.4.

M40 House 6, Inv. No. H23 3/m20, Fig. 3.35. Dim.: 1) 6.0; 2) 1.3×3.0; 3) 0.2.

M41 House 6, Inv. No. H23 5/m1. Dim.: 1) 6.3; 2) 1.4×3.3; 3) 0.8×0.7, point 0.4.

M42 House 6, Inv. No. H23 5/m2. Dim.: 1) 6.1; 2) 1.6/1.4×2.3; 3) 0.8×0.7, point 0.1.

M43 House 6, Inv. No. H23 5/m3. Dim.: 1) 4.6; 2) 1.3/0.9×2.0; 3) 0.7, point 0.3.

M44 House 6, Inv. No. H23 5/m4. Dim.: 1) 6.2; 2) 1.6/1.2×too heavily corroded to determine; 3) 1.1×0.9, point 0.3.

M45 House 6, Inv. No. H23 5/m5. Dim.: 1) 6.6; 2) 1.8/1.5×too heavily corroded to determine; 3) 0.7×0.6, point broken off.

M46 House 6, Inv. No. H23 6/m15. Dim.: 1) 8.1; 2) 2.0×4.8; 3) 1.2×0.9. NB appears to be intact.

M47 House 5, Inv. No. HA 59/m1, Fig. 3.35. Dim.: 1) 13.4; 2) 2.0×3.2; 3) 0.7×0.7.

M48 House 5, Inv. No. HA 60/m2. Dim.: 1) 4.8; 2) 1.6×1.2; 3) 0.3.

M49 House 5, Inv. No. HA 73/m4. Dim.: 1) 18.7; 2) 1.5×too heavily corroded to determine; 3) 0.6.

M50 House 4, Inv. No. HA 153/8A, Fig. 3.35. Dim.: 1) 6.6; 2) 1.5×2.3; 3) 0.4/0.2.

M51 House 3, Inv. No. HA 227/4. Dim.: 1) 9.0; 2) 1.3×3.2; 3) 0.3.

M52 House 3, Inv. No. HA 246/3. Dim.: 1) 5.6; 2) 1.5×2.1; 3) 0.5.

M53 House 3, Inv. No. HA 248/2. Dim.: 1) 5.6; 2) 1.2×2.5; 3) 0.5.

M54 House 3, Inv. No. HA 262/7. Dim.: 1) 4.8; 2) 1.2×1.9; 3) 0.4.

M55 House 3, Inv. No. HA 266/5. Dim.: 1) 5.0; 2) 1.1×2.0; 3) 0.1.

M56 House 3, Inv. No. HA 271/4. Dim.: 1) 6.0; 2) 1.1×2.2; 3) 0.4.

Incomplete

M57 House 3, 216/8. Dim.: 1) 4.8; 2) 0.7×0.8; 3) 0.3.

M58 House 3, 271/2. Dim.: 1) 4.8; 2) 1.1×1.4; 3) 0.4.

DOMESTIC UTENSILS

Knives

Type 1, iron and bronze. Long, straight, rectangular blade, in section tapering to the cutting edge. At one end, the upper part of the blade thickens and appears to consist of three layers. A small strip of bronze at this end, attached to the corrosion, may be a remnant of a clamp fastening the handle to the blade. References: Raubitschek, 1998: p. 113, no. 371 (dated archaic-classical, handle attachment not with bronze band but with decorative pattern of 12 rivets); Robinson 1941: pp. 337–338, 1603–1604, pl. CII (larger). Measurements: 1. Length×width×thickness blade; 2. Length×width×thickness bronze strip.

M59 House 1, Inv. No. X 715, Fig. 3.36. Dim.: 1) 18.8×1.8–2.4×0.3–0.9; 2) 1.8×0.7×unknown = Reinders, 1988: 34.21.

Type 2, iron. Flat, fairly wide blade, in section flat, tapering towards a point; the wide end bent to a slight semicircle. No trace of handle or handle attachments. References: Robinson, 1941: p. 337, no. 1601, pl. CII. Measurements: Length× width×thickness.

M60 House 1, Inv. No. X 708. Dim.: 13.4×3.9–1.4×0.2–0.4; Point broken off? = Reinders, 1988: 34.24 ('fragment of iron').

Type 3, iron. Flat strip of iron, slightly curved, one end tapering towards a straight tang for the handle, other end tapering towards a blunt point. References: Raubitschek, 1998: p. 113: no. 373. Dim.: Length×width×thickness.

M61 House 6, Inv. No. H23 3/m16. Dim.: 16.2×1.6–0.8×0.4–0.5; Point (if any) broken off. Identification uncertain.

M62 House 6, Inv. No. H23 6/m17. Dim.: 14.0×3.3–1.9 0.2–0.4; Identification uncertain. There appears to be a second plate of iron (4.8×1.3×0.2) along the upper half of the shaft end, although heavy corrosion makes this difficult to establish this with certainty; cf. knife XXX above (X 715).

Type 4, iron. Chopping knife. Very wide, almost rectangular blade, flat ridge along the top, presumably straight tang, most of which is broken off. References: Manning, 1976: pp. 175–176, no. 40; Henig, 1994: p. 270, no. 8, fig. 14,9. Measurements: Length× width×thickness.

M63 House 5, Inv. No. HA 61/m2. Dim.: 12.3×5.1–4.3 (blade), 1.5–0.6 (handle attachment)×0.3 (blade), 1.4 (flat ridge and handle attachment). Most of handle attachment broken off.

Ladle

M64 House 6, Inv. No. H23 5/m49, Fig. 3.36. Fragments of a small bronze ladle. The shank, in section rectangular(?) curves inwards from the ladle and then outwards towards a hook. Only fragments of the ladle remain, making it impossible to give accurate measurements reflecting its original size. The fragments are very brittle and corroded. Dim.: Shank >11.5×1.0–0.7×0.2–0.4; ladle: unknown (fragments only), thickness 0.1.

Pins and Needles

M65 House 5, Inv. No. HA 40/m1, Fig. 3.36. Bronze pin, one end bent around to form a round eye; pin in section round, tapering to a point. Bent. Dim.: 10.7×0.1–0.2; diameter eye 0.8.

M66 House 4, Inv. No. HA 97/7, Fig. 3.36. Straight pin with a small, thick, flat, round head, sides hammered; long, thin shank in section round/angular, tapering to a sharp point. Dim.: 11.8×0.4–<0.1; head 1.1×0.3.

M67 House 4, Inv. No. HA 171/2. Point of an iron pin or needle, in section round. Dim.: 2.1×0.3–0.1.

Strigil

M68 House 6, Inv. No. H23 5/m23, Fig. 3.36. Small, bronze strigil, handle fused to a rectangular, iron plate. Curved, concave blade; simple handle, bent back sharply to form a rectangle, decorated with two longitudinal grooves. Dim.: 14.0 (total length); handle: 5.7×1.4–0.4×0.2; blade: 8.7×2.2× 0.5; iron plate: 5.3×4.2×0.2; rivets 0.8×0.2; 0.3×0.2.

Weights

Type 1. Rectangular weight, in section flat rectangular. References: Robinson, 1941: pp. 463–464, nos. 2421–2439, pls. CXLV–CXLVII. Measurements: Length×width×thickness; weight.

M69 House 6, Inv. No. H23 5/m46, Fig. 3.36. Dim.: 6.4×6.0× 1.7.

M70 House 6, Inv. No. H23 5/m48a. Dim.: 9.9×7.1×0.3–0.8; original weight unknown. Very irregular in form as a large part of the weight has melted and is now missing.

Type 2. Discoid weight, one side flat, one side convex. References: Robinson, 1941: p. 469, no. 2464, pl. CL.
Measurements: diameter×thickness; weight.
M71 House 6, Inv. No.H23 6/m14. Dim.: 4.0×0.4–1.1.

Type 3. Discoid loom weight, in section rectangular, pierced by two holes near the top. Measurements: diameter×thickness
M72 House 4, Inv. No. HA 149/5B, Fig. 3.36. Dim.: 5.4–5.2×0.4–0.5.

'Meathook'

M73 House 6, Inv. No. H23 5/m24, Fig. 3.36. Long rod, in section irregular/rectangular, one end bent round to form a small eye, the other end a five-pronged, star-shaped hook, concave. The hook was formed by flattening the rod near the end, and bending the very end (point broken off) inwards; two additional short rods, consisting of a flattened central section and an inwardly curving hook at either end, are fastened crosswise to the flat part of the main rod (all points broken off). References: Petrie, 1917: p. 57, par. 165, pl. LXIII, no. W 66; Robinson, 1941: pp. 198–199, no. 623, pl. L. Dim.: 39.1× 0.7; diameter eye (inside) 0.4; two crossing hooks: 4.5×0.7; 4.5×0.7.

Fork

M74 House 5, Inv. No. HA 85/m1, Fig. 3.36. Long straight shank, in section rectangular, ending in three prongs; the central prong is the continuation of the shank, the two lateral ones are the ends of a U-shaped piece of iron, attached to the shank just below the central prong at a point where the shank is flattened and widened. Dim.: 29.7×0.6; 3.0 (distance between lateral prongs), 2.8×0.4 (lateral prong A), 3.3×0.3 (central prong), 2.5×0.4 (lateral prong B).

Stylus

References: Robinson, 1941: pp. 357–359, nos. 1725–1735, pl. CXIV & fig. 22 (with useful refs.); Corinth XII: pp. 185–187, nos. 1348–1376, pls. 83–84 (bronze, ivory, and bone, Classical to Byzantine; no pre-Roman bronze styli).
M75 House 6, Inv. No. H23 5/m44, Fig. 3.36. Bronze stylus. Long, flat shaft; sharp, narrow, lozenge-shaped point. Dim.: 15.2×0.5×0.1–0.2; point: 1.9×0.3–0.1.
M76 House 4, Inv. No. HA 148/3. Long, straight iron shank, in section round, twisted like a cable, flattening and widening at one end, broken at the other. Identification as a stylus not certain. References: Carapanos, 1878: pl. LIII no. 13, cf. 8 & 11 (bronze); Robinson, 1941: p. 359, no. 1731, pl. CXIV. Dim.: 9.5×0.4–0.9/0.2.

JEWELLERY AND COSMETICS

Hairpin, 'Glasinac' type

M77 House 1, Inv. No. X 73. Bronze, ornamental (hair?)pin of the 'Glasinac' type, in section round, bent to an intricate, symmetrical form. From both ends the pin moves outwards to a shoulder, where both sides bend sharply inwards towards each other, and then loop upwards and outwards in a small circle, connecting in a much larger loop above. References: Philipp, 1981: pp. 97–99, 101, nos. 323–324, pl. 37; Maier,

1956: pp. 69–70, 73 (dates this type to 6th–5th c. BC), with refs; Jacobsthal, 1956: pp. 135–141. Lost in the earthquake of 1980; recorded dim.: length 5.0; max. width 1.7 (Reinders 1988, 35.30, fig. 115).

Fibulas

Large, round bow of a fibula, swelling in the middle. Three lengthwise parallel grooves along the middle. Both fibulas lack the pin and catch.
M78 House 5, Inv. No. HA 77, Fig. 3.37. Dim.: 7.4×3.0–0.6; 0.5 (diameter knob); 0.6×1.0×0.2 (flattened vertical continuation).
M79 House 3, Inv. No. HA 216/31. Dim.: 3.6×1.2–0.6×0.6

OTHER JEWELLERY

Bronze bracelet

M80 House 2, Inv. No. H21 5/15, Fig. 3.37. Bronze bracelet, in section round, both ends ending in flat, stylized snake's heads decorated with a pattern of dots and lines. The bracelet is twisted out of shape. References: Philipp, 1981: pp. 222–251 (overview of snake-head bracelets, no direct parallel); Robinson, 1941: pp. 68–72, nos. 182–224 (photographs and descriptions insufficiently detailed to allow identification of close parallels). Dim.: total length approx. 20.0, present diameter 13.4 (original diameter cannot be determined), section 0.5. Heads: 1.4×0.7×0.3; 1.3×0.7×0.3.

Decorative element in bronze relief

M81 House 5, Inv. No. HA 55/m6, Fig. 3.37. Decorative element in bronze relief, with a bent over lip at the top; frontal face of a woman, long, wavy, Medusa-like hair. References: For production technique (with matrixes) cf. Marić, 1995: pp. 60–61, figs. 17–18, pls. 21–23. Dim.: 5.7×5.0×0.1.

Bronze inlay of a pyxis

M82 House 3, Inv. No. HA 215/2. Bronze inlay of a pyxis, originally forming a rectangle in one piece, one short side now detached; bent; otherwise undamaged. Lid, lid-hinges and bottom missing. Each side consists of a rectangle with two corner legs; the two short sides (B&D) are equal in shape and dimension; the front (A) is the same height as the short sides, but the back (C) is lower, allowing space for the hinges of the lost lid of the pyxis. Each side is decorated with two vertical lines defining the legs, and two horizontal lines just below the middle. References: For production technique (with matrixes) cf. Marić, 1995: pp. 60–61, figs. 17–18, pls. 21–23; Marić, 1995: pp. 60, pl. 18,6; cf. Marić, 1978: p. 27, no. 9, pp. 51–54, pls. XXI–XXII. A very similar pyxis, almost twice as large, found in the smith's hoard of the 2nd century BC discovered in 1977 at Daors, Bosnia. This pyxis also lacks a bottom, but has a decorated bronze lid. Dim.: length 4.8; width 3.5; height 2.4 (A,B,D).

RITUAL OBJECTS

Snakes

M83 House 3.
M84 House 3.

Miniature bed

Thin, flat, rectangular plaque of lead, a narrow rectangle cut away very close to one short side; decorated with a chevron pattern on one side. Bent.
M85 House 2, Inv. No. H21 4/7, Fig. 3.37. Dim.: 6.4×3.4×0.2–0.3.

STRUCTURAL ELEMENTS

Nails

To avoid confusion or overly detailed descriptions, we have divided the nails into two basic groups: A. small nails and tacks; B. large nails. Small nails will basically not exceed a length of 5.0, and the diameter of the shank will generally be 0.4 or less. All other nails are large. We believe that this subdivision reflects a real division between nails used for building and construction purposes (large) and nails used for fine carpentry and metalwork (small). References: Briggs, 1956: p. 443 (clinching, 13th century AD); Robinson, 1941: pp. 309–329, pls. LXXXIX–XCVI.

A. Small nails and tacks

Type 1, iron. Small flat head, rectangular or irregularly rectangular in shape; shank, in section rectangular or irregular, set in the middle of the head. Measurements: 1. Head: length×width×thickness: 2. Shank: total length×section; 3. (only if the shank is not straight): Bent/clinched, distance to 1st clinch (from outside to outside), distance to 2nd clinch (outside to outside).
Complete
M86 House 1, Inv. No. X 4. Dim.: 1) 1.1×1.3×0.3–0.5; 2) 3.5×0.2–0.5; 3) clinched, 2.0, 0.9; = Reinders, 1988: 37.27.
M87 House 2, Inv. No. H21 4/3, Fig. 3.38. Dim.: 1) 1.7×0.9×0.3; 2) 2.3×0.6–0.3; 3) clinched (?), point curving inward like a hook.
M88 House 6, Inv. No. H23 4/m4, Fig. 3.38. Dim.: 1) 1.2×1.0× 0.3; 2) 3.6×0.4 (width shank)×0.2 (thickness shank); 3) clinched (?), 3.5; point bent sharply to one side.

Incomplete
M89 House 2, Inv. No. H21 5/-. Dim.: 1) 1.2×1.0×0.2, 2) 1.7× 0.4–0.2; 3) bent, point broken off.

Type 2, iron. Small flat head, rectangular or irregular in shape; shank, in section rectangular or irregular, forms the continuation of one side of the head. Measurements: 1. Head: length×width× thickness; 2. Shank: total length×section; 3. (only if the shank is not straight): Bent/clinched, distance to 1st clinch (from outside to outside), distance to 2nd clinch (outside to outside).
Complete
M90 House 1, Inv. No. X 72. Dim.: 1) 2.4×0.9×0.5; 2) 3.4×0.5–0.9; 3) clinched, 1.9; = Reinders, 1988: 32.09.
M91 House 4, Inv. No. HA 98/2, Fig. 3.38. Dim.: 1) 1.6×1.4×0.5; 2) 2.1×0.4–0.3.

M92 House 5, Inv. No. HA 30/m1. Dim.: 1) 1.3×1.8×0.3; 2) 3.9×0.8–0.2.

Type 3, iron. Small, sometimes very small, flat, round head, short shank in section rectangular or irregular. Shank set more or less in the middle of the head. It appears that the shank was sometimes 'preforged' for clinching at a specific point. In the case of HA145/ 2, for instance, the shank is of a uniform thickness right up to the clinch, while the bent over point is much thinner. This would imply that such nails were 'made to order'. Measurements: 1. Head: diameter×thickness; 2. Shank: total length×section; 3. (only if the shank is not straight): Bent/clinched, distance to 1st clinch (from outside to outside), distance to 2nd clinch (outside to outside).
Complete
M93 House 1, Inv. No. X 345. Dim.: 1) 1.0×0.3; 2) 0.8×0.2; = Reinders, 1988: 37.31.
M94 House 5, Inv. No. HA 39/m2, Fig. 3.38. Dim.: 1) 0.9×0.2; 2) 1.4×0.3–0.1; 3) clinched, 0.9.
M95 House 4, Inv. No. HA 109/6. Dim.: 1) 1.6–1.7×0.3; 2) 1.3×0.4.
M96 House 4, Inv. No. HA 127/61, Fig. 3.38. Dim.: 1) 1.9–2.2×0.1–0.3; 2) 2.6×0.5–0.2; 3) clinched, 1.4, 0.8.
M97 House 4, Inv. No. HA 145/2. Dim.: 1) 1.7–1.9×0.4; 2) 2.1×0.6–0.2; 3) clinched, 1.2.
M98 House 3, Inv. No. HA 247/28. Dim.: 1) 2.1–2.0×0.3–0.4; 2) 3.6×0.7–0.4.
M99 House 3, Inv. No. HA 247/30. Dim.: 1) 1.9×0.5; 2) 4.8×1.0–0.5; 3) bent just below the head.

Incomplete. Unless stated otherwise, the head is more or less complete and the end of the shank is broken off.
Note that some of these could also originally have been large nails!
M100 House 6, Inv. No. H23 2/m24a. Dim.: 1) 1.5×0.2–0.3; 2) 1.0×0.5.
M101 House 6, Inv. No. H23 5/m30. Dim.: 1) 1.8×0.2–0.3; 2) 1.4×0.4.
M102 House 6, Inv. No. H23 6/m5. Dim.: 1) 2.0/1.8×0.4; 2) 1.5×0.4.
M103 House 5, Inv. No. HA 13/m2. Dim.: 1) 1.0×0.2; 2) 3.4× 0.5–0.3; 3) clinched, 2.4, point broken off.
M104 House 5, Inv. No. HA 65/m3. Dim.: 1) 2.0×0.4; 2) 1.1×0.4.
M105 House 5, Inv. No. HA 73/m3. Dim.: 1) 1.7×0.3–0.4; 2) -.
M106 House 4, Inv. No. HA 126/20. Dim.: 1) 1.3–1.5×0.1–0.4; 2) -.
M107 House 4, Inv. No. HA 127/69. Dim.: 1) 2.1×0.3; 2) 1.2× 0.4.
M108 House 4, Inv. No. HA 141/5. Dim.: 1) 1.7–1.9×0.2; 2) 1.2×0.4.
M109 House 4, Inv. No. HA 143/1. Dim.: 1) 1.5×0.4; 2) -.
M110 House 4, Inv. No. HA 166/2. Dim.: 1) 1.6×0.3 0.4; 2) 1.2×0.4.

Type 4, iron. Shank, in section rectangular, tapering to a point, other end thickening to form a hardly differentiated, irregular globular head. Measurements: 1. Head: diameter; 2. Shank: total length×section; 3. (only if the shank is not straight): bent/clinched, distance to 1st clinch (from outside to outside), distance to 2nd clinch (outside to outside).
Complete
M111 House 6, Inv. No. H23 6/m2, Fig. 3.38. Dim.: 1) 1.1; 2) 3.0×0.5–0.2.
M112 House 6, Inv. No. HA 4/m3. Dim.: 1) 1.0; 2) 4.0×0.3–0.2; 3) clinched, 1.5.

M113   House 3, Inv. No. HA 232/28.2. Dim.: 1) 0.9/1.1; 2) 3.2×0.7–0.4; 3) clinched, 2.1, 2.8.

Type 1, bronze. Large, flat, circular head; shank, in section round, set in or near the middle of the head. References: Sahm 1994, 116, pl. 41 (with references). Measurements: 1. Head: diameter×thickness; 2. Shank: total length×section.

Incomplete

M114   House 1, Inv. No. X 630, Fig. 3.38. Dim.: 1) 1.6–1.7×0.1; 2) 1.4×0.3 (Reinders, 1988: 32.08).

Type 2, bronze. Small convex round or irregular head, shank in section round, tapering to a point. Measurements: 1. Head: diameter×thickness; 2. Shank: total length×section; 3. Bent/clinched, distance to 1st clinch (from outside to outside).

Complete

M115   House 6, Inv. No. H23 5/m46a, Fig. 3.38. Dim.: 1) 0.5×0.4; 2) 4.8×0.3–0.2; 3) Clinched, 3.7. A small flat, roughly rectangular piece of iron; directly below the head, remains of the object fastened by this nail: 1.8×0.5×0.2.

Type 3, bronze. Inclined square head, shank, in section flat rectangle, tapering to a point. Measurements: 1. Head: length×width× thickness; 2. Shank: total length×section.

Complete

M116   House 3, Inv. No. HA 198/5. Dim.: 1) 0.4×0.4×0.2; 2) 1.3×0.2/0.3 – 0.1/0.04.

B. Large nails

Type 1, iron. Small round head, shank, in section rectangular or irregular, set in the middle of the head. References: Marić, 1995: pp. 47, pl. 1, 11–13 (all clinched twice); Raubitschek, 1998: pp. 176–177, app. L, IM 5120, fig. 35 (clinched twice); Waldbaum & Knox, 1983: p. 68. Measurements: 1. Head: diameter×thickness; 2. Shank: total length×section; 3. (only if the shank is not straight): Bent/clinched, distance to 1st clinch (from outside to outside), distance to 2nd clinch (outside to outside).

Complete

M117   House 1, Inv. No. X 647 D, Fig. 3.38. Dim.: 1) 2.4–2.1×0.6; 2) 12.8×1.0–0.5; 3) Clinched, 8.4, 3.2 = Reinders, 1988: 33.60.

M118   House 4, Inv. No. HA 80/m1. Dim.: 1) 2.0×0.4–0.5; 2) 9.4×0.6–0.4.

M119   House 4, Inv. No. HA 60/m3. Dim.: 1) 2.0×0.3–0.5; 2) 9.6×0.6; 3) Clinched (or bent?), 7.2.

M120   House 4, Inv. No. HA80/m4. Dim.: 1) 2.0×0.4–0.5 (damaged); 2) 9.4×0.6.

M121   House 3, Inv. No. HA 240/15. Dim.: 1) 1.9–2.1×0.4; 2) 7.2×0.8–0.3; 3) Clinched, 4.4.

M122   House 3, Inv. No. HA 247/32. Dim.: 1) 1.8×0.4; 2) 5.5×0.9–0.4; 3) -.

Incomplete; unless stated otherwise, the head is more or less complete, and the end of the shank is broken off.

M123   House 1, Inv. No. X 115. Dim.: 1) 2.0×0.5; 2) 3.7×0.7; 3) slightly bent. = Reinders, 1988: 33.67.

M124   House 1, Inv. No. X 126. Dim.: 1) 1.8×0.3; 2) 1.1×0.5. Reinders, 1988; not listed.

M125   House 1, Inv. No. X 625 A. Dim.: 1) 1.7–1.8×0.4; 2) 11.8×0.7–0.4; 3) clinched, 8.1, 3.5. = Reinders, 1988: 33.49.

M126   House 1, Inv. No. X 647 E. Dim.: 1) 1.9×0.3; 2) 7.3×0.7–0.5; 3) bent. = Reinders, 1988: 33.61.

M127   House 1, Inv. No. X 655 B. Dim.: 1) 2.5×0.5; 2) 7.8×1.0–0.6. = Reinders, 1988: 33.64.

M128   House 1, Inv. No. X 655 C. Dim.: 1) 2.0×0.4; 2) 8.0×0.8–0.4. = Reinders, 1988: 33.65.

M129   House 1, Inv. No. X 655 D. Dim.: 1) 2.0×0.4; 2) 11.8×0.9–0.4; 3) clinched, 7.6. = Reinders, 1988: 33.66.

M130   House 1, Inv. No. X 734. Dim.: 1) 2.0×0.4; 2) -. = Reinders, 1988: 34.27.

M131   House 1, Inv. No. X 647 F. Dim.: 1) 3.0×0.5; 2) 13.5×0.5–0.4. = Reinders, 1988: 33.62.

M132   House 1, Inv. No. X 35. Lost in the earthquake of 1980. Recorded dim.: diameter head 1.8; section shank 0.5. = Reinders, 1988: 39.12.

M133   House 6, Inv. No. H23 2/m23. Dim.: 1) 1.9×0.3; 2) 6.4× 0.6; 3) clinched? If so, broken at clinch.

M134   House 6, Inv. No. H23 3/m19. Dim.: 1) 3.1/2.5×0.3–0.4; 2) 2.3×0.8; 3) part of head broken off.

M135   House 6, Inv. No. H23 5/m37, Fig. 3.38. Dim.: 1) 2.8×0.5; 2) 7.2×0.6–0.5.

M136   House 6, Inv. No. H23 5/m39. Dim.: 1) 1.7×0.6; 2) 7.6×0.6.

M137   House 6, Inv. No. H23 6/m7. Dim.: 1) 1.9/1.5×0.4; 2) 1.3×0.9; 3) bent to one side.

M138   House 6, Inv. No. HA 43/m1. Dim.: 1) 1.9/1.7×0.4–0.5; 2) 3.1×0.8–0.6; 3) head bent to one side.

M139   House 4, Inv. No. HA 106/26. Dim.: 1) 2.1×0.5–0.8; 2) 1.4×0.6.

M140   House 4, Inv. No. HA 115/28. Dim.: 1) 1.1×0.9×0.3 (part broken); 2) 3.3×0.5; 3) end of shank bent, rest broken off.

M141   House 4, Inv. No. HA 120/12. Dim.: 1) 2.7–3.1×0.3; 2) 1.4×0.7.

M142   House 4, Inv. No. HA 157/2. Dim.: 1) 1.6×0.3; 2) 4.4×0.6.

M143   House 3, Inv. No. HA 191/2. Dim.: 1) 2.0×0.5; 2) 5.9×0.8.

M144   House 3, Inv. No. HA 194/1. Dim.: 1) 1.4–1.5×0.3; 2) 4.5×0.6–0.4.

M145   House 3, Inv. No. HA 210/1. Dim.: 1) 1.5×0.4; 2) 5.5×0.6.

M146   House 3, Inv. No. HA 210/13. Dim.: 1) 2.1×0.4; 2) 4.0×0.7–0.4.

M147   House 3, Inv. No. HA 216/12. Dim.: 1) 1.9×0.3; 2) 5.9×0.6.

M148   House 3, Inv. No. HA 220/4. Dim.: 1) 2.1×0.3; 2) 3.2×0.5.

M149   House 3, Inv. No. HA 224/6. Dim.: 1) 2.5×0.3 (half broken off; 2) 3.2×0.7.

M150   House 3, Inv. No. HA 224/11. Dim.: 1) 2.2×0.4; 2) 2.8×0.8.

M151   House 3, Inv. No. HA 232/6. Dim.: 1) 1.5×0.4; 2) 3.3×0.6.

M152   House 3, Inv. No. HA 236/3. Dim.: 1) 1.5×0.4; 2) 3.7×0.4.

M153   House 3, Inv. No. HA 247/60. Dim.: 1) 1.5×0.3; 2) 6.4×0.6–0.4.

M154   House 3, Inv. No. HA 267/25. Dim.: 1) 1.6×0.3; 2) 1.3×0.6.

Type 2, iron. Large round head, shank, in section rectangular or irregular, set in the middle of the head. We assume that the shank does not necessarily taper to a point; it would otherwise be quite remarkable to have so many double-clinched nails of this type with only the point missing. References: Raubitschek, 1998: p. 134 (refs. in n. 20), p. 138, nos. 492–493, p.175 app. K2, fig. 35. Measurements: 1 Head: diameter×thickness; 2. Shank: total length×section; 3. (only if the shank is not straight): Bent/clinched, distance to 1st clinch (from outside to outside), distance to 2nd clinch (outside to outside).

Complete

M155 House 1, Inv. No. X 639, Fig. 3.38. Dim.: 1) 4.5×0.4; 2) 13.4×0.7–0.3; 3) clinched, 7.7, 5.5. = Reinders, 1988: 33.55.

M156 House 1, Inv. No. X 640. Dim.: 1) 4.5×0.4; 2) 13.2×0.6–0.5; 3) clinched, 8.1, 4.8. = Reinders, 1988: 33.50.

M157 House 1, Inv. No. X 643 A. Dim.: 1) 4.4×0.4; 2) 13.4×0.7–0.5; 3) clinched, 7.9, 5.0. Reinders, 1988; not listed.

M158 House 1, Inv. No. X 645. Dim.: 1) 4.6×0.4; 2) 12.9×0.6–0.5; 3) clinched, 7.9, 4.0, second clinch not marked. = Reinders, 1988: 33.57.

M159 House 1, Inv. No. X 647 B. Dim.: 1) 4.3×0.4; 2) 13.2×0.7–0.4; 3) clinched, 7.9, 4.5. = Reinders, 1988: 33.58.

M160 House 1, X 647 A. Lost in earthquake of 1980. Recorded dim.: diameter head 4.4. = Reinders, 1988: 33.51 (fig. 114).

M161 House 5, HA80/m7. Dim.: 1) 5.2×0.4; 2) 9.3×0.6–0.5; 3) clinched, 7.0.

Incomplete; unless stated otherwise, the head is more or less complete, and the end of the shank is broken off.

M162 House 1, Inv. No. X 633. Dim.: 1) 4.4×0.4; 2) 11.1×0.7–0.4; 3) clinched, 7.7. = Reinders, 1988: 33.54.

M163 House 1, Inv. No. X 641. Dim.: 1) 4.5×0.4; 2) 12.2×0.8–0.5; 3) clinched, 7.8. = Reinders, 1988: 33.56.

M164 House 1, Inv. No. X 647 C. Dim.: 1) 4.6×0.4; 2) 7.7×1.0–0.6. = Reinders, 1988: 33.59.

M165 House 1, Inv. No. X 655 A. Dim.: 1) 4.5×0.5; 2) 12.2×0.6–0.5; 3) clinched, 7.9. = Reinders, 1988: 33.63.

M166 House 2, Inv. No. H21 4/5. Dim.: 1) 3.8–4.0×0.3; 2) 3.1×0.8–0.5.

M167 House 6, Inv. No. H23 1/m4. Dim.: 1) 4.6×3.6×0.4; 2) 5.2×0.6; 3) part of head broken.

M168 House 6, Inv. No. H23 1/m8. Dim.: 1) 3.4×2.5×0.3; 2) 2.9×0.4; 3) clinched, 2.8, rest of shank broken off.

M169 House 6, Inv. No. H23 3/m3. Dim.: 1) 4.2×0.2–0.5; 2) 1.3×0.6–0.8.

M170 House 6, Inv. No. H23 3/m8. Dim.: 1) 4.4–4.8×0.2–0.4; 2) 2.6×0.7.

M171 House 6, Inv. No. H23 3/m18. Dim.: 1) 4.2×0.2–0.4; 2) 3.3×0.7–0.5; 3) bent.

M172 House 6, Inv. No. H23 3/m26. Dim.: 1) 2.4×0.8×0.2; 2) 1.6×0.7; 3) head broken off, only small rectangular fragment remaining.

M173 House 6, Inv. No. H23 4/m3. Dim.: 1) 4.3×0.2; 2) 1.3×0.6; 3) shank bent sharply to one side.

M174 House 5, Inv. No. HA80/m14. Dim.: 1) 5.1×0.4; 2) -.

M175 House 3, Inv. No. HA 197/1. Dim.: 1) 5.2×0.3; 2) -.

M176 House 3, Inv. No. HA 202/20. Dim.: 1) 4.6×0.4; 2) 2.1×0.9.

M177 House 3, Inv. No. HA 202/21. Dim.: 1) 4.1×0.3 (part broken); 2) 8.0×0.5.

M178 House 3, Inv. No. HA 246/8. Dim.: 1) 5.9×3.6×0.3 (broken, originally round?); 2) 4.0×0.9–0.4.

M179 House 3, Inv. No. HA 247/13. Dim.: 1) 4.4–5.1×0.3 (broken); 2) 5.3×0.8.

M180 House 3, Inv. No. HA 250/1a. Dim.: 1) 4.7×0.3; 2) 1.8×0.9.

Type 3, iron. Pinched, wedge-shaped head, shank in section rectangular or irregular. The head of this type of nail is made by squeezing the end of the shank with a pair of pincers during forging. Measurements: 1. Head: width (thickness and height range from 0.2–0.5); 2. Shank: total length×section; 3) (only if the shank is not

straight): Bent/clinched, distance to 1st clinch (from outside to outside), distance to 2nd clinch (outside to outside).

Complete

M181 House 1, Inv. No. X 152, Fig. 3.38. Dim.: 1) 1.2; 2) 9.9×0.7–0.3; 3) head bent slightly back. = Reinders, 1988: 35.29.

M182 House 6, Inv. No. H23 5/m36, Fig. 3.38. Dim.: 1) 1.3; 2) 12.0×0.7; 3) bent (6.8), no point (broken off?).

M183 House 6, Inv. No. H23 6/m10. Dim.: 1) 0.9; 2) 9.4×0.7–0.3; 3) clinched (?), 3.1.

M184 House 6, Inv. No. HA 74/m1. Dim.: 1) 1.4; 2) 13.6×0.7–0.3; 3) clinched, 6.0, very sharp angle.

M185 House 4, Inv. No. HA 126/27. Dim.: 1) 1.1; 2) 10.0×0.8–0.3.

M186 House 4, Inv. No. HA 130/10. Dim.: 1) 0.9; 2) 4.8×0.6–0.3; 3) lower part of shank twisted, almost broken.

M187 House 3, Inv. No. HA 216/23.1. Dim.: 1) 1.1; 2) 8.2×0.8–0.3; 3) bent sharply just below the head.

M188 House 3, Inv. No. HA 232/15. Dim.: 1) 0.7; 2) 6.8×0.5–0.2; 3) slightly bent just below the head.

M189 House 3, Inv. No. HA 232/28.1. Dim.: 1) 0.9; 2) 4.7×0.7–0.3.

Incomplete; unless stated otherwise, the head is more or less complete and the end of the shank is broken off.

M190 House 1, Inv. No. X 145–3. Dim.: 1) 0.9; 2) 5.7×0.4–0.6; 3) slightly bent. Reinders, 1988: not listed.

M191 House 1, Inv. No. X 655 e+f-III. Dim.: 1) 1.4; 2) 11.7×0.6–0.5; 3) clinched, 7.4. Reinders, 1988: not listed.

M192 House 1, Inv. No. X 655 e+f-IV. Dim.: 1) 1.3; 2) 11.0×0.6–0.5; 3) clinched, 7.6. Reinders, 1988: not listed.

M193 House 6, Inv. No. H23 5/m41. Dim.: 1) 0.9; 2) 6.7×0.6.

M194 House 6, Inv. No. H23 6/m3. Dim.: 1) 1.1; 2) 8.1×0.5.

M195 House 6, Inv. No. H23 6/m3. Dim.: 1) 0.9; 2) 7.9×0.5.

M196 House 6, Inv. No. H23 6/m13. Dim.: 1) 1.0; 2) 4.9×0.7–0.6.

M197 House 3, Inv. No. HA 232/10. Dim.: 1) 1.8; 2) 6.0×0.6–0.5; 3) Bent sharply just below the head.

M198 House 3, Inv. No. HA 274/1. Dim.: 1) 1.3; 2) 5.1×0.7–0.6.

Type 4, iron. Fairly large triangular, irregular, or irregular rectangular head; shank, in section rectangular or irregular, forms the continuation of one side of the head. The head of this type of nail is made by hammering flat the end of the shank during forging and then bending the flat end over at a right angle. Measurements: 1. Head: length×width×thickness; 2. Shank: total length×section; 3) (only if the shank is not straight): Bent/clinched, distance to 1st clinch (from outside to outside), distance to 2nd clinch (outside to outside).

Complete

M199 House 1, Inv. No. 79 6 B. Dim.: 1) 1.4×1.1×0.3; 2) 5.4×1.0–0.3. Reinders 1988, not listed.

M200 House 2, Inv. No. H21 3/37. Dim.: 1) 1.4×1.1×0.1; 2) 8.4×0.7–0.3; 3) small part of head broken off.

M201 House 6, Inv. No. H23 1/m5. Dim.: 1) 1.7×1.8×0.3–0.4; 2) 8.4×0.8–0.4; 3) clinched, 4.5; point bent downward.

M202 House 6, Inv. No. H23 2/m24. Dim.: 1) 1.1×1.0×0.3–0.4; 2) 6.4×0.5–0.3; 3) clinched? or sharply bent near point; 5.3.

M203 House 6, Inv. No. H23 4/m5, Fig. 3.38. Dim.: 1) 1.6×1.0×0.5; 2) 6.4×0.7–0.2.

M204 House 6, Inv. No. H23 6/m4, Fig. 3.38. Dim: 1) 1.8×1.3×0.4; 2) 9.3×0.8–0.2; 3) bent at 6.0.

M205 House 6, Inv. No. HA 4/m2. Dim.: 1) 1.8×1.0–0.3×0.3–0.4; 2) 12.2×0.6–0.3; 3) clinched, 9.9.

M206 House 5, Inv. No. HA 71/m1. Dim.: 1) 1.5×1.1×0.3; 2) 8.0×0.8–0.3.

M207 House 4, Inv. No. HA 104/10. Dim.: 1) 2.3×1.8×1.4; exceptionally large, irregular head; 2) 13.7×0.8; 3) clinched. 7.5.

M208 House 3, Inv. No. HA 198/19. Dim.: 1) 2.4×1.9×0.6; exceptionally large, roughly oval head; 2) 12.7×1.1–0.5.

M209 House 3, Inv. No. HA 202/7. Dim.: 1) 1.1×1.1×0.4 (part of head missing); 2) 8.2×0.7–0.3.

M210 House 3, Inv. No. HA 202/12. Dim.: 1) 1.1×0.8×0.3 (part of head missing); 2) 6.6×0.5–0.2; 3) clinched, 5.1.

M211 House 3, Inv. No. HA 215/1. Dim.: 1) 1.3×0.4×0.3 (large part broken); 2) 9.0×0.7–0.3; 3) bent twice, first inwards, then outwards.

M212 House 3, Inv. No. HA 216/10. Dim.: 1) 1.9×1.2×0.4; 2) 11.3×0.8–0.3; 3) clinched, 6.2, 2.9.

M213 House 3, Inv. No. HA 216/18.1. Dim.: 1) 2.2×1.4×0.5; 2) 10.2×1.1–0.4.

M214 House 3, Inv. No. HA 216/20.2. Dim.: 1) 1.3×1.0×0.2; 2) 4.3×0.6–0.3.

M215 House 3, Inv. No. HA 232/7. Dim.: 1) 2.1×1.7×0.3; 2) 7.2×0.9–0.3.

M216 House 3, Inv. No. HA 232/9. Dim: 1) 2.2×1.6×0.9 (damaged, corroded); 2) 8.4×0.7–0.4.

M217 House 3, Inv. No. HA 232/26. Dim.: 1) 1.3×1.3×0.4; 2) 4.8×0.8–0.4.

M218 House 3, Inv. No. HA 247/55. Dim.: 1) 1.5×1.3×0.3; 2) 6.9×0.7–0.3.

M219 House 3, Inv. No. HA 253/20. Dim.: 1) 1.6×1.2×0.5; 2) 10.3×0.8–0.3; 3) clinched, 3.2, point bent outwards.

Incomplete; unless stated otherwise, the head is more or less complete and the end of the shank is broken off.

M220 House 1, Inv. No. X 146-1. Dim.: 1) 1.5×1.5×0.3; 2) 7.7×1.0–0.5; 3) part of head broken. = Reinders, 1988: 35.27.

M221 House 1, Inv. No. X 607. Dim.: 1) 1.0×0.9×0.4; 2) 7.2×0.7–0.4; 3) part of head broken. = Reinders, 1988: 33.52.

M222 House 1, Inv. No. X 613. Dim.: 1) 2.1×1.4×0.4; 2) 13.9×0.9–0.6; 3) clinched, 8.6. = Reinders, 1988: 33.53.

M223 House 1, Inv. No. X 655 e+f-I. Dim.: 1) 1.4×1.6×0.4; 2) 15.9×0.7–0.4; 3) clinched, 8.9; part of head broken off. Reinders 1988, not listed.

M224 House 1, Inv. No. X 655 e+f-II. Dim.: 1) 1.0×2.0×0.5; 2) 12.6×0.9–0.4; 3) clinched, 9.2; part of head broken off. Reinders 1988, not listed.

M225 House 1, Inv. No. X 655 e+f-V. Dim.: 1) -; 2) 8.6×0.6–0.5; 3) clinched, 7.0; most of head broken off. Reinders, 1988: not listed.

M226 House 1, Inv. No. X 655 e+f-VII. Dim.: 1) 1.9×1.0×0.4; 2) 5.8×1.0–0.7; 3) part of head broken off. Reinders, 1988: not listed.

M227 House 2, Inv. No. H21 2/53. Dim.: 1) 1.3×1.7×? (measurement impossible due to corrosion); 2) 6.6×0.6; 3) part of head broken off.

M228 House 2, Inv. No. H21 3/14sep. Dim.: 1) 1.4×1.0×0.3; 2) 2.3×0.4.

M229 House 6, Inv. No. H23 2/m21A. Dim.: 1) 1.5×1.0×0.2; 2) 9.9×0.7–0.4; 3) Clinched, 7.7; part of head broken off.

M230 House 6, Inv. No. H23 2/m26. Dim.: 1) 1.3×0.9×0.3; 2) 2.8×0.6; 3) part of head missing.

M231 House 6, Inv. No. H23 3/m27. Dim.: 1) 2.3×1.7×0.3; 2) 1.5×0.6; 3) part of head missing.

M232 House 6, Inv. No. H23 5/m40. Dim.: 1) 1.8×1.7×0.3; 2) 4.9×0.4–0.3.

M233 House 6, Inv. No. H23 6/m1. Dim.: 1) 1.2×0.7×0.3; 2) 9.0×0.6.

M234 House 6, Inv. No. H23 6/m6. Dim.: 1) 1.7×1.5×0.3–0.7; 2) 5.5×0.6.

M235 House 6, Inv. No. HA74/m3. Dim.: 1) 1.4×1.2×0.4; 2) 7.4×0.7–0.4.

M236 House 5, Inv. No. HA39/m3. Dim.: 1) 1.1×1.1×0.3 (broken, irregular); 2) 6.3×0.6.

M237 House 3, Inv. No. HA 224/5. Dim.: 1) 2.1×1.7×0.5; 2) 2.6×0.6.

M238 House 3, Inv. No. HA 224/22. Dim.: 1) 1.4×1.3×0.3; 2) 6.7×0.7–0.4.

M239 House 3, Inv. No. HA 226/11. Dim.: 1) 1.7×0.7×0.3; 2) 4.0×0.5.

M240 House 3, Inv. No. HA 228/7. Dim.: 1) 2.4×2.2×0.4; 2) 7.7×0.7.

M241 House 3, Inv. No. HA 232/14. Dim.: 1) 1.3×1.0×0.4; 2) 4.8×0.5–0.3.

M242 House 3, Inv. No. HA 249/1. Dim.: 1) 1.2×0.9×0.3; 2) 4.5×0.5–0.3.

M243 House 3, Inv. No. HA 267/2. Dim.: 1) 1.6×1.5×0.4; 2) 6.8×0.8–0.7.

M244 House 3, Inv. No. HA 271/9. Dim: 1) 1.1×0.9×0.4 (broken); 2) 5.5×0.5–0.4.

Type 5, iron. Broad, flaring, flat, semicircular head; the straight shank, in section rectangular, tapering slightly, is the continuation, without transition, of the lower end of the head. Measurements: 1. Head: length×width×thickness; 2. Shank: total length×section.
Complete

M245 House 2, Inv. No. H21 3/24, Fig. 3.38. Dim.: 1) 1.7×0.9×0.7; 2) 3.5×0.5–0.3.

Type 6, iron. Round, flattened head, very thick shank, in section irregular or rectangular, tapering to an often blunt point. This type of nail is characterized by the thickness of the shank; the head is formed by hammering the end of the shank so that it is widened. Measurements: 1. Head: diameter×thickness; 2. Shank: total length×section; 3. (only if the shank is not straight): Bent/clinched, distance to 1st clinch (from outside to outside), distance to 2nd clinch (outside to outside).
Complete

M246 House 6, Inv. No. H23 2/m25. Dim.: 1) 1.7×0.6; 2) 4.3×1.0–0.7; 3) sharply bent below the head.

M247 House 6, Inv. No. H23 5/m35, Fig. 3.38. Dim.: 1) 2.0×0.7; 2) 9.9×1.1–0.4.

M248 House 5, Inv. No. HA 50/m1. Dim.: 1) 2.2×0.5; 2) 3.5×1.1.

M249 House 3, Inv. No. HA 250/1b. Dim.: 1) 1.6/1.1×0.4; 2) 6.9×1.5–0.4; 3) clinched, 3.0.
Incomplete

M250 House 4, Inv. No. HA 146/1. Dim.: 1) 2.2×0.4; 2) 3.2×1.0; 3) head bent to one side.

M251 House 4, Inv. No. HA 204/1. Dim.: 1) 2.2×0.5; 2) 5.8×1.0; 3) bent near the head.

Type 7, iron. Straight shank, in section lozenge-shaped, tapering to a point; irregular head, rolled to one side. The head of this type of nail is formed by flattening the upper end of the nail and then rolling it over roughly into a tube. Measurements: 1. Head: width×height×thickness; 2. Shank: total length×section.

Complete

M252 House 6, Inv. No. H23 3/m2, Fig. 3.8. Dim.: 1) 1.2×0.6×
0.5; 2) 4.6×0.3–0.5.

*Type 8, iron. Shank, in section rectangular, tapering to a point, other
end thickening to form a hardly differentiated, irregular circular
head. This type often does not have a pronounced head at all, and
in view of the wedge-like thickness of some of the shanks, their identification as nails must remain tentative. Measurements: 1. Head:
diameter; 2. Shank: total length×section; 3. (only if the shank is not
straight): Bent/clinched, distance to 1st clinch (from outside to outside), distance to 2nd clinch (outside to outside).*

Complete

M253 House 6, Inv. No. H23 3/m4. Dim.: 1) 1.1–1.2; 2) 5.3×0.9–
0.4.

M254 House 4, Inv. No. HA 153/5, Fig. 3.38. Dim.: 1) 2.0; 2)
3.8×1.3–0.4.

M255 House 3, Inv. No. HA 199/7.1. Dim.: 1) 1.1; 2) 7.0×1.1–
0.4.

M256 House 3, Inv. No. HA 199/7.2. Dim.: 1) 1.2; 2) 6.8×1.2–
0.4.

M257 House 3, Inv. No. HA 199/9. Dim.: 1) 1.7/1.2; 2) 6.7×1.2–
0.4.

Incomplete; unless stated otherwise, the head is more or less complete and the end of the shank is broken off.

M258 House 6, Inv. No. H23 3/m7. Dim.: 1) 0.7; 2) 3.3×0.6–0.4.

M259 House 6, Inv. No. H23 5/m53. Dim.: 1) 0.8/0.7; 2) 3.7×
0.7–0.3; 3) bent.

*Type 9, iron. Straight shank, in section rectangular, tapering to a
point; the irregular, triangular head was formed by hammering a
piece of iron across the shank just below the top during the forging
of the nail. Measurements: 1. Head: width×height×thickness; 2.
Shank: total length×section.*

Complete

M260 House 6, Inv. No. HA 4/m1, Fig. 3.38. Dim.: 1) 2.8×1.1–
0.3×0.7–0.5; 2) 8.2×0.8–0.3.

*Type 10, iron. Shank, in section rectangular, tapering to a point;
very large, convex head, the two long sides curving to a point at
each end, forming a pointed oval. Measurements: 1. Head: length×
max width×max. thickness; 2. Shank: total length×section; 3. (only
if the shank is not straight): Bent/clinched, distance to 1st clinch
(from outside to outside), distance to 2nd clinch (outside to outside).*

Complete

M261 House 5, Inv. No. HA 26/m1, Fig. 3.38. Dim.: 1) 8.9×
2.8×0.9; 2) 7.5×1.5–0.5; 3) clinched (?), 6.0.

*Type 11, iron. Straight shank, in section irregular, flaring outwards
at the end to a sharply defined, irregular conical head. Measurements: 1. Head: diameter×thickness; 2. Shank: total length×section.*

Complete

M262 House 6, Inv. No. HA 48/m1, Fig. 3.38. Dim.: 1) 1.2×0.7;
2) 4.4×0.6–0.5.

Incomplete

M263 House 3, Inv. No. HA 201/1. Dim.: 1) 1.8×0.9; 2) 5.6×0.6–
0.3.

M264 House 3, Inv. No. HA 247/40. Dim.: 1) 0.9×0.7; 2)
3.7×0.5.

*Type 12, iron. Straight, wedge-shaped shank, in section flat rec-
tangular, tapering to a point; flat irregular head to one
side. Measurements: 1. Head: Length×width×thickness; 2.
Shank: total length×width×thickness.*

Complete

M265 House 5, Inv. No. HA 59/m2, Fig. 3.38. Dim.: 1) 2.4×1.9×
0.3; 2) 7.7×2.4–0.6×1.3–0.6.

*Type 13, iron. T-shaped nail, straight shank, in section rectangu-
lar, tapering to a point, head convex crossbar. Head and shank are
made out of one piece. Measurements: 1. Head: Length×width×
thickness; 2) Shank: total length×width×thickness.*

Complete

M266 House 4, Inv. No. HA 113/16, Fig. 3.38. Dim.: 1) 2.9×0.6×
0.8–0.2; 2) 8.4×0.9–0.3.

*Type 14, iron. Straight, wedge-shaped shank, in section irregular;
roughly oval, 'gabled' head. Measurements: 1. Head: Length×
width×thickness; 2. Shank: total length×width×thickness.*

Complete

M267 House 4, Inv. No. HA 130/17, Fig. 3.38. Dim.: 1) 3.6×2.4×
0.9–0.2; 2) 3.1×1.4–0.5.

Cramps, clamps

*Type 1, iron. Small rectangular clamp made out of one piece of iron,
bent at right angles. The long sides are flat, narrow at the ends,
widening towards the middle. The shorter, upright shanks are in
section irregular or rectangular. Measurements: 1. Long side a:
length×width×thickness; 2. Long side b: length×width×thickness;
3. Shank a: length (outside to outside)×section; 4. Shank b: length
(outside to outside)×section.*

M268 House 1, Inv. No. X 77. Dim.: 1) 2.9×1.1×0.1–0.2; 2) missing; 3) trace only; 4) missing; = Reinders 1988 31.23
('fragment, iron').

M269 House 1, Inv. No. X 655 G. Dim.: 1) 3.3×0.4–1.0×0.2; 2)
1.6×0.4–0.6×0.2 (broken); 3) 2.4×0.2; 4) 1.9×0.2; = Reinders, 1988: 33.70.

M270 House 1, Inv. No. H23 5/m32. Dim.: 1) 2.8×0.3–0.4×0.3
(broken); 2) 2.0×0.4–1.2×0.2 (broken); 3) 2.2×0.3; 4) missing; identification uncertain.

M271 House 1, Inv. No. H23 5/m57. Dim.: 1) 5.0×0.4–1.5×0.4;
2) 2.3×0.4 and 1.6×0.4 (middle part broken off); 3) 3.1×
0.3; 4) 3.4×0.3.

*Type 2. Two flat, horizontal, parallel sheets of iron, connected by
one or two vertical rivets. Measurements: 1. Sheet a: shape; length×
width×thickness; 2. Sheet b: shape; length×width×thickness; 3.
Rivet a: length×section; diameter head (if visible); 4) Rivet b:
length×section; diameter head (if visible).*

Complete

M272 House 6, inv. No. H23 5/m54 + 5/m55, Fig. 3.38. Dim.:
1) rectangular, flat; 7.4×2.6×0.4–0.5; 2) rectangular, flat;
7.1×2.7×0.4–0.5; 3) 5.4×0.8; 2.0, 1.1; 4) 6.1×0.8; 2.2, 1.2.

Incomplete

M273 House 6, Inv. No. H23 3/m14. Dim.: 1) rectangular; 3.8×
2.8×0.3; 2) irregular; 1.6×1.3×0.3; 3) 2.2×0.8; 4) -. Broken on all sides.

*Type 3. π-shaped cramp consisting of two parallel shanks and a
crossbar, made out of one piece; in section rectangular.
Measurements: Shank A length×section, shank B length×
section, crossbar length×section.*

M274 House 5, Inv. No. HA 81/m1, Fig. 3.38. Dim.: 10.6×0.5–0.9 (curving outwards), 8.5×0.5–0.7 (end broken off), 5.5×0.6–0.9.

M275 House 5, Inv. No. HA81/m2, Fig. 3.38. Dim.: 9.1×0.4–0.7, 9.1×0.4–0.7, 4.1×0.6–0.8.

M276 House 3, Inv. No. HA 202/13. Dim.: 10.8×0.4–0.8, 10.1×0.4–0.8, 5.2×0.8.

Type 4. Flat band of iron, omega-shaped, i.e. central part bent to form a vertical semicircle, ends extending outwards horizontally; a hole pierces each end.

M277 House 3, Inv. No. HA 214/8, Fig. 3.38. Dim.: 15.2×2.9–3.4×0.4–0.5; 4.8 (height semicircle), 0.7 (diameter holes).

Bronze clamps

Type 1. Two parallel thin, flat, rectangular strips, connected through a hole at either end by two rivets, in section round, ends flaring. References: Robinson, 1941: p. 302, p. 305, nos. 1315, 1330 (rounded corners), 1331 (rectangular), pls. LXXXVIII–LXXXIX. Measurements: 1. Strips: length×width×thickness (strip 1); length×width×thickness (strip 2); 2. rivets: length×section (rivet 1); lengh×section (rivet 2).

M278 House 2, Inv. No. H21 4/8, Fig. 3.38. Dim.: 1) 4.2×1.8×0.2; 4.7×1.7×0.2; 2) 1,7×0.3–0.4; 1.5×0.3–0.4.

Lead clamps

Type I. Two narrow parallel strips of lead, flat, outward-facing side generally convex, inward-facing side flat, connected by two shanks, in section round, at either end. References: Robinson, 1941: pp. 332–333, nos. 1566–1582, pls. XCVII–XCIX. All lead clamps from Olynthos appear to belong to this type; Branigan, 1992: p. 366, M72, pl. 304,13 ('purpose unknown'); Waldbaum & Knox, 1983: pp.67–68, nos. 292–294 (293 with sherd). Measurements: 1. Strip 1: L×W×T; 2. Strip 2: L×W×T; 3) Pin 1: L (from outside to outside)×S; 4) Pin 2: L (from outside to outside)×S.

Complete or virtually complete

M279 House 2, Inv. No. H21 2/13, Fig. 3.38. Dim.: 1) 5.0×0.9–1.1×0.4–0.5; 2) 3.9×1.0–0.7×0.4; 3) 3.1×0.7–0.5; 4) 3.1×0.9–0.6.

M280 House 6, Inv. No. H23 4/m1, Fig. 3.38. Dim.: 1) 6.1×1.8×0.4; 2) 6.5×1.7×0.4; 3) 1.3×0.5; 4) 1.6×0.5.

M281 House 4, Inv. No. HA 127/68. Wall fragments of an amphora with clamp, approx. same size as HA149/7B.

M282 House 4, Inv. No. HA 149/5A. Dim.: 1) 8.0×1.1–1.8×0.5; 2) 6.7×0.8–1.6×0.5; 3) 2.5×0.6; 4) 2.4×0.6.

M283 House 4, Inv. No. HA 149/6A. Dim.: 1) 6.4×1.0–1.5×0.3–0.5; 2) 6.0×0.8–1.5×0.3–0.5; 3) 2.5×0.6; 4) 1.5×0.6 (broken).

M284 House 4, Inv. No. HA 149/7A, Fig. 3.38. Dim.: 1) 7.6×1.3–1.5×0.4–0.5; 2) 7.6×1.1–1.5×0.3–0.4; 3) 3.0×0.7; 4) 2.5×0.6.

M285 House 3, Inv. No. HA 202/30. Dim.: 1) 5.6×1.3×0.4; 2) 5.6×1.3×0.4; 3) 3.1×0.7; 4) 3.6×0.8.

M286 House 3, Inv. No. HA 214/1. Dim.: 1) 4.5×1.8×0.2; 2) 4.5×1.8×0.2; 3) 2.0×0.4; 4) 2.0×-. Attached to sherd, middle coarse red, 2.3×1.6×0.5.

M287 House 3, Inv. No. HA 253/1. Dim: 1) 5.0×0.9×0.3; 2) 4.0×0.8×0.2; 3) 1.2×0.4; 4) 1.2×0.4.

Incomplete

M288 House 1, Inv. No. X 134. Dim.: 1) 5.3×0.9–1.1×0.3–0.4; 2) 2.5×1.0×0.3–0.5 (broken); 3) 2.8×0.6–0.8; 4) -; Reinders, 1988; not listed.

M289 House 1, Inv. No. X 635A. Dim.: 1) 3.1×0.6×0.2 (broken); 2) 3.3×0.7×0.2 (broken); 3) 1.1×unknown (in sherd); 4) 0.6×0.2 (broken). Attached to base fragment; = Reinders, 1988: 33.30.

M290 House 1, Inv. No. X 716. Dim.: 1) 2.4×2.1×0.3 (broken); 2) -; 3) 3.3×0.9–1.0 (broken); 4) -; = Reinders, 1988: 34.28.

M291 House 2, Inv. No. H21 2/34, Fig. 3.38. Dim.: 1) 3.6×1.3–1.4×0.3–0.4 (broken); 2) -; 3) 4.0×0.8–0.6 (broken); 4) -.

M292 House 2, Inv. No. H21 2/59. Dim.: 1) 3.6×1.3×0.3 (broken); 2) 3.0×1.0×0.2 (broken); 3) 1.6×? (in sherd); 4) -. In sherd; connecting sherd with other half of clamp missing.

M293 House 2, Inv. No. H21 2/60. Dim.: 1) 1.7×1.2×0.4 (broken); 2) 1.0×1.3×0.4 (broken); 3) 2.4×0.6; 4) -.

M294 House 2, Inv. No. H21 3/32. Dim.: 1) 1.8×1.3×0.2–0.4 (broken); 2) 1.9×1.2×0.3 (broken); 3) 4.3×0.6–0.5; 4) -.

M295 House 5, Inv. No. HA 27/m1. Dim.: 1) 1.9×1.4×0.3 (broken); 2) -; 3) 2.6×0.7–0.5 (broken); 4) -.

M296 House 5, Inv. No. HA 36/m13. Dim.: 1) 2.4×0.9–1.1×0.3 (broken); 2) 2.5×1.0–1.1×0.3 (broken); 3) 1.0×- (complete, but in sherd); 4) -. In sherd, medium coarse orange, 4.2×2.9×0.5.

M297 House 4, Inv. No. HA 126/14. Dim.: 1) 7.4×1.2–1.3×0.3 (broken); 2) -; 3) 1.1×0.7 (broken); 4) -.Partially molten; molten accretion on one end: 3.0×2.3×1.1.

M298 House 4, Inv. No. HA 126/34. Dim.: 1) 4.3×1.2–1.4×0.3 (broken); 2) 2.8×1.0–1.4×0.3 (broken); 3) 2.8×0.6; 4) -.

M299 House 4, Inv. No. HA 149/4A. Dim.: 1) 5.5×0.8–1.4×0.3 (broken); 2) -; 3) 3.4×0.9; 4) -.

M300 House 4, Inv. No. HA 149/4B. Dim.: 1) 4.6×0.8–2.1×0.5 (broken); 2) -; 3) 0.8×0.9 (broken); 4) -. Various molten accretions attached to this fragment.

M301 House 4, Inv. No. HA 149/6B. Dim: 1) 6.5×1.1–1.5×0.4–0.6 (broken); 2) 4.4×1.1–1.5×0.5–0.6 (broken); 3) 3.3×0.7; 4) -.

M302 House 4, Inv. No. HA 149/7B. Dim: 1) 7.6×1.2–1.5×0.4–0.6; 2) 2.4×1.4×0.4 (broken); 3) 3.0×0.8–1.0; 4) 1.9×1.0 (broken).

M303 House 4, Inv. No. HA 159/1. Dim.: 1) 3.6×1.1–1.2×0.3 (broken); 2) 2.8×1.4–1.5×0.4–0.5 (broken); 3) 2.9×0.7–0.8; 4) -.

M304 House 3, Inv. No. HA 240/14. Dim.: 1) 6.4×0.9–1.3×0.5 (broken); 2) -; 3) 1.0×0.5 (broken); 4) -.

Type 2. Type 2 differs from type 1 in that it does not fasten two sherds of the same pot together, but fixes a sherd – not necessarily of the same pot as the one which is being repaired – in place into a hole in the pot. This calls for a more elaborate construction, which can be described as follows: Sherd, usually more or less circular, encased in a roughly formed lead ring, concave on both sides, and fitted into the hole of the repaired pot. The sherd is held into place by lead strips - two per side, crossing at right angles – which pass beyond the break to a neighbouring sherd to which they are fastened by means of lead rivets which pass through a hole in the sherd from one strip to the other. Measurements: 1. Sherd: length×width / diameter×thickness: 2. Side A: strip 1: length×width×thickness; strip 2: length×width×thickness. Side B: strip 1: length×width×thickness; strip 2: length×width×thickness; 3. Pin 1: length×section; Pin 2: length×section; Pin 3: length×section; Pin 4: length×section.

M305 House 6, Inv. No. H23 1/m2, Fig. 3.38. Dim.: 1) 6.4×5.5×1.5; 2) side A: 13.0×1.3–3.3×0.7–0.3; 11.0×1.3–1.6×0.6–0.3. Side B: 11.4×1.4–1.6×0.6; missing; 3) 2.8×0.9; 3.2×0.8; 1.3×0.8 (broken); missing.

Fragments

Measurements: 1. Form/type of fragment; 2. Measurements conform appropriate segment.

M306   House 6, Inv. No. HA 24/m3. Dim.: 1) sherd, encased in lead ring; 2) 3.9×1.8×0.4 (sherd); 6.4×4.0×0.6 (incl. lead).

M307   House 6, Inv. No. HA 155/1. Dim.: 1) part of the lead ring encasing the inserted sherd; 2) 5.2×0.5–1.8×1.3–1.9.

Bolts or rivets

Type 1, iron. Straight shank, in section round, flattened and flaring at both ends. Measurements: Length×section shank×diameter end a, end b.

M308   House 6, Inv. No. H23 6/m8. Dim.: 2.3×0.4×0.6, 0.6.

Type 2. Straight shank, not tapering, in section irregular; on one end a flat head, irregular rectangular in shape; on the other end a rove fits onto the shank, which abruptly narrows and forms a point which pierces the rove, beyond which it is clinched. The rivet of this type is very corroded, so that the description must remain tentative. Measurements: Length×section shank; length×width×thickness head, length×width×thickness rove.

M309   House 5, Inv. No. HA 66/m2. Dim.: 6.9×1.9; 2.1×1.5×0.3–0.6, 3.0×2.6×0.3–0.7. The fragments of metal HA66/m2a were found in context with this rivet.

Type 3. As type 1, but with a rectangular or round head on one end. References: Raubitschek, 1998: p. 105, nos. 345A, B. Measurements: Head: length×width×thickness; shank: length×section, diameter end.

M310   House 4, Inv. No. HA 127/37. Dim.: 1.3×1.1×0.2; 3.1×0.5, 0.6.

M311   House 3, Inv. No. HA 216/22. Dim: 2.1 (diameter)×0.4; 5.2×0.8; 2.4×2.1×0.4 (piece of iron into which other end is fitted).

M312   House 3, Inv. No. HA 216/34. Dim.: 2.8 (diameter)×0.3; 5.0×0.7; 2.4×1.9×0.2 (piece of iron into which other end is fitted).

Type 1, bronze. Straight shank, in section round, flattened and flaring at both ends. References: Robinson, 1941: pp. 309–310, nos. 1354–1359, pl. LXXXIX. Measurements: Length×section shank×diameter end a, end b.

M313   House 6, Inv. No. H23 5/m42, Fig. 3.38. Dim.: 3.0×0.6×0.9, 0.8.

Keyhole reinforcements

Rings fastened by two, three, or four prongs onto a (wooden) surface are regularly identified as keyhole reinforcements. References: Robinson, 1941: pp. 253–260, pl. LXIX (rings with two, three, or four prongs, all of bronze).

Ring, in section flat with a prong at either end on one side, tapering to a point. Measurements: 1. Diameter (outer)×width×thickness ring; 2. Prong 1: length×section; prong 2: length×section.

M314   House 6, Inv. No. H23 5/m6, Fig. 3.38. Dim.: 1) 2.8×0.8×0.5; 2) 1.4×0.5; 1.6×0.4.

M315   House 5, Inv. No. HA 73/m1. Dim.: 1) 2.8–3.0×0.6–0.8×0.6; 2) 1.3×0.5; 1.3×0.3.

M316   House 3, Inv. No. HA 204/4. Dim.: 1) 3.5–3.2×0.8–1.0× 0.5; 2) 2,0×0.7/0.4–0.2; broken off.

Latch

M317   House 3, Inv. No. HA 202/14, Fig. 3.38. Dim.: Flat rectangular plate of iron with two rectangular openings, one horizontal, one vertical, one narrower than the other. The door handle passes through the narrow opening; it consists of a long shank, clinched, and a crossbar handle, both rectangular in section. The opening is such that the shank can move a significant distance from one end of the opening to the other. The second, wider opening is presumably the keyhole. Very slight traces remain of the very corroded nails in the four corners where the plate was fastened to the door; a few minute fragments of the shanks of these nails were also found (not described seperately). A small rectangular sheet of bronze (M 598) was found in context with this latch. References: Marić, 1995: p. 47, pl. 1, 2. Measurements: 9.7×8.4×0.3; keyhole 2.5×1.2; door handle hole 3.1×0.5; shank of door handle 11.0 (clinched 6.3)×0.4; handle 7.4×0.8.

Key

M318   House 5, Inv. No. HA 68/m1, Fig. 3.38. Straight shank, in section rectangular, broken at one end, ending in a ring, in section irregular, at the other. Dim.: 9.0×0.7 (shank), 2.6 (diameter ring)×0.5. Perhaps the end of a key? Reference: Robinson, 1941: pp. 507–509, nos. 2577–2583, pl. CLXV.

Pivot hole lining

M319   House 6, Inv. No. H23 5/m33. Hollow, convex, conical object, flat base with concave centre; hole in the middle. Dim.: 4.0 (diameter)×2.0 (height) 1.0 (width base).

OBJECT PARTS

Lids

M320   House 6, Inv. No. H23 5/m25, Fig. 3.39. Flat, round, lead lid with downturned rim, round knob in centre, flaring towards the top. Bent somewhat out of shape. Dim.: 9.5 (diameter)×1.5–1.0 (rim)×0.4–0.7; 1.7 (diameter)×2.3 (height) knob.

M321   House 6, Inv. No. H23 5/m48c. Presumably similar to the previous one, though much larger; only part of the knob (diameter 3.9) remains, the rest is melted out of shape or missing. Dim.: 14.0×7.5×0.2–1.8 (molten remains plus knob).

M322   House 3, Inv. No. HA 251/3, Fig. 3.39. Small, rectangular bronze lid, with a ring-handle near one short side. Three sides have a downturned rim, the fourth (back) does not. The rims of the long sides widen slightly near the back, where they are pierced by small holes for the hinge nails. Dim.: 3.1×2.5×0.1; rim: 0.3–0.4, holes (diameter) 0.2; ring-handle 0.7 (diameter)×0.1–0.2.

Rings

Iron rings. Circular ring, fully closed, in section rectangular or irregular. References: Henig, 1994: pp. 270, no. 7 fig 14,9; Sahm, 1994: 117 M10, pl. 42 (with reference); Robinson, 1941: p. 523, nos. 2632–2636, pl. CLXIX.

Measurements: Diameter (external)×section.
Complete
M323   House 1, Inv. No. X 305. Dim.: 3.5×0.4–0.6. = Reinders, 1988: 37.30.
M324   House 2, Inv. No. H21 1/22, Fig. 3.38. Dim.: 8.4×0.8–1.4.
M325   House 5, Inv. No. HA 60/m1. Dim.: 3.3×0.8–0.3.
M326   House 5, Inv. No. HA 80/m5. Dim.: 5.3×0.8.

Bronze rings
Type 1. Circular ring, fully closed, in section round. Most bronze rings are quite small. References: Sahm, 1994: p. 118, M29, pl. 42 (with references). Measurements: Diameter (external)×section.
Complete
M327   House 1, Inv. No. X 649. Dim.: 1.3×0.2. = Reinders, 1988: 33.71.
M328   House 1, Inv. No. X 735. Dim.: 2.0–2.2×0.3; irregular circle (damaged). = Reinders, 1988: 34.29.
M329   House 6, Inv. No. H23 5/m17, Fig. 3.39. Dim.: 0.9×0.2.
M330   House 6, Inv. No. H23 5/m18, Fig. 3.39. Dim.: 1.0×0.2.
M331   House 5, Inv. No. HA 18/m1, Fig. 3.39. Dim.: 1.4×0.2.
M332   House 5, Inv. No. HA 73/m2. Dim.: 1.7–1.8×0.3.
M333   House 4, Inv. No. HA 127/64. Dim.: 1.7×0.3.
M334   House 4, Inv. No. HA 129/7. Dim.: 1.7×0.3.
M335   House 4, Inv. No. HA 155/5. Dim.: 1.4×0.2–0.3.

Type 2. Bronze wire, bent around to form a ring, ends overlapping. Raubitschek (1998: p. 68) tentatively classifies similar bronze rings as earrings when the ends do not (completely) overlap. I prefer her alternative suggestion that these rings had a more practical purpose. References: Raubitschek, 1998: p. 66, p. 69, nos. 254–258, pl. 40. Measurements: Diameter (external)×section.
Complete
M336   House 6, Inv. No. H23 5/m16, Fig. 3.39. Dim.: 1.6×0.2.
M337   House 3, Inv. No. HA 238/7. Dim.: 2.7–2.8×0.3–0.4.

Chain

M338   House 3, Inv. No. HA 208/2. Dim.: Link of a chain (or hook?), in section rectangular, bent into an 8-form, one side open (small opening) 6.0×1.0, 0.2–0.4 (section).

Rods

Type 1. Long straight shank, in section irregular, small round head. Measurements: 1. Head: diameter×thickness; 2. Shank: length×section.
M339   House 6, Inv. No. H23 2/m2, Fig. 3.39. Dim.: 1) 1.2×0.4; 2) 22.9×0.7.
M340   House 5, Inv. No. HA 86/m1, Fig. 3.39. Dim.: 1) 2.7×0.5; 2) 22.0×0.8; bent in a gentle curve.

Hooks

Type 1. Shank, in section rectangular or irregular, bent twice at sharp angles to form a hook. This hook is roughly formed, and differs little from the (clinched) shanks of nails. Note that the distance from the first angle to the second is very small however, and that a sharp corner at the first bend shows that the hook was forged with that bend, not bent later. For this reason it is classified as a hook, rather than as a nail shank. Measurements: 1. Total length×section; 2. Distance to first angle; distance to second angle.
M341   House 6, Inv. No. H23 3/m1. Dim.: 1) 7.2×0.5–0.6; 2) 4.2; 1.9.

Type 2. Shank, in section irregular/rectangular, bent in the form of an 8, one circle closed, the other open. The form is similar to that of the link of a chain (cf. M338 above), but the hole is rather small, relative to the thickness of the shank. This, combined with the lack of any subsequent links, suggests that it is a hook meant to be attached to a rope or a leather thong. Measurements: Length×width, section, diameter hole.
M342   House 3, Inv. No. HA 214/9. Dim.: 5.2×2.6, 0.6, 0.9–1.3.

Type 3. Shank, in section rectangular or irregular, bent round to form a hook. Measurements: 1. Total length×section; length to bend
M343   House 6, Inv. No. H23 5/m31. Dim.: 1) 5.2×0.5; 3.6.
M344   House 6, Inv. No. HA 2/m1. Dim.: 1) 10.9×1.0×0.5; 7.4, 8.3.

Type 4. Large hook shaped as a lazy S, in section flat rectangular, end bent back to form an eye. This hook would be suitable for a wide range of uses. It is not dissimilar from butcher's meat-hooks. Measurements: 1. Total length×section; length×width eye.
M345   House 6, Inv. No. H23 3/m13, Fig. 3.39. Dim.: 1) 31.2× 0.5/0.3–0.9/0.7; 1.3×0.5.

Handles

M346   House 6, Inv. No. H23 5/m15, Fig. 3.39. Swing handle, iron, omega-shaped, in section square, ends bent back outwards to form two rings for the attachment. Dim.: 10.7 (width from end to end)×9.5 (diameter)×0.8. References: Robinson, 1941: pp. 207–221, pp. 668–769, pls. LVI–LVII (all bronze rather than iron, smaller, ends usually open, ends sometimes decorative knobs).
M347   House 6, Inv. No. HA 1/m1, Fig. 3.39. Fragment of a bronze swing handle, in section round, curving, ending in a tapering knob with fillet, broken at the other end just beyond a sharp bend. Dim.: 3.5×0.5. References: Robinson, 1941: pp. 217–220, nos. 737, 740, 743, 746, 750, 758, pl. LVIII; Marić, 1995: p. 47, fig. 16.
M348   House 6, Inv. No. HA 55/m4, fig. 3.39. Bronze handle, horizontal, rectangular in shape with upturned end, in section round. Dim.: 4.3×5.7×0.5–0.8. References: Stillwell, 1948: p. 115, no. 3, pl 47, with references; Carapanos, 1878: pl. XLVI, no. 11; Robinson 1941: pp. 203–204, nos. 644–647, pl. LIII.
M349   House 3, Inv. No. HA 247/33, Fig. 3.39. Straight, narrow iron band handle (part missing), curving sharply inwards near either end and then sharply outwards again, taking the form of a flat, leaf-shaped ornament. Each ornamental end was attached to the surface of the object by means of two very small bronze nails, clinched. The size of the nails shows that the handle was unsuitable for heavy objects; one could imagine it as the handle of a flat lid of a wooden pyxis. Dim.: Handle 3.5×1.0×0.4 (part missing); distance from ornament to curve unknown. Ornament 1: 2.9×2.8× 0.4; nail a: 1.2×0.1–0.025, clinched 0.5 (measured from inside); nail b: 0.4×0.1 (broken). Ornament 2: 3.0×3.0×0.3; nail c: 1.1×0.1–0.07, clinched, 0.5 (measured from the inside); nail d: 1.1×0.1–0.04, clinched, 0.5. Parallels: Robinson, 1941, 205, pl. liv, nos. 655–656 (bronze), with ref.

METAL FRAGMENTS

Iron fittings, plating, reinforcements, structural elements

Iron plaques of uniform thickness, flat or curving, generally rectangular, shapes ranging from almost square to long narrow shanks; often with vertical shanks, holes for nails, etc. The potential range of functions is vast. Description and measurements: 1. Shape, flat/bent; 2. Length×width×thickness; 3. Shanks (1–x) length×section. All objects listed here are incomplete and broken, unless stated otherwise.

M350 House 1, Inv. No. X 40. Irregular, slightly curved. Dim.: 3.0×2.4×0.3; hole in the middle – two pieces stuck together. = Reinders, 1988: 39.13.

M351 House 1, Inv. No. X 59. Irregular, flat. Dim.: 3.6×3.0×0.3; two pieces stuck together. Reinders, 1988; not listed.

M352 House 1, Inv. No. X 84. Irregular, flat. Dim.: 10.6×6.2×1.3. = Reinders, 1988: 35.24.

M353 House 1, Inv. No. X 135. Irregular, flat. Dim.: 1.7×1.3×0.5; shank through hole 1.7×0.5. = Reinders, 1988: 35.25.

M354 House 1, Inv. No. X 304. Rectangular, flat. Dim.: 3.6×3.2×0.4; hole at one end. = Reinders, 1988: 37.29.

M355 House 1, Inv. No. X 625 B+D III. Rectangular strip, flat. Dim.: 3.3×0.6×0.2–0.4. = Reinders, 1988: 33,69?

M356 House 1, Inv. No. X 668. Two roughly rectangular fragments. Dim.: 1) 4.7×0.6×0.6; 2) 2.0×0.6×0.3. = Reinders, 1988: 31.21.

M357 House 1, Inv. No. X 726. Small, roughly horseshoe shaped, flat. Dim.: 3.5×0.8–1.3×0.5. = Reinders, 1988: 34.25.

M358 House 2, Inv. No. H21 4/31c. Irregular, flat, 3.2×2.1×0.5; straight shank. Dim.: 2.0×0.8. Very corroded, fragmentary.

M359 House 2, Inv. No. H21 1/12. Irregular, flat, shattered into 10 fragments. Dim. largest fragm.: 5.0×2.2×0.4. Possibly some sort of implement or tool with handle attachment, e.g. a shovel.

M360 House 2, Inv. No. H21 2/47. Semicircular band, flat. Dim.: 5.5×1.0–1.1×0.4–0.6.

M361 House 2, Inv. No. H21 3/27. Triangular, flat. Dim.: 9.0×6.9×6.6 (sides); 0.4 (thickness).

M362 House 2, Inv. No. H21 4/31A. Rectangular, slightly curved. Dim.: 3.9×2.9×0.5.

M363 House 6, Inv. No. H23 1/m3, Fig. 3.40. Long shank, in section rectangular, bent, widening and flattening at one end to an irregular oval with a hole in the centre. Dim.: 12.3×0.5–0.8 (section shank); 2.3×0.3 (end).

M364 House 6, Inv. No. H23 1/m9. Rectangular, flat. Dim.: 6.8×1.8×0.5–0.6.

M365 House 6, Inv. No. H23 1/m10. Rectangular, flat. Dim.: 3.1×2.0×0.3.

M366 House 6, Inv. No. H23 2/m27. Straight shank, in section irregular, flaring and flattening towards one end. Dim.: 6.0×0.6 (section shank)–1.1×0.3 (width and thickness flat end).

M367 House 6, Inv. No. H23 3/m9. Wedge-like, in section rectangular. Dim.: 5.2×1.4–0.6 (width)×0.6 (thickness).

M368 House 6, Inv. No. H23 3 Dim.: 8.9×0.6×0.3–0.4; head: 1.1×0.5.

M369 House 6, Inv. No. H23 3/m12. Rectangular, flat, curving slightly lengthwise; fragment of a second flat, rectangular piece attached to one corner with a rivet (? very corroded). Both pieces broken at both ends. Dim.: 8.6×4.4×0.3; 3.2×3.4×0.3.

M370 House 6, Inv. No. H23 3/m15. Rectangular, flat, corner missing; perforated. Dim.: 7.8×5.8×1.4–0.3; hole 1.2×0.6.

M371 House 6, Inv. No. H23 4/m8. Rectangular, flat, rivet through one corner. Dim.: 5.0×2.8×0.2–0.3; 1.5×0.4; one corner missing, broken at one side?

M372 House 6, Inv. No. H23 4/m9. Rectangular, flat, rivet through one corner. Dim.: 3.7×2.8×0.1–0.3; 1.2×0.4; broken at one side. Does not fit to H23 4/m8.

M373 House 6, Inv. No. H23 5/m7. Irregularly rectangular, flat; three straight sides, one rounded; small hole (damage?) near the centre, large round hole (part broken away) near rounded side; two rivets along one long side. Dim.: 9.2×6.8×0.5; rivet 1 0.8×1.2–0.5; rivet 2 0.8×0.4.

M374 House 6, Inv. No. H23 5/m8–5/m11. Four fragments together forming about half of a circular band, in section flat. Such ring-shaped metal bands could serve as hoops around barrels. Dim. pails, etc.: 17.0 (diameter)×1.5 (width)×0.3–0.5 (thickness).

M375 House 6, Inv. No. H23 5/m12. Rectangular, curved lengthwise to a semicircle, forming half a cylinder. Dim.: 3.5×2.7 (diameter)×0.2.

M376 House 6, Inv. No. H23 5/m20. Shank, in section square, bent horizontally at a right angle at one end, vertical rivet at the other end. Very heavily corroded, damaged. Dim.: 15.0×0.8; rivet 3.8×1.0.

M377 House 6, Inv. No. H23 5/m34. Shank, in section flat rectangular, slightly bent. Dim.: 9.7×1.1×0.6.

M378 House 6, Inv. No. H23 5/m50. Straight shank, in section flat rectangular, flaring and thinning at one end, bent slightly. Dim.: 5.5×1.1–2.5×0.7–0.4.

M379 House 6, Inv. No. H23 5/m51. Flat, irregular fragment, tapering, narrow end bent over at a right angle. Dim: 4.1 (total length)×0.8–0.5×0.2–0.3, bent at 2.6.

M380 House 6, Inv. No. H23 5/m56. Rectangular, flat, two short sides bent downwards, shank, broken, in the middle. Dim.: 3.8×1.9×0.3; 1.5×0.6.

M381 House 6, Inv. No. H23 6/m19. Irregular, flat. Dim.: 5.9×2.2×0.3–0.7.

M382 House 6, Inv. No. HA 1/m2. Irregular, flat, curving shank from one side. Dim.: 2.0×1.3×0.2; 2.6×0.2.

M383 House 6, Inv. No. HA 19/m1. Rectangular, flat, long sides bent upwards and inwards. Dim.: 7.5×1.9×0.2–0.4.

M384 House 6, Inv. No. HA 68/m2. Irregular, flat/irregular. Dim.: 4.2×2.9×1.7.

M385 House 5, Inv. No. HA 13/m1. Rectangular, flat, one side curving slightly lengthwise. Dim.: 5.5×3.0×0.3. Broken.

M386 House 5, Inv. No. HA 13/m3. Rectangular, flat, both sides curving slightly lengthwise. Dim: 5.1×2.9×0.3. Broken.

M387 House 5, Inv. No. HA 14/m1. Rectangular, flat, long sides curved around to form a tube. Dim.: 3.8×2.2 (diameter)×0.4–0.6.

M388 House 5, Inv. No. HA 37/m2. Rectangular, flat, curved round lengthwise to slightly more than a semicircle. A nail, small round head, shank in section irregular, clinched, passes through a hole from one short end to the other. Dim.: 6.3–6.6 (diameter)×4.2–5.0×0.2–0.3; 1.6 (diameter)×0.2 (head); 9.7 (total length shank)×0.4–0.3.

M389 House 5, Inv. No. HA 46/m1. Straight shank, in section irregular, flattening and broadening on one end. Dim.: 7.4×0.5–0.3×0.3–1.1.

M390 House 5, Inv. No. HA 46/m3. Rectangular, flat, long sides curved around to form a tube. Dim.: 4.2×1.7–2.1 (diameter)×0.2–0.4.

M391 House 5, Inv. No. HA 49/m2. Rectangular, flat. Dim.: 7.0×2.3×0.4–0.5.

M392 House 5, Inv. No. HA 50/m2. Rectangular, flat, curving slightly, in section irregular. Dim.: 7.3×0.9×0.5.

M393 House 5, Inv. No. HA 55/m3. Straight shank, in section irregular, flattening and broadening on one end. Dim.: 10.4×0.4–0.3×0.6–2.0.

M394 House 5, Inv. No. HA 55/m5. Rectangular, flat, curving slightly in horizontal plane. Dim.: 18.5×1.5–1.8×0.4. Cf. Robinson, 1941: pp. 647–648.

M395 House 5, Inv. No. HA 60/m4. Very narrow rectangular, flat. Dim.: 6.3×0.3–0.5×0.3. Heavily corroded, broken into two pieces.

M396 House 5, Inv. No. HA 66/m2a. Irregular, flat, 6 fragments. Dim.: 2.0×1.8×0.4 (largest) – 0.8×0.5×0.2 (smallest).

M397 House 5, Inv. No. HA 71/m3. Irregular, flat. Dim.: 1.8× 0.6×0.3.

M398 House 5, Inv. No. HA 80/m2. Crescent-shaped, flat. Dim.: 6.9×2.0×0.3. Probably fragment of a sickle.

M399 House 5, Inv. No. HA 80/m6. Irregular, flat, folded over once. Dim: 5.5×2.4×0.7.

M400 House 5, Inv. No. HA 80/m8. Irregular, flat. Dim.: 3.5× 1.0×0.5.

M401 House 5, Inv. No. HA 86/m5. Rectangular, flat. Dim.: 2.8×1.0×0.2.

M402 House 4, Inv. No. HA 95/1. Rectangular, tapering, flat. Dim.: 7.9×1.4–1.0×0.4–0.5.

M403 House 4, Inv. No. HA 104/13A+B. Straight shank, flattening and flaring at one end. Dim.: 5.1×0.4/0.7 (shank), 1.6/0.4 (flattened end).

M404 House 4, Inv. No. HA 106/31. Straight shank, flattening and flaring at one end. Dim: 7.0×0.5/0.6 (shank)×1.4/0.4 (flattened end).

M405 House 4, Inv. No. HA 46/m3. Rectangular, flat, long sides curved around to form a tube. Dim.: 4.5×1.5 (diameter) ×0.2–0.3.

M406 House 4, Inv. No. HA 115/18. Shank, in section rectangular, one end flattened and bent sharply around. Dim.: 7.5×0.5–0.5/0.2.

M407 House 4, Inv. No. HA 115/19. Thin shank, widening and flattening in the middle, then tapering again to a somewhat thicker shank. Dim.: 7.5×0.2, 0.3/0.8, 0.4/0.5.

M408 House 4, Inv. No. HA 116/1. Rectangular, flat, curving slightly at one short end, bent sharply in the opposite direction at the other end. Dim.: 8.0×2.8×0.2.

M409 House 4, Inv. No. HA 120/10. Irregular, flat. Dim.: 3.0× 1.5×0.3.

M410 House 4, Inv. No. HA 120/14. Rectangular, thick, flat, hole in middle; one short end curving, other end broken away at hole; fragment begins to bend at hole. Dim.: 4.8×4.6× 0.9–1.0.

M411 House 4, Inv. Nos. HA 126/19 + HA 126/32. Narrow flat strip, bent downwards halfway, at one end dividing into two (ends broken off) or widening for a hole in the middle. Two fragments, not certain that they fit. Dim.: 19.: 12.4×1.1–2.2×0.3–0.5; 32.: 10.9×1.0–1.2×0.4.

M412 House 4, Inv. No. HA 126/24. Rectangular, flat, one short side folded over to form an eye. Dim.: 6.3×3.1×1.0–0.4; 0.6×1.7 (diameter eye side a), 0.5×1.1 (diameter eye side b). Possibly fragment of a hinge?

M413 House 4, Inv. No. HA 127/41. Rectangular, flat, one end bent over sharply. Dim.: 3.4×0.6×0.2.

M414 House 4, Inv. No. HA 136/10A+B. Thin wide band of iron, bent halfway to form a rounded corner (90?); two pieces, fit together. Dim.: A. 4.1×3.1×0.2; B. 2.5×2.3×0.2.

M415 House 4, Inv. No. HA 146/7. Flat, two long sides curving to a point, short side broken. Dim.: 1.7×1.3×0.2; point of a knife?

M416 House 4, Inv. No. HA 152/4. Rectangular, flat, very small hole in the middle. Dim.: 1.8×1.6×0.4.

M417 House 4, Inv. No. HA 153/8C. Irregular, flat. Dim.: 3.1×0.7×0.2.

M418 House 4, Inv. No. HA 166/17. Rectangular, flat, curved lengthwise, shank. Dim.: 3.5×2.4×0.2–0.5; 1.9×0.6 (shank).

M419 House 4, Inv. No. HA 178/8. Rectangular, flat, long sides curved around to form a tube. Dim.: 4.7×1.6 (diameter)×0.3. Apparently intact but for one small corner.

M420 House 3, Inv. No. HA 198/16. Irregular, flat, bent sharply at one end. Dim.: 3.8×2.9×0.3.

M421 House 3, 208/4: Irregular rectangular, flat, rivet (broken) through a hole (broken) at one side. Dim.: 5.0×3.2×0.4; rivet: 2.0×0.8, head 1.6×1.2×0.3.

M422 House 3, Inv. No. HA 216/35. Rectangular, flat, shank. Dim.: 1.9×1.5×0.5; 5.2×0.7.

M423 House 3, Inv. No. HA 224/25. Irregular, one long straight side, flat, curving gently; found with fifteen small fragments belonging to the same object. Dim.: 19.0×3.5–7.5×0.2.

M424 House 3, Inv. No. HA 226/4. Rectangular, flat, narrow strip of iron, broken into many pieces; remnants of a small nail through one end. Dim.: 5.8×1.3×0.2. Found in connection with 10 other, very small fragments.

M425 House 3, Inv. No. HA 228/1. Irregular, flat. Dim.: 2.2×1.8×0.3.

M426 House 3, Inv. No. HA 231/1. Irregular, one straight side, flat. Dim.: 7.8×4.2×0.3.

M427 House 3, Inv. No. HA 235/2. Irregular, flat, strip flaring at one end. Dim.: 6.3×0.8–1.2×0.3. Could be the handle of some instrument (knife, small sickle?).

M428 House 3, Inv. No. HA 253/30. Shank, in section irregular, bent, flattening and flaring at one end. Dim.: 6.7×0.4 (shank), 0.9×0.2 (end).

M429 House 3, Inv. No. HA 253/46. Straight shank, in section irregular, bent, flattening and flaring at one end. Dim.: 4.3×0.4 (shank), 0.9×0.2 (end).

M430 House 3, Inv. No. HA 259/2. Rectangular, flat, long sides curved around to form a tube. Dim.: 4.5×1.6 (diameter) ×0.4. Apparently intact but for one small corner.

Iron shanks

Type 1. Shank in section irregular. Measurements: 1. Total length× section; 2. (only if shank is not straight) Bent/clinched, longest distance to 1st clinch (from outside to outside), distance to 2nd clinch (outside to outside). All shanks are broken at both ends, unless stated otherwise.

M431 House 1, Inv. No. X 27. Dim.: 1) 3.8×0.3–0.4; 2) clinched, 2.0. = Reinders, 1988: 38.06.

M432 House 1, Inv. No. X 145-4. Dim.: 1) 3.6×0.5; 2) slightly bent. Reinders, 1988: 35.26.

M433 House 1, Inv. No.X 501. Dim.: 1) 4.1×0.6–0.3. = Reinders, 1988: 31.20.

M434 House 1, Inv. No. X 625 B&D-I. Dim.: 1) 9.4×0.9–0.4; 2) clinched, 5.0, 3.8. = Reinders, 1988: 33.68.

M435 House 1, Inv. No. X 647 f. Dim.: 1) 3.5×0.6–03. Reinders, 1988; not listed.

M436 House 1, Inv. No. X 655 e+f-VI. Dim.: 1) 8.0×0.7–0.5; 2) clinched, 7.0. Reinders, 1988; not listed.

M437   House 2, Inv. No. H21 1/24. Dim.: 1) 2.9×1.7.

M438   House 2, Inv. No. H21 3/31. Dim.: 1) 3.4×0.7–0.4; 2) head broken off, point bent.

M439   House 2, Inv. No. H21 4/31B. Dim.: 1) 4.0×1.2–0.9.

M440   House 6, Inv. No. H23 3m/10. Dim.: 1) 7.9×0.5; 2) clinched, 0.5.

M441   House 6, Inv. No. H23 3/m22. Dim.: 1) 5.5×0.9–0.6.

M442   House 6, Inv. No. H23 3/m23. Dim.: 1) 5.0×1.0–0.6.

M443   House 6, Inv. No. H23 5/m38. Dim.: 1) 8.0×0.7.

M444   House 6, Inv. No. H23 5/m52. Dim.: 1) 4.9×0.4.

M445   House 6, Inv. No. H23 5/m58. Dim.: 1) 3.4×0.4–0.7; 2) bent.

M446   House 6, Inv. No. H23 6/m16. Dim.: 1) 20.6×0.3–0.5; 2) bent slightly.

M447   House 6, Inv. No. HA 28/m1. Dim.: 1) 4.8×0.7–0.3; 2) clinched, 3.3; head missing, point intact.

M448   House 6, Inv. No. HA 43/m2. Dim.: 1) 3.7×0.9–0.5; 2) bent slightly.

M449   House 5, Inv. No. HA 36/m1. Dim.: 1) 2.8×0.6.

M450   House 5, Inv. No. HA 44/m1. Dim.: 1) 3.1×0.8–0.6.

M451   House 5, Inv. No. HA 50/m3. Dim.: 1) 6.3×0.6.

M452   House 5, Inv. No. HA 54/m1. Dim.: 1) 5.5×1.2–0.6.

M453   House 5, Inv. No. HA 60/m3. Dim.: 1) 5.1×0.6–0.4.

M454   House 5, Inv. No. HA 65/m1. Dim.: 1) 6.8×1.9–0.5; 2) clinched, 4.9; head broken off, point intact.

M455   House 5, Inv. No. HA 75/m1. Dim.: 1) 8.3×0.7–0.3.

M456   House 5, Inv. No. HA 86/m2. Dim.: 1) 3.2×0.6.

M457   House 5, Inv. No. HA 86/m3. Dim.: 1) 2.4×0.6.

M458   House 5, Inv. No. HA 88/m1. Dim.: 1) 3.9×0.8.

M459   House 4, Inv. No. HA 96/3. Dim.: 1) 2.8×0.5–0.2.

M460   House 4, Inv. No. HA 104/6. Dim.: 1) 5.0×0.7–0.5.

M461   House 4, Inv. No. HA 104/11A. Dim.: 1) 2.2×0.4.

M462   House 4, Inv. No. HA 104/11B. Dim.: 1) 1.8×0.4.

M463   House 4, Inv. No. HA 104/14. Dim.: 1) 11.7×1.2–0.9; 2) clinched, 5.3; head broken off, point intact.

M464   House 4, Inv. No. HA 104/16. Dim.: 1) 4.0×0.4.

M465   House 4, Inv. No. HA 106/20. Dim.: 1) 1.7×1.0–0.5.

M466   House 4, Inv. No. HA 106/32. Dim.: 1) 5.8×1.4–1.0; 2) clinched, 3.8.

M467   House 4, Inv. No. HA 120/15A. Dim.: 1) 3.4×0.9.

M468   House 4, Inv. No. HA 120/15B. Dim.: 1) 2.4×0.7–0.4.

M469   House 4, Inv. No. HA 129/5. Dim.: 1) 6.4×0.9–0.5.

M470   House 4, Inv. No. HA 145/3. Dim.: 1) 2.6×0.4.

M471   House 4, Inv. No. HA 157/1. Dim.: 1) 5.2×0.3–0.5.

M472   House 4, Inv. No. HA 166/5. Dim.: 1) 6.3×0.7.

M473   House 4, Inv. No. HA 166/14. Dim.: 1) 4.2×0.4–0.7.

M474   House 3, Inv. No. HA 198/24. Dim.: 1) 2.8×0.4.

M475   House 3, Inv. No. HA 202/11. Dim.: 1) 2.6×0.4–0.2.

M476   House 3, Inv. No. HA 210/7. Dim.: 1) 3.1×0.4.

M477   House 3, Inv. No. HA 210/8. Dim.: 1) 3.0×0.6.

M478   House 3, Inv. No. HA 210/16. Dim.: 1) 3.7×0.7.

M479   House 3, Inv. No. HA 216/16. Dim.: 1) 2.6×0.4; 2) bent.

M480   House 3, Inv. No. HA 216/23.2. Dim.: 1) 2.6×0.9.

M481   House 3, Inv. No. HA 223/1. Dim.: 1) 5.6×0.7; 2) beginning of a clinch/bend at one end.

M482   House 3, Inv. No. HA 224/7. Dim.: 1) 2.7×0.8.

M483   House 3, Inv. No. HA 224/12. Dim.: 1) 3.0×0.6.

M484   House 3, Inv. No. HA 226/3. Dim.: 1) 3.7×0.6–0.2; 2) very corroded and brittle; found with eight small, irregular fragments of iron.

M485   House 3, Inv. No. HA 226/12. Dim.: 1) 4.0×0.7; 2) slightly bent.

M486   House 3, Inv. No. HA 232/11. Dim.: 1) 1.9×0.4–0.2; 2) point intact, head broken off.

M487   House 3, Inv. No. HA 232/13. Dim.: 1) 2.1×0.5–0.2; 2) point intact, head broken off.

M488   House 3, Inv. No. HA 232/16.2. Dim.: 1) 3.5×0.4.

M489   House 3, Inv. No. HA 232/16.3. Dim.: 1) 2.4×0.4; 2) point intact, head broken off.

M490   House 3, Inv. No. HA 232/27. Dim.: 1) 3.9×0.6; 2) clinched, 2.2.

M491   House 3, Inv. No. HA 236/5. Dim.: 1) 4.0×0.5.

M492   House 3, Inv. No. HA 240/18. Dim.: 1) 3.1×0.7.

M493   House 3, Inv. No. HA 247/6. Dim.: 1) 2.1×0.6.

M494   House 3, Inv. No. HA 247/8. Dim.: 1) 5.5×0.4.

M495   House 3, Inv. No. HA 247/44. Dim.: 1) 2.1×0.3–0.2.

M496   House 3, Inv. No. HA 247/58. Dim.: 1) 3.8×0.5–0.2; 2) point intact, head broken off; clinched, broken at clinch.

M497   House 3, Inv. No. HA 247/61. Dim.: 1) 6.6×0.8–0.5; 2) clinched, 3.3; very heavily corroded, but virtually complete. The 'head' is an indeterminate lump of corrosion.

M498   House 3, Inv. No. HA 248/8b. Dim.: 1) 1.1×0.4; 2) point intact, head broken off.

M499   House 3, Inv. No. HA 267/13. Dim.: 1) 3.3×0.6–0.3; 2) clinched, 1.8.

M500   House 3, Inv. No. HA 261/3. Dim.: 1) 5.0×0.7–0.3.

M501   House 3, Inv. No. HA 262/8. Dim.: 1) 2.4×0.5.

Type 2. Shank in section rectangular. Measurements: 1. Total length×section; 2. (only if shank is not straight): Bent / clinched, longest distance to 1st clinch (from outside to outside), distance to 2nd clinch (outside to outside).

M502   House 1, Inv. No. X 146–2. Dim.: 1) 6.7×0.8–0.3; 2) broken near head, point intact. Reinders, 1988: 35.27.

M503   House 1, Inv. No. X 625 B&D–II. Dim.: 1) 4.8×1.1–0.4; Reinders, 1988: 33.68.

M504   House 1, Inv. No. X 655 e+f-VIII. Dim.: 1) 4.7×0.6–0.5. Reinders, 1988; not listed.

M505   House 1, Inv. No. X 66. Lost in earthquake of 1980; Reinders, 1988: 39.14.

M506   House 6, Inv. No. H23 1/m6. Dim.: 1) 6.3×0.6–0.4; 2) clinched, 3.0, broken below the head, point intact.

M507   House 6, Inv. No. H23 3/m5, Fig. 3.40. Dim.: 1) 17.7×0.9–0.4; 2) clinched 3-dimensionally (one angle in vertical, one in horizontal plane); 5.2; 8.1; both ends broken.

M508   House 6, Inv. No. H23 3/m6. Dim.: 1) 9.3×0.5–0.3; 2) clinched 5.4.

M509   House 6, Inv. No. H23 3/m17. Dim.: 1) 8.4×0.5–0.8; 2) clinched 6.9.

M510   House 6, Inv. No. H23 3/m21. Dim.: 1) 5.7×0.8–0.4.

M511   House 6, Inv. No. H23 4/m6. Dim.: 1) 9.4×0.5–0.4 ; 2) clinched, 7.1.

M512   House 6, Inv. No. H23 4/m7. Dim.: 1) 4.5×0.9–0.4.

M513   House 6, Inv. No. H23 5/m45. Dim.: 1) 10.5×0.6–0.5.

M514   House 6, Inv. No. H23 6/m11. Dim.: 1) 5.6×1.1–0.8.

M515   House 6, Inv. No. H23 6/m12. Dim.: 1) 6.5×0.8–0.4 ; 2) bent.

M516   House 6, Inv. No. HA 12/m1. Dim.: 1) 4.4×0.6–0.5; 2) bent.

M517   House 6, Inv. No. HA 24/m1. Dim.: 1) 3.3×0.4–0.3; 2) clinched (?), 3.0; head broken off, point intact.

M518   House 6, Inv. No. HA 24/m2. Dim.: 1) 9.5×0.9–0.4; 2) clinched, 8.3.

M519   House 6, Inv. No. HA 43/m1. Dim.: 1) 5.1×0.5.

M520   House 6, Inv. No. HA 74/m2. Dim.: 1) 7.6×0.8–0.2; 2) head missing, point intact.

M521   House 5, Inv. No. HA 21/m1. Dim.: 1) 6.8×0.5–0.3; 2) clinched, 1.5; head broken off (?).

M522   House 5, Inv. No. HA 46/m2. Dim.: 1) 4.1×0.5.

M523 House 5, Inv. No. HA 49/m1. Dim.: 1) 11.2×0.4–0.7; 2) clinched 3-dimensionally (one angle in vertical, one in horizontal plane); 2.0, 3.5; head broken off, point intact.

M524 House 5, Inv. No. HA 49/m3. Dim.: 1) 8.3×0.4–0.2; 2) head broken off, point intact.

M525 House 5, Inv. No. HA 49/m4. Dim.: 1) 6.9×0.4–0.3.

M526 House 5, Inv. No. HA 55/m1. Dim.: 1) 8.5×0.7–0.6.

M527 House 5, Inv. No. HA 55/m2. Dim.: 1) 7.4×0.9–0.3; 2) bent twice, head broken off, point intact.

M528 House 5, Inv. No. HA 61/m1. Dim.: 1) 2.4×0.3–0.4.

M529 House 5, Inv. No. HA 85/m2. Dim.: 1) 3.0×0.4–0.2; 2) head broken off, point intact.

M530 House 4, Inv. No. HA 96/2. Dim.: 1) 4.1×0.5–0.3; 2) head broken off, point intact.

M531 House 4, Inv. No. HA 115/13. Dim.: 1) 5.3×0.5.

M532 House 4, Inv. No. HA 125/1. Dim.: 1) 7.5×0.7–0.2; 2) head broken off, point intact.

M533 House 4, Inv. No. HA 126/37. Dim.: 1) 7.1×0.6–0.3.

M534 House 4, Inv. No. HA 130/14. Dim.: 1) 6.4×0.6–0.5.

M535 House 4, Inv. No. HA 135/5. Dim.: 1) 4.0×0.6–0.4; 2) clinched, 2.5.

M536 House 4, Inv. No. HA 153/8B. Dim.: 1) 4.1×0.5/0.3.

M537 House 4, Inv. No. HA 156/3. Dim.: 1) 7.1×1.7–1.0; 2) point intact, head broken off.

M538 House 4, Inv. No. HA 166/1. Dim.: 1) 6.0×1.0–0.7; 2) clinched, 4.8.

M539 House 4, Inv. No. HA 178/2. Dim.: 1) 7.0×0.5–0.6; 2) clinched, 4.9.

M540 House 3, Inv. No. HA 193/11. Dim.: 1) 4.0×0.6.

M541 House 3, Inv. No. HA 199/4. Dim.: 1) 5.9×1.1–0.3; 2) point intact, head broken off.

M542 House 3, Inv. No. HA 199/10. Dim.: 1) 5.9×1.1–0.5; 2) point intact, head broken off.

M543 House 3, Inv. No. HA 208/10. Dim.: 1) 4.0×0.8–0.3.

M544 House 3, Inv. No. HA 216/18.2. Dim.: 1) 10.2×0.9–0.3; 2) bent, point intact, head broken off.

M545 House 3, Inv. No. HA 216/20.1. Dim.: 1) 4.0×0.8–0.4.

M546 House 3, Inv. No. HA 224/24. Dim.: 1) 9.8×0.6; 2) clinched, 4.8, bent outward beyond the clinch.

M547 House 3, Inv. No. HA 232/16.1. Dim.: 1) 3.6×0.6–0.3; 2) point intact, twisted, head broken off.

M548 House 3, Inv. No. HA 232/16.4. Dim.: 1) 2.7×0.6–0.3; 2) point intact, head broken off.

M549 House 3, Inv. No. HA 232/35. Dim.: 1) 6.5×0.6–0.5.

M550 House 3, Inv. No. HA 247/39. Dim.: 1) 7.0×0.6.

M551 House 3, Inv. No. HA 248/8a. Dim.: 1) 1.8×0.5–0.3; 2) point intact, bent, head broken off.

M552 House 3, Inv. No. HA 254/2. Dim.: 1) 2.9×0.3; 2) bent.

M553 House 3, Inv. No. HA 267/37. Dim.: 1) 4.7×0.4; 2) clinched, 2.4.

M554 House 3, Inv. No. HA 271/10. Dim.: 1) 6.4×0.7–0.3; 2) point intact, twisted at other end, head broken off.

M555 House 3, Inv. No. HA 279/3. Dim.: 1) 8.5×1.8/1.4–0.4/0.2; 2) point intact, bent halfway, head broken off (?).

Iron shapeless lumps

Measurements: Length×width×thickness / length×section (s).

M556 House 1, Inv. No. X 24. Dim.: 2.7, s. 1.3–0.6. = Reinders, 1988: 31.22.

M557 House 1, Inv. No. X 145-1. Dim.: 3.5×2.1×1.6. = Reinders, 1988: 35.26.

M558 House 1, Inv. No. X 145-2. Dim.: 3.1×1.6×1.3. = Reinders, 1988: 35.26.

M559 House 1, Inv. No. X 151 B. Dim.: 4.1×1.5×0.6. = Reinders, 1988: 35.28.

M560 House 1, Inv. No. X 729 A. Dim.: Lost in the earthquake of 1980; Reinders, 1988: 34.26.

M561 House 1, Inv. No. X 67. Lost in the earthquake of 1980; Reinders, 1988: 39.15.

M562 House 2, Inv. No. H21 4/31D. Dim.: 2.8×1.8×0.5 plus second, smaller lump.

M563 House 6, Inv. No. H23 2/m23A. Dim.: 3.6×1.2–0.4.

M564 House 6, Inv. No. H23 3/m24. Dim.: 4.2×2.0×0.5–1.6.

M565 House 6, Inv. No. H23 3/m25. Dim.: 3.1×0.9–0.4.

M566 House 5, Inv. No. HA 80/m9. Dim.: 4.3×0.8×0.5.

M567 House 5, Inv. No. HA 80/m10. Dim.: 2.2×0.6×0.2.

M568 House 5, Inv. No. HA 80/m11. Dim.: 2.2×0.8×0.5.

M569 House 5, Inv. No. HA 80/m12. Dim.: 1.8×0.5×0.2.

M570 House 5, Inv. No. HA 86/m4. Dim.: 2.4×1.2×0.7.

M571 House 4, inv. No. HA 106/11. Dim.: 4.2×1.4×1.3–0.5. Possibly modern.

M572 House 4, Inv. No. HA 109/18. Dim.: 1.4×1.0×0.3.

M573 House 4, Inv. No. HA 115/-. Dim.: 7×1.1×0.2.

M574 House 4, Inv. No. HA 120/5. Dim.: 3.0×2.7×0.6.

M575 House 4, Inv. No. HA 126/3. Three fragments. Dim. largest: 4.5×1.7×0.7–1.1.

M576 House 4, Inv. No. HA 126/38. Dim.: 2.4×1.5×0.5.

M577 House 4, Inv. No. HA 141/8. Dim.: 3.6×1.4×0.9.

M578 House 3, Inv. No. HA 198/18. Three fragments. Dim. each: <1.0×<1.0×<0.5.

M579 House 3, Inv. No. HA 199/8. Dim.: 9.8×0.9×0.8.

M580 House 3, Inv. No. HA 202/31. Dim.: 7.4×1.1×0.8.

M581 House 3, Inv. No. HA 210/5. Three minute fragments. Dim. each: <1.0×<1.0×<0.5.

M582 House 3, Inv. No. HA 210/9. Dim.: 0.8×0.6×0.2.

M583 House 3, Inv. No. HA 215/10. Seven fragments. Dim. largest: 3.9×2.6×1.8.

M584 House 3, Inv. No. HA 220/1. Three fragments. Dim. each: <1.0×<1.0×<0.5.

M585 House 3, Inv. No. HA 224/15. Two fragments. Dim. both: <1.0×<1.0×<0.5.

M586 House 3, Inv. No. HA 226/5. Five fragments. Dim. each: <1.0×<1.0×<0.5.

M587 House 3, Inv. No. HA 244/1. Dim.: 6.5×1.8×1.7.

M588 House 3, Inv. No. HA 253/17. Seven fragments. Dim.: 2.3×1.1×1.0 (largest) – 1.4×1.0×1.0 (smallest).

M589 House 3, Inv. No. HA 253/19. Two fragments. Dim. both: <1.0×<1.0×<0.5.

M590 House 3, Inv. No. HA 253/25. Dim.: 1.5×1.4×0.6.

M591 House 3, Inv. No. HA 253/36. Dim.: 6.5×4.8×1.9.

Other unidentifiable objects of iron

M592 House 4, Inv. No. HA 126/21. Flat object, one long side straight, two short sides flaring outwards slightly to other long side which is concave. In section the object tapers from the straight long side to the concave one. Dim. 5.0× 2.8×0.7–0.1.

M593 House 3: Straight shank, in section irregular, one end flattened to a leaf-shaped point. Dim. 12.2×0.4–0.1×0.4–0.8.

Sheets of bronze (for inlay work and the like?)

Measurements: Shape, flat/curved, length×width×thickness.

M594 House 1, Inv. No. X 64. Rectangular, flat. Dim. 6.4× 4.6×0.1. = Reinders, 1988: 38.05.

M595 House 6, H23 5/m19, Fig. 3.40. Flat band, horizontal semicircle, twisted, broken at both ends. Dim. 24.0 (approx. total length)×2.1×0.1; diameter approx. 20.0.

M596 House 4, Inv. No. HA 146/6. Irregular, flat. Dim.: 1.7×1.1×<0.1.

M597 House 4, Inv. No. HA 150/13. Five wafer-thin fragments of bronze, very irregular in shape, not flat but in bas relief; perhaps a fragment of bronze relief decoration (folds of clothing)? Dim. largest fragment: 2.3×1.8×0.1.

M598 House 3, Inv. No. HA 202/14A. Rectangular, flat. Dim.: 4.3×1.3×0.1.

Wire

Measurements: Shape (straight/bent/twisted); total length×section.

M599 House 3, Inv. No. H23 2/m21. Straight. Dim.: 1.4×0.2.

M600 House 5, Inv. No. HA 23/m1. Bent, twisted. Dim.: 6.7×0.2–0.3; topsoil find, possibly modern.

M601 House 5, Inv. No. HA 65/m2. Bent. Dim.: 2.0×0.1–0.2.

M602 House 4, Inv. No. HA 148/2. Bent. Dim.: 4.2×0.4.

M603 House 3, Inv. No. HA 267/9. Bent. Dim.: 5.3×0.3–0.2.

Shapeless fragments

Measurements: Length×width×thickness / length×section (s).

M604 House 1, Inv. No X 8. Dim.: 2.0×1.6×0.3. Reinders, 1988; not listed

M605 House 6, Inv. No. HA 74/m4. Dim.: 1.6×1.1×0.1–0.2.

M606 House 5, Inv. No. HA 80/m13. Dim.: 1.1×1.3×0.1.

Fragments of unidentified objects

M607 House 4, Inv. No. HA 127/59. Central part of an unidentified object; flat, with rounded edges; from one end, narrow, broken off, the object widens rapidly to two 'shoulders' from where it gradually tapers to the other end (broken off). In profile the object curves gently inwards from the narrow end, then sharply outwards just before the other end. Dim. 6.8×0.9/3.1/2.4×0.5.

Unidentifiable objects of lead

'Clamps'

Straight, narrow strip of lead, folded over once lengthwise. References: Robinson, 1941: p. 333, nos. 1583–1589, pl. XCIX. Measurements: Shape (flat, semicircle, twisted); total length×width (after folding)×thickness (before folding).

M608 House 2, Inv. No. H21 3/25. Irregular semicircle. Dim.: 8.0×0.7–1.1×0.3.

M609 House 2, Inv. No. H21 2/9. Flat, straight. Dim.: 7.8×0.7–0.5×0.2–0.3.

M610 House 2, Inv. No. H21 2/36. Flat, straight. Dim.: 7.0×1.6–1.0×0.2–0.3.

M611 House 2, H21 4/18: Bent. Dim.: 5.9×0.9×0.2–0.3.

Psimythion?

Short bar of lead, in section round, both ends flat

M612 House 6, Inv. No. HA 48/m2. Dim.: 1.7×1.1.

Shapeless fragments

Measurements: Length×width×thickness / length×section (s).

M613 House 6, Inv. No. H23 5/m47. Two large and seven small fragments of molten lead. Dim.: 11.6×9.4×0.8 (largest)– 1.3×1.3×0.3 (smallest).

M614 House 6, Inv. No. H23 5/m48. Six larger and six smaller fragments of molten lead, very irregular in form. Dim.: 21.0×13.8×0.7 (largest)–1.3×1.2×0.4 (smallest).

M615 House 6, Inv. No. H23 5/m48a. Large piece of lead which melted in or into a basket, preserving its woven pattern on one side. Dim.: 11.7×6.7×0.3–1.1. The basket was presumably of soft basketry as there is no trace of a framework, only of the woven rushes. References: White, 1975: pp. 52–104; cf. esp. 88–91, fig. 29.

M616 House 6, Inv. No. H23 5/m48d. Dim.: 5.6×4.0×0.3–0.5 (molten lead).

M617 House 6, Inv. No. H23 5/m48e. Semicircular strip, wavy line on one side. Dim.: 11.7×2.5×0.3–0.7.

M618 House 6, Inv. No. H23 6/m18. Irregular 'head', straight 'shank'. Dim. 1.5×1.3×0.3 (head); 2.1×0.5–0.3 (shank); perhaps a fragment of a pot repair?

M619 House 6, Inv. No. HA 6/m1. Irregular, flat. Dim.: 3.2×2.0×0.3.

M620 House 5, Inv. No. HA 25/m1. Irregular strip of lead, remnant of a shank (?) protruding from one end; conceivably a fragment of a very large pot repair. Dim.: 8.7×1.0–2.6×0.4; 'shank' 1.5×0.4.

M621 House 5, Inv. No. HA 81/m3. Irregular, flat. Dim.: 9.3×7.3×0.6.

M622 House 4, Inv. No. HA 147/12. Irregular, flat. Dim.: 4.6×2.9×0.1–0.4.

M623 House 4, Inv. No. HA 149/4C. Irregular, flat. Dim.: 5.2×3.9×0.5.

M624 House 4, Inv. No. HA 149/4D. Irregular, flat. Dim.: 4.2×4.0×0.5.

M625 House 4, Inv. No. HA 149/4E. Irregular, flat. Dim.: 3.1×2.9×0.5.

M626 House 4, Inv. No. HA 149/4F. Irregular, flat. Dim.: 4.0×2.5×0.6.

M627 House 4, Inv. No. HA 149/6C. Irregular, flat. Dim.: 2.9×2.2×0.5.

M628 House 4, Inv. No. HA 149/7C. Irregular, flat. Dim.: 2.7×1.7×0.4.

M629 House 4, Inv. No. HA 149/7D. Irregular, flat. Dim.: 2.5×1.0×0.2.

M630 House 4, Inv. No. HA 155/3. Irregular. Dim.: 11.2×2.6×2.4.

M631 House 3, Inv. No. HA 224/14. Roughly rectangular, flat. Dim.: 2.3×2.1×0.2.

M632 House 3, Inv. No. HA 247/3. Irregular. Dim.: 4.5×3.6×2.2.

MODERN FINDS

M633 House 1, Inv. No. X 5. Large, shapeless. Dim.: 9.7×1.9–0.4×1.3–0.7. = Reinders, 1988: 37.28

M634 House 1, Inv. No. X 346. Fragment of a pair of pliers. Dim.: 3.1×1.1×0.9–0.4. = Reinders, 1988: 37.32

M635 House 5, Inv. No. HA 37/m1. Nail, section round. Dim.: 4.3×0.3.

M636 House 5, Inv. No. HA 66/m1. Spoke; straight shank in section round. Dim.: 17.9×0.5.

M637 House 4, Inv. No. HA 105. Fragment of a pair of pliers.

M638 House 3, Inv. No. HA 197/.. Large, shapeless. Dim.: 8.9×1.5×1.3.

M639 House 3, Inv. No. HA 254/.. Bolt.

APPENDIX 6. CATALOGUE OF COINS

H. REINDER REINDERS

Abbreviations

Obv.	= Obverse		AR	= silver
Rev.	= Reverse		AE	= bronze
Leg.	= Legend		W.	= Weight in grams
Inv. No.	= Inventory number		D.	= Diameter in mm
Cat. No.	= Catalogue number (Reinders, 1988)		D.P.	= Die Position
W.	= Weight in grams			
D.	= Diameter in mm			
D.P.	= Die Position			

THRAKE, Orthagoreia
Obv.   Apollo laureate
Rev.   Helmet, ΟΡΘΑΓΟΡΕΩΝ
Ref.   SNG Cop: no. 691
C1   House 4, Inv. No. HA 93/1   AE; W. 2.17; D. 13; D.P. 8

MAKEDONIA, Kassandros (316–297 BC)
Obv.   Herakles in lion's skin, border of dots
Rev.   Naked youth on horse, pacing towards the r., ΒΑΣΙΛΕΩΣ ΚΑΣΣΑΝΔΟΥ
Ref.   SNG Cop: nos. 1160–1162
C2   House 3, Inv. No. HA 260/7   AE; W. 5.70; D. 19/20; D.P. 6
C3   House 3, Inv. No. HA 204/2   AE; W. 6.60; D. 18/20; D.P. ?
C4   House 3, Inv. No. HA 247/19   AE; W. 6.45; D. 20; D.P. 6
C5   House 3, Inv. No. HA 222/4   AE; W. 6.50; D. 18/19; D.P. 6 (countermark, bird)
C6   House 4, Inv. No. HA 102/1   AE; W. 1.01 (?), D. 18; D. P. ?
C7   House 5, Inv. No. 1989/21   AE; W. 6.91; D. 19; D.P. 5
C8   House 5, Inv. No. 1989/55   AE; W. 4.93; D. 20–22; D.P. 10
C9   House 5, Inv. No. 1989/59   AE; W. 5.59; D. 18; D.P. 11
C10   House 5, Inv. No. 1989/80   AE; W. 5.49; D. 19; D.P. 12
C11   House 6, Inv. No. 1987/17   AE; W. 6.50; D. 18; D.P. 9

Obv.   Herakles in lion's skin
Rev.   Lion seated; ΚΑΣΣΑ-Ν-ΔΡΟΥ
Ref.   SNG Cop. 1140
C12   House 6, Inv. No. 1989/48   AE; W. 3.55; D. 17; D.P. ?

MAKEDONIA, Demetrios Poliorketes (294-288 BC)
Obv.   Monogram ΔΗΜΗΤΔΗ on the boss of a Macedonian shield
Rev.   Crested Macedonian helmet between BA ΣΙ
Ref.   Newell, 1927: pp. 119–120, nos. 125–132; SNG Cop: 1424–1425
C13   House 2, Inv. No. H21 B 4   AE; W. 4.11; D. 16/17; D.P. 5
C14   House 3, Inv. No. HA 208/1   AE; W. 4.45; D. 14/16; D.P. 3
C15   House 3, Inv. No. HA 220/5   AE; W. 3.97; D. 14/15; D.P. 5
C16   House 4, Inv. No. HA 183/3   AE; W. 3.77; D. 15; D.P. ? (almost illegible)
C17   House 5, Inv. No. 1989/25   AE; W. 5.08; D. 13; D.P. 5
C18   House 6, Inv. No. 1987/7   AE; W. 4.31; D. 16–18; D.P. 2

Obv.   Male head, r., wearing Corinthian helmet
Rev.   Prow r., BA above, monogram below
Ref.   Newell, 1927: pl. II.10; SNG Cop 1194
C19   House 2, Inv. No. H21 B 41   AE; W. 3.30; D. 14–17; D.P. 12
C20   House 5, Inv. No. 1989/44   AE; W. 3.60; D. 14–17; D.P. 11

MAKEDONIA, Antigonos Gonatas (277-239 BC)
Obv.   Monogram ANTI on the boss of a Macedonian shield
Rev.   Crested Macedonian helmet between BA ΣΙ; Monogram r. above, l. and below
Ref.   SNG Sweden 1141; Furtwängler
C21   House 2, Inv. No. H21 B 28   AE; W. 3.93; D. 17–19; D.P. 6
C22   House 3, Inv. No. HA 195/2   AE; W. 4.92; D.P. 7

EPEIROS, Pyrrhos (295–272 BC)
Obv.   Head of Zeus
Rev.   Thunderbolt within olive wreath
Ref.   SNG Cop: no. 101
C23   House 5, Inv. No. 1989/36   AE; W. 5.60; D. 18; D.P. ?
C24   House 5, Inv. No. 1989/60   AE; W. 3.98; D. 19; D.P. ?

KORKYRA, Korkyra
Obv.   Amphora
Rev.   Bunch of grapes
Ref.   SNG Cop: no. 167–168
C25   House 4, Inv. No. HA 93/1   AE; W. 2.17; D. 13; D.P. 6

PERRHAIBIA, Orthe
Obv.   Head of Athena, r., wearing crested helmet
Rev.   Forepart of horse, springing towards the r. from rock on which olive bushes grow; ΟΡΘΙΕΩΝ
Ref.   Rogers, 1932: no. 315
C26   House 4, Inv. No. HA 113/3   AE; W. 2.19; D. 14; D.P. 12

PELASGIOTIS, Gyrton
Obv.   Laureate head of Zeus, l.
Rev.   Horse trotting l.; ΓΥΡΤΩΝΙΩΝ, monogram below
Ref.   SNG Cop: no. 61
C27   House 3, Inv. No. HA 222/3   AE; W. 6.50; D. 20; D.P. 8

PELASGIOTIS, Larisa
Obv.   Head of nymph, three quarters facing l.
Rev.   Horseman r.; monogram M below; ΛΑRΙ ΣΑΙΩΝ
Ref.   SNG Cop: no. 141; Rogers, 1932: no. 285
C28   House 1, Cat. No. 39.20   AE; W. 6.33; D. 18; D.P. 6 (monogram M)
C29   House 3, Inv. No. HA 216/6   AE; W. 5.76; D. 17; D.P. 9 (monogram between legs of horse)
C30   House 5, Inv. No. HA 1989/24   AE; W. 7.06; D. 18; D.P. 6

PELASGIOTIS, Skotoussa
Obv.   Head of Athena or Ares in close-fitting helmet, r.
Rev.   Horse prancing; line of ground; SKOTOY above, ΣΑΙΩΝ below
Ref.   SNG Cop: nos. 257–258; Rogers, 1932: no. 547; Reinders, 1988
C31   House 1, Cat. No. 35.32   AE; W. 5.13; D. 19; D.P. 12 (ΣΑΣΑΙΟΝ)
C32   House 1, Cat. No. 39.18   AE; W. 5.76; D. 19; D.P. 12
C33   House 3, Inv. No. HA 226/14   AE; W. 4.76; D.P. 12
C34   House 6, Inv. No. 1987/3   AE; W. 5.21; D. 18; D.P. 12

HESTIAIOTIS, Trikka
Obv.   Head of nymph Trikka, r., hair in sphendone
Rev.   Asklepios seated, r., feeding serpent with bird held in outstretched r. hand; ΤΡΙΚΚΑΙΩΝ
Ref..   SNG Cop: nos. 266–267
C35   House 3, Inv. No. HA 218/3   AE; W. 6.35; D. 19/21; D.P. 7

ACHAIA PHTHIOTIS, Thebai
Obv.   Head of Demeter
Rev.   Protesilaos armed with shield and sword, leaping on shore from prow of galley; in field r. XA
Ref.   Rogers, 1932: no. 551; SNG Cop. 259–260
C36   House 4, Inv. No. HA 120/2   AE; W. 1.60; D. 14; D.P. ?

ACHAIA PHTHIOTIS, Peuma
Obv.   Male head
Rev.   Monogram XA; ΠΕΥΜΑΤΙΩΝ
Ref.   Rogers, 1932
C37   House 2, Inv. No. H21 B 1   AE; W. 2.50; D. 13; D.P. 12
C38   House 3, Inv. No. HA 199/2   AE; W. 2.12; D. 13; D.P. 11
C39   House 4, Inv. No. HA 110/4   AE; W. 2.11; D. 13; D.P. 12
C40   House 5, Inv. No. 1989 13   AE; W. 1.47; D. 13; D.P. 9
C41   House 6, Inv. No. 1987/2   AE; W. 2.00; D. 13; D.P. 9 (helmet in the field r.)
C42   House 6, Inv. No. 1987/11   AE; W. 1.99; D. 13; D.P. 3 (helmet in the field r.)

C43   House 6, Inv. No. 1987/44   AE; W. 2.61; D. 14; D.P. 12

ACHAIA PHTHIOTIS, Halos
Obv.   Head of Zeus Laphystios, border of dots
Rev.   Phrixos clinging to ram, flying r., with chlamys in the wind; monogram XA, AΛEΩN
Ref.   Reinders, 1988: p. 241, series 6, fig. 113
Series 6
C44   House 3, Inv. No. HA 243/1   AE; W. 6.41; D. 18/19; D.P. 6
C45   House 3, Inv. No. HA 208/6   AE; W. 5.93; D. 17/18; D.P. 6

Obv.   Head of Zeus Laphystios, r., border of dots
Rev.   Phrixos clinging to ram, flying r., with chlamis in the wind; monogram XA; AΛEΩN below
Ref.   Reinders, 1988: pp. 241–243, series 7, fig. 113
Series 7
C46   House 1, Cat. No. 35.35   AE; W. 1.92; D. 14; D.P. 6
C47   House 2, Inv. No. H21 B 69   AE; W. 3.18; D. 13; D.P. 12
C48   House 4, Inv. No. HA 103/8   AE; W. 2.04; D. 15; D.P. 6
C49   House 4, Inv. No. HA 97/10   AE; W. 1.38; D. 14; D.P. 12
C50   House 6, Inv. No. 1987/16   AE; W. 2.60; D. 14; D.P. 6
Series 7.1
C51   House 1, Cat. No. 39.16   AE; W. 2.19; D. 14; D.P. 12
C52   House 1, Cat. No. 39.17   AE; W. 1.85; D. 13.5; D.P. 12
C53   House 3, Inv. No. HA 243/2   AE; W. 2.28; D. 14; D.P. 12
C54   House 3, Inv. No. HA 199/6   AE; W. 2.25; D. 16; D.P. 7
Series 7.2
C55   House 6, Inv. No. 1987/1   AE; W. 2.38; D. 14; D.P. 12
C56   House 6, Inv. No. 1987/5   AE; W. 2.19; D. 15; D.P. 12
Series 7.3
C57   House 3, Inv. No. HA 231/3   AE; W. 2.04; D. 13/14; D.P. 6
Series 7.4
C58   House 2, Inv. No. H21 B 66   AE; W. 2.20; D. 14; D.P. 7
C59   House 3, Inv. No. HA 224/4   AE; W. 2.27. D. 15; D.P. 12
C60   House 3, Inv. No. HA 207/10   AE; W. 2.58; D. 14/16; D.P. 6
C61   House 3, Inv. No. HA 199/3   AE; W. 2.46; D. 13/14; D.P. 12
Series 7.5
C62   House 2, Inv. No. H21 B 33   AE; W. 2.41; D. 14; D.P. 6
C63   House 2, Inv. No. H21 B 62   AE; W. 1.90; D. 14; D.P. 12
C64   House 3, Inv. No. HA 269/7   AE; W. 2.52; D. 14/15; D.P. 7
C65   House 3, Inv. No. HA 214/3   AE; W. 1.91; D. 14/15; D.P. 12
C66   House 5, Inv. No. 1989/55   AE; W. 2.26; D.P. 12

Obv.   Head of Zeus Laphystios
Rev.   Phrixos clinging to ram, flying r., his right hand holds horn
Ref.   Reinders, 1988: p. 243, series 8, fig. 113
Series 8
C67   House 1, Cat. No. 33.74   AE; W. 1.63; D. 1.36; D.P. 12

Obv.   Head of Zeus Laphystios
Rev.   Phrixos clinging to ram, flying r.; AΛE below
Ref.   Reinders, 1988: p. 243, series 9, fig. 113
Series 9
C68   House 1, Cat. No. 31.26   AE; W. 2.06 gr; D. 13; D.P. 9

Obv.   Head of Zeus Laphystios, r.
Rev.   Phrixos clinging to ram, with chlamys in the wind; waves below; AΛ-E-Ω-N
Ref.   Reinders, 1988: p. 245, series 19, fig. 113
Series 19
C69   House 2, Inv. No. H21 B 14   AE; W. 1.51; D. 14; D.P. 7

Obv.   Head of Zeus Laphystios
Rev.   Phrixos clinging to ram, flying r.
Ref.   Reinders, 1988, series ?
C70   House 1, Cat. No. 33.73   AE; W. 2.58 gr; D. 13; D.P. 12
C71   House 4, Inv. No. HA 128/5   AE; W. 1.84; D. 14; D.P. 8

ACHAIA PHTHIOTIS, Ekkara

Obv.	Head of Zeus laureate, r.; border of dots	
Rev.	Artemis standing, l., her r. hand resting on hunting spear; ΕΚΚΑΡΕΩΝ	
Ref.	Rogers, 1932: no. 208; SNG Cop: nos. 47–48	
C72	House 4, Inv. No. HA 106/19	AE; W. 2.23; D. 14; D.P. 10
C73	House 4, Inv. No. HA 126/17	AE; W. 1.95; D. 12; D.P. 1.95
C74	House 5, Inv. No. 1989/13	AE; W. 1.86; D. 13; D.P. 2

ACHAIA PHTHIOTIS, Larisa Kremaste

Obv.	Male head, l.	
Rev.	Thetis seated on hippocampus, holding in l. hand shield, bearing monogram XA; ΛΑΡΙ	
Ref.	Rogers, 1932: no. 315	
C75	House 2, Inv. No. H21 B 37	AE; W. 5.11; D. 16/19; D.P. 11
C76	House 4, Inv. No. HA 106/7	AE; W. 5.49; D. 19; D.P. 6
C77	House 4, Inv. No. HA 129/2	AE; W. 5.39; D. 18; D.P.

Obv.	Head of nymph, r.	
Rev.	Harpa with hook l., upright within olive wreath; ΛΑΡΙ r. upwards	
Ref.	Rogers, 1932: p. 104	
C78	House 2, Inv. No. H21 B 6	AE; W. 2.40; D. 14; D.P. 4
C79	House 2, Inv. No. H21 B 29	AE; W. 1.82; D. 13/14; D.P. 6 (Rogers, 1932: no. 317?)
C80	House 2, Inv. No. H21 B 31	AE; W. 2.01; D. 13; DP 11 (Rogers, 1932: no. 318)
C81	House 3, Inv. No. HA 220/3	AE; W. 2.09; D.P. 12
C82	House 4, Inv. No. HA 92/1	AE; W. 2.14; D. 15; D.P. 12
C83	House 5, Inv. No. 1989/79	AE; W. 2.27; D.P. ?
C84	House 6, Inv. No. 1987/9	AE; W. 1.49; D. 13; D.P. 3
C85	House 6, Inv. No. 1989/57	AE; W. 1.99; D. 12; D.P. 12

MALIS, Lamia

Obv.	Head of Athena, r., with Corinthian helmet	
Rev.	Philoktetes standing and shooting birds; ΜΑΛΙΕΩΝ, l.	
Ref.	SNG Cop: nos. 87–88; Rogers, 1932: no. 384	
C86	House 1, Cat. No. 39.19	AE; W. 2.52; D. 14; D.P. 6
C87	House 6, Inv. No. 1987/10	AE; W. 2.29; D. 15; D.P. 1
C88	House 6, Inv. No. 1987/13	AE; W. 1.71; D. 13; D.P. 1

Obv.	Female head r.	
Rev.	Philoktetes kneels on r. knee, discharging arrow; ΛΑΜΙ-Ε-ΩΝ	
Ref.	SNG Cop: no. 86	
C89	House 5, Inv. No. 1989/40	AE; W. 2.66; D. 12; D.P. 11

LOKRIS, Lokris

Obv.	Head of Demeter or Persephone, wearing earring and necklace, hair bound with wreath	
Rev.	Ajax in fighting attitude, armed with shield, sword and helmet; ΟΠΟΝΤΙΩΝ	
Ref.	SNG Cop: 48–54	
C90	House 3, stray find	AR; W. 2.49; D. 13/15; D.P. 12
C91	House 5, Inv. No. 1989/36	AR; W. 2.46; D. 15; D.P. 5 (on ground two spears)
C92	House 6, Inv. No. 1989/57	AR; W. 2.53; D. 16; DP 6 (on ground helmet and spear)

Obv.	Apollon Laureate	
Rev.	Bunch of grapes, between and O	
Ref.	SNG Cop, no. 72, 338–300 BC; Reinders 1988: no. 31.27, fig. 119	
C93	House 1, Cat. No. 31.27	AE; W. 2.25 gr.; D. 13; D.P. 12

Obv.	Athena with helmet, r.	
Rev.	Bunch of grapes; ΚΡ ΕΠ, ΛΟ, ΛΟΚΡ ΕΠΙΚΝΑ or ΛΟΚΡΩΝ	
Ref.	SNG Cop, nos. 65–75; Reinders, 1988	
C94	House 1, Cat. No. 33.72	AE; W. 1.75; D. 13; D.P. 12 (ΚΡ ΕΠ)
C95	House 1, Cat. No. 35.34	AE; W. 1.68; D. 13; D.P. 5
C96	House 4, Inv. No. HA 105/2	AE; W. 1.70; D. 14; D.P. 11 (ΛΟΚ Π ΟΝ)
C97	House 5, Inv. No. 1989/55	AE; W. 1.80; D. 14; D.P. ? (ΛΟΚΡ ΕΠΙΚΝΑ?)
C98	House 6, Inv. No. 1987/6	AE; W. 0.80; D. 12; D.P. 12 (L O)
C99	House 6, Inv. No. 1987/18	AE; W. 2.00; D. 14; D.P. 12

EUBOIA, Chalkis
Obv.   Illegible
Rev.   Illegible
Ref.   Picard, 1979
C100   House 1, Cat. No. 35.31   AE; W. 1.36; D. 12; D.P. ?

Obv.   Head of Hera facing
Rev.   Eagle to l., holding serpent in beak and claws
Ref.   Picard, 1979; Reinders, 1988
C101   House 1, Cat. No. 31.24   AE; W. 3.41; D. 17; D.P.
C102   House 2, Inv. No. H21 B 26   AE; W. 1.81; D. 12; D.P. 12
C103   House 2, Inv. No. H21 B 63   AE; W. 1.52; D. 11; D.P. 6
C104   House 2, Inv. No. H21 B 64   AE; W. 1.71; D. 13; D.P. 7
C105   House 2, Inv. No. H21 B 65   AE; W. 1.91; D. 13; D.P. 10
C106   House 2, Inv. No. H21 E 20   AE; W. 1.00; D. 12; D.P. 1
C107   House 3, Inv. No. HA 240/21   AE; W. 2.05; D. 11/12; D.P. ?
C108   House 3, stray find   AE; W. 1.23; D.P. ?
C109   House 6, Inv. No. 1987/12   AE; W. 2.13; D. 12; D.P. 7
C110   House 6, Inv. No. 1987/15   AE; W. 1.31; D. 12; D.P. ?
C111   House 6, Inv. No. 1987/19   AE; W. 1.80; D. 13; D.P. 12
C112   House 6, Inv. No. 1989/32   AE; W. 1.21; D. 13; D.P. 12
C113   House 6, Inv. No. 1987/69   AE; W. 2.06; D. 13; D.P. 12

Obv.   Head of Hera facing, placed on Ionic capital
Rev.   Eagle flying r., holding serpent in beak and claws
Ref.   Picard, 1979: series 15; Reinders, 1988
C114   House 1, Cat. No. 31.25   AE; W. 1.71; D. 13; D.P. 7

EUBOIA, Histiaia
Obv.   Female head, r.
Rev.   Bull, walking r.; IΣ-TI
Ref.   SNG Cop: nos. 74–76, Reinders 1988
C115   House 1, Cat. No. 35.33   AE; W. 1.68; D. 13; D.P. 5
C116   House 2, Inv. No. H21 D 1   AE; W. 2.30; D. 14; D.P. 10
C117   House 4, Inv. No. HA 106/3   AE; W. 1.96; D. 13; D.P. 11
C118   House 6, Inv. No. 1989/89   AE; W. 2.05; D. 14; D.P. 12

Obv.   Female head, r., wearing earring, necklace and vine wreath, hair in sphendone
Rev.   Head and neck of bull, three-quarter face towards r., bound with sacrificial fillet; to l. bunch of grapes; IΣTI
Ref.   BMC Histiaia: no. 31, pl. 24.8
C119   House 2, Inv. No. H21 B 38   AE; W. 3.63; D. 14; D.P. 12
C120   House 2, Inv. No. H21 B 51   AE; W. 2.40; D. 15; D.P. 6
C121   House 2, Inv. No. H21 C 18   AE; W. 3.11; D. 14/17; D.P. 12
C122   House 3, outside house   AE; W. 2.26; D. 14/15; D.P. 7
C123   House 6, Inv. No. 1987/8   AE; W. 1.82; D. 15; D.P. 1

EUBOIA, Euboian League
Obv.   Bull standing r.
Rev.   Bunch of grapes; EYBO l. upwards
Ref.   SNG Cop: nos. 488–491; Picard, 1979: p. 171;
C124   House 2, Inv. No. H21 B 69   AE; W. 1.33; D. 14; D.P. 12

PHOKIS, Phokis
Obv.   Head of Athena (?), three-quarter face to the left, wearing triple-crested helmet
Rev.   Φ within olive wreath
Ref.   BMC Phocis: nos. 66–74, pl. 3.17; SNG Cop: 113–118
C125   House 2, Inv. No. H21 E 12   AE; W. 2.21; D. 14/15; D.P. 12

BOEOTIA, Thebai
Obv.   Boeotian shield
Rev.   Kantharos; BO
Ref.   SNG Cop: nos. 274–275
C126   House 3, Inv. No. HA 260/6   AE; W. 2.55; D.P. 9
C127   House 5, Inv. No. 1989/23   AR; W. 2.48; D. 13. D.P. 12

BOEOTIA, Thespiai
Obv.   Boiotian shield
Rev.   Aphrodite Melainis, r.; in front crescent
Ref.   SNG Cop: no. 403
C128   House 6, Inv. No. 1987/4   AR; W. 2.61; D. 12–15; D.P. 12

BOEOTIA, Federal mint
Obv.   Boeotian shield
Rev.   Trident; ΒΟΙΩΤΩΝ
Ref.   SNG Cop: 179–181
C129   House 4, Inv. No. HA 104/2   AE; W. 1.10; D. 13; D.P. 9

BOEOTIA, Uncertain mint
Obv.   Boeotian shield
Rev.   Ornamented trident
Ref.   SNG Cop: nos. 179–181
C130   House 2, Inv. No. H21 C 17   AE; W. 1.80; D. 13/14; D.P. 12

KORINTHIA, Korinthos
Obv.   Pegasos, l.
Rev.   Head of nymph, l., hair in sakkos, Λ (?) below chin
Ref.   SNG Cop: no. 114
C131   House 2, Inv. No. H21 B 45   AR; W. 2.41; D. 15; D.P. 11

SYKIONIA, Sykion
Obv.   Dove flying
Rev.   Olive wreath; ΣΙ
Ref.   Head, 1977: p. 410
C132   House 4, Inv. No. HA 135/3   AE; W. 3.15; D. 17; D.P. ?

AIGYPTOS, Ptolemaios II (285–246 BC)
Obv.   Laureate head of Zeus r., border of dots
Rev.   Eagle standing on thunderbolt; shield and monogram; ΠΤΟΛΕΜΑΙΟΥ ΒΑΣΙΛΕΩΣ
Ref.   Svoronos, 1904-8; Reinders 1988
C133   House 1, Cat. No. 34.30   AE; W. 14.58; D. 25; D.P. 12 (Λ)
C134   House 1, Cat. No. 34.32   AE; W. 14.56; D. 27; D.P. 12 (Λ)
C135   House 4, Inv. No. HA 127/34   AE; W. 14.63; D. 25; D.P. 1 (Λ)
C136   House 1, Cat. No. 34.31   AE; W. 16.73; D. 27; D.P. 12 (Τ)
C137   House 4, Inv. No. HA 126/26   AE; W. 16.94; D. 29; D.P. 12 (Τ)
C138   House 1, Cat. No. 34.33   AE; W. 13.96; D. 26; D.P. 12 (Φ)
C139   House 2, Inv. No. H21 B 23   AE; W. 14.95; D. 27; D.P. 12
C140   House 3, Inv. No. HA 238/12   AE; W. 12.87; D. 27; D.P. 12
C141   House 3, Inv. No. HA 219/1   AE; W. 13.69; D. 25; D.P. 12 (Φ)
C142   House 3, Inv. No. HA 224/37   AE; W. 15.67; D. 28; D.P. 12
C143   House 4, Inv. No. HA 170/3   AE; W. 14.42; D. 26; D.P. 12
C144   House 4, Inv. No. HA 133/2   AE; W. 14.15; D. 26; D.P. 12
C145   House 5, Inv. No. 1989/83   AE; W. 14.64; D. 26; D.P. 12

Obv.   Laureate head of Zeus r., border of dots
Rev.   Eagle standing on thunderbolt, shield and monogram; ΠΤΟΛΕΜΑΙΟΥ ΒΑΣΙΛΕΩΣ
Ref.   Svoronos, 1904-8
C146   House 3, Inv. No. HA 218/2   AE; W. 6.18; D. 18; D.P. 12

Illegible
C147   House 3, Inv. No. HA 216/4   AE; W. 2.69; D. 15; D.P. 12
C148   House 3, Inv. No. HA 218/1   AE; W. 3.66; D. 14/15; D.P. 7
C149   House 5, Inv. No. 1989/23   AE; W. 3.30; D. 15; D.P. ?
C150   House 6, Inv. No. 1987/14   AE; W. 1.51; D. 14; D.P. ?

Roman coin
C151   House 3   AE; W. 2.65; D.P. 12

Denier Tournois
ACHAIA, William II of Villehardouin (1246-1278)
Obv.   Cross; G.PRINCEPS
Rev.   Chatel Tournois; CLARENTIA
Ref.   Metcalf, 1983: pp. 70–71
C152   House 3, Inv. No. HA 204/3   AR; W. 0.58; D. 17; D.P. 10

Printed and bound by CPI Group (UK) Ltd, Croydon, CR0 4YY

23/10/2024

01777711-0001